W9-CLC-577

Engines

PUBLISHER

Fundamentals of Service (FOS) is a series of manuals created by Deere & Company. Each book in the series is conceived, researched, outlined, edited, and published by Deere & Company. Authors are selected to provide a basic technical manuscript which is edited and rewritten by editors.

PUBLISHER: DEERE & COMPANY SERVICE PUBLICATIONS, Dept. FOS/FMO, John Deere Road, Moline, Illinois 61265-8098; DEPT. MANAGER: Alton E. Miller.

FUNDAMENTAL WRITING SERVICES EDITORIAL STAFF

Managing Editor: Louis R. Hathaway
Editor: John E. Kuhar
Publisher: Lori J. Lees
Promotions: Cindy S. Calloway

TO THE READER

PURPOSE OF THIS MANUAL

The main purpose of this manual is to help the reader understand and service engines with speed and skill. Starting with "how it works," we build up to "why it fails" and "what to do about it." This manual is also an excellent reference for the trained mechanic who wants to refresh knowledge of engines. It is written in a simple form using many illustrations, so that it can be easily understood.

APPLICATION OF ENGINES IN THIS MANUAL

"Engines" is a broad term. But in this manual, the prime interest is in engines (gas reciprocating, diesel, and LP fuel) as they are commonly used to produce work on the farm and in industry. Automotive, truck and bus applications are often covered as well.

HOW TO USE THIS MANUAL

This manual can be used by anyone—experienced mechanics, shop trainees, vocational students, and lay readers.

By starting with the basics, build your knowledge step by step. Chapter 1 covers the basics—"how it works;" Chapter 2 covers the basic engine components, while Chapters 3 through 9 explain the various systems—fuel, lubrication, cooling, etc. Chapters 10, 11, and 12 return to the complete engine in terms of testing, diagnosis, and tune-up.

Persons not familiar with engines should start at Chapter 1 and study the chapters in sequence. The experienced person can find what is needed on the "Contents" page.

OTHER MANUALS IN THIS SERIES

Other manuals in the FOS series are:

- **Hydraulics**
- **Electrical Systems**
- **Power Trains**
- **Shop Tools**
- **Welding**
- **Air Conditioning**
- **Fuels, Lubricants, and Coolants**
- **Tires and Tracks**
- **Belts and Chains**
- **Bearings and Seals**
- **Mowing and Spraying Equipment**
- **Fiber Glass and Plastics**
- **Fasteners**
- **Identification of Parts Failures**

Each manual is backed up by a set of 35 mm color slides for classroom use.

FOR MORE INFORMATION

Write for a free *Catalog of Educational Materials.* Send your request to:

John Deere Service Publications
Dept. FOS/FMO
John Deere Road
Moline, Illinois 61265-8098

ACKNOWLEDGEMENTS

John Deere gratefully acknowledges help from the following groups: Dana Corporation, Allen Electric and Equipment Co., Bacharach Industrial Instrument Co., Bendix Corp., Central Tool Co., Federal-Mogul Corp., F. W. Dwyer Co., Garrett Corp., General Motors Corp., J. H. Williams and Co., Kent-Moore Corp., K. O. Lee Co., Koppers Co., L. S. Starrett Co., Marquette Manufacturing Co., Marvel-Schebler, Division of Borg-Warner Corp., National LP-Gas Assn., Nuday Co., Owatonna Tool Co., Roosa Master, Hartford Division of Standard Screw Co., Rottler Boring Bar Co., Shell Oil Co., Standard Oil Co., Sterling Products Co., Sun Electric Corp., Sunnen Products Co., Taylor Dynamometer and Machine Co., Texaco, Inc., TRW Replacement Div., Union Carbide Corp., United Tool Process Corp., Westberg Manufacturing Co.

FOS — 30 **Litho in U.S.A.**

ISBN 0-86691-137-5

CONTENTS

Litho in U.S.A.

CONTENTS

Litho in U.S.A.

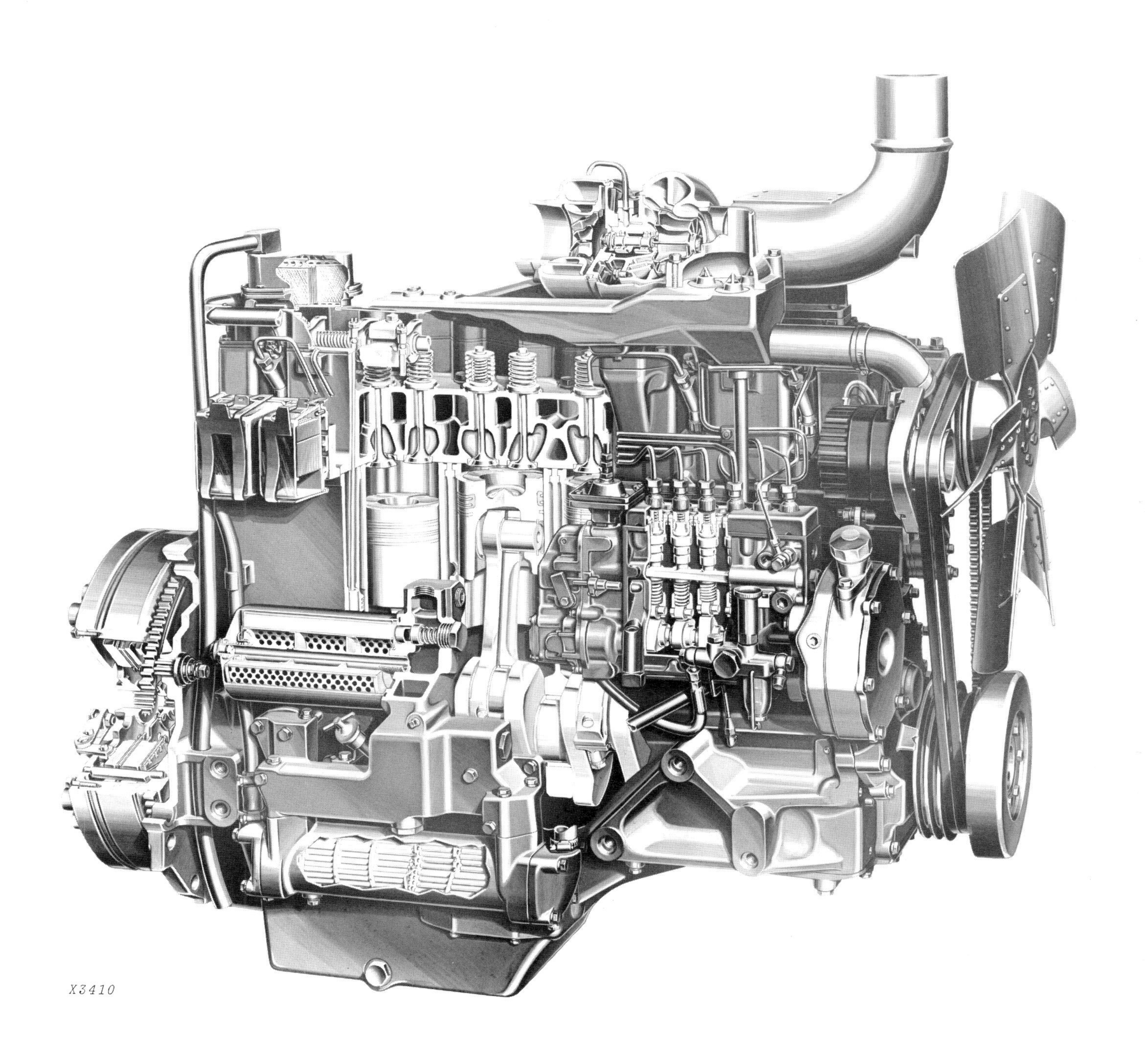

X3410

ENGINES—How They Work / CHAPTER 1

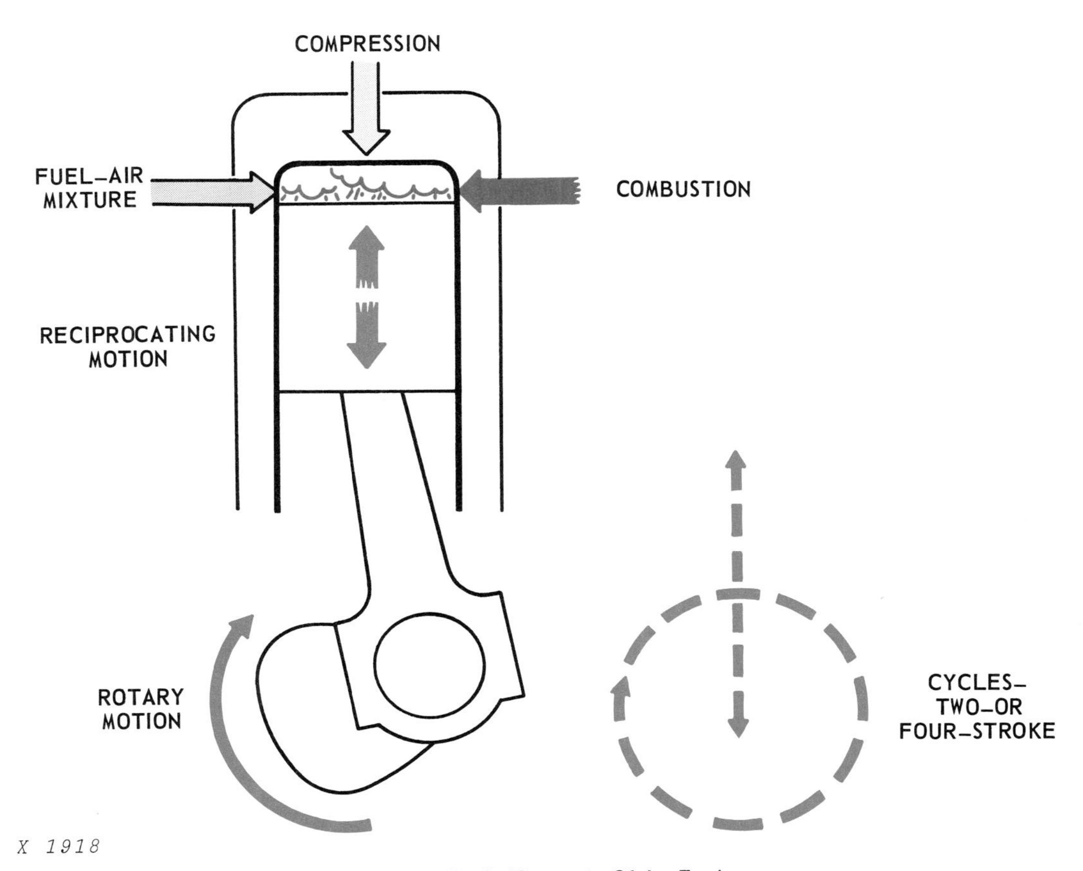

Fig. 1 — Basic Elements Of An Engine

WHAT IS AN "INTERNAL COMBUSTION" ENGINE?

Let's see what the term "internal combustion" means:

- **"Internal" means "inside" or "enclosed"**
- **"Combustion" is the "act of burning"**

Thus an internal combustion engine is one that *burns fuel internally.*

Basically this engine is a container in which we put fuel and air and start them burning.

The mixture expands rapidly while burning and pushes outward. This push can be used to move a part of the engine, and transmitted to drive the machine.

In summary, an engine is a device which converts heat energy into mechanical energy to do work.

WHAT ELEMENTS ARE NEEDED FOR AN ENGINE?

These elements are needed to construct a simple engine:

- **Air, Fuel, and Combustion**
- **Reciprocating and Rotary Motion**
- **Compression of Fuel-Air Mixture**
- **Engine Cycles—Two-or Four-Stroke**

Let's discuss these items one by one.

AIR, FUEL AND COMBUSTION

Three basic elements are needed to produce heat energy in the engine:

- **Air**
- **Fuel**
- **Combustion**

 Litho in U.S.A.

AIR is needed to combine with fuel and give it oxygen for fast burning. Air also has two other properties which affect the engine:

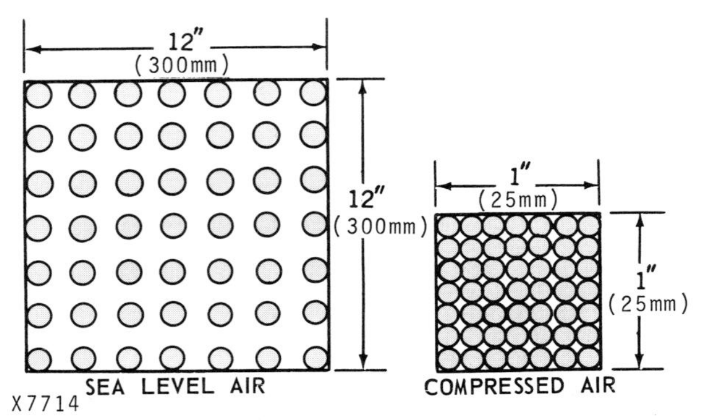

Fig. 2 — Air Can Be Compressed

(1) Air will compress; one cubic foot (28 L) of air can be packed into one cubic inch (16 cm^3) or less (Fig. 2).

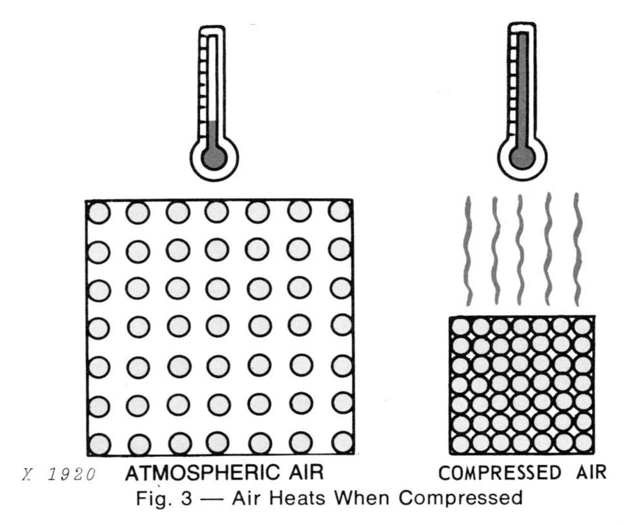

Fig. 3 — Air Heats When Compressed

(2) Air heats when it is compressed. The molecules of air rub against each other and produce heat (Fig. 3).

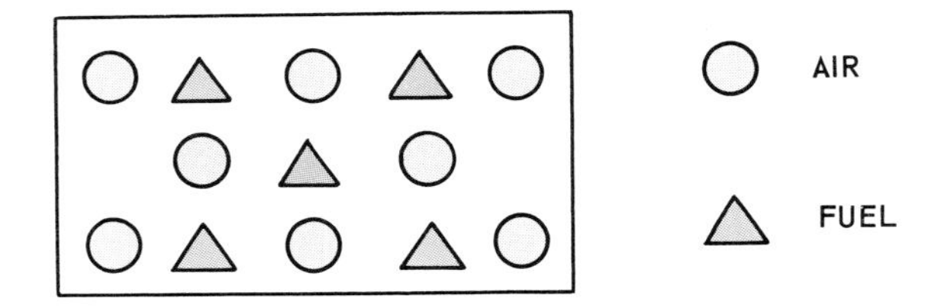

Fig. 4 — Fuel Must Mix Readily With Air And Ignite Easily

FUEL must mix readily with air and ignite easily (Fig. 4). The three we will cover are gasoline, LP-gas, and diesel fuel.

These fuels ignite easily and are readily broken down or vaporized.

Why do we want to vaporize the fuel? To help each particle of fuel contact enough air to burn fully.

COMBUSTION is the actual igniting and burning of the fuel-air mixture. It is the *oxygen* in the air that combines with the fuel for combustion.

What is important here is *how fast* the fuel burns, for this force must be "explosive" to get full power from the engine.

If a container of gasoline is ignited in calm outside air, it burns rather lazily (Fig. 5). This is because the air contacts only the surface of the fuel. To make the fuel burn faster, two things can be done:

1) Heat up the fuel

2) Vaporize the fuel (Fig. 5)

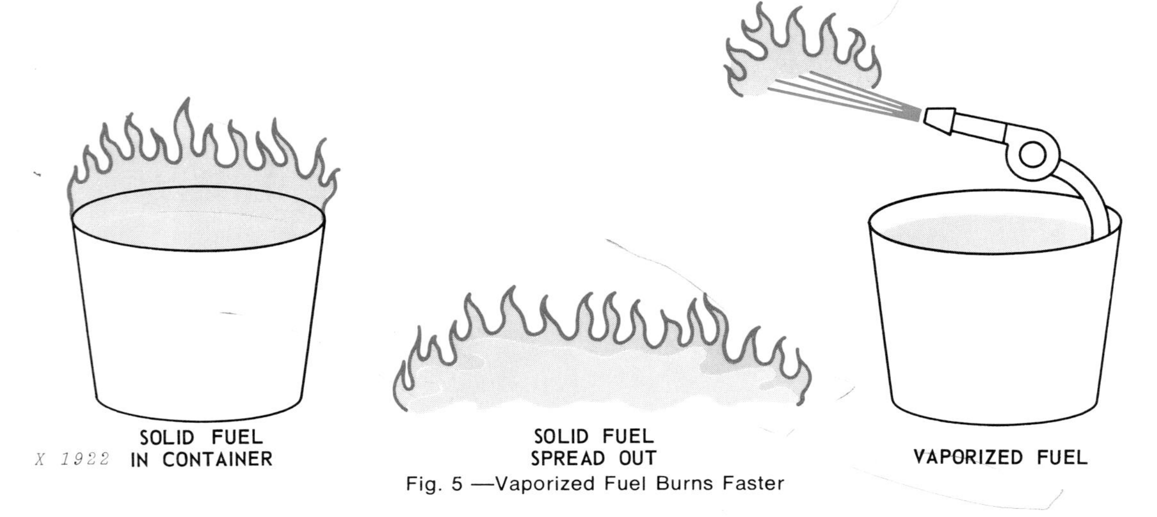

Fig. 5 —Vaporized Fuel Burns Faster

However, too powerful an explosion would destroy an engine, since combustion takes place in a closed container.

We can control the rate of burning by 1) how far we compress the air (and so heat it up), 2) how much fuel we use and 3) how volatile it is.

RECIPROCATING AND ROTARY MOTION

The engine uses two forms of motion to transmit energy:

- **Reciprocating Motion—up-and-down or back-and-forth motion**
- **Rotary Motion—circular motion around a point**

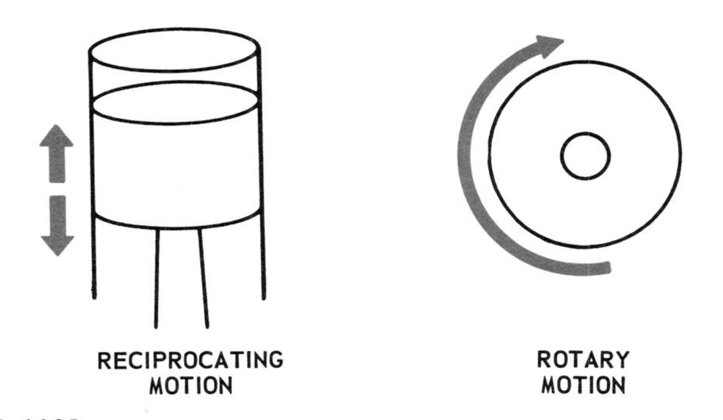

Fig. 6 — Reciprocating And Rotary Motion

The engine converts reciprocating motion into rotary motion (Fig 6).

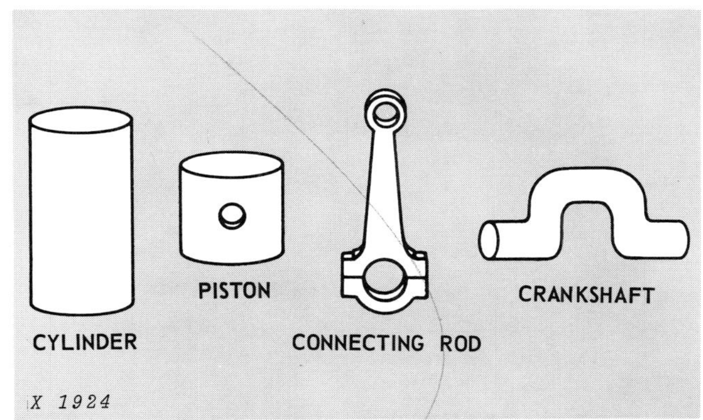

Fig. 7 — Basic Parts Of The Engine

Four basic parts are needed to make the engine work in this way:

- **Cylinder**
- **Piston**
- **Connecting Rod**
- **Crankshaft**

The *piston* and *cylinder* are mated parts, fitted closely so that the piston glides easily in the cylinder but with little clearance at the sides. The

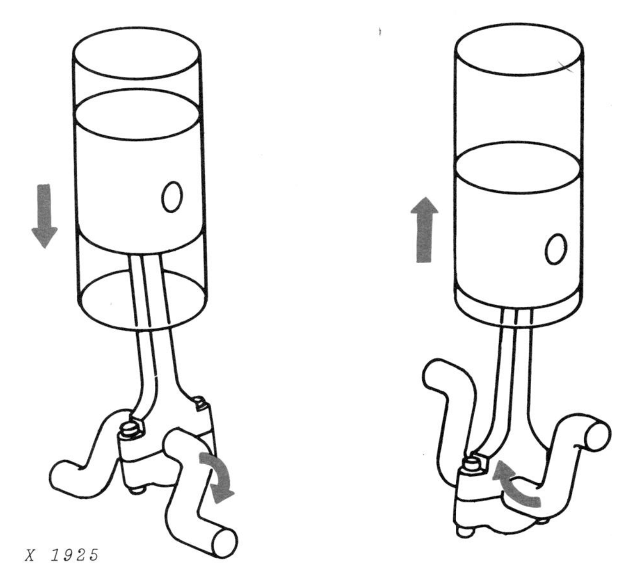

Fig. 8 — How Reciprocating Motion Is Transmitted To The Crankshaft As Rotary Motion

top of the cylinder is closed, but has extra space for the combustion chamber. A cylinder head of the engine generally closes the end of the cylinder.

The *connecting rod* is the link which transmits the motion of the piston to the crankshaft.

A simple *crankshaft* has a section offset from the center line of the shaft so that it "cranks" when the shaft is turned.

The motion is basically the same as when you pedal a bicycle. Your leg is like the connecting rod while the pedal crank and sprocket are like the crankshaft.

As a result, we have a way of converting the reciprocating motion of the piston into useful rotary motion (Fig. 8).

The stroke of the piston (how far it travels in the cylinder) is set by the "throw" of the crankshaft (how far it is offset).

COMPRESSION OF THE FUEL-AIR MIXTURE

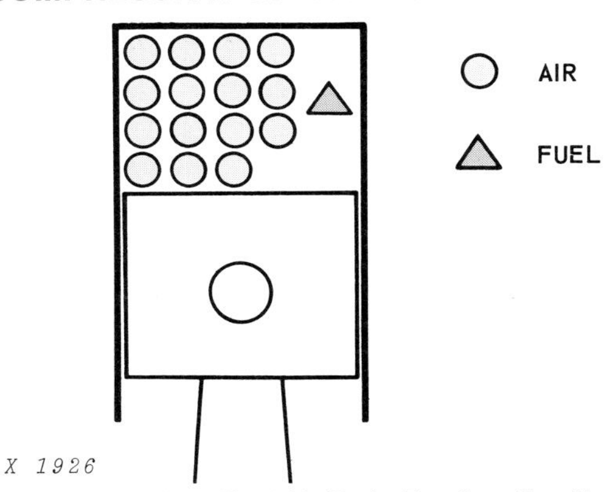

Fig. 9 — Fuel-Air Ratio For Gasoline Engine

The modern gasoline engine works best when about 15 parts of air are mixed with 1 part of fuel (Fig. 9).

Fig. 10 — Volume Of Fuel-Air Needed For Gasoline Engine

Fig. 10 shows how much greater the *volume* of air needed is than the volume of fuel. The gasoline engine mixes one gallon (4 L) of gasoline with about 9,000 (34 000 L) gallons of air.

As a result, we must **compress** the fuel-air mixture to get the desired ratio.

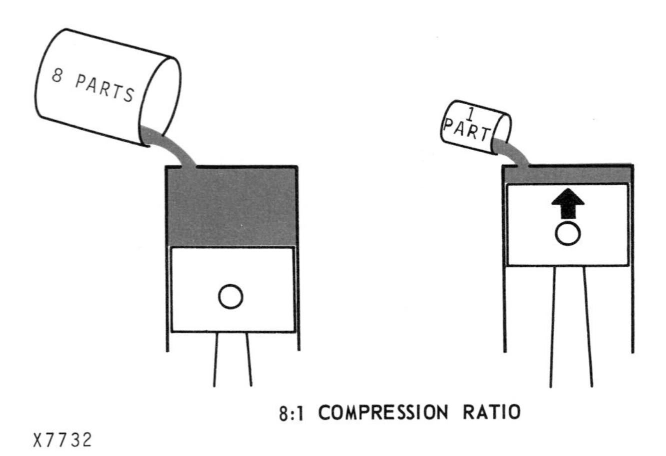

Fig. 11 — Compression Ratios Tell How Much The Fuel-Air Mixture Is Compressed By Volume

Compression ratios tell us how much the fuel-air mixture is compressed by volume. Fig. 11 shows an imaginary case.

When the piston is at the bottom of its stroke, let's measure the amount of liquid the cylinder will hold and say it takes 8 pints (3.75 L).

Now if we remove all the liquid and move the piston to the top of its stroke, we can again pour the cylinder full of liquid. Let's say it only holds one pint (0.5 L).

The ratio is then 8 to 1, which is the *compression ratio.*

In other words, air in this engine is compressed to one-eighth of its former volume by the moving piston. Later we'll see how compression affects the engine.

ENGINE CYCLES

For an engine to operate, a definite series of events must occur in sequence. They are:

1. *Fill the cylinder with a combustible mixture.*

2. *Compress this mixture into a smaller space.*

3. *Ignite the mixture and cause it to expand, producing power.*

4. *Remove the burned gases from the cylinder.*

The sequence above is generally called:

- **Intake**
- **Compression**
- **Power**
- **Exhaust**

To produce sustained power, the engine must repeat this sequence over and over again.

One complete series of these events in an engine is called a *cycle.*

Most engines have one of two types of cycles:

- **Two-Stroke Cycle**
- **Four-Stroke Cycle**

In the TWO-STROKE CYCLE engine, there are two strokes of the piston, one up and one down, during each cycle (Fig. 12). Then it starts over again on another cycle of the same two strokes. This whole cycle occurs during one revolution of the crankshaft.

In the FOUR-STROKE CYCLE engine, there are four strokes of the piston, two up and two down, during each cycle (Fig. 12). Then it starts over again on another cycle of the same four strokes. This cycle occurs during *two* revolutions of the crankshaft. Most engines today operate on the four-stroke cycle.

Let's see how each type of cycle works in detail.

TWO-STROKE CYCLE ENGINE

In the **two-stroke cycle** engine, the complete cycle of events—intake, compression, power, and exhaust—takes place during two piston strokes.

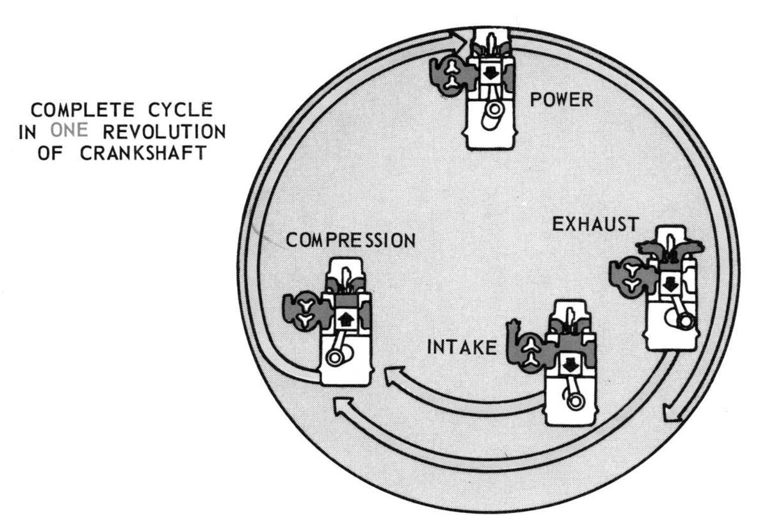

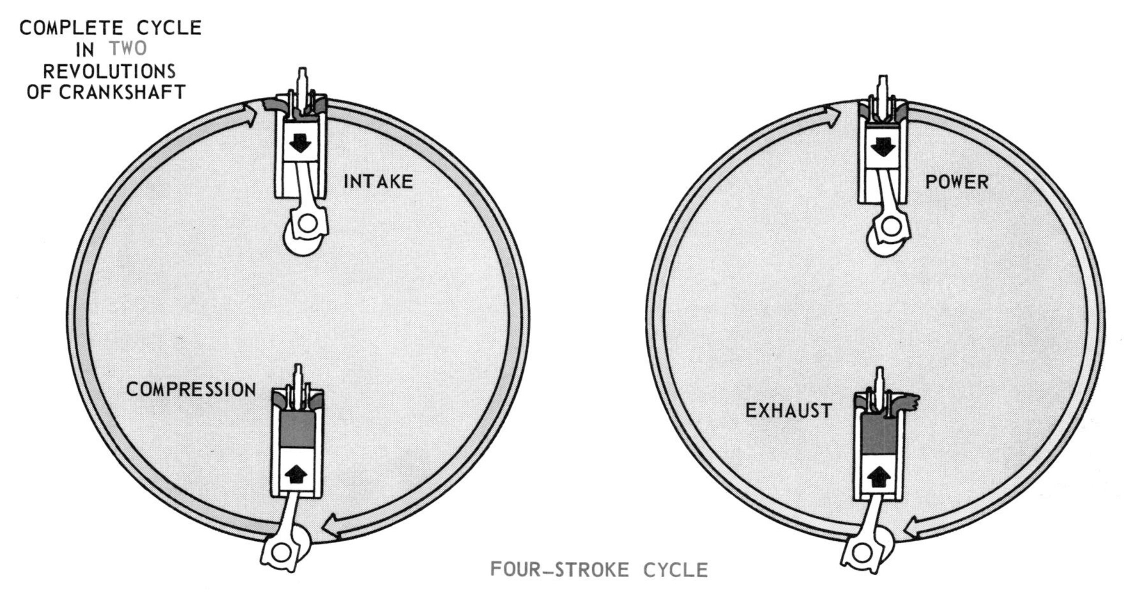

Fig. 12 — Two-Stroke And Four-Stroke Cycles Compared

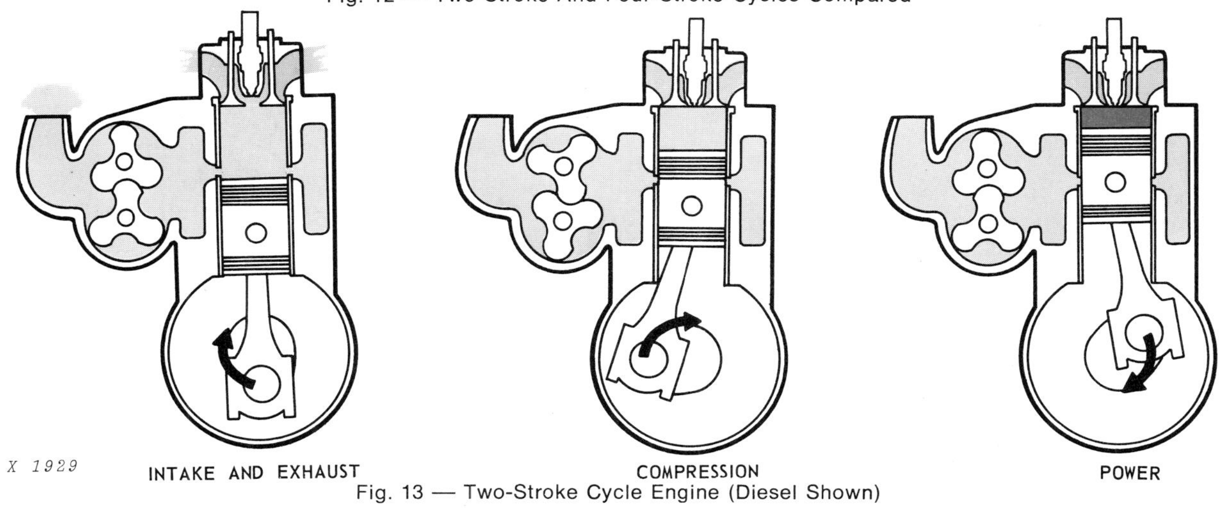

Fig. 13 — Two-Stroke Cycle Engine (Diesel Shown)

Litho in U.S.A.

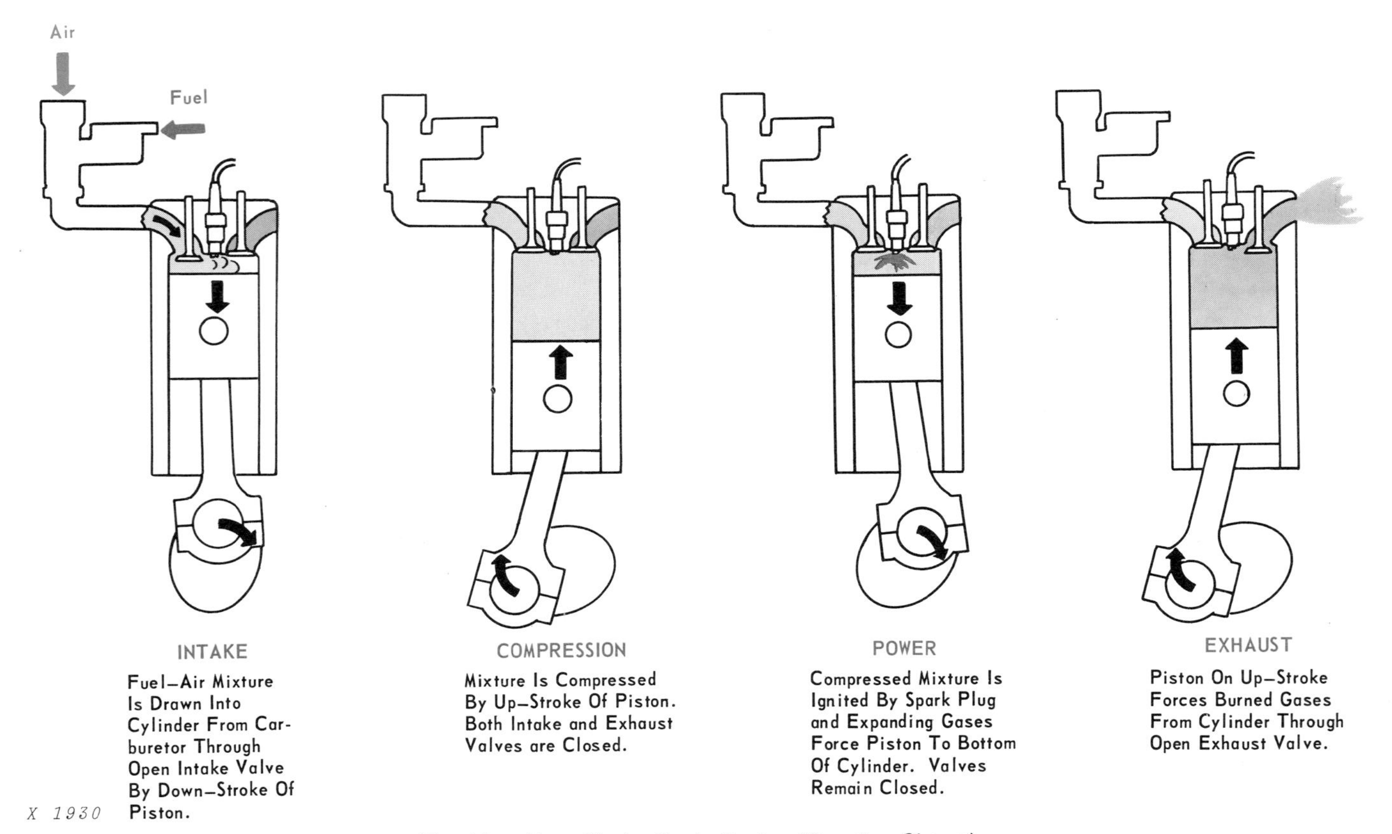

Fig. 14 — Four-Stroke Cycle Engine (Gasoline Shown)

Fig. 13 shows a two-cycle diesel engine in operation. Every other stroke is a power stroke; each time the piston moves down it is a power stroke.

In the diesel engine shown, air alone is compressed in the cylinder. A charge of fuel is then sprayed into the cylinder and ignites from the heat of compression.

In the two-cycle engine, intake and exhaust take place during part of the compression and power strokes.

A blower is sometimes used to force air into the cylinder for expelling exhaust gases and to supply fresh air for combustion. The cylinder wall contains a row of ports which are above the piston when it is at the bottom of its stroke. These ports admit air from the blower into the cylinder when they are uncovered (during *intake).*

The flow of air toward the exhaust valves pushes the *exhaust* gases out of the cylinders and leaves them full of clean air when the piston again rises to cover the ports (during *compression).*

At the same time, the exhaust valves close and the fresh air is compressed in the closed cylinder.

When the piston almost reaches the top of its *compression* stroke, fuel is sprayed into the combustion area as shown. The heat of compression ignites the fuel and the resulting pressure forces the piston down on its *power* stroke.

As the piston nears the bottom of its stroke, the *exhaust* valves are again opened and the burned gases escape.

The piston then uncovers the *intake* ports and the cycle begins once more.

This entire cycle is completed in *one* revolution of the crankshaft or *two strokes* of the piston—one up and one down.

FOUR-STROKE CYCLE ENGINE

In **four-stroke cycle** engines, the same four operations occur—intake, compression, power, and exhaust. However, four strokes of the piston—two up and two down—are needed to complete the cycle. As a result, the crankshaft will rotate *two* complete turns before one cycle is completed (Fig. 14).

Intake Stroke

The intake stroke starts with the piston near the top and ends shortly after the bottom of its stroke. The intake valve is opened, allowing the cylinder

 Litho in U.S.A.

as the piston moves down to receive the fuel-air mixture. The valve is then closed, sealing the cylinder.

Compression Stroke

The compression stroke begins with the piston at bottom and rising up to compress the fuel-air mixture. Since the intake and exhaust valves are closed, there is no escape for the fuel-air and it is compressed to a fraction of its original volume.

Power Stroke

The power stroke begins when the piston almost reaches the top of its stroke and the fuel-air mixture is ignited. As the mixture burns and expands, it forces the piston down on its power stroke. The valves remain closed so that all the force is exerted on the piston.

Exhaust Stroke

The exhaust stroke begins when the piston nears the end of its power stroke. The exhaust valve is opened and the piston rises, pushing out the burned gases. When the piston reaches the top, the exhaust valve is closed and the piston is ready for a new four-stroke cycle of intake, compression, power, and exhaust.

As it completes the cycle, the crankshaft has gone all the way around twice.

TWO-CYCLE VS. FOUR-CYCLE ENGINES

It might seem that the two-cycle engine can produce twice as much power as a four-cycle engine.

However, this is not true. With the two-cycle engine, some power may be used to drive the blower that forces the fuel-air charge into the cylinder under pressure. Also, the burned gases are not completely cleared from the cylinder, this results in less power per power stroke.

The actual gain in power with a two-cycle engine is about 75 percent (over a four-cycle engine of the same displacement).

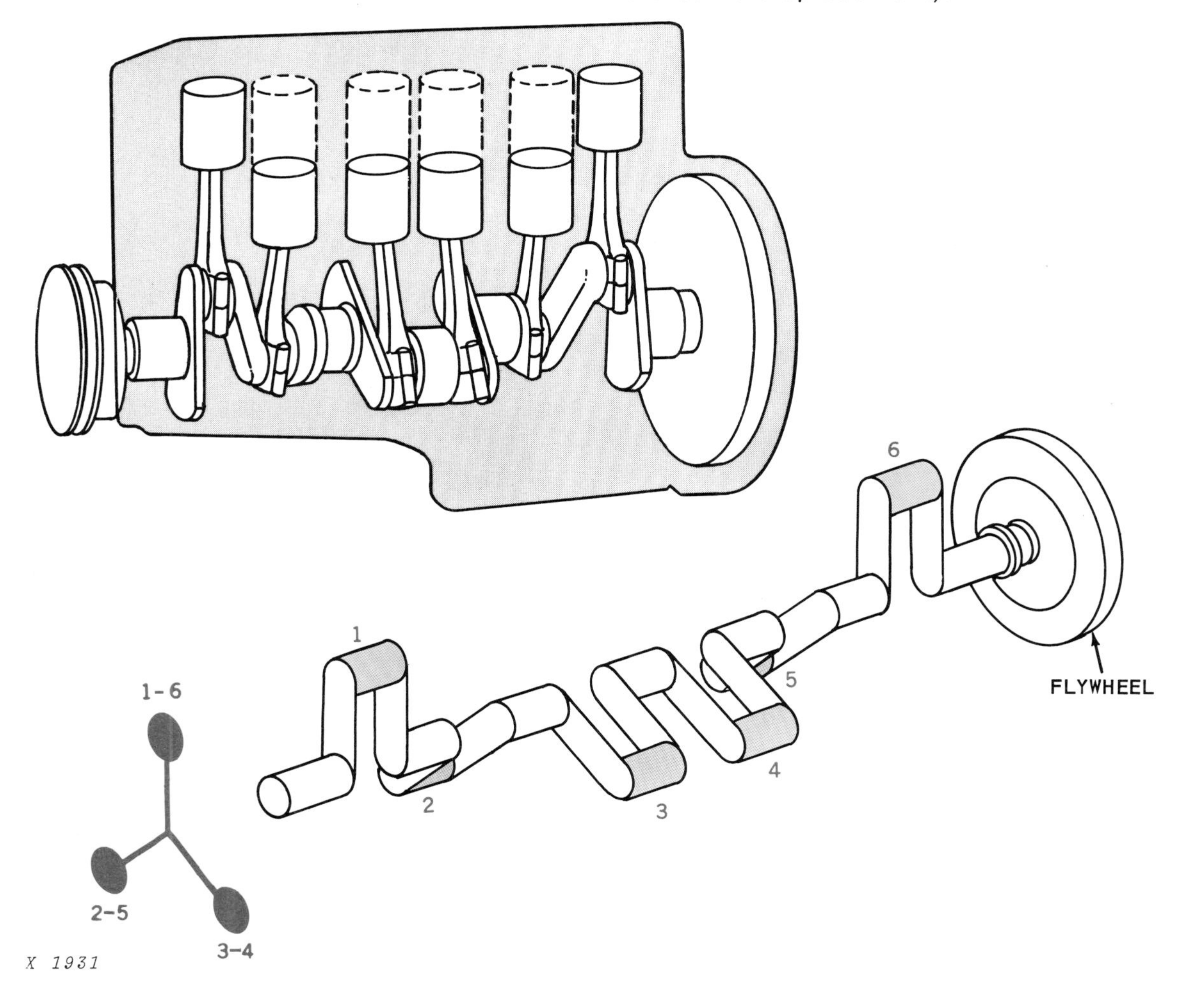

Fig. 15 — Crankshaft For A Six-Cylinder Engine

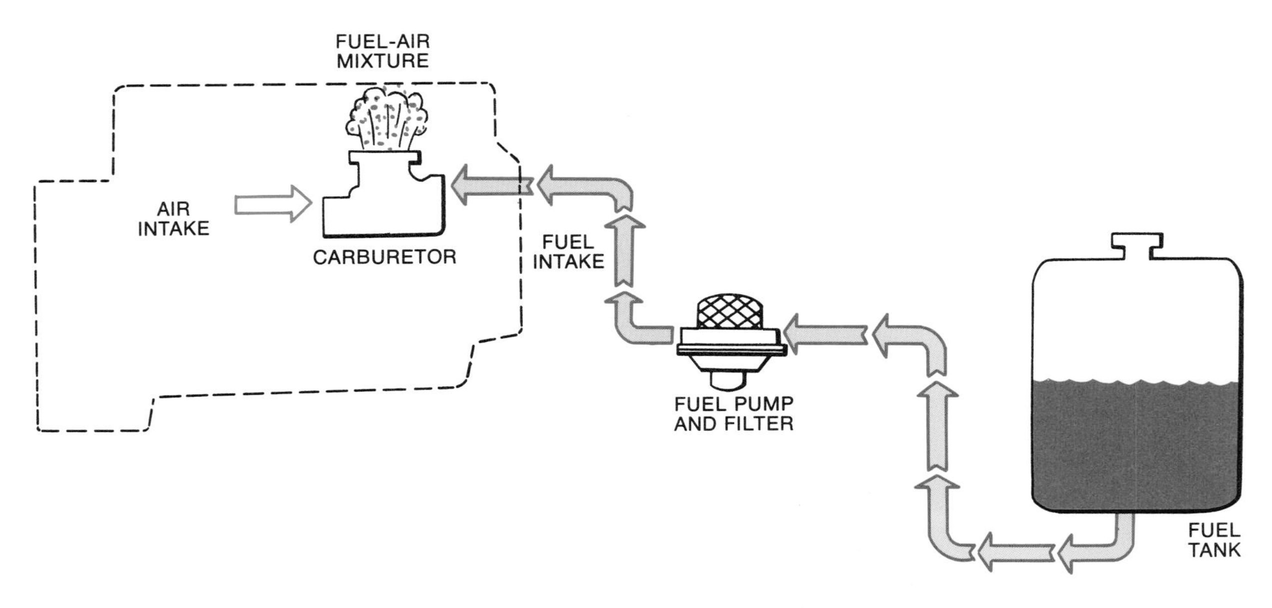

Fig. 16 — Gasoline Fuel System

MULTIPLE-CYLINDER ENGINES

So far we have covered only basic one-cylinder engines.

A single cylinder gives only one power impulse every two revolutions of the crankshaft in a four-cycle engine. Thus it is producing power only one-fourth of the time.

For a more continuous flow of power, modern engines use four, six, eight, or more cylinders. The same series of cycles takes place in each cylinder.

For example, in a typical four-stroke cycle engine having six cylinders, the cranks on the crankshaft are set 120 degrees apart (Fig. 15). The cranks for cylinders 1 and 6, 2 and 5, 3 and 4 are in line with each other as shown.

The cylinders normally fire and deliver their power strokes in the following order: 1-5-3-6-2-4. Thus the power strokes follow each other so closely that there is a fairly continuous and smooth delivery of power to the crankshaft.

The heavy *flywheel* attaches to the rear of the crankshaft and gives it momentum to return the pistons to the tops of the cylinders after each power stroke. Weights on the crankshaft are used to help balance the forces created in the engine by the rapidly moving parts.

For more details on construction of the basic engine, see Chapter 2.

Now let's look at some of the auxiliary systems that help the engine to operate.

ENGINE SYSTEMS

Now that we've put together a basic engine, let's look at some other systems that are required for good operation:

- **Fuel System**
- **Intake and Exhaust System**
- **Lubrication System**
- **Cooling System**
- **Governing System**

Let's discuss these systems one by one.

FUEL SYSTEMS

A fuel system must deliver clean fuel, in the quantity required, to the fuel intake of an engine. It must provide for safe fuel storage and transfer.

The three fuel systems of concern to us are:

- **Gasoline**
- **LP-Gas**
- **Diesel**

GASOLINE FUEL SYSTEMS

The gasoline fuel system supplies a combustible mixture of fuel and air for the engine.

The basic gasoline fuel system (see Fig. 16) has three parts.

- **Fuel Tank—stores fuel**
- **Fuel Pump—moves fuel to carburetor**
- **Carburetor—atomizes fuel and mixes with air**

In operation, the *fuel pump* moves gasoline from the tank to the carburetor bowl.

The *carburetor* is basically an air tube which atomizes fuel and mixes it with air by a difference in air pressure. It meters both the fuel and air for the engine.

On its intake stroke, the engine creates a partial vacuum. This allows outside air pressure to force the fuel-air vapor mixed in the carburetor into the engine cylinder.

Fuel Supply Systems

Fuel can be supplied to the carburetor in two ways:

- **Gravity-Feed**
- **Force-Feed**

The GRAVITY-FEED system has the fuel tank placed *above* the level of the carburetor. This system does *not* use a fuel pump. Instead, the fuel flows by *gravity* to the carburetor.

The FORCE-FEED system allows the fuel tank to be located *below* the carburetor if necessary. A fuel pump moves the fuel from the tank to the carburetor as shown in Fig. 16.

More details on gasoline fuel systems are given in Chapter 3.

LP-GAS FUEL SYSTEMS

The LP-gas fuel system (Fig. 17) also supplies a combustible mixture of fuel and air to its engine. However, LP-gas vaporizes at low temperatures.

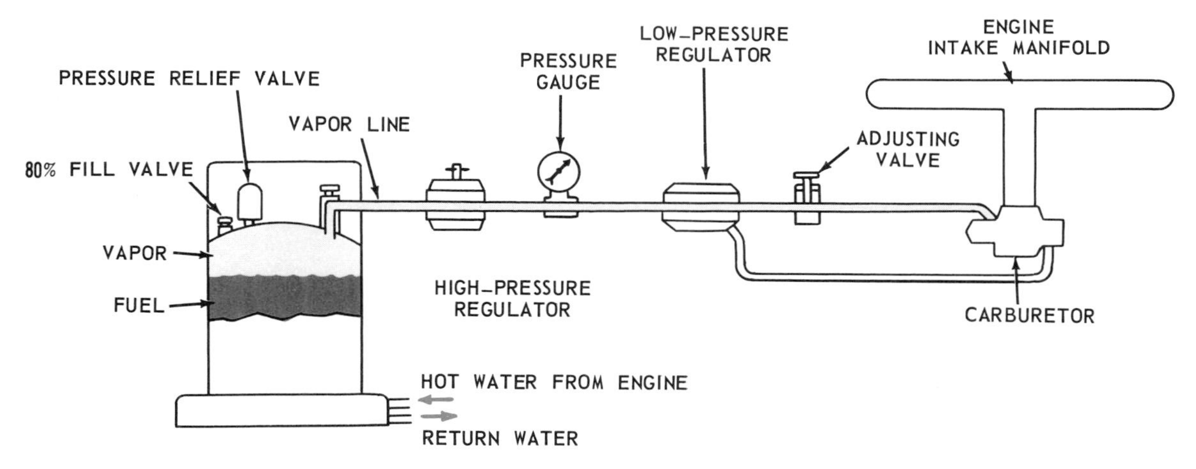

LP-GAS SYSTEM USING VAPOR WITHDRAWAL

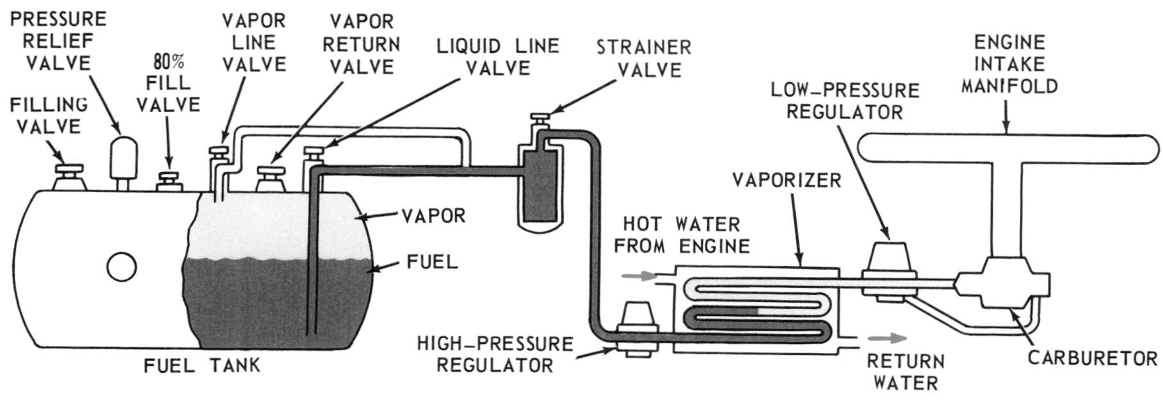

LP-GAS SYSTEM USING LIQUID WITHDRAWAL

Fig. 17 — LP-Gas Fuel System

Thus the fuel tank must be a closed unit to prevent vapor from escaping.

To withdraw fuel from the tank, two methods are used:

- **Liquid Withdrawal**
- **Vapor Withdrawal**

VAPOR withdrawal of fuel is used in starting; the fuel system is later switched to liquid withdrawal after warm-up. This is because in a cold engine the heat exchanger cannot change the liquid fuel to vapor, and the carburetor operates only on vapor.

The liquid and vapor line valves shown in Fig. 17 provide for safety and for selection of fuel—liquid or vapor. The filters remove moisture and dirt. The pressure regulators keep a constant pressure of fuel at the carburetor for accurate fuel metering.

In the LIQUID withdrawal system, a heat exchanger converts the liquid fuel to vapor. The heat exchanger does this by circulating hot water from the engine cooling system around the fuel line. As the fuel heats up and pressure is reduced, it vaporizes. The *liquid* withdrawal system is most common today.

The LP-gas carburetor is simpler than the gasoline type, since the fuel is already vaporized. It meters the vapor and mixes it with the proper amount of air for the engine.

Chapter 4 of this manual gives more details on LP-gas systems.

DIESEL FUEL SYSTEMS

In the diesel fuel system, fuel is sprayed directly into the engine combustion chamber where it mixes with hot compressed air and ignites. No electrical spark is used to ignite the mixtures (as in gasoline and LP-gas engines).

Instead of a carburetor, a fuel injection pump and spray nozzle are used.

The major parts of the diesel fuel system are:

- **Fuel Tank—stores fuel**
- **Fuel Pump—moves fuel to injection pump**
- **Fuel Filters—help clean the fuel**
- **Injection Pump—times, measures, and delivers fuel under pressure**
- **Injection Nozzles—atomize and spray fuel into cylinders**

Fig. 18 shows these major parts of the diesel system.

In operation, the *fuel pump* moves fuel from the tank and pushes it through the *filters.* Clean fuel

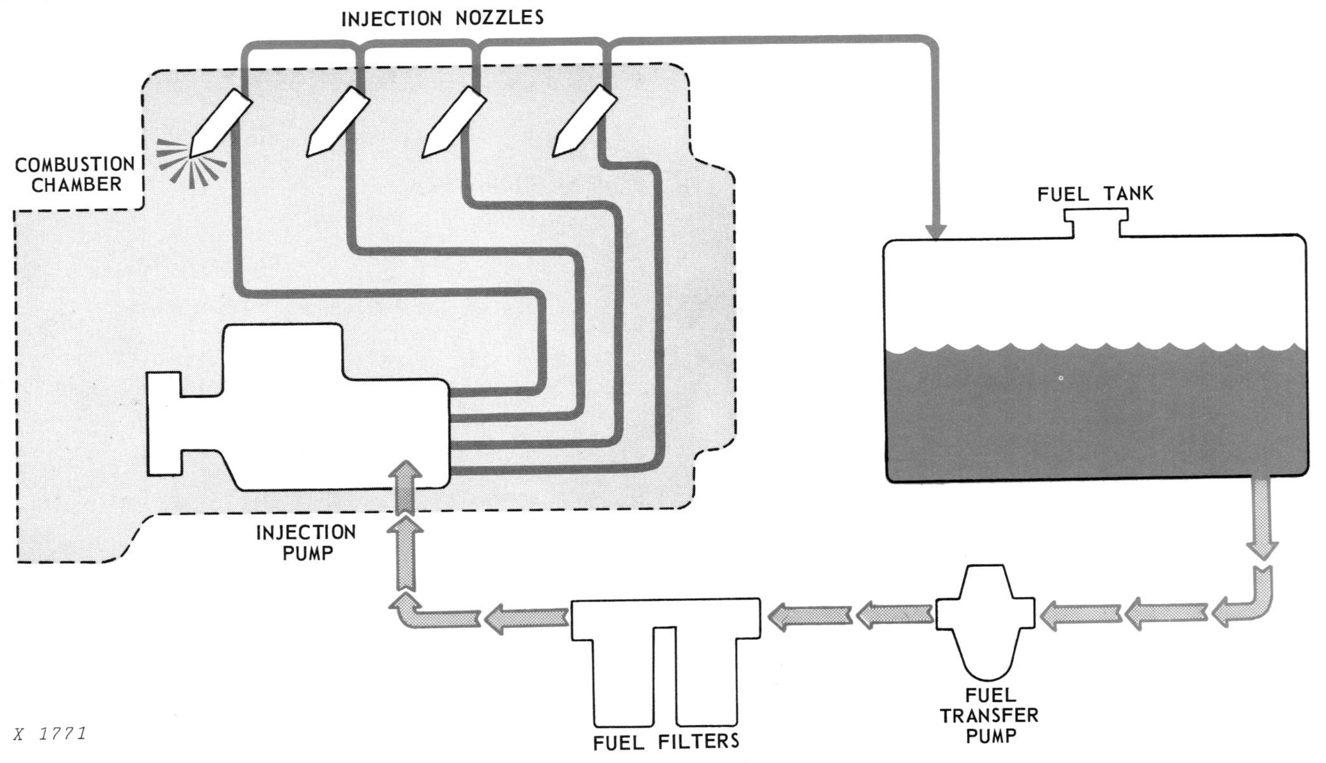

Fig. 18 — Diesel Fuel System

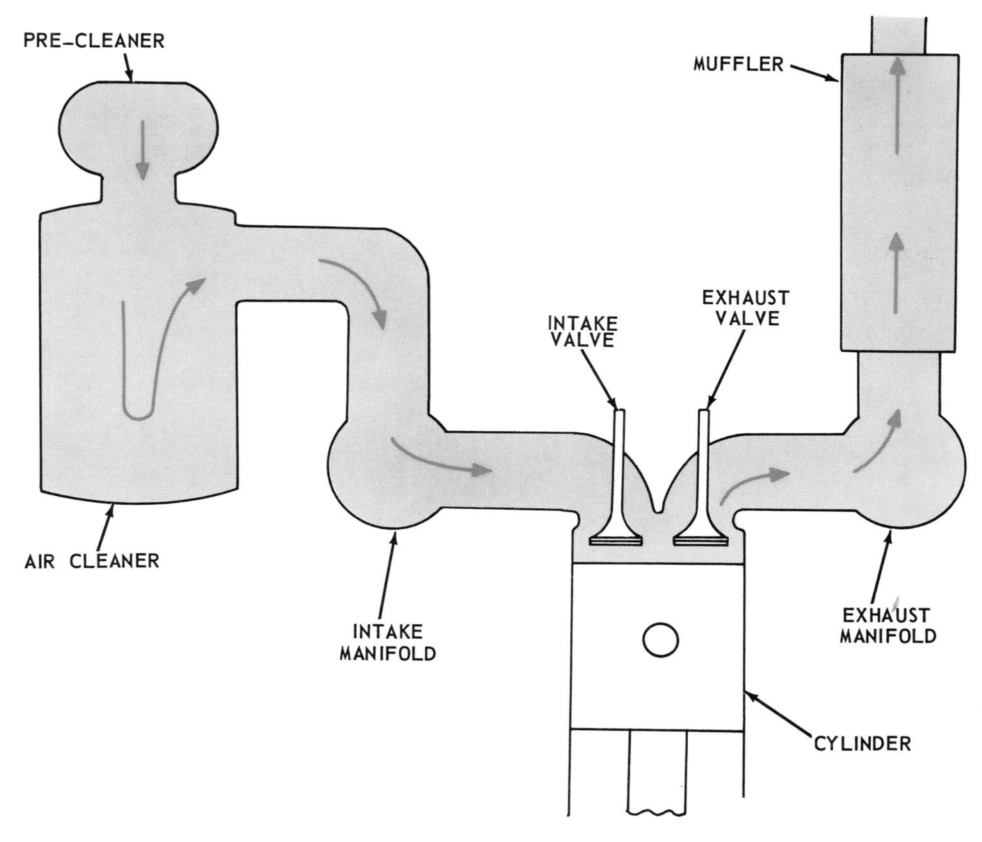

Fig. 19 — Intake And Exhaust Systems

free of water is very vital to the precision parts of the diesel injection system. Extra filters are often used to assure clean fuel, but buying clean fuel and storing it properly are also prime needs.

The fuel is then pushed on to the *injection pump* where it is metered, put under high pressure, and delivered to each *injection nozzle* in turn.

The nozzles each serve one cylinder; they atomize the fuel and spray it under controlled high pressure into the combustion chamber at the proper moment.

High-pressure fuel is needed at each nozzle to get a fine spray of fuel. This assures good mixing of fuel with the hot compressed air for full combustion.

Chapter 5 of this manual gives more details on the diesel fuel system.

INTAKE AND EXHAUST SYSTEMS

Intake and exhaust systems carry the fuel-air mixture into the engine and remove the exhaust gases after combustion (Fig. 19).

INTAKE SYSTEM

The intake system supplies the engine with clean air of the proper quantity, temperature, and mix for good combustion.

The intake system has five parts:

- **Air cleaners**
- **Blower or turbocharger (optional)**
- **Intake manifold**
- **Carburetor air inlet**
- **Intake valves**

Air cleaners filter dust and dirt from the air passing through them enroute to the carburetor. Precleaners prevent larger particles from reaching the air cleaners and plugging it.

Blowers can be used on two-cycle engines to force air into the cylinder while exhaust gases are driven out. The blower is an air pump which pressurizes air.

Turbochargers increase horsepower by packing more air or fuel-air mixture into the engine cylinders than the engine could take in by natural aspiration.

Intake manifolds transport the air-fuel mixture (pure air on diesel engines) to the engine cylinders.

Carburetors mix incoming air with fuel in the proper proportion for combustion, and control engine speed.

Intake valves admit air to diesel engines and the fuel-air mixture to spark-ignition engines. They are normally opened and closed by mechanical linkage from the camshaft.

For more details on intake systems, see Chapter 6.

EXHAUST SYSTEMS

The exhaust system collects the exhaust gases after combustion and carries them away. This is really three jobs:

1) Removing heat

2) Muffling engine sounds

3) Carrying away burned and unburned gases

The exhaust system has these basic parts:

- **Exhaust valves**
- **Exhaust manifold**
- **Muffler**

Exhaust valves open to release the burnt gases on four-cycle engines. The valves are normally operated by the camshaft.

The *exhaust manifold* collects the exhaust gases and conducts them away from the cylinder.

The *muffler* reduces the sounds of the engine during the exhaust period.

See Chapter 6 for details on exhaust systems.

LUBRICATION SYSTEMS

The lubrication system does these jobs for the engine:

1) Reduces friction between moving parts

2) Absorbs and dissipates heat

3) Seals the piston rings and cylinder walls

4) Cleans and flushes moving parts

5) Helps deaden the noise of the engine

With lubricating oil, the system is able to do all these jobs at once (Fig. 20). Without oil, the engine would soon wear out, burn up, or seize. For oil not only reduces friction by forming a film between parts, it also conducts heat away from these parts.

The lubrication system may work by splashing oil on the moving parts or it may feed oil under pressure to the parts via internal oil passages as shown in Fig. 20. In some cases, both methods are used at the same time.

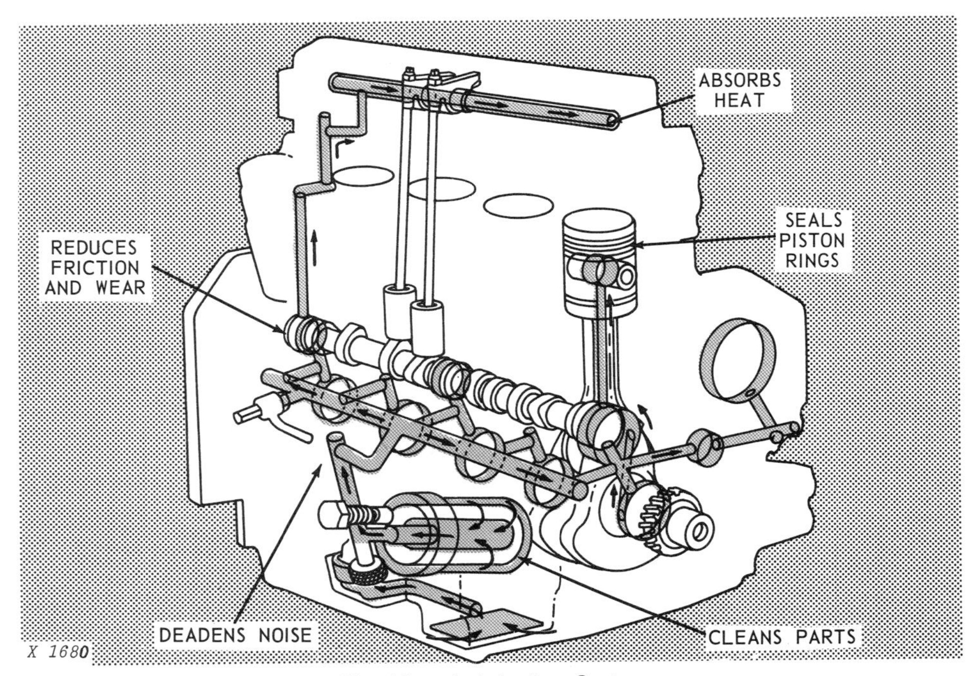

Fig. 20 — Lubrication System

Litho in U.S.A.

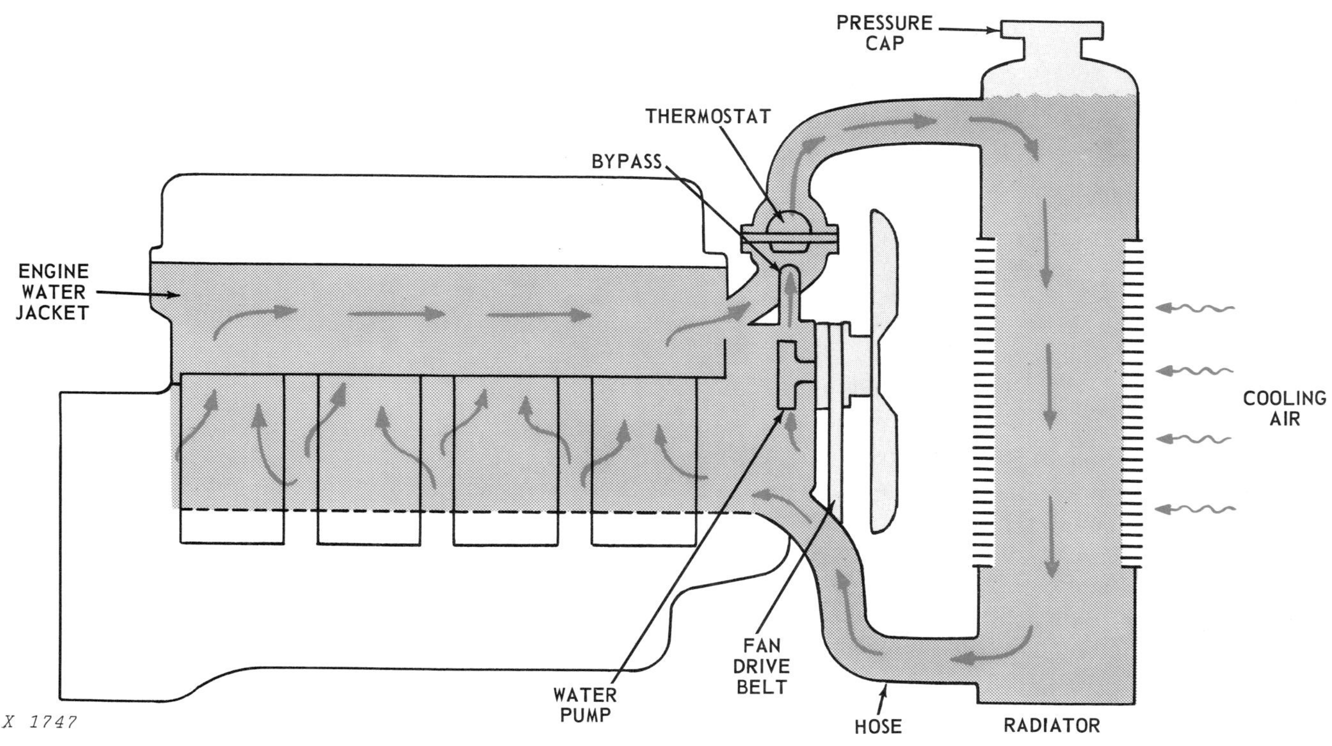

Fig. 21 — Cooling System (Liquid Type Shown)

The engine crankcase forms an oil reservoir where oil is stored and also cooled.

The crankcase must be vented to prevent pressure build-ups from the blow-by of gases past the pistons.

Modern venting sometimes includes a system which routes crankcase vapors back to the intake system to reduce air pollution.

See Chapter 7 for further details on lubrication systems.

COOLING SYSTEMS

The cooling system prevents overheating of the engine. Some heat is necessary for combustion, but the working engine generates too much heat. The cooling system carries off this excess heat.

Cooling systems are designed to use parts that are *matched* in capacity. A matched cooling system will provide adequate heat rejection. If one part is replaced that is under or overcapacity, the effectiveness of the system will be decreased. Parts include: water pump, radiator, coolant, piping, thermostat, and fan.

TYPES OF COOLING SYSTEMS

Two types of cooling systems are used on modern engines:

• *Air Cooling—uses air passing around the engine to dissipate heat*

• *Liquid Cooling—uses water around the engine to dissipate heat*

AIR COOLING is used primarily on small engines or aircraft as it is difficult to route air to all the heat points of larger engines. Metal baffles, ducts, and blowers are used to aid in distributing air.

LIQUID COOLING normally uses water as a coolant. In cold weather, anti-freeze solutions are added to the water to prevent freezing. The water circulates in a jacket around the cylinders and cylinder head. As heat radiates, it is absorbed by the water, which then flows to the radiator. Air flow through the radiator cools the water and dissipates heat into the air. The water then recirculates into the engine to pick up more heat (Fig. 21).

See Chapter 8 for more details on cooling systems.

 Litho in U.S.A.

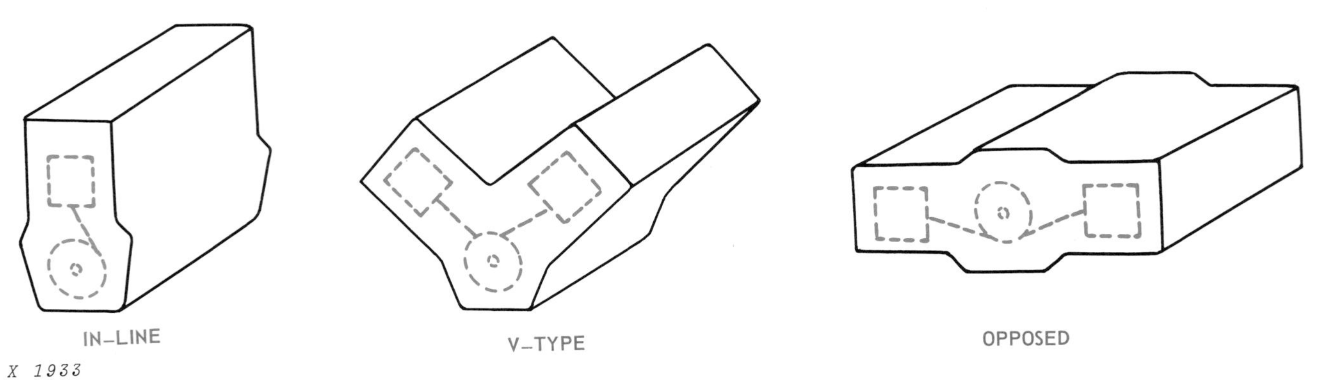

Fig. 22 — Arrangement Of Cylinders—Three Types

GOVERNING SYSTEMS

The governing system keeps the engine speed at a constant level. It does this by varying the amount of fuel on fuel-air mixture supplied to the engine, according to the demands of the load. The level of engine speed is controlled by the position of the speed control lever, connected by linkage to the governor.

The object is to get the engine's power to match the load at all times, to keep the speed at a steady level.

Governors can be either mechanical, hydraulic, or electrical. See Chapter 9 for details.

TYPES OF ENGINES

Engines can be typed in three ways:

- **Cylinder arrangement**
- **Valve arrangement**
- **Type of fuel used**

ARRANGEMENT OF CYLINDERS

Multi-cylinder engines are classified according to the arrangement of cylinders:

- *In-Line—all cylinders in straight line above crankshaft*
- *V-Type—two banks of cylinders in V-shape above crankshaft*
- *Opposed—two rows of cylinders opposite the crankshaft*

The IN-LINE model (Fig. 22) is most popular on farm and industrial machines. The V-TYPE is most popular on automobiles although its use in farm and industrial machines is increasing, while the OPPOSED is limited primarily to small cars and aircraft. The cylinders are normally numbered. With in-line models, the No. 1 cylinder is normally at the end opposite the flywheel. The others are 2, 3, 4 etc. from front to rear. In V-and opposed types, the sequence varies with the manufacturer.

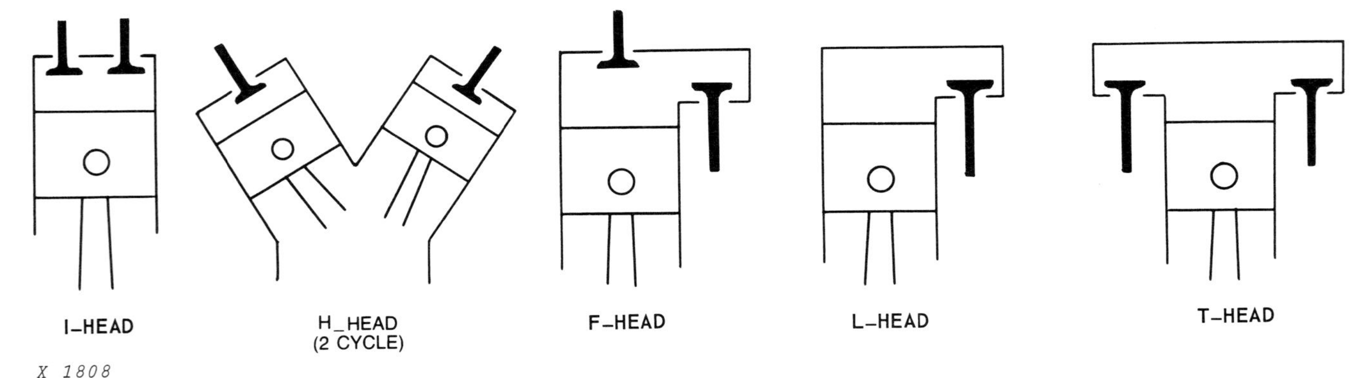

Fig. 23 — Valve Arrangement — Five Types

 Litho in U.S.A.

VALVE ARRANGEMENT

Engines can also be classified by the position and arrangement of the intake and exhaust valves. This normally depends on whether the valves are located in the cylinder block or the cylinder head. The most common types are the *I-head, H-head* (2 cycle), *F-head, L-head* and *T-head* (Fig. 23). See Chapter 2 for details.

FUEL TYPES OF ENGINES

The most common way to type engines is by the type of fuel used. Three fuel types are most common:

- **Gasoline Engine**
- **LP-Gas Engine**
- **Diesel Engine**

The basic operation of each engine is the same and we have already compared the methods of fueling. But now let's look at the overall performance of each one while comparing gasoline and diesel.

What Are The Main Differences Between Gasoline And Diesel Engines?

1. *The method of supplying and igniting fuel.*
2. *The higher compression ratio in diesels.*
3. *The generally more rugged design of diesels.*
4. *The grade and type of fuel used.*

Let's look at each of these differences.

METHODS OF SUPPLYING AND IGNITING FUEL: GASOLINE VS. DIESEL

In *gasoline engines,* fuel and air are mixed *outside* the cylinders, in the carburetor and manifold. The mixture is forced in due to the partial vacuum of the pistons' intake stroke.

In *diesel engines,* there is no premixing of air outside the cylinder. Air only is taken into the cylinder through the intake manifold and compressed. Fuel is then sprayed into the cylinder and mixed with air as the piston nears the top of its compression stroke. See Fig. 24.

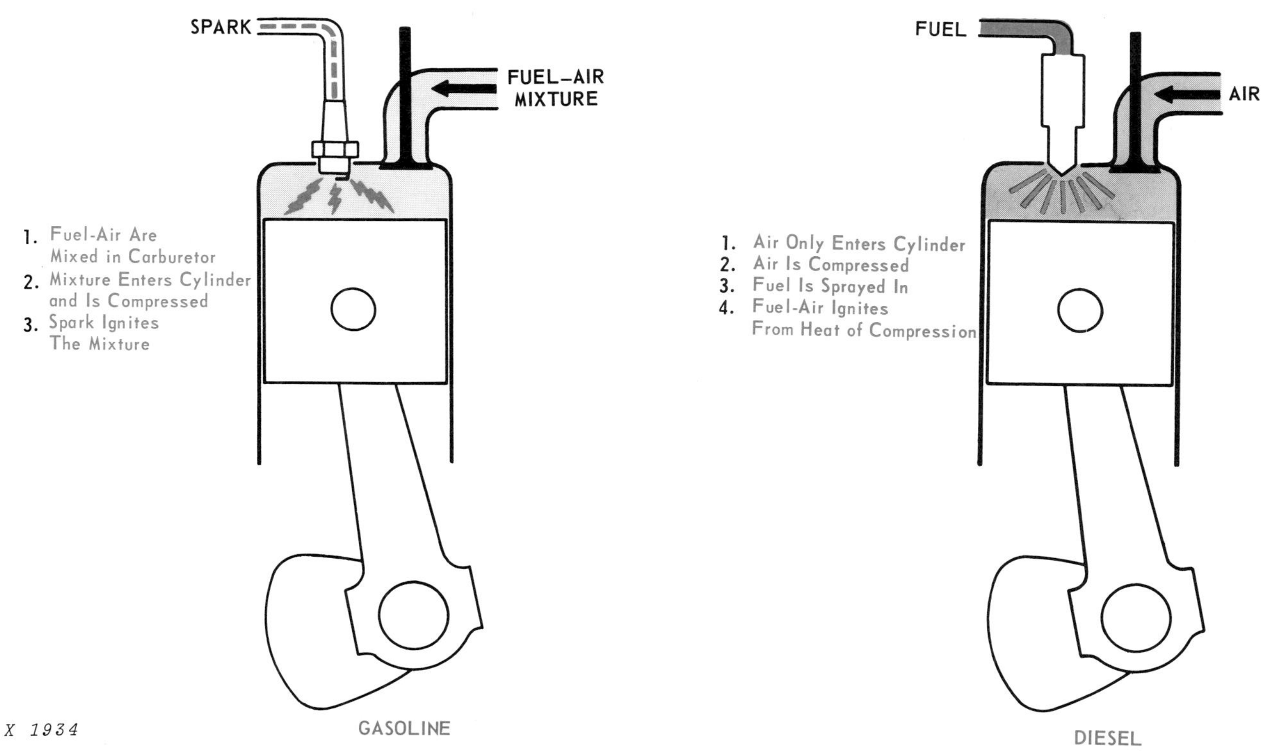

Fig. 24 — Methods Of Supplying And Igniting Fuel (Power Stroke)

Litho in U.S.A.

Gasoline engines use an electric spark to ignite the fuel-air mixture, while diesels use the heat of the compressed air for ignition.

COMPRESSION RATIOS: GASOLINE VS. DIESEL

Compression ratio compares the volume of air in the cylinder before compression with the volume after compression.

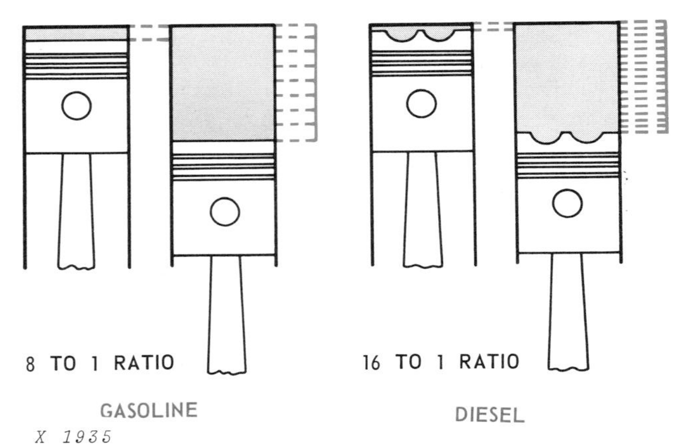

Fig. 25 — Compression Ratios Compared

An 8 to 1 compression ratio is typical for gasoline engines, while a 16 to 1 ratio is common for diesels (Fig. 25).

The higher compression ratio of the diesel raises the temperature of the air high enough to ignite the fuel without a spark.

This also gives the diesel more efficiency because the higher compression results in greater expansion of gases in the cylinder following combustion. Result: a more powerful stroke.

The higher efficiency which results from diesel combustion must be offset by the need for sturdier, more expensive parts to withstand the greater forces of combustion.

DESIGN OF ENGINE PARTS: GASOLINE VS. DIESEL

We have just touched on the next point: diesels must be built sturdier to withstand the greater forces of combustion. This is generally done by "beefing up" the pistons, pins, rods, and cranks, and by adding more main bearings to support the crankshaft.

GRADES AND TYPES OF FUEL: GASOLINE VS. DIESEL

Fuel energy is measured in standard heat units or "British Thermal Units" (BTU) or watts and gives a comparison of the power possible from each fuel.

Diesel fuel has more heat units (BTU) watts per gallon, and so gives more work per gallon of fuel. In addition, diesel fuel is normally cheaper than gasoline.

However, diesel fuel injection equipment is more expensive than gasoline equipment.

When selecting the fuel type for the engine, the deciding factor is how much fuel is consumed per year in the engine operation.

LP-Gas Engines

The LP-gas engine is similar to the gasoline model, but requires special fuel handling and equipment.

LP-gas engines have higher compression ratios than gasoline engines but not as high as diesels.

In areas where LP-gas fuel is available at low prices, these engines are very popular. However, in many areas, LP-gas fuel is not competitive with the other fuels.

Summary: Comparing Engines

The chart below compares gasoline, LP-gas, and diesel engines.

The comparisons assume that each fuel is available at reasonable prices. Performance is based on general applications which are suited to the engine and fuel type. It is also assumed that the engines are all in good condition.

COMPARING THE ENGINES

	Gasoline	LP-Gas	Diesel
Fuel Economy	Fair	Good	Best
Hours Before Maintenance	Fair	Good	Good
Weight per Horsepower	Low	Low	High
Cold Weather Starting	Good	Fair	Fair
Acceleration	Good	Good	Fair
Continuous Duty	Fair	Fair	Good
Lubricating Oil Contamination	Moderate	Lowest	Low

 Litho in U.S.A.

Fig. 26 — Stationary Engine

USES OF ENGINES

There are two basic uses for engines:

- **Stationary**
- **Mobile**

STATIONARY ENGINES

Stationary engines supply power from a fixed location (Fig. 26). Couplers, belts, chains and drive shafts transfer the engine power to other machines.

Because they are in a "fixed" location, they can be designed for one application. Stationary engines, or "power units," drive such machines as compressors, motor driven pumps, and generators.

MOBILE ENGINES

Mobile engines supply power on the move. They power a wide variety of vehicles from road graders to race cars. Mobile engines can be subdivided into two basic types:

- **Structural**
- **Non-structural**

A *structural* engine is mounted to and becomes part of the vehicle frame. It helps to support and carry the load of the vehicle. The structural engine *block* must be strong enough to withstand the load and road stress.

Structural engines are used on some construction equipment and other off-the-road vehicles. An example of this type is a farm tractor engine that is bolted directly to the sides of the main machine frame. The engine helps support and hold the tractor together.

A *non-structural* engine is not a part of the vehicle frame. The vehicle frame in this example is just as strong with or without the engine. Non-structural engine *blocks* do not have to withstand load and road stress like a structural engine. The non-structural engine block can be made of lighter metals, such as aluminum.

These "lighter" engines are usually mounted on rubber pads on the vehicle frame. The main application for non-structural engines is over-the-road vehicles, such as cars and trucks. The lighter engine block weight helps boost vehicle fuel economy and load carrying ability.

A non-structural engine may supply the same performance and reliability of a comparable structural engine. However, when a new engine is used to replace original equipment, make sure you do not replace a structural with a non-structural engine. Structural stress may severely damage the engine block.

Also, if you are replacing original equipment with a different type of engine, be sure the fuel and cooling systems are compatible. These systems must be matched with the engine performance and load carrying ability.

THE BASICS OF ENGINES

We have seen how the engine works. Now let's look at some of the basics which go into the design and operation of engines.

First the laws of mechanics:

- **Matter**
- **Mass**
- **Energy**
- **Inertia**
- **Force**
- **Momentum**
- **Torque**
- **Work**
- **Mechanical Power**

Let's discuss these basics one by one.

MATTER

The substances we encounter in engines are: **solids, liquids** and **gases;** they are the three physical states of matter.

If you look at Fig. 27, you will see the following truths:

1. *Solids* have a definite volume and shape.
2. *Liquids* have a definite volume but no definite shape.
3. *Gases* have no definite volume or shape.

All matter can be changed from one state to another by heating or cooling. Water is a liquid which can be changed to ice (solid) or steam (gas) by changing its temperature. However, if the temperature is returned to the original point, the water will become liquid again; the water has been subjected to physical change only, because its characteristics stayed the same.

Fig. 27 — Three Physical States Of Matter

In summary, matter can be changed, but it cannot be destroyed.

MASS

Mass is often confused with weight. *Mass is the measure of how much matter is in a body.* Weight is the measure of Earth's gravitational pull. A body has the same mass on Earth as it has 2,000 miles (3200 km) out in space, but its weight is much less out in space.

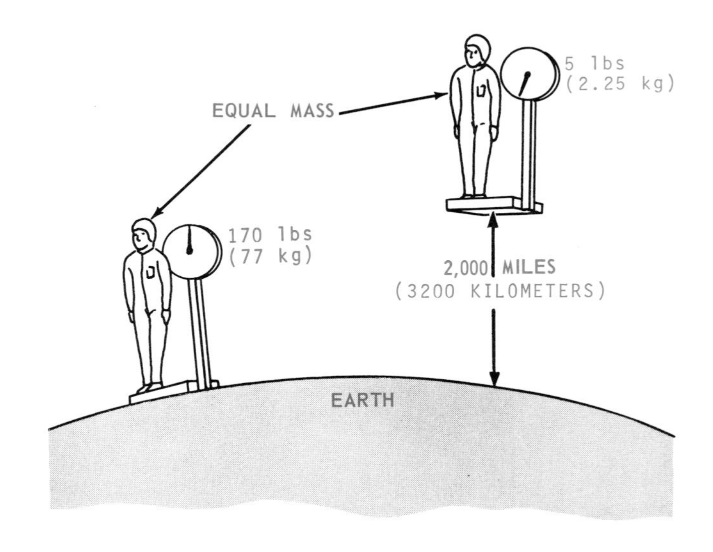

Fig. 28 — Mass Versus Weight

In Fig. 28, the man weighs 170 pounds (77 kg) while standing on the earth. Out in space he may weigh only 5 pounds (2.25 kg). However, in either location he has the same amount of *matter* (or mass) in his body.

 Litho in U.S.A.

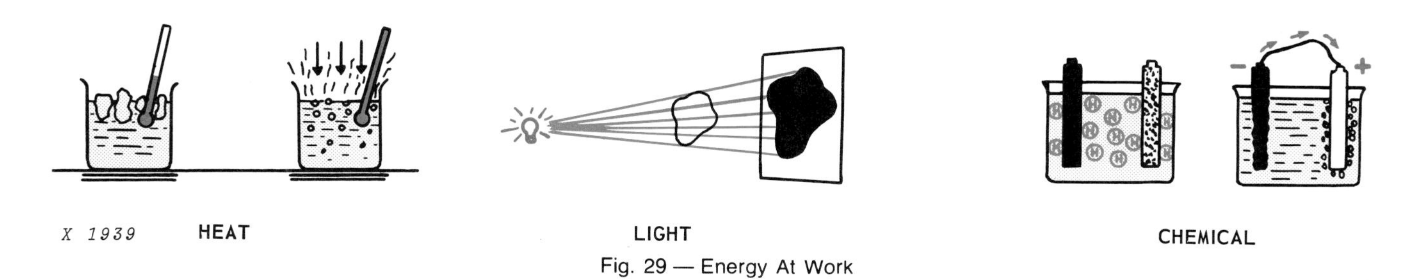

Fig. 29 — Energy At Work

ENERGY

Things like electricity, light, sound and heat are forms of **energy.** They do not occupy space or have weight in the usual sense. *Energy is the thing that produces changes in matter.*

Fig. 29 shows heat energy converting water, light energy forming an image on film, and electrical energy working in a chemical cell. Chemical energy heats your home and runs an engine, and mechanical energy does work.

INERTIA

Inertia is the tendency of a body to keep its state of rest or motion.

If you're sitting in the wagon shown in Fig. 30 and someone gives you a push from behind, your body will fall backward. Nothing actually pushed you backward, your body just tried to stay at rest.

If someone stops you while you're in motion, you will pitch forward. This is because your body wants to keep moving at the same speed. The larger the mass of your body, the more you will be affected by inertia.

FORCE

A **force** is a push or pull which starts, stops, or changes the motion of a body.

From this we conclude that if all the forces acting on a body are equal from all directions, the body will be at rest. If any one of the forces is greater, the body will be set into motion in the direction of the force.

If six men push equally hard where shown in Fig. 31, the box will retain its position. When the top man pushes harder, the box will move downward.

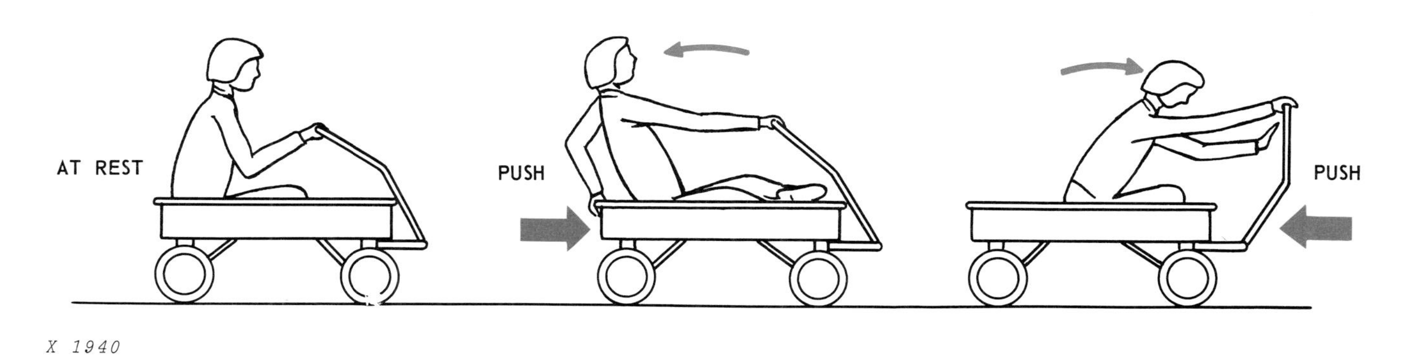

Fig. 30 — Inertia

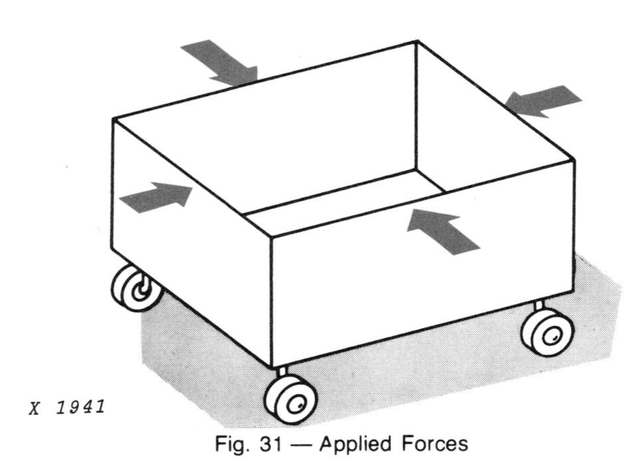

Fig. 31 — Applied Forces

MOMENTUM

Fig. 32 — Momentum Forces

When a body is in motion, it is said to have **momentum** which is the product of its mass and velocity (speed). A body moving in a straight line will keep going in a straight line at the same speed forever if no other forces act upon it (Fig. 32). The laws of momentum are equally effective when a body is rotating; it would continue to rotate. Momentum and inertia are sources of energy because of their mass.

TORQUE

If the forces applied to a body do not all act at a single point, the body will tend to rotate. The turning effect of any force applied to a body is found by multiplying the amount of the force by the distance from the pivot point to the line of the force.

This *turning* effect of a force is called **torque** and the distance mentioned is the torque arm.

Consider the torque wrench shown in Fig. 33. If we apply more force at the point shown or apply the force farther out, the torque will be increased. This increased torque will cause the wrench to rotate faster or give us more turning force at the pivot point.

An engine crankshaft reacts to the pushing force of the piston and connecting rod in the same manner.

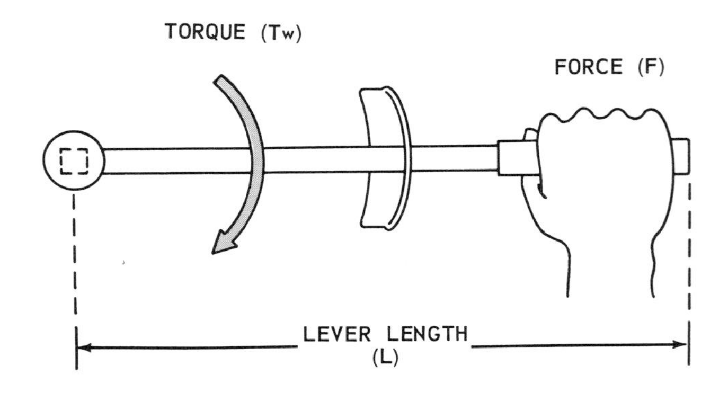

Fig. 33 — Torque (Or Turning Force)

WORK

When you're pushing on a large rock and it fails to move, you feel like you're working hard. In Physics, **work** is accomplished only when the rock is moved by the pushing you're doing. Work is expressed as a force unit multiplied by a distance unit.

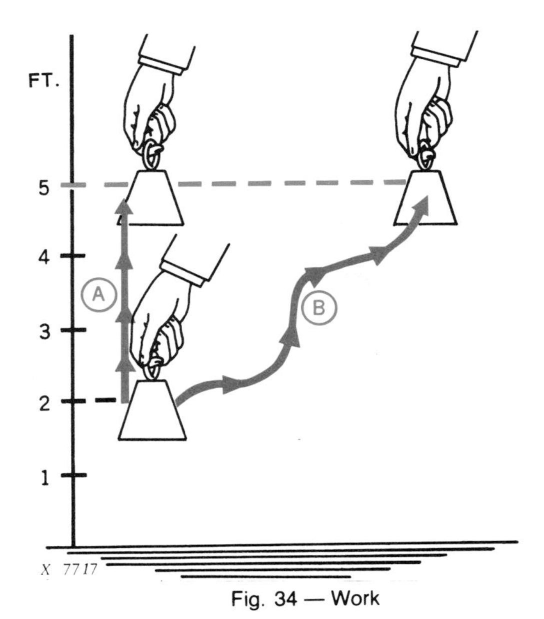

Fig. 34 — Work

If you stand and hold a weight at the 2-foot (600 mm) level as shown in Fig. 34, you aren't really doing any work, as the weight is stationary. However, if you move it to the 5-foot (1.5 m) level, you have moved the weight 3 feet (900 mm), and work was done. The path you take to get to the 5-foot (1.5 m) level is not important; the amount of work done is 20 lbs (9 kg) times 3 feet (900 mm) = 60 lb. ft. (80 J) as shown, whether you take path A or B.

MECHANICAL POWER

Power is a rate of doing work and the term **mechanical** is the energy method used. Other energy methods include chemical, electrical, heat and sound.

Because we are mainly concerned with engines, our unit of power is the **horsepower (watt-metric).** One horsepower is equal to the lifting of a weight of 550 lbs. (250 kg) one foot (300 mm) in one second [or 33,000 lbs (15,000 kg), one foot (300 mm) in one minute].

Energy conversion allows us to compare different types of power. A diesel engine produces *mechanical power* by chemical change (Fig. 35). Fuel (liquid) and air (gas) is burned (chemical change) in the combustion chamber and the engine crankshaft rotates.

An electric motor also produces mechanical power when connected to a battery (Fig. 35). The electrolyte in the battery reacting on the battery plates produces electricity by chemical change and the electric motor runs.

When diesel fuel burns and produces heat at the rate of 2545 BTU per hour (746 watts), the fuel energy is being expended in the engine at the rate of one horsepower per hour.

If a battery delivers electricity to a motor for one hour at the rate of 746 watts, the motor will consume 746 watt-hours or 1 hp-hr of energy.

It is interesting to note that by simple conversion, 746 watts of electrical power equals 2545 BTU per hour when converted to heat.

This completes our discussion of the basic **laws of mechanics.**

HORSEPOWER

The term horsepower is a unit of measurement in rating engines and motors. There are several categories of horsepower, all very necessary for the design of an efficient engine. To sum them up quickly, we talk about **theoretical horsepower** and **net horsepower (useful)** and of those in between.

The most common horsepower terms are:

- **Indicated (IHP)**
- **Friction (FHP)**
- **Flywheel or Brake (BHP)**
- **Drawbar**
- **Power Take-Off (PTO)**
- **Rated**

Let's see what each type of horsepower means.

INDICATED HORSEPOWER (IHP)

Indicated horsepower (IHP) is measured in the combustion chamber of a cylinder by special instruments. The instrument measures the actual gas pressure developed. Using this measurement, an engineer can calculate the amount of energy that is released in the cylinder.

However, here we're more interested in the measurements we can do something about, for *indicated* horsepower neglects such things as friction and the type of power we actually need to do our work. Indicated horsepower is *theoretical* horsepower (Fig. 36).

.02 GALLON (0.08 LITER) 2545 B.T.U. 746 WATTS } PER HOUR = 1 HP

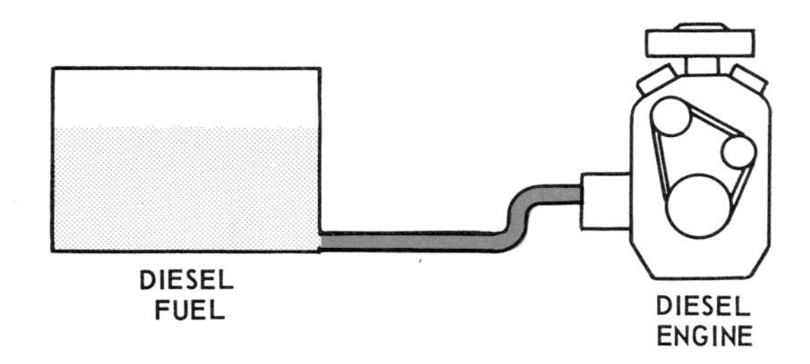

MECHANICAL POWER

746 WATT-HRS. 2545 B.T.U. } PER HOUR = 746 Watts = 1 hp

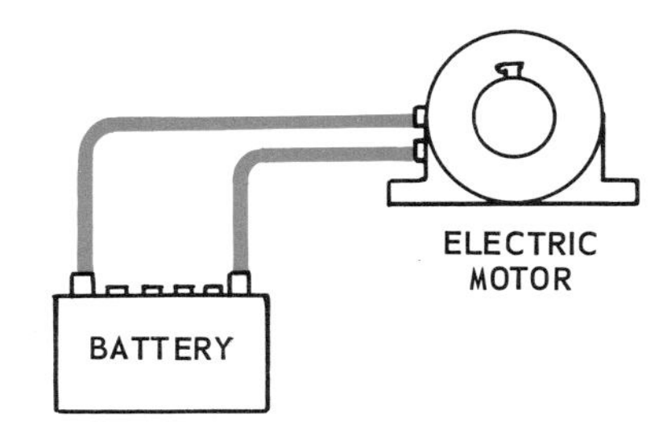

ELECTRICAL POWER

X7718

Fig. 35 — Mechanical And Electrical Power Compared

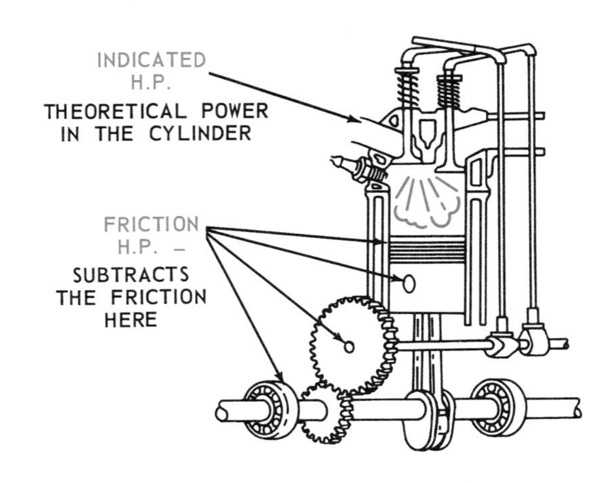

Fig. 36 — Indicated And Friction Horsepower

Friction horsepower (FHP) allows for the friction between engine parts such as pistons and cylinder walls, and the power needed for compression. Friction is a loss factor and a producer of heat.

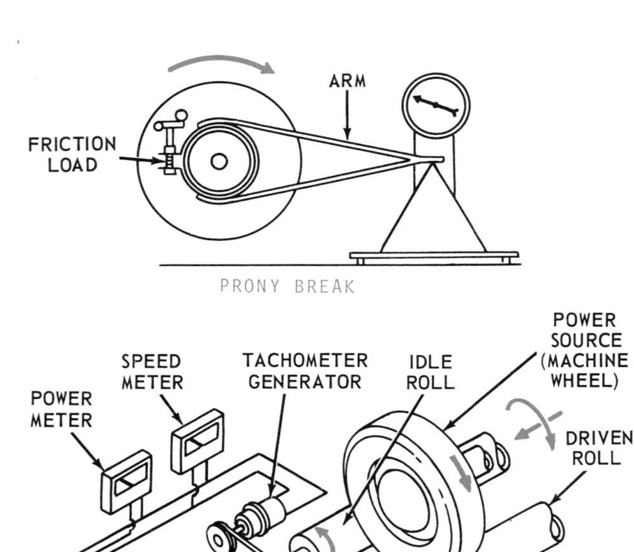

Fig. 37 — Units Which Measure Horsepower Of An Engine

Remember, energy cannot be destroyed, merely converted or divided. If the bearings are heating up while the mechanical energy is working, some of that mechanical energy is being converted to heat and is lost into the cooling system.

Friction horsepower then is the difference between *indicated* horsepower and *usable* horsepower (Fig. 36). So it is a factor in engine *efficiency.*

Lubricating oils place a thin film between two surfaces to reduce friction. In most engines, bearings are lubricated with oil under pressure to make the shafts float on an oil film and so reduce friction.

FLYWHEEL HORSEPOWER OR BRAKE HORSEPOWER (BHP)

Now we come to the first really practical unit of measurement for an engine. This is the point where we can couple to the engine and actually draw power. **Flywheel horsepower** is the FHP (friction losses) subtracted from the IHP (theoretical horsepower). For simplicity, FHP is all the engine losses; friction, etc. If the losses equal 10 horsepower (7.5 kW), and if IHP equals 50 horsepower (37 kW), the result is 50-10=40 (37-7.5=29.5 kW) which is the flywheel horsepower.

Flywheel horsepower is also called brake horsepower (BHP). It is the maximum horsepower the engine can produce without alteration.

Flywheel or brake horsepower is measured by a *Prony brake or a dynamometer* (Fig. 37). Both test instruments apply a load to the engine which is measured in pounds (kilograms). If the length of the lever arm is known in feet (meters), we can measure foot-pounds (newton-meters) of force as engine load. Speed is measured with a tachometer in revolutions per minute (rpm).

DRAWBAR HORSEPOWER

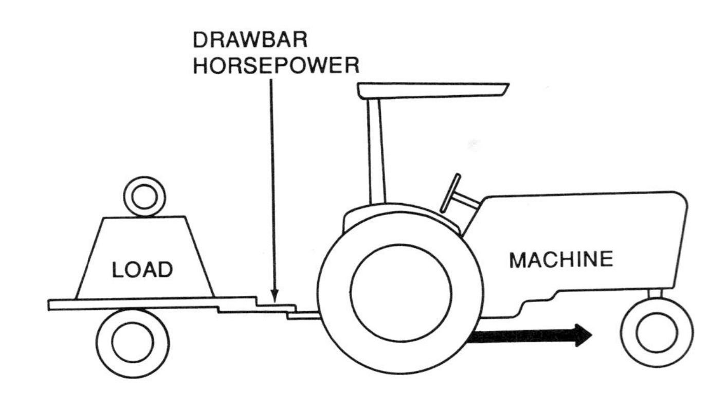

Fig. 38 — Drawbar Horsepower

Drawbar horsepower is the measure of pulling power an engine can produce when mounted in a moving machine. The load is attached to the machine and the horsepower required to move the machine is calculated by knowing the force required to pull the machine and the speed with which it is moved (Fig. 38).

PTO HORSEPOWER

PTO or power take-off horsepower is a function of torque and speed (rpm) and is measured at the machines' power take-off shaft.

A power take-off usually has some gear reduction between the engine and the PTO shaft. This reduction increases the torque value but reduces the speed. When measuring PTO horsepower, the speed is usually held constant at 1,000 rpm, so the horsepower can be read directly on a gauge measuring torque but having its scale calibrated in horsepower.

RATED HORSEPOWER

Rated horsepower is a value used by engine manufacturers to indicate the horsepower an engine should produce under normal operating conditions. This rating takes into account the maximum pressure forces in the engine as well as the speed and torsional forces. If these values are exceeded, the engine can be damaged.

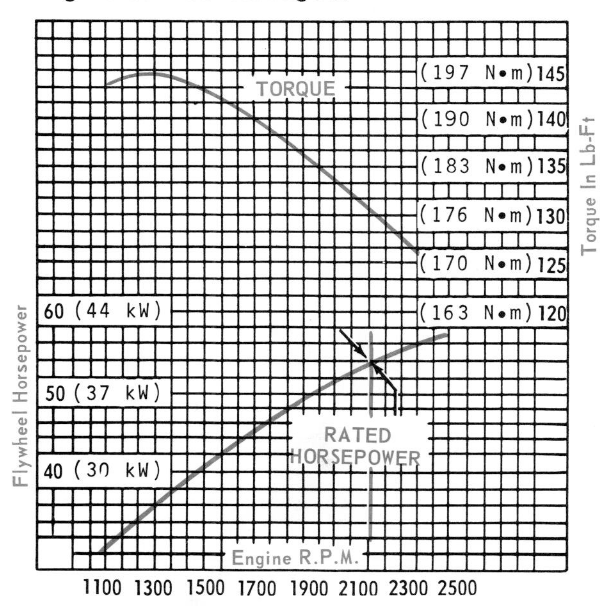

Fig. 39 — Typical Rated Horsepower Chart

Rated horsepower depends in part on the total cubic inches of piston displacement in the engine. From this, the manufacturer determines the maximum pressure stresses and rpm the engine can tolerate without internal damage. He then tests and develops the engine for long life and reliability at a given rated horsepower (Fig. 39).

Rated horsepower may not be the most efficient operating point for the best fuel consumption; this rating is expressed in terms of recommended operating horsepower and rpm.

ENGINE EFFICIENCY

We talk a lot about efficiency, and its importance can't be overstressed. It represents more than fuel economy; it also means the ability to do work at a constant rate with low maintenance.

Here are the prime efficiency factors in engines:

- **Mechanical Efficiency**
- **Volumetric Efficiency**
- **Thermal Efficiency**
- **Effective Pressure**
- **Fuel Consumption**
- **Compression Efficiency**
- **Load Effects**

Let's take a look at each type of engine efficiency.

MECHANICAL EFFICIENCY

We've mentioned some losses in horsepower between IHP and BHP—friction losses, etc. But in a functional engine other things also absorb horsepower:

1) *Fuel Pump*
2) *Water Pump*
3) *Cooling Fan*
4) *Generator or Alternator*
5) *Ignition System*
6) *Valves*
7) *Oil Pump*
8) *Blowers and Superchargers*
9) *Hydraulic Pumps (on standby)*
10) *Air Conditioner Compressor*

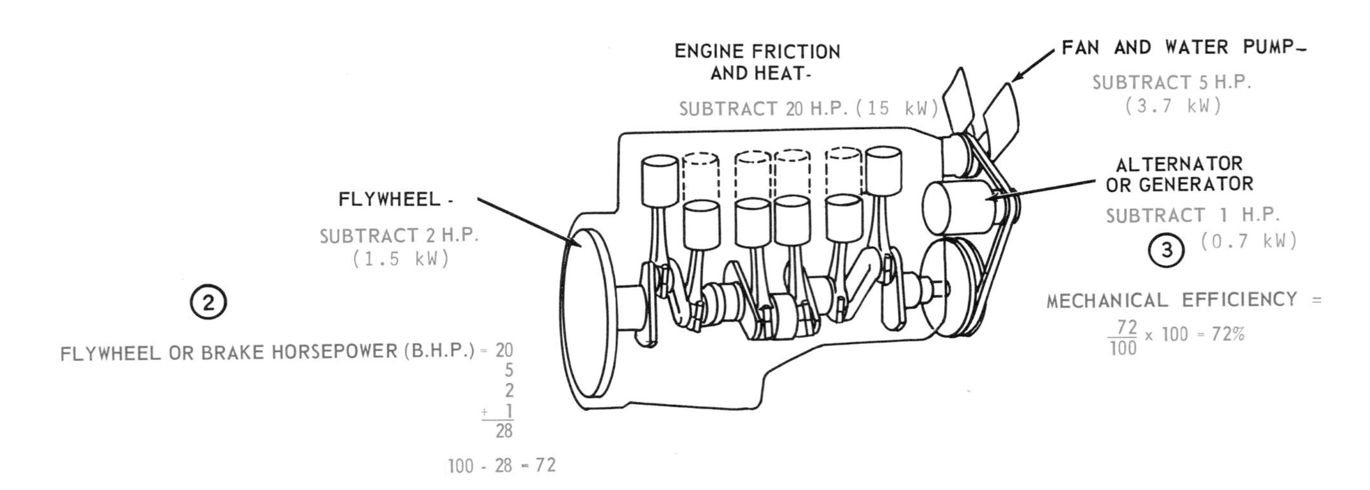

Fig. 40 — Mechanical Efficiency

Driving all these extra devices draws power from the engine and reduces its efficiency.

Mechanical efficiency then takes into account all these losses as well as frictional losses. To get a true value for mechanical efficiency the engine must be operating at its rated output, with all accessories performing their normal functions when we measure flywheel or brake horsepower (BHP). Then we divide the BHP by the IHP and multiply by 100 to get the mechanical efficiency of the engine (Fig. 40).

VOLUMETRIC EFFICIENCY

Most engines get intake air into the cylinders by creating a partial vacuum as the piston travels down on the intake stroke. However, the intake manifold, carburetor, air cleaner and intake system restrict the amount of air that can actually get into the cylinder.

Volumetric efficiency (Fig. 41) is calculated by dividing the actual amount of engine air taken in by the piston displacement and multiplying by 100.

Volumetric efficiency is one of the main factors governing the maximum torque output of an engine. The rpm at which an engine "breathes" the best will often determine the point of maximum torque.

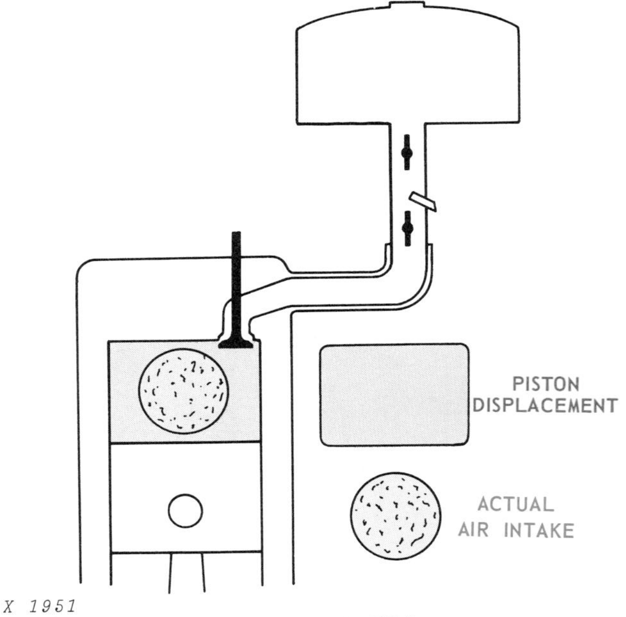

Fig. 41 — Volumetric Efficiency

THERMAL EFFICIENCY

The ratio of the work done by the gases in a cylinder (indicated work) to the heat energy (thermal energy) of the fuel is called **thermal efficiency.** Thermal efficiency is a laboratory value. We want something more practical—**brake thermal efficiency.**

Brake thermal efficiency is the brake horsepower, converted to BTU, divided by the fuel heat input in BTU, and multiplied by 100.

Thus, the brake thermal efficiency tells us how effectively an engine converts heat energy into usable power.

Brake thermal efficiency takes into account all engine losses and is sometimes called overall efficiency.

1. ENGINE DEVELOPS 100 FLYWHEEL OR BRAKE H.P. (75 kW) PER HOUR
2. 2545 x 100 B.H.P. = 254,500 B.T.U. (75 kW)
3. FUEL BURNED PER HOUR = 800,000 B.T.U. (234 kW)
4. BRAKE THERMAL EFFICIENCY = $\frac{254,500}{800,000} \times 100 = 31.8\%$

X7722

Fig. 42 — Brake Thermal Efficiency For A Typical Engine

Fig. 42 shows the formula applied to a typical engine.

MEAN EFFECTIVE PRESSURE

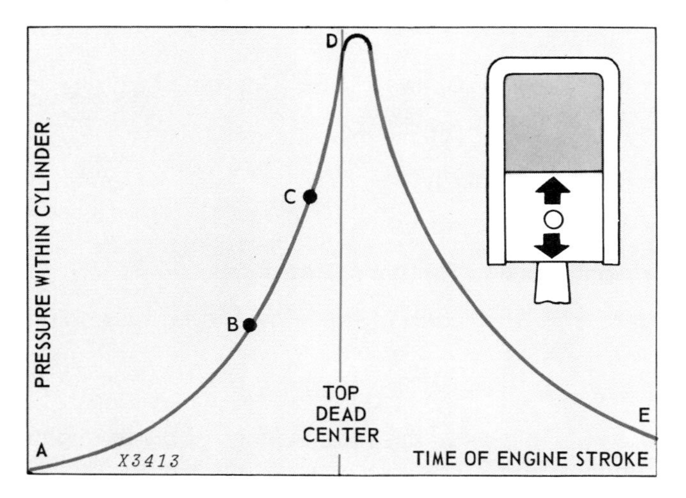

Fig. 43 — Typical Diesel Engine Pressure Indicator Tracing

Mean effective pressure consists of two types:

- **Indicated mean effective pressure (IMEP)**
- **Brake mean effective pressure (BMEP)**

The elements shown in Fig. 43 are:

A) Beginning of compression stroke

B) Start of injection

C) Beginning of injection

D) Peak firing pressure

*BC) Ignition delay**

DE) Power stroke

**Ignition delay is the period of time from the beginning of injection until the actual burning starts.*

These pressures are those developed within the cylinder. Pressures increase from the beginning of the compression stroke become maximum at peak firing pressure, and decrease as the power stroke progresses until the exhaust valve opens (Fig. 43).

Indicated (IMEP) mean effective pressure is the average pressure in the cylinder. This is measured by an instrument which records a tracing on a calibrated chart (Fig. 43). From this tracing the actual horsepower developed within the cylinder can be calculated.

Calculating the indicated mean effective pressure from the tracing is a complicated process requiring instruments not commonly available outside of engineering establishments. For those interested in more detail, refer to engineering publications on the subject.

Indicated (IMEP) mean effective pressure is used to calculate the horsepower the engine develops without taking into consideration the losses due to friction and other losses (heat, volumetric efficiency, etc.). Obviously this calculated horsepower is much higher than the usable horsepower obtained from the engine.

Brake (BMEP) mean effective pressure is calculated from the actual horsepower developed by the engine as measured by a dynamometer.

COMPRESSION RATIO

Compression ratio is the ratio of the volume of the combustion chamber at the beginning of the upstroke of the piston to the volume at the end of that stroke.

Thus, if the volume of the combustion chamber at the end of the upstroke is one-tenth (1/10) of the volume at the bottom of the upstroke, the compression ratio of the engine would be ten to one—commonly expressed 10:1.

FUEL CONSUMPTION

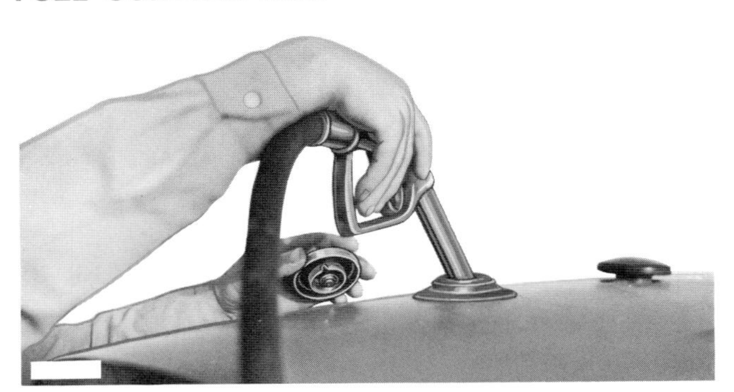

Fig. 44 — The Amount Of Fuel Used Is Measured By Pounds (kilograms) Per Horsepower-hour (kilowatt-hour)

Fuel consumption is generally measured by pounds (kilograms) of fuel per horsepower-hour (kilowatt-hour).

In a gasoline engine, fuel-air ratio is considered best at 15 pounds (7 kg) of air to one pound (0.5 kg) of fuel. Pounds (kilograms) of fuel per horsepower-hour (kilowatt-hour) is the measure of engine efficiency.

Each engine manufacturer can furnish a chart which indicates fuel consumption at various speeds and horsepower outputs.

COMPRESSION EFFICIENCY

Increased compression ratio permits the fuel-air mixture to be compressed more, which in turn gives greater heat and expansion during combustion.

Also, the higher compression and the resulting extra heat and pressure gives better burning and more energy.

There are limiting factors in the thermal efficiency produced by increasing the compression ratio. These include mechanical stresses, temperatures, and combustion chamber pressures.

Increased compression ratio may reduce fuel consumption, but certain fuels do not permit high compression ratios.

LOAD EFFECTS

Every phase of engine efficiency is affected by the nature of the load the engine drives.

A fluctuating load can set up serious vibration and, in extreme cases, "lugging" of the engine.

"Lugging" slows an engine while the throttle is still wide open. The result is excessive heat and firing pressures. This may cause the engine to be damaged.

Lean fuel mixtures result in high cylinder temperatures and possible detonation with certain fuels.

USES OF THERMODYNAMICS

Thermodynamics is the science that deals with heat and mechanical energy and their conversion one to the other.

The basics are as follows:

- **Laws of Thermodynamics**
- **Boyle's Law**
- **Charles' Law**
- **Boyle's and Charles' Law Combined**
- **Conservation of Energy**
- **Heat and Energy**
- **Adiabatic Compression**
- **Compression Ratio Vs. Pressure**

Let's see what each of these means to the engine.

LAWS OF THERMODYNAMICS

Two basic laws of thermodynamics apply to engine fundamentals: *The first states that energy is conserved. The second deals with the irreversibility of this process.*

Mechanical energy can be converted completely to heat but heat energy can never be converted completely to mechanical energy.

These fundamental truths are based upon the fact that heat naturally travels to a lower temperature object and seeks equilibrium.

An engine burns fuel and develops heat to produce mechanical energy for doing work. However, a major portion of the heat in the combustion chamber is absorbed by the cylinder walls or blown out the exhaust system; it is lost as far as mechanical energy resulting in BHP output is concerned.

Therefore, the efficiency of an engine is limited in part by the engine temperature range.

BOYLE'S LAW

Boyle's Law says that a gaseous mass can be compressed and that its volume is inversely proportional to the pressure on it, as long as the temperature stays the same.

This law is a governing factor in engine design, relative to piston displacement and compression ratio.

CHARLES' LAW

Charles' Law says that temperature changes on a gaseous mass result in direct changes of volume and pressure.

If the temperature of the gas is raised and the volume is held constant, the pressure will increase; raising the temperature and holding the pressure constant will increase the volume of the gas.

The heat energy in an engine combustion chamber causes the fuel-air mixture to expand. The combustion chamber is somewhat restricted in volume (compression ratio) and the change in volume is quite large over the entire power stroke. However, the change is not as great as the gas would need for full expansion at combustion, so the pressure is raised appreciably.

BOYLE'S AND CHARLES' LAWS COMBINED

The diesel engine takes advantage of the laws just discussed.

Air (a gas) is fed into the combustion chamber. Reducing the volume during the compression stroke raises the air temperature and the pressure.

Fuel injected into this mixture will then ignite due to the high temperatures of compression. Ignition further raises the temperature and the gases expand to force the piston down on the power stroke.

The pressure applied is dependent on the rate of burning and the heat retained in the gases and not lost through the cylinder walls and exhaust system.

This same principal is the cause of detonation or the preignition in gasoline engines. Gasoline will ignite at much lower temperatures than diesel fuel. If the temperature is raised too much by compression, spontaneous ignition will take place before the compression stroke is complete and the engine will tend to run backwards.

Engine design must, therefore, take into account all known conditions resulting from the three laws.

CONSERVATION OF ENERGY

Energy can neither be created nor destroyed. The different forms (work, heat, etc.) are mutually convertible. When, in some thermodynamic change, a quantity of one form disappears, an equivalent quantity must necessarily appear.

HEAT AND ENERGY

Heat is the greatest source of energy loss, as most engines operate in the varying temperature of the atmosphere.

Remember that heat energy lost to the engine cannot be recovered. This includes the normal losses of heat through the engine cooling and exhaust systems.

COMPRESSION RATIO VS. PRESSURE

The compression ratio of an engine is fixed by the design engineer. He then determines the combustion pressures which result under normal operation of the engine.

Tampering with the fuel rate or combustion rate can change the cylinder pressure and may damage the engine.

The higher the compression ratio of gasoline and LP-gas engines, the more critical become fuel quality and operating temperatures.

Detonation is the stray, unwanted explosions that take place in the engine as a result of too-high compression or heat, or low-quality fuels. These explosions cause piston and cylinder damage by creating excess pressure.

"Lugging" a high-compression engine at low rpm can cause serious detonation as the charge is in the combustion chamber a longer interval.

The laws of thermodynamics cannot be ignored when efficient, trouble-free engine performance is required.

SAFETY

After a shop has been planned to be as hazard-free as possible, *it must be managed* to keep it that way. Consider these key management procedures:

Make sure someone knows you are working in the shop and will check on you and render aid if you are injured.

1. Keep all tools and service equipment in good condition.

2. Use personal protective equipment, goggles, face shields, gloves.

3. Keep floors and benches clean to reduce fire and tripping hazards.

4. Clean up as you go while doing a job, and clean the area completely after the job is done.

5. Empty trash containers regularly.

6. Keep lighting, wiring, heating, and ventilation systems in good shape.

7. Lock your shop to prevent accidents. A shop is an attraction to a child.

8. Don't let anyone use tools or service equipment unless they've had adequate instruction.

9. Keep guards and other safety devices in place and functioning.

10. Use tools and service equipment for the jobs they were designed to do.

11. Supervise children carefully when they are in the shop.

12. Keep the fire extinguishers serviced, and the first aid kit replenished with supplies.

CHEMICALS AND CLEANING EQUIPMENT

In service operations, cleaning is needed in order to:

- **Keep dirt from entering the machine when parts are disassembled**
- **Inspect parts for wear and damage**
- **Install and adjust parts properly during reassembly**

Let's discuss the safe use of solvents, steam cleaners, and high-pressure washers.

SOLVENTS

Most solvents are toxic, caustic, and flammable. Be careful, therefore, to keep them from being taken internally, from burning the skin and eyes, and from catching on fire. Read the manufacturer's instructions and precautions before using any commercial cleaner or solvent.

Fig. 45—Use Commercial Cleaning Solvents—Not Gasoline

Whenever possible, use a commercially available solvent (Fig. 45). Many types are available for general-purpose cleaning and for specific cleaning jobs. Always read and follow the manufacturer's instructions to get the best results and to be able to handle each product safely.

Never clean with gasoline. It vaporizes at a rate sufficient to form a flammable mixture with air at temperatures as low as 50° F below zero. It is *always* unsafe to use. If you don't use a commercial solvent, use diesel fuel or kerosene. These *will* burn, but not as easily or as explosively as gasoline (Fig. 46).

Follow these general safety practices when using chemical cleaners:

1. *Follow the manufacturer's instructions.* Cleaning agents intended for the same purpose may be quite different in chemical composition. Read the label on each container you buy and follow instructions carefully.

Fig. 46—One Gallon Of Gasoline Mixed With The Right Amount Of Air Can Have The Explosive Force Of 83 Pounds of Dynamite

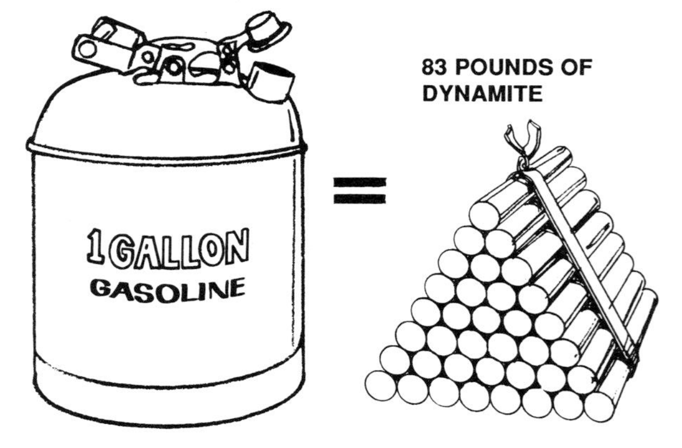

X4589

Fig. 47—Use Cleaning Solvents Only In A Ventilated Area, Outdoors If Possible

2. *Work in a well-ventilated area (Fig. 47).* Do the cleaning outdoors if possible. If you can't, provide ventilation by opening doors and windows.

3. *Keep solvents away from sparks and flame.* Don't let anyone smoke in the immediate area. Don't use solvents near heaters, sparks or open flames. Some solvent tanks and tubs should be "grounded." Contact your solvent supplier for instructions on how to ground your solvent tank or tub.

4. *Never heat solvents unless instructed to do so.* And don't mix them, because one might vaporize more readily and act as a fuse to ignite the other.

5. *Wipe up spills promptly.* Keep soaked rags in closed metal containers and dispose of them promptly.

6. *Store solvents in their original containers or in sealed metal cans properly labeled.* In a sealed container, the solvent will be kept clean from further contamination, toxic fumes will be controlled, the fire hazard will be reduced, and there will be less likelihood of spillage. Never use open pans (Fig. 48).

7. *If a commercially made parts washer is used, close the lid when you're finished cleaning.* Never destroy the fusable link that closes the lid automatically in case of fire.

8. *Protect your skin and eyes.* Wear a face shield and rubber gloves when working with strong, concentrated cleaning solutions (Fig. 49). Some are caustic, and they can destroy the natural oils of the skin and burn it severely. Always read and follow the manufacturer's instructions carefully. Even when gloves are not called for, avoid long exposures to solvents, and wash your hands when the job is done. Wear an apron if it's needed to keep your clothing dry.

9. *Avoid accidental poisoning.* Wash your hands and arms before eating or smoking. Keep all solutions in labeled containers. Never use empty containers, no matter how thoroughly cleaned, for carrying food or beverages. Keep poison containers sealed even when they are empty. Keep them out of the reach of children.

10. *Be prepared for emergencies.* Keep fire-fighting equipment near your cleaning area. Save the original containers for all solvents until they are completely used. In case of accidental poisoning, follow the instructions provided by the manufacturer immediately, and when going for medical aid, take the container with you so the chemical can be quickly identified.

Fig. 48—Store Solvents In Sealed And Labeled Cans, Never In Open Pans

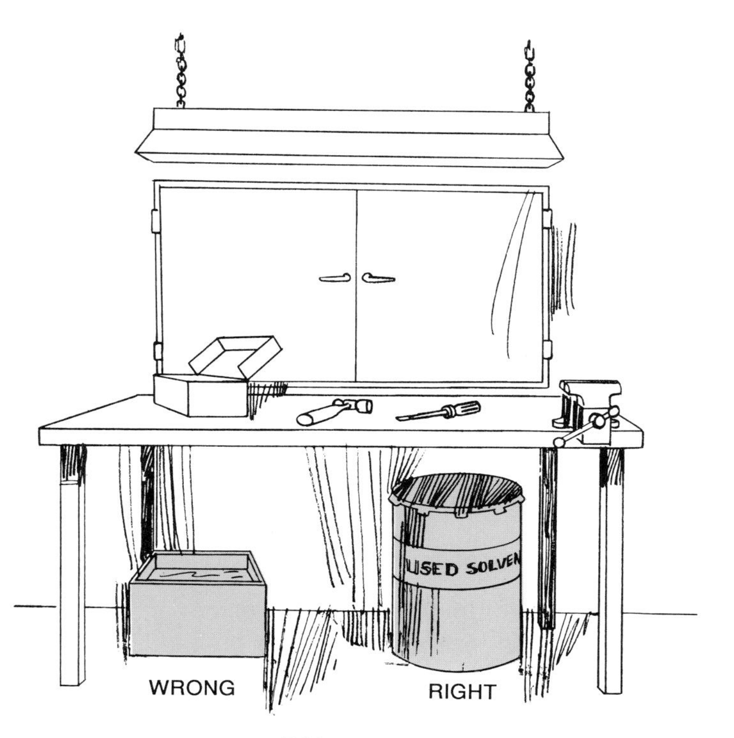

NOTE: If a solvent is splashed into your eyes, flush them thoroughly with water, keeping your eyelids open. Then get medical attention immediately. Some solvents may cause skin irritation or dizziness. If that happens, stop. Wash thoroughly with mild soap and water, and get some fresh air. If possible, use solvents in a location where a clean water source is available and ready for immediate use if needed.

Fig. 49—Wear Face Shield And Rubber Gloves When Cleaning With Cleaning Solutions

STEAM CLEANERS

A number of cleaners are on the market. Operating instructions vary with type and model. Before using any steam cleaner, become thoroughly familiar with its operation, and read its instruction manual carefully.

NOTE: Do not steam carburetors, electrical components, diesel injection pumps, or air-conditioning lines as damage to the components may result. Avoid oversteaming hydraulic lines, wiring, and bearing seals. And don't use cleaning compounds on painted surfaces where appearance is important. Many cleaning compounds will dull the paint.

Certain safe practices apply to the operation of all steam cleaning machines:

1. *Work in a ventilated area.* Clean equipment outdoors when possible (Fig. 50). Otherwise, provide adequate ventilation for your indoor cleaning area. Steam must be ventilated away for good vision, and to prevent damage and rust to your shop facilities and tools caused by condensing moisture.

2. *Protect yourself and others from burns.* Wearing gloves is not essential when using a well-insulated gun, but gloves are recommended if you light the burner by hand. When cleaning, warn others to stay away. Watch for unexpected bystanders when swinging the gun around. Always hold the gun securely and never lay it down until the cleaner has been shut down. When shutting it down, turn off the burner and keep the water running until all steam disappears from the nozzle. Then you can be sure that the equipment will not burn or scald anyone.

3. *Keep the cleaner in good condition.* Avoid damage to the steam gun and hose. Remember that tape does not make safe repairs. Keep fuel connections tight. If your cleaner does not ignite readily, burn cleanly, or maintain steam pressure within safe limits, have a qualified repairman service it.

Fig. 50—Clean Outdoors

HIGH-PRESSURE WASHERS

Cleaning with high-pressure washers is becoming common on farms. Take these precautions when using this equipment:

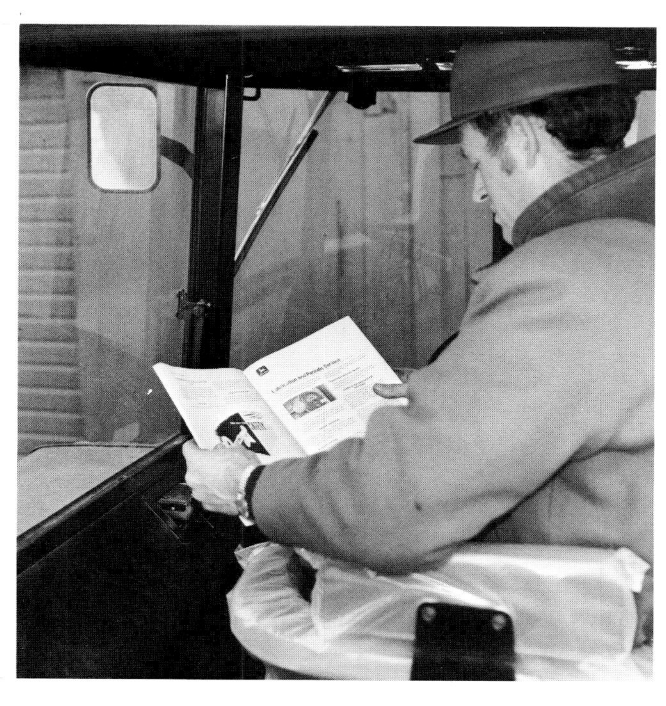

Fig. 51—Follow the Safety Instructions for Your High Pressure Washer. Read Your Operator's Manual Carefully, Use Only the Chemicals Recommended By the High Pressure Washer Manufacturer

1. *Wear eye protection.* The high pressure stream may cause dirt to be flung into your eyes. In addition, your eyes need protection in case the high pressure stream should become directed toward your face.

2. *Read the installation instructions (Fig. 51).* In addition to an adequate water supply, you'll need a grounding-type circuit or one protected with a ground-fault interrupter. The electrical circuit must have sufficient capacity to handle the load of the motor, and it should be equipped with a safety switch that can be locked in the *off* position. If you need a new circuit for your washer, have a qualified electrician install one for you.

3. *Use extension cords of adequate capacity.* Conductors must be large enough to prevent excessive voltage drop. And three-wire cords must be used if the circuit is not protected with a ground-fault interrupter. Refer to your owner's manual for specific recommendations.

4. *Always direct the spray away from the body.* The pressurized stream can penetrate skin. If the water is heated, it can burn. When cleaning, warn others to stay away, and watch for unexpected onlookers when moving the spray wand around. Hold the wand securely and don't put it down unless the machine is turned off. Even if the wand is equipped with a shut-off valve, the valve could fail. Then, the high pressure discharge would violently whip the wand around and possibly cause a serious injury. If injured by the spray, see a doctor at once.

5. *Protect yourself from splatter and throwback.* Always keep yourself out of line of the sludge and water thrown back from the surface you're cleaning. Hold the gun at an angle to the surface, and stand to one side when cleaning corners from which the sludge is thrown straight backward.

PROTECT AGAINST HIGH PRESSURE SPRAY

Spray from high pressure nozzles can penetrate the skin and cause serious injury. Keep spray from contacting hands or body.

Any high pressure spray injected into the skin must be surgically removed within a few hours by a doctor familiar with this type of injury or gangrene may result.

Fig. 52—Escaping High Pressure Fluid

ELECTRICAL SYSTEM

Avoid these hazards when servicing an electrical system:

- **Battery explosions**
- **Being burned by the battery electroyte**
- **Electrical shock**
- **Bypass start hazard**

NOTE: When servicing the electrical system, always follow the steps outlined in your operator's or service manual to avoid damaging the electrical system. Lock out or disconnect the electrical power source from the electrical system before you begin your service procedure.

BATTERY EXPLOSIONS

When charging and discharging, a lead-acid storage battery generates hydrogen and oxygen gas. Hydrogen will burn, and is very explosive in the presence of oxygen. A spark or flame near the battery could ignite these gases, rupturing the battery case and throwing acid all over.

To prevent battery explosions:

1. *Maintain the electrolyte at the recommended level.* Check this level frequently. When properly maintained, less space will be available in the battery for gases to accumulate. Refer to your operator's manual. Put only water or battery electrolyte in the battery.

2. *Use a flashlight to check the electrolyte level.* Never use a match or lighter, because these could set off an explosion.

3. *Do not short across the battery terminals.* If the battery isn't completely dead, the resulting spark may set off an explosion if hydrogen gas is present.

4. *Remove and replace battery clamps in the right order (Fig. 53).* This is very important. If your wrench touches the ungrounded battery post and the tractor chassis at the same time, the heavy flow of current will produce a dangerous spark. To prevent this, follow these rules:

a) *Battery removal:* disconnect the grounded battery clamp first. *Note: Some systems may be positive ground. Make sure you know which post is grounded.*

b) *Battery installation:* connect the grounded battery clamp *last*.

5. *Prevent sparks from battery charger leads.* Turn the battery charger off or pull the power cord before connecting or disconnecting charger leads to battery posts (Fig. 54). If you don't, the current flowing in the leads will spark at the battery posts. These sparks could ignite the explosive hydrogen gas which is always present when a battery is being charged.

Fig. 53—Remove And Replace Battery Clamps In The Correct Order

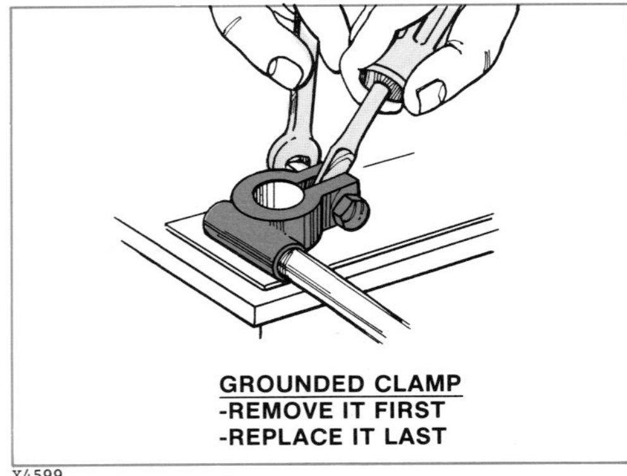

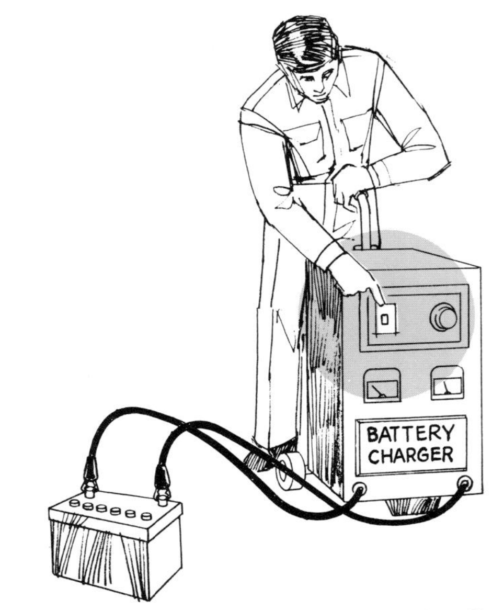

Fig. 54—Turn Charger Off Before Disconnecting Clamps From Battery Posts

CONNECTING A BOOSTER BATTERY

Improper jump-starting of a dead battery can be dangerous. Follow these procedures when jump starting from a booster battery.

1. Remove cell caps (if so equipped).

2. Check for a frozen battery. Never attempt to jump start a battery with ice in the cells.

3. Be sure that booster battery and dead battery are of the same voltage.

4. Turn off accessories and ignition in both vehicles.

5. Place gearshift of both vehicles in neutral or park and set parking brake. Make sure vehicles do not touch each other.

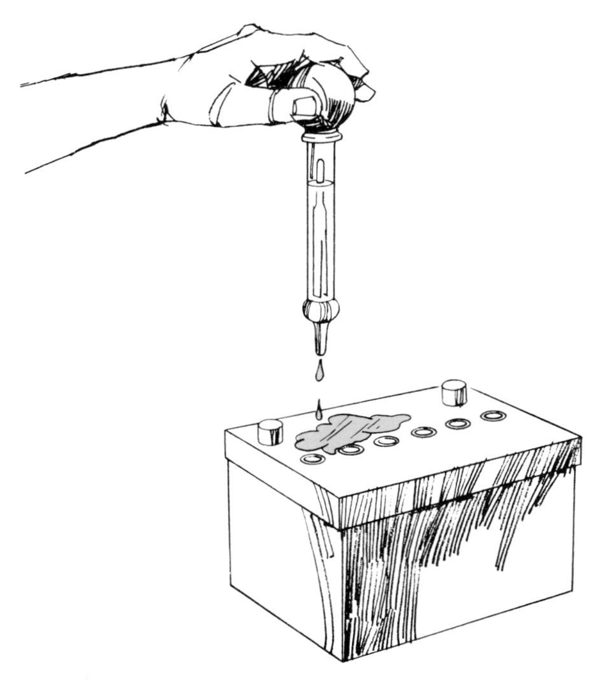

Fig. 55—Avoid Dripping Electrolyte When Reading Specific Gravity

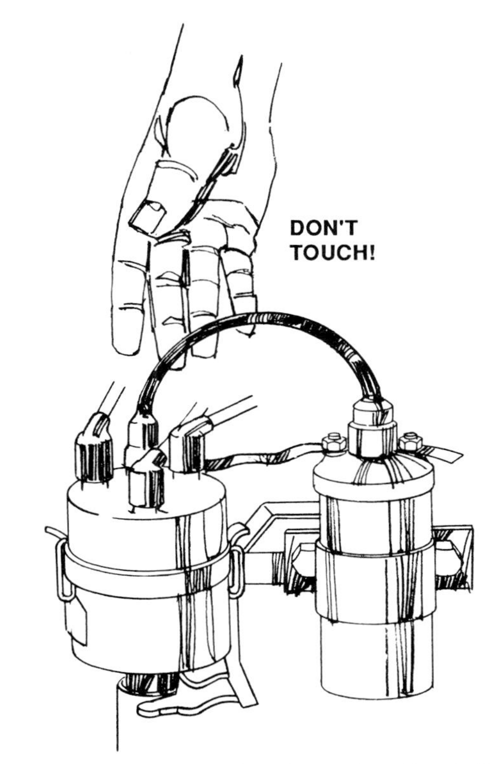

Fig. 56—Never Touch Wires In The Secondary Circuit Or Spark Plug Terminals While The Ignition Switch Is Turned On

6. Remove vent caps from both batteries (if so equipped). Add electrolyte if low. Cover the vent holes with a damp cloth, or if caps are safety vent type, replace the caps before attaching jumper cables.

7. Attach one end of one jumper cable to the booster battery positive terminal. Attach other end of the same cable to the positive terminal of the dead battery. Make sure of good, metal-to-metal contact between cable ends and terminals.

8. Attach one end of the other cable to the booster battery negative terminal. Make sure of good, metal-to-metal contact between the cable end and the battery terminal.

CAUTION: Never allow ends of the two cables to touch while attached to batteries.

9. Connect other end of second cable to engine block or vehicle metal frame *below* dead battery and as far away from dead battery as possible. That way, if a spark should occur at this connection, it would not ignite hydrogen gas that may be present above dead battery.

10. Try to start vehicle with dead battery. Do not engage the starter for more than 30 seconds or starter may overheat and booster battery will be drained of power. If vehicle with dead battery will not start, start the other vehicle and let it run for a few minutes with cables attached. Try to start second vehicle again.

11. Remove cables in exactly the reverse order from installation. Replace vent caps.

ACID BURNS

Battery electrolyte is approximately 36 percent full-strength sulfuric acid and 64 percent water. Even though it's diluted, it is strong enough to burn skin, eat holes in clothing, and cause blindness if splashed into eyes. Fill new batteries with electrolyte in a well ventilated area, wear eye protection and rubber gloves, and avoid breathing any fumes from the battery when the electrolyte is added. Avoid spilling or dripping electrolyte when using a hydrometer to check specific gravity readings (Fig. 55).

If you spill acid on yourself, flush your skin immediately with lots of water. Apply baking soda or lime to help neutralize the acid. If acid gets in your eyes, flush them *right away* with large amounts of water, and see a doctor at once.

ELECTRICAL SHOCK

The voltage in the secondary circuit of an ignition system may exceed 25,000 volts. For this reason, don't touch spark plug terminals, spark plug cables, or the coil-to-distributor high-tension cable when the ignition switch is turned on or the engine is running (Fig. 56). The cable insulation should protect you, but it could be defective.

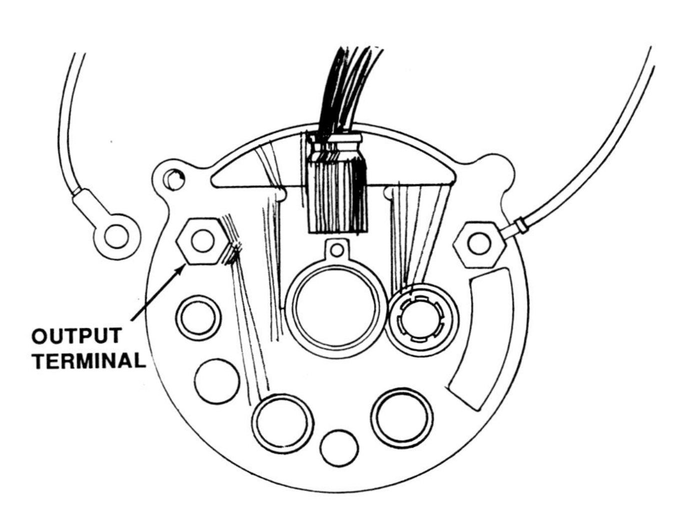

X4603
Fig. 57—Don't Run The Engine When The Wire Has Been Disconnected From The Alternator Or Generator Output Terminal

Never run an engine when the wire connected to the output terminal of an alternator or generator is disconnected (Fig. 57). If you do, and if you touch the terminal, you could receive a severe shock. When the battery wire is disconnected, the voltage can go dangerously high, and it may also damage the generator, alternator, regulator, or wiring harness.

TEST YOURSELF

QUESTIONS

1. (Fill in the blanks.) "An internal combustion engine converts ________ energy into ________ energy."

2. What three basic elements are needed to produce heat energy in the engine cylinder?

3. Air is compressed in the engine cylinder. What else happens to this air?

4. Fuel must be in what form for good combustion in the engine?

5. Match the two items at left with the definitions at the right.

a. Rotary Motion	*1. Up-and-down movement*
b. Reciprocating Motion	*2. Circular movement*

6. (Fill in the blanks.) "The engine converts ________ motion to ________ motion."

7. If an engine compresses eight parts of air into a one-part volume, what is its compression ratio?

8. Name the four parts of an engine combustion cycle in the proper sequence.

9. True or false? "In a four-stroke cycle engine, the crankshaft turns one complete revolution during a complete cycle."

10. Match each item on the left with the correct one on the right.

a. Diesel engine	*1. Fuel is mixed with air in the cylinder.*
b. Gasoline engine	*2. Fuel is vaporized before mixing with air.*
c. LP-gas engine	*3. Fuel is mixed with air before it goes into the cylinder.*

11. What are the two major types of engine cooling systems?

12. Which normally has a higher compression ratio—gasoline or diesel engines?

13. Why is the compression ratio higher on diesel engines?

(Answers in back of text.)

BASIC ENGINE / CHAPTER 2

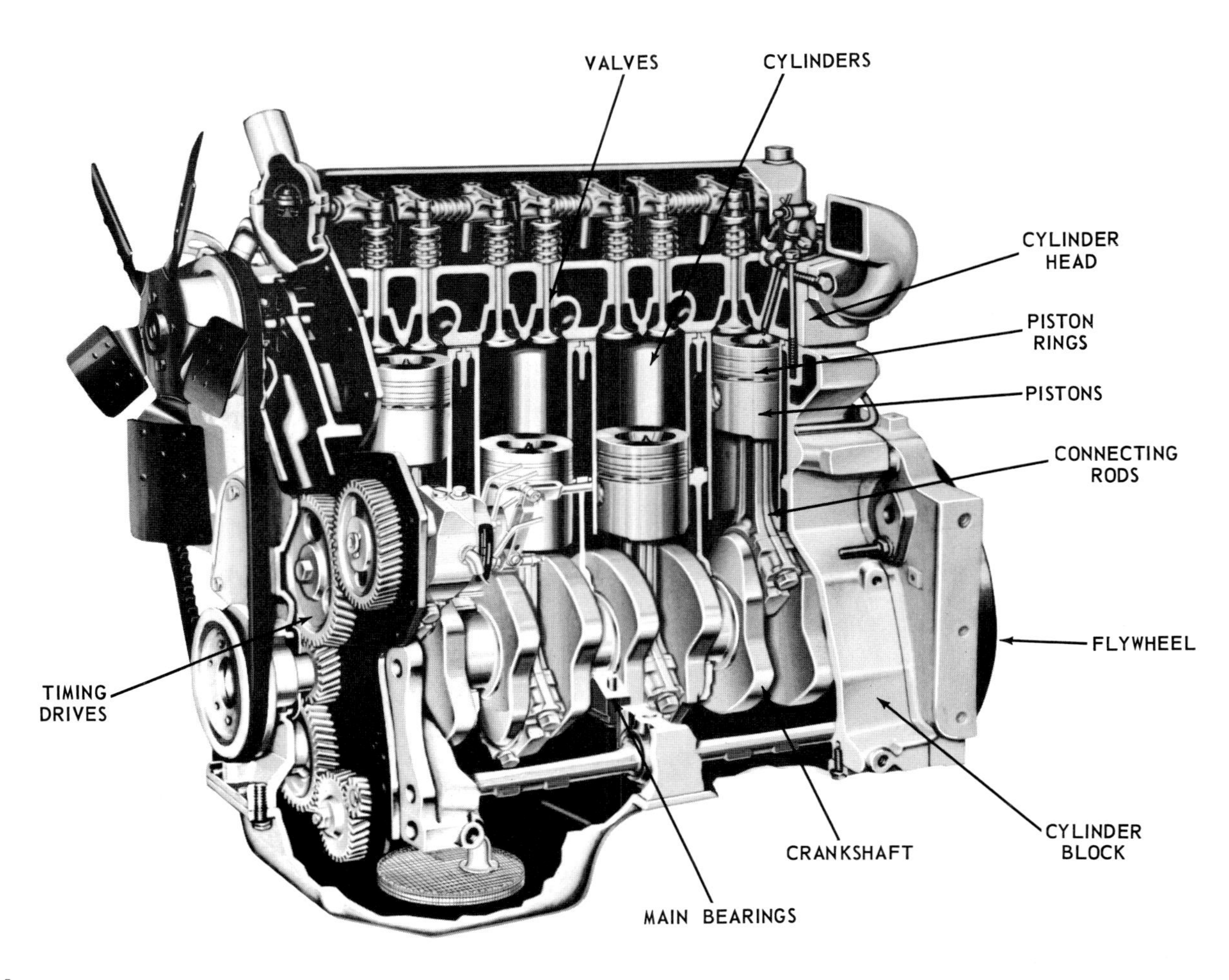

Fig. 1 — Cutaway of Basic Engine

INTRODUCTION

This chapter covers the *basic* engine—parts common to *all* internal combustion engines.

Chapter 1 told how the engine works, while later chapters cover the fuel systems and other accessories.

General testing and diagnosis of the engine is outlined in Chapter 11.

In this chapter you will learn about these parts of the engine:

• **Cylinder Head** is at the top of the engine and houses the valves and the intake and exhaust passages.

• **Valves** open and close to let fuel in and exhaust gases out of each cylinder.

• **Camshaft** rotating in the engine block opens the valves by cam action.

• **Cylinder Block** is the main housing of the engine and supports the other main parts.

• **Cylinders** are hollow tubes in which the piston works. They may be cast into the cylinder block or made of liners or sleeves.

• **Pistons** move up and down in the cylinders by the force of combustion.

• **Piston Rings** seal the compression in the combustion chamber and also help to transfer heat.

 Litho in U.S.A.

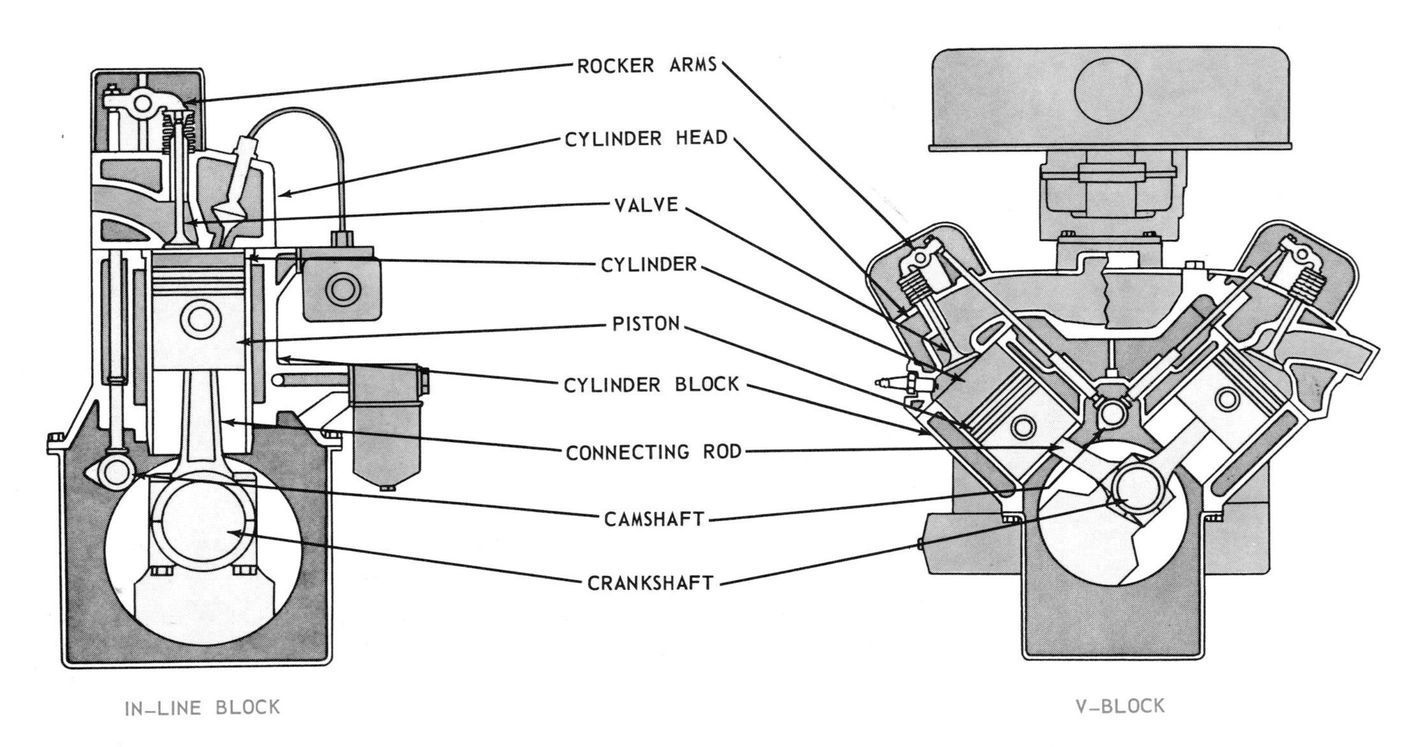

Fig. 2 — Parts Of A Basic Engine

• **Connecting Rods** transmit the motion of the pistons to the crankshaft.

• **Crankshaft** receives the force from the pistons and transmits it as rotary driving power.

• **Main Bearings** support the crankshaft in the cylinder block.

• **Flywheel** attaches to the crankshaft and gives it momentum to return the pistons to the top of the cylinders after each downward thrust.

• **Balancers** such as shafts or dampers, if used, balance the vibrations in the engine.

• **Timing Drives** link the crankshaft, camshaft, and other key parts together to assure that each is doing its job at the right time.

Engine parts are most commonly arranged in one of two ways (Fig. 2):

1) In-Line Block

2) V-Block

A typical V-block engine is shown in Fig. 3.
Now let's look at each part of the engine in detail — how it works and how to service it.

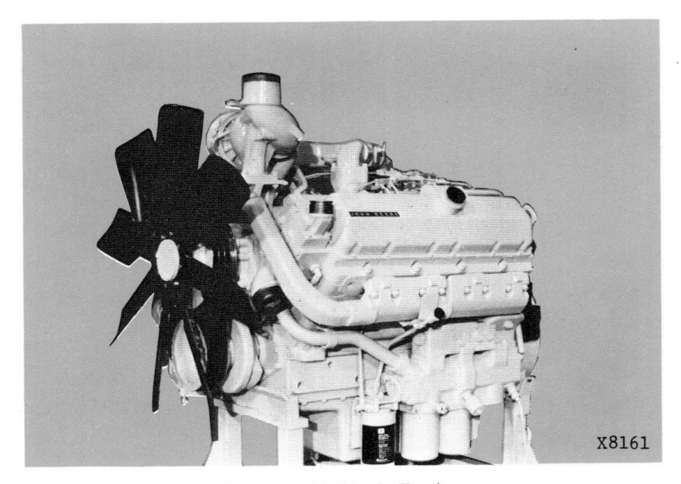

Fig. 3 — V-Block Engine

CYLINDER HEADS

Most cylinder heads today are separate one-piece castings (Fig. 4) made of an alloy of iron and copper or aluminum for greater strength and durability. The intake and exhaust passages are cast or bored into the head.

The cylinder head is sealed to the engine block with a gasket and careful tightening of the attach-

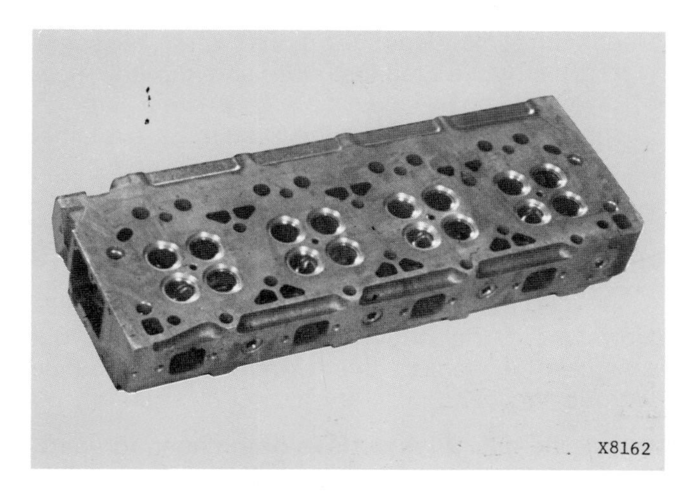

Fig. 4 — Engine Cylinder Head (2 used on V-block Engine)

ing cap screws or studs. This is critical in preventing distortion or gas leaks that lose power from the cylinders.

All heads are designed for swirling the fuel to get the fullest combustion.

This is done by having combustion occur when there is the least space between the piston crown and the cylinder head (Fig. 5). Some pistons are recessed as shown to help swirl the fuel-air mixture.

VALVE ARRANGEMENTS

Valves are located in five major ways (see page 1-14, Fig. 23 for illustration of different arrangements):

- **I-Head—both valves above the cylinder (parallel)**
- **H-Head (2 cycle)—both valves above angled cylinders (V-block)**
- **F-Head—one valve above and other to side of cylinder.**
- **L-Head—both valves to one side of cylinder (parallel)**
- **T-Head—one valve on each side of cylinder (parallel)**

I-HEAD valves are widely used on modern engines.

H-HEAD valves are found on V-block engines. Only one rocker arm is needed to operate two intake or two exhaust valves.

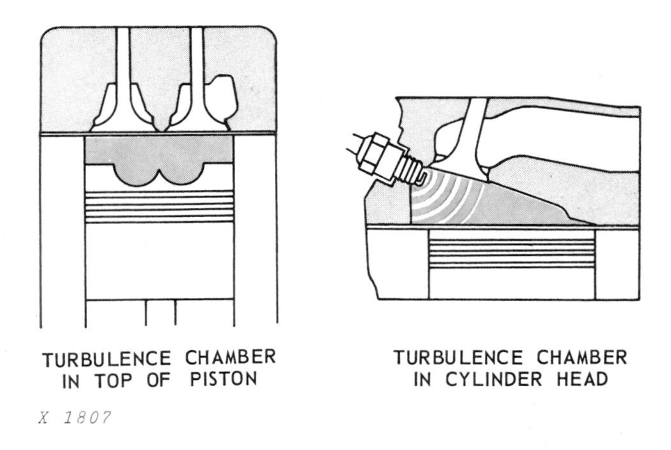

Fig. 5 — Turbulence Chambers

F-HEAD valves are sometimes used to allow room for larger valves.

L-HEAD valves are not used much because of the trend toward valve-in-head engines.

T-HEAD was used on early engines but required two camshafts.

CYLINDER HEAD GASKETS

Joints between the cylinders and the head must be gas- and water-tight.

To get this tight seal, head gaskets are made of copper-asbestos, plain copper, asbestos encased in steel, or special composition with a steel core.

These head gaskets firmly seal by "crushing" when the cylinder head is tightened down.

SERVICING CYLINDER HEADS

Removing The Head

Check for oil or gas leaks around the base of the head at the gasket line.

Drain coolant from engine before removing the head.

Steam clean the engine to prevent dirt from getting on the internal parts.

Litho in U.S.A.

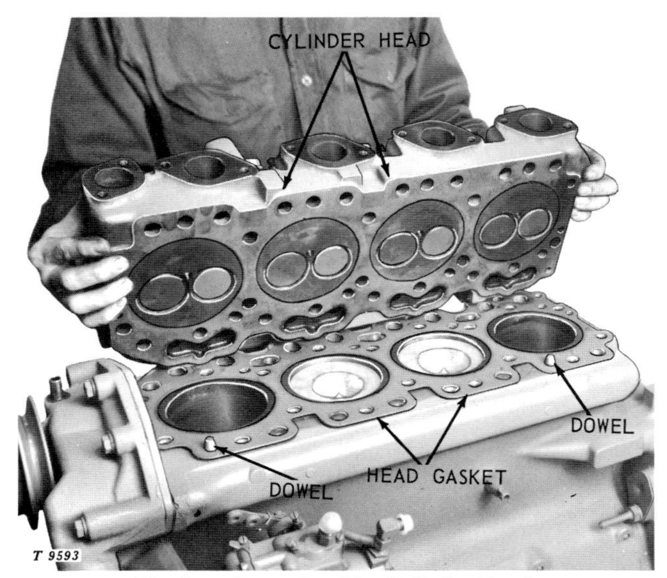

Fig. 6 — Removing The Cylinder Head

When removing a cylinder head (Fig. 6) *do not pry on the contact surfaces.*

If necessary, tap the head lightly with a *soft* hammer to loosen it.

Inspecting The Head

Engines with I-head valves have an oil gallery leading from the bottom to the top of the engine block to carry oil to the rocker arms. When removing the old head gasket, check for leaks around the area.

Replace the gasket to correct the leak.

Check for damage to the sealing surfaces of the head or cylinder block. Clean all carbon deposits from the head by scraping or brushing with a wire brush.

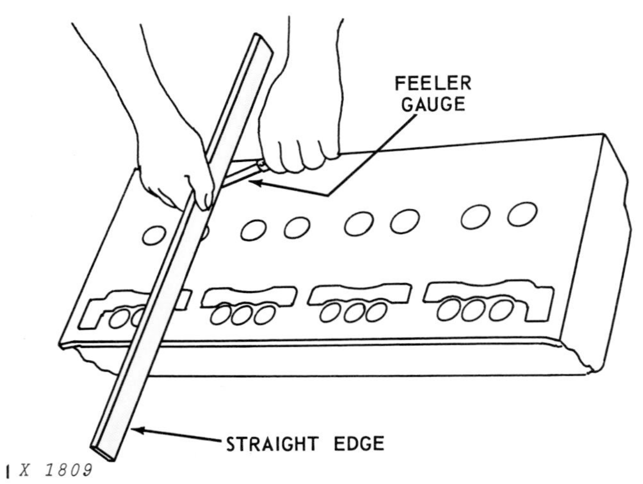

Fig. 7 — Checking The Head For Warpage

Check for lime deposits in the water passages. Use a recommended solution and dip the head to clean out scale and lime. (To prevent lime deposits, condition the cooling system as explained in Chapter 8.)

CHECKING HEAD FOR WARPAGE

After long operation, the head may contour to match the block—this is normal.

However, if the engine has overheated or had compression leaks, the extra heat may have warped the head.

Check the machined surface of the head for warpage as follows:

1. Clean off the machined surface of the head.

2. Use a heavy, accurate straightedge and feeler gauge to check for warpage at each end and between all cylinders (Fig. 7). Also check for end-to-end warpage in at least six places.

3. Decide whether to reinstall the head or to reface it. This will of course be determined by the allowable amount of metal that can be removed. Consult the engine Technical Manual for refacing limits.

Two methods can be used to check a head for cracks or leaks:

1) Water-and-air pressure method

2) Magnetic crack detector method

Using the *water-* and *air-pressure method*, the head is sealed and connected to an air hose. The head is then immersed in hot water (180-200°F) (82-93°C) for 15 minutes. Leaks are detected by any air bubbles which appear in the water.

The *magnetic crack detector* is placed over the suspected area, setting up a magnetic field (Fig. 8). Fine white metallic powder is then sprinkled over the area and the tool is rotated 90 degrees. After the excess powder is blown off, any cracks are clearly shown in white.

Installing The Cylinder Head

Install the head as follows to prevent leaks and blow-by:

1. Check for defects such as scratches or nicks on the sealing surfaces of the head and block.

2. Clean the cylinder block and head contact surfaces. Install a new cylinder head gasket. Follow the engine manufacturer's recommendations for

Fig. 8 — Magnetic Crack Detector

applying a sealing compound to one or both sides of the head gasket.

3. Carefully set the head on the engine block without disturbing the head gasket.

4. Be sure the head cap screws or bolts are clean and lightly oiled.

5. Draw the cylinder head down gradually and uniformly to be sure of a good seal between the cylinder head and block. Tightening the head bolts in one step may distort the head or cylinder liners and result in compression leaks.

6. Turn the head bolts or nuts down finger tight. Tighten each bolt or nut about one-half turn at a time.

7. Fig. 9 shows one sequence for tightening cylinder heads. Some manufacturers use this method for "older" engines. Start at the center of the head and work out to both ends of the head until the specified torque is reached.

8. Fig. 10 shows another sequence for tightening cylinder heads. Tighten head bolt No. 17 first. This prevents the cylinder head from tipping during the tightening sequence. Tighten remaining screws beginning with No. 1.

9. Most engines require a "break-in" or "run-in" after an engine overhaul. The cylinder head should generally be retightened to the specific torque after this run-in. Retightening cylinder head bolts is not recommended for some newer engines. Follow engine manufacturer's specifications.

10. Most engine heads must be retightened after a warm-up. See "Engine Break-In" at the end of this chapter.

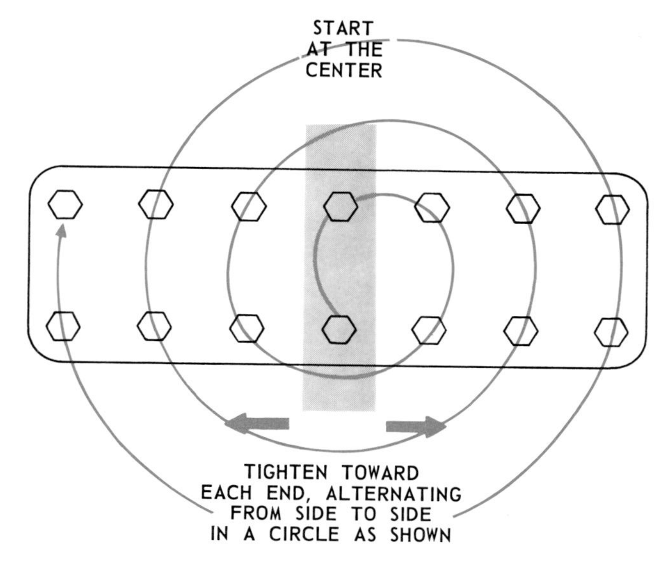

Fig. 9—Guide To Proper Sequence For Tightening Cylinder Heads—Older Engines

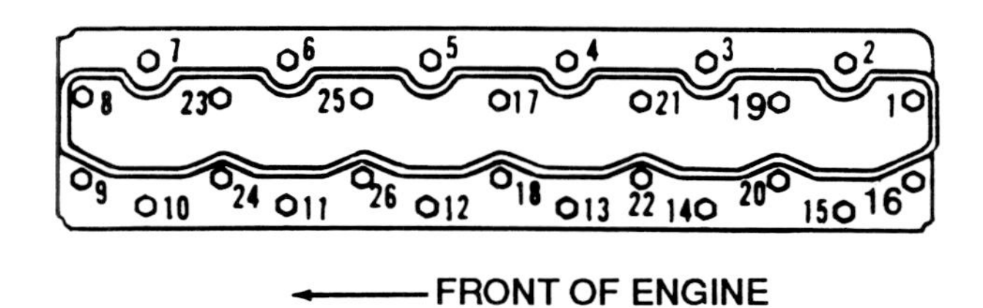

Fig. 10—Sequence For Tightening Cylinder Heads—Newer Engines

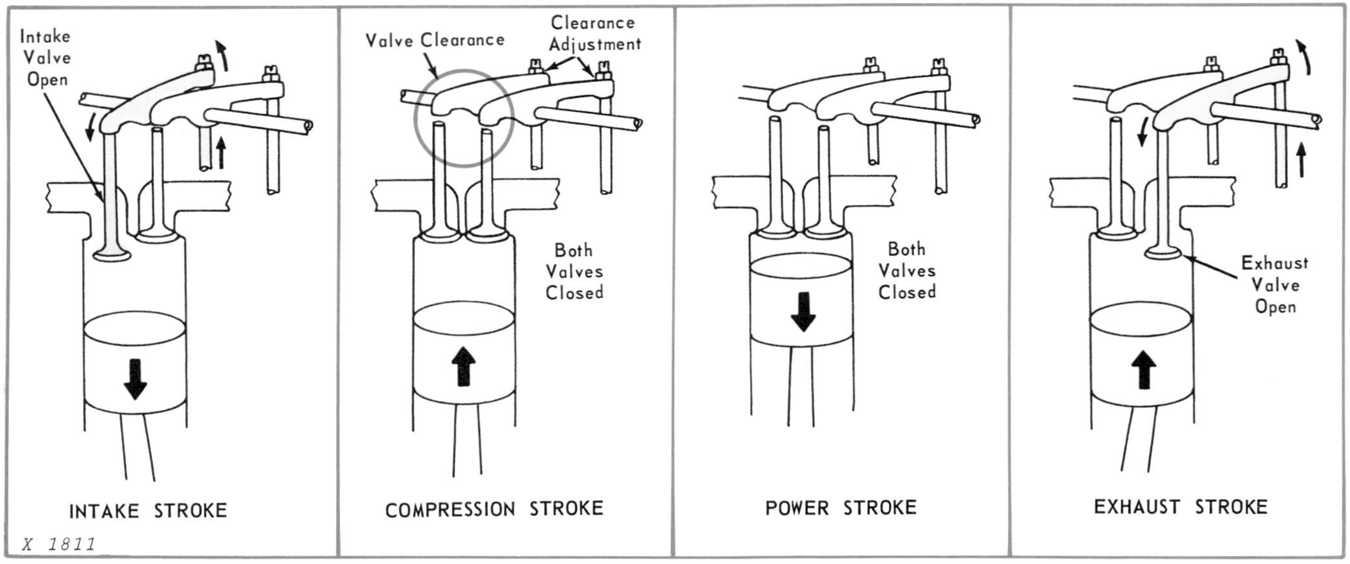

Fig. 11 — Valve Operation In A Four-Cycle Engine

VALVES

The engine must take in fuel-air and exhaust spent gases at precise intervals. The **valves** do this job by opening and closing the intake and exhaust ports to the cylinder.

Fig. 11 shows the sequence of valve operation for a typical four-cycle engine. Each cylinder may have two or four valves — one intake and one exhaust or two of each (Fig. 12).

Fig. 12—Two Intake And Two Exhaust Valves In One Cylinder

During the intake stroke, the intake valve(s) opens as shown.

During the exhaust stroke, the exhaust valve(s) opens.

All valves are closed to seal in the combustible mixture during the other two strokes, compression and power.

HOW VALVES ARE COOLED

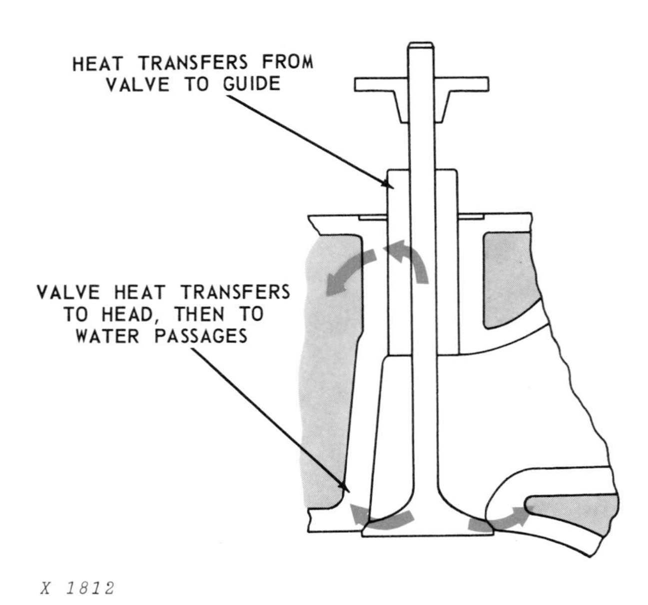

Fig. 13—How Valves Are Cooled

Fig. 13 shows how valves are cooled during the heat of engine operation.

When the valve is closed, heat in the valve head is transferred to the cylinder head, then to the water passages as shown.

At all times, heat is transferred from the valve stem to the guide.

 Litho in U.S.A.

TYPES OF VALVES

Modern valves are shaped like a poppet or mushroom. The valve is made up of the **head** and the **stem.** The sealing edge is called the **valve face.**

The valve head is conical in shape so that it centers itself when it closes.

Fig. 14 shows three forms of conical engine valves.

The STANDARD VALVE is commonly used in American engines.

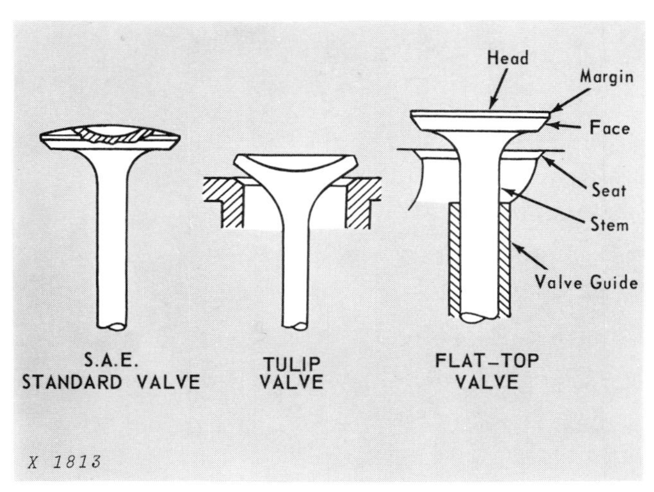

Fig. 14 — Forms Of Engine Valves

The TULIP VALVE is used more in aircraft and racing engines for better flow of gases.

The FLAT-TOP VALVE combines the standard and tulip valve features. The larger taper under the head helps in better flow of gases and also strengthens the valve.

Valves are usually constructed from one or two pieces of special alloy steel—chrome-nickel for intake valves, silichrome for exhaust valves (because of the greater heat).

SERVICING VALVES

Analyzing Valve Troubles

When rebuilding an engine, some servicemen replace almost every part in the valve train while others replace some of the exhaust valves and reface the others. These are extreme cases. The best method is somewhere between the two.

What is best for one valve overhaul may not be best for another.

MAJOR CAUSES OF VALVE FAILURES ARE:

- **Distortion of the Valve Seat**
- **Deposits on Valve**
- **Too little Tappet Clearance—Burned Valves**
- **Preignition—Burned Valves**
- **Erosion**
- **Heat Fatigue**
- **Breaks**
- **Worn Valve Guides**

Distortion Of Valve Seat

Fig. 15 — Valve Face Burned From Distortion Of Seat

The valve shown in Fig. 15 is burned because of distortion of the valve seat.

Major causes of seat distortion are:

1. *Failure in the cooling system.*
2. *Out-of-round or loose seat (see "Seat Inserts" later). This may stop the transfer of heat between the insert and the head or block.*
3. *Warped sealing surfaces on heads or blocks. This often distorts the seats when the head is tightened. Improper tightening, too much torque, and the wrong sequence can also distort valve seats.*
4. *Failure to grind the valve seat concentric with the valve guide hole.*

 Litho in U.S.A.

Deposits On Valve Stems

Fig. 16 — Valve Damaged Because Of Face Deposits Breaking Off

Fig. 16 shows a valve that failed because face deposits built up and then broke off. This damaged the seat and the resulting "blow-by" burned the valve.

Other factors that may cause this type of failure are:

1. Weak valve springs. They cause a poor seal between the seat and the face, allowing deposits to form.

2. Too little tappet clearance. This also causes a poor valve-to-seat seal.

3. Valves sticking in the valve guide. This allows deposits to build up on the valve face and seat.

4. Valve seats that are too wide. They cut down the seating pressure and reduce the crushing of deposits when the valve closes.

5. Lack of valve rotation (which is needed to "scrub" the valve).

Fig. 17 — Valve Stem Deposits

Too Little Tappet Clearance—Burned Valves

Fig. 18 — Valve Burned From Too Little Tappet Clearance

Failure of the valve in Fig. 18 was caused by too little tappet clearance.

The valve was held off its seat and blow-by caused face burning.

Causes of too little tappet clearance are:

1. Tappet clearance not set to specifications.

2. Failed valve rotators and, as a result, burned valves.

3. Cooling system not operating properly or wrong thermostat installed.

(Heat affects tappet clearance.)

4. Tappet clearance not rechecked after break-in and retorquing of head as recommended by engine manufacturers.

Preignition—Burned Valves

Fig. 19 — Valve Burned By Preignition

Fig. 19 shows a valve that has burned and failed as a result of preignition. Valve temperatures can reach the point where portions of the valve face melt away. Preignition can also cause excessive cylinder wear, and piston ring and land failures.

General causes of preignition are:

1. **Improper timing.** *Timing for maximum power is not always recommended, as preignition cannot always be heard. Follow the recommendations in the engine Technical Manual on timing.*

2. **Combustion chamber deposits.** *Long idling periods, rich fuel-air mixtures and cold engine operation can result in excessive deposits. Oil burning can also cause this.*

3. **High heat range spark plugs or cracked porcelains.** *Replace defective plugs as recommended for the operating conditions.*

4. **Too-high compression ratios.** *When cylinder heads or blocks are resurfaced, the fuel octane requirements also increase.*

Erosion Of Valves

Fig. 20 — Erosion Under Head Of Valve

The valve in Fig. 20 is eroded but has not failed. However, it would have broken after much more service because of the erosion under the head.

Causes of valve erosion are:

1. *Wrong type of fuel.*
2. *Faulty combustion.*
3. *Too-high valve temperatures.*
4. *Lean fuel air mixtures which overheat valves and erode them.*

Causes of cracked valves from overheating are:

1. *Worn guides*
2. *Distorted seats*
3. *Preignition*
4. *Lean fuel-air mixtures*

Valves Cracked From Heat

Fig. 21 — Valve Cracked By Heat

Overheating of valves can cause cracks in the valve head (Fig. 21). This is sometimes called "Thermal Fatigue." More cracking may cause parts of the valve to break off.

Broken Valves

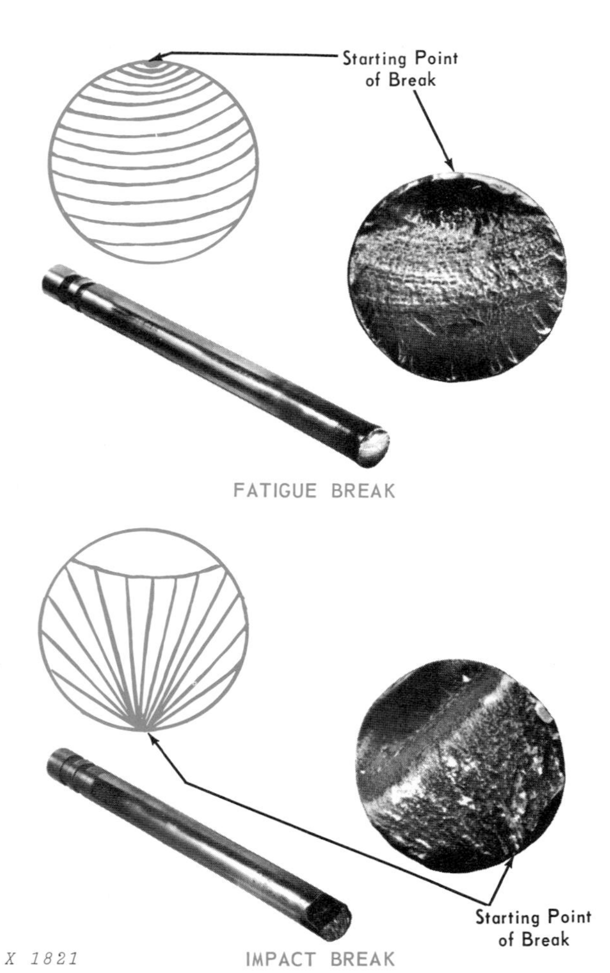

Fig. 22 — Broken Valves — Two Causes

The two leading causes of broken valves are:

1. **Fatigue Break.** *This is the gradual breakdown of the valve due to high heat and pressure. A fatigue break usually shows lines of progression as shown at the top in Fig. 22.*

2. **Impact Break.** *The mechanical breakage of the valve. The cause of this breakage is seating the valve with too much force, often caused by too much valve clearance. An impact break does not show the lines of progression but rather the familiar crow's-feet (see bottom, Fig. 22).*

Broken valves are not always clearly one type or the other. Combinations of heat and high seating force can produce failures of varying degree and appearance. Fig. 22 shows the differences between fatigue and impact breaks.

Worn Valve Guides

Fig. 23 — Effect Of Worn Guide On Valve Stem

The valve shown in Fig. 23 has failed because of face burning. Wear on the stem and carbon projecting into the guide shows that the valve guide was worn and was the likely cause of the failure.

Worn valve guides lead to valve failures:

1. *Worn guides prevent concentric grinding of the valve seat. This leads to out-of-square seating, which in turn allows burning gas to leak out and burn the valves.*

2. Worn guides cause the valves to strike at an angle. This damages the sealing surface and leads to blow-by and burning.

3. Excessive stem-to-guide clearance allows too much oil to run down the stem, resulting in excessive carbon deposits that cause valve sticking.

4. When the inside edges of the valve guides wear, they can no longer act as carbon scrapers.

Installing new guides when new valves are installed does not mean they will be trouble-free.

Other factors can also cause valve guides to wear out too fast:

Worn Rocker Arms—cause excessive side thrust on the valve stem.

Poor Lubrication—results in scoring.

Carbon Deposits—deposits on the valve stem wear the valve guide into a bell-mouthed shape.

Cocked Valve Springs—places side thrust on the valve stem and results in excessive wear.

Dimensions Needed When Servicing Valves

A—I.D. of Valve Guide

B—Length of Valve Guide

C—Distance of Guide from Head

D—O.D. of Valve Stem

E—Angle of Valve Face

F—Angle of Valve Seat

G—Width of Valve Head

H—Width of Valve Seat

Fig. 24 shows the dimensions which must be known to recondition valves, seats, and guides.

Refer to the engine Technical Manual for the correct dimension for each engine.

Reworking Or Replacing Valves

When removing valves from the engine, place them in a numbered rack or otherwise mark them. This returns each valve to its own valve guide after servicing.

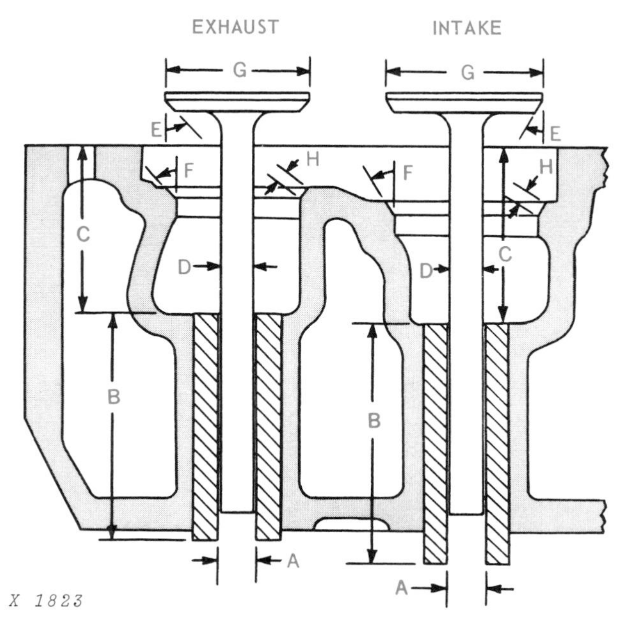

Fig. 24 — Valve Dimensions Needed When Servicing Valves

Special tools are a must when reworking engine valves. *Valves can't be refaced on a shop grinding wheel.*

1. CLEANING

Hold each valve firmly against a wire wheel on a bench grinder.

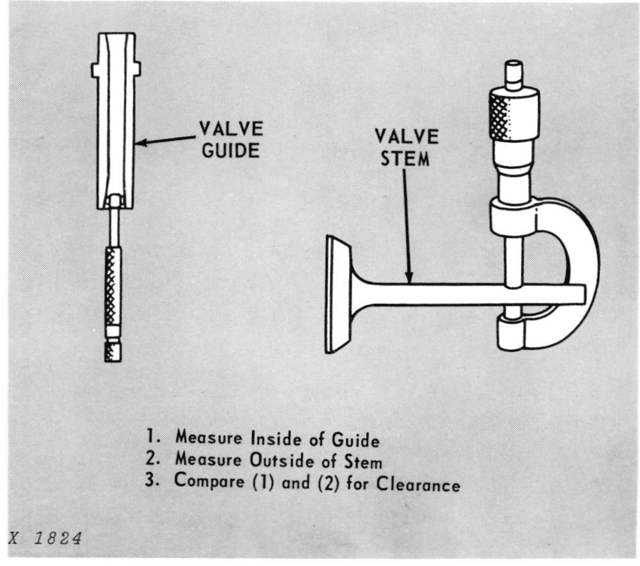

Fig. 25 — Measuring Wear On Valve Stem And Guide

Remove **all** carbon from the valve head, face and stem. Any carbon left on the stem will affect alignment in the valve refacer. Polish the valve stems with steel wool or crocus cloth to remove any scratch marks left by the wire brush.

2. INSPECTION

Use micrometers to measure the clearance of valve stems and guides (Fig. 25). If the measurements exceed the specified clearance, replace either the valve or guide or both.

3. TESTING VALVES FOR BREAKS

Test all used valves as follows:

Hold the valve by its stem, head down, and strike it sharply on end of stem with a hammer. If a fracture exists, the head will break off.

4. REFACING THE VALVES

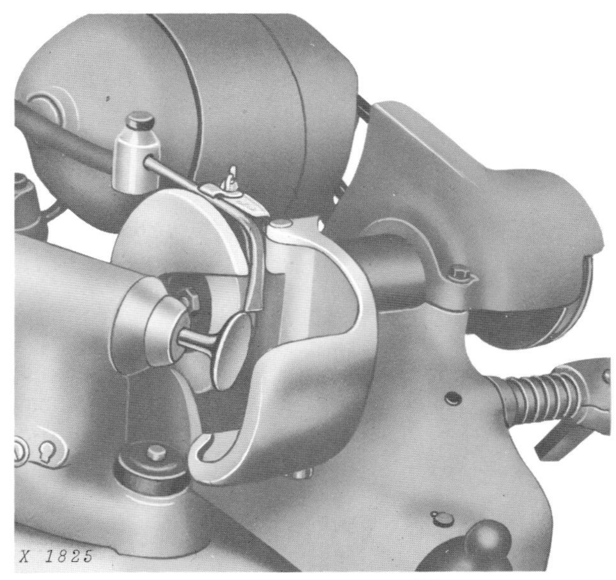

Fig. 26 — Refacing The Valves

Locate the working head of the valve refacer at the specified angle with the valve head (Fig. 26). For quicker seating, an interference angle is normally used.

First be sure the chamfer on the end of the stem is uniform and then be sure the valve is positioned all the way in the chuck and is seated against the conical centering stop.

When grinding, the first cut from face of valve must be light. Observe if there is an unevenness of the metal being removed. If only 1/3 or 1/2 of the valve's face has been touched, check to see if the machine is dirty or the valve is warped, worn, or distorted. When the cut is even around the

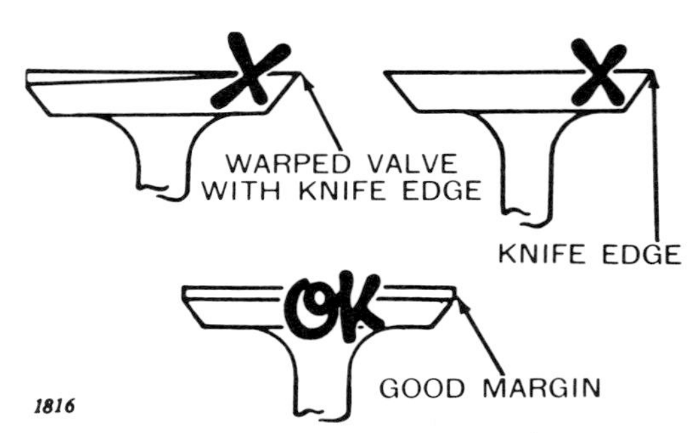

Fig. 27 — Valve Head Margin

whole valve, keep on until the complete face is ground clean. Be sure the correct angle is maintained.

Scrap and replace any valves that cannot be entirely refaced while keeping a good valve margin (Fig. 27). The amount of grinding necessary to true a valve tells whether the head is worn or warped.

Avoid a knife edge around part or all of the valve head (Fig. 27). Heavy valve heads are required for strength and good heat dissipation. Knife edges lead to breakage, burning, and pre-ignition because heat localizes on the edge.

5. GRINDING END OF VALVE STEM

If the end of the valve stem is pitted or worn, true it and clean it up on the refacer attachment—not the grinding wheel. A very light grind is usually enough to square the stem and remove any pits or burrs.

6. ASSEMBLING VALVES

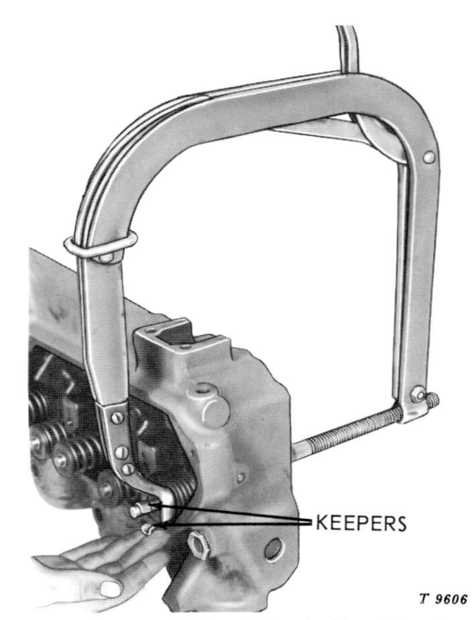

Fig. 28 — Installing Valves (Valve-In-Head Engine Shown)

Apply oil to valve stems and return the valves to same ports from which they were removed. (Commercial valve stem lubricants are available for this purpose). Work the valves back and forth to make sure they slip through easily and seat properly. A properly seated valve will "bounce" when dropped on its seat (without oiling it).

Be sure to seat the valve keepers (locks) and springs seat properly (Fig. 28). Improper seating of these parts may cause valve breakage.

It is a good practice to use **new** valve keepers when reinstalling the valves.

With valve-in-head models, "pop" each spring and valve assembly three or four times by tapping on the end of each valve stem with a soft mallet to insure good seating of the keepers.

If a failed valve is replaced, also inspect the rocker arm and push rod for that valve. The rocker arm or push rod may be bent or otherwise damaged when the valve fails, even though this may not be noticeable.

7. ADJUSTING VALVE CLEARANCES

Refer to page 2—23 of this chapter for adjusting the valve clearances.

VALVE SEATS

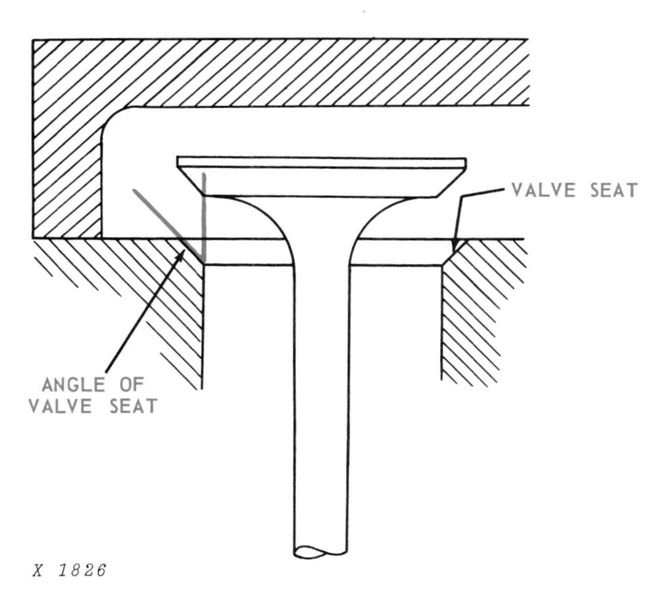

Fig. 29 - Valve Seat

The machined surface of the block or the cylinder head on which the valve rests when closed is the **valve seat.** This seat normally makes an angle of 20, 30, or 45 degrees with the plane of the valve head or bore (Fig. 29).

The valve seat in the block or head is normally of the same material as the block or head and is ground for a tight seal with the face of the valve.

SEAT INSERTS

Valve seat inserts are used on some engines, often only for exhaust valves. The valve insert reduces wear, prevents leakage, and reduces valve grinding frequency. It also allows the seat to be replaced.

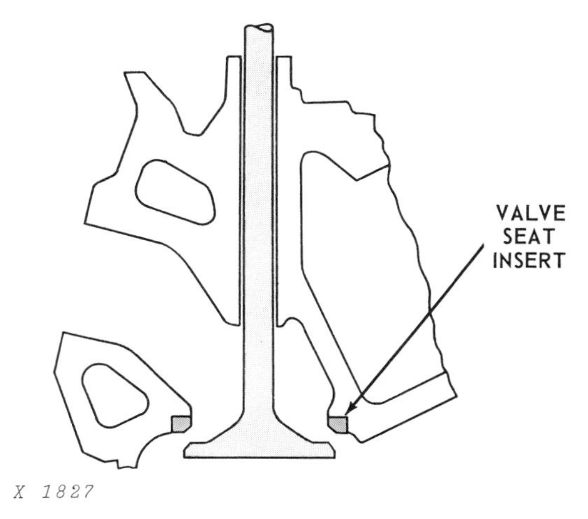

Fig. 30 — Valve Seat Insert

The insert is a ring of special high-grade alloy metal placed in the block or head to serve as a valve seat (Fig. 30).

The seat insert is generally a shrink-fit, although screw-in types are sometimes used. In most applications the inserts are shrunk in place by using dry ice to get the necessary fit without distorting the insert during installation.

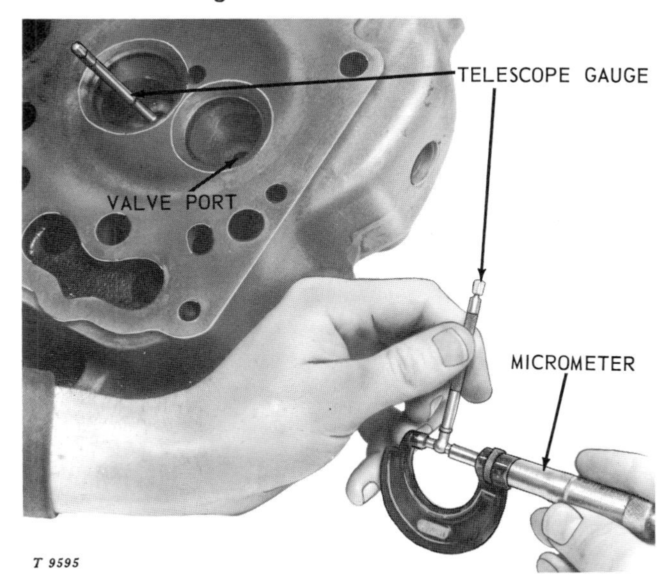

Fig. 31 — Measuring Valve Guides

SERVICING VALVE SEATS

1. **Cleaning Valve Seats**—*Use an electric hand drill with wire cleaning brush and remove all carbon. Use kerosene to help loosen very hard carbon.*

2. **Cleaning Valve Guides**—*Clean valve guides as instructed under "Valve Guides" later in this chapter.*

3. **Measuring Valve Guides** — *Measure the inside diameter of the guide for wear before machining the seat or counterbore (Fig. 31). Replace the guide or ream the guide hole for oversized valves if necessary.*

4. **Servicing Valve Seat Inserts**—*Recut or clean up the counterbore if it is out-of-round or is rough or damaged. Measure the counterbore after recutting and follow recommendations for press fits or other methods. Normally, chill each insert and the driver in dry ice. Use a driver that pilots in the guide hole. Blow out any chips from under the seat before installation. Also check for any nicks or raised edges.*

5. **Servicing Valve Seats**—*The tendency to cut seats too wide can be corrected by using a narrowing stone. This stone is dressed with a sharp face and is used very lightly to reduce the width of the seat face until it is narrower than the valve face.*

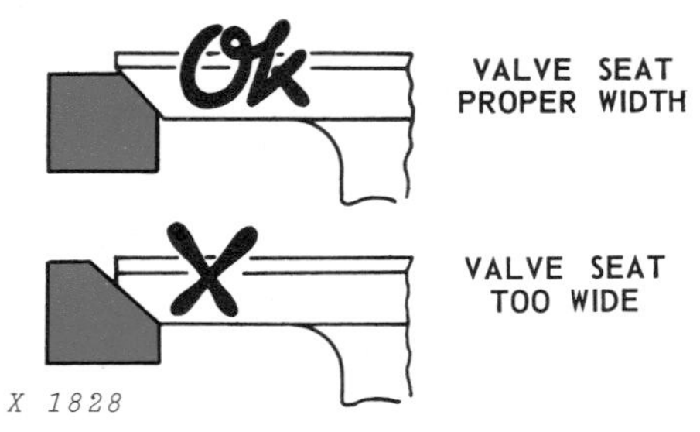

Fig. 32 — Width Of Valve Seat

Fig. 32 shows the right and wrong widths of the valve seat in relation to the valve.

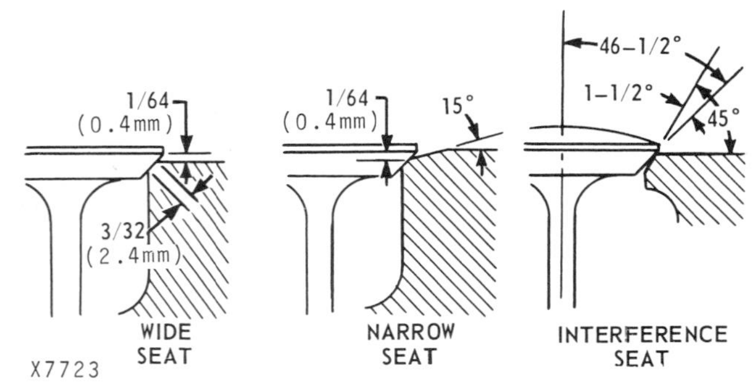

Fig. 33 — Typical Valve Seating Locations

Wide seats improve the heat transfer, while **narrower seats** give better crushing of valve deposits. For even better seating forces, an **interference angle** of the seat to the valve face is used. This is best for high deposits and in some seat distortion conditions, but it will also cause the valve to run hotter.

Grind the valve seat so that little of the valve face is exposed to the combustion chamber. See Fig. 33 for three examples.

Fig. 34 — Grinding Valve Seats

PRECAUTIONS FOR GRINDING VALVE SEATS

1. **Do Not Grind Too Long.** *Only a few seconds are required to recondition the average valve seat. Avoid the natural tendency to grind off too much.*

2. **Do Not Use Too Much Pressure.** *While grinding, support the weight of the driver to avoid excess pressure on the stone.*

3. **Keep The Work Area Clean.** *There can be no precision workmanship in the presence of dirt. Always keep tools and work clean.*

4. **Check The Seat Width and Contact Pattern With Bluing.** *If there are any uneven spots, regrind the seat. Do not lap the seats with grinding compound since this may ruin a good grinding job.*

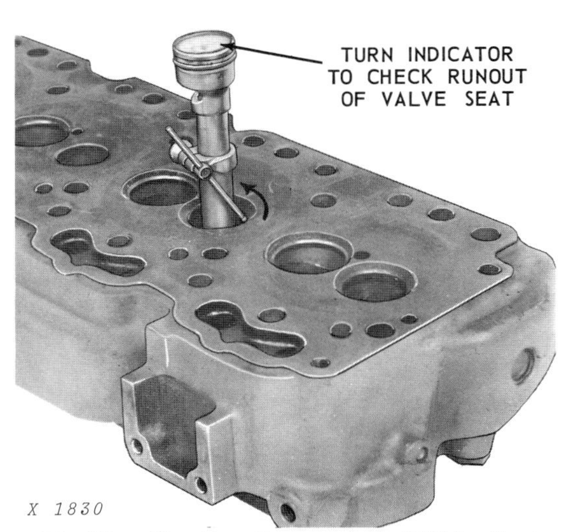

Fig. 35 — Checking Concentricity Of Valve Seat

5. **Check the runout (concentricity) of the valve seat with a dial indicator (Fig. 35).** *Hold the indicator reading within the specifications shown in the engine Technical Manual.*

This is perhaps the most important valve check. If the indicator reading is low, the grinding equipment and stone are in good condition, and together with the bluing check, assures a good seal of the valve to the seat.

Rotate the pilot 90 degrees in the guide and take a second runout reading. If this reading is also low, the pilot is reasonably straight and the guide is within limits.

VALVE GUIDES

The valve guide holds the valve centered for proper seating. It also transfers heat from the valve stem to help cool the valve.

The guide may be a drilled bore in the head or block or it may be an insert (Fig. 36).

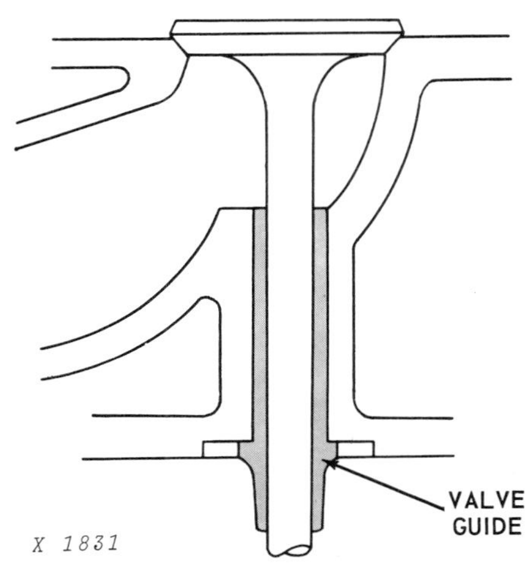

Fig. 36 — Valve Guide

The guide is usually as long as space permits to reduce any "cocking" of the valve by the rocker arm.

The valve guide should be lubricated enough to prevent scuffing of the valve stem. Too much lubrication leads to heavy deposits at the hot end of the valve stem. In some engines, an oil seal is provided on the intake valve stem to prevent this.

SERVICING VALVE GUIDES

It is a good policy to replace valve guides or knurl or ream the integral type. However, this is not always practical. But do this when guide-to-stem clearance is more than 50 percent above the specifications.

IMPORTANT: Never grasp the valve guides when turning over or handling the cylinder head as you may bend or loosen the guides.

Cleaning Valve Guides

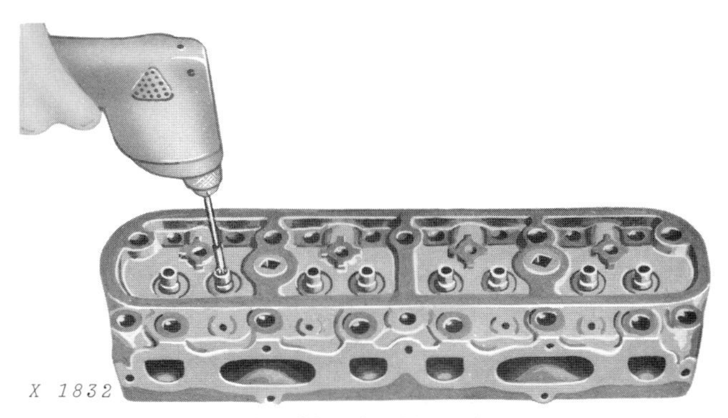

Fig. 37 — Cleaning Valve Guides

Use the correct size wire brush in an electric drill (Fig. 37). Run the brush up and down the full length of the guide. A few drops of light oil or kerosene will help to fully clean the guide.

Measuring Valve Guides

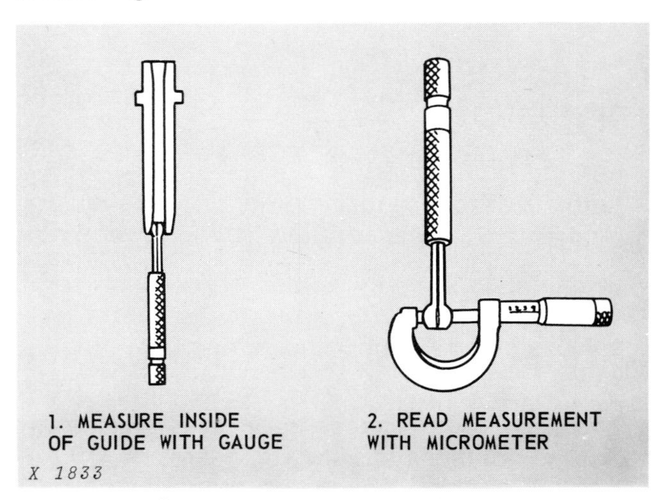

Fig. 38 — Measuring Valve Guides

Guides do not wear round or uniformly. Therefore, don't use plugs, valve stems, or pilots to measure them. Measure the guides at different points within the guide as shown in Fig. 38.

Replacing Valve Guides

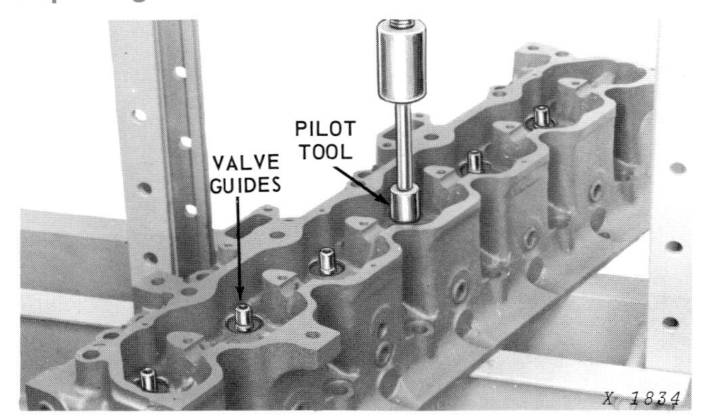

Fig. 39 — Installing Valve Guides

Remove old guides and install new guides as needed (Fig. 39).

Some new valve guides will compress slightly when installed. They must be precision-reamed to specifications after they are installed.

Knurling Valve Guides

"Knurling" a valve guide is upsetting metal on the inside wall to make the bore smaller. Some manufacturers recommend this as a way of reconditioning valve guides using a special tool.

VALVE ROTATORS

Valve rotators are devices that cause the valve to turn during operation. This removes deposits that form on the valve stem and face. The result: an improved gas seal, cooler valves, and less valve erosion.

Two types of valve rotators are:

- **Release Type**
- **Positive Type**

RELEASE-TYPE VALVE ROTATORS

In the release-type rotator (Fig. 40), valve tension is momentarily released, allowing the valve to rotate.

During each valve cycle, the tappet or rocker arm forces the locks away from the shoulder on the valve stem, releasing the spring load from the valve.

The valve then rotates freely for a moment from engine vibrations and moving gases.

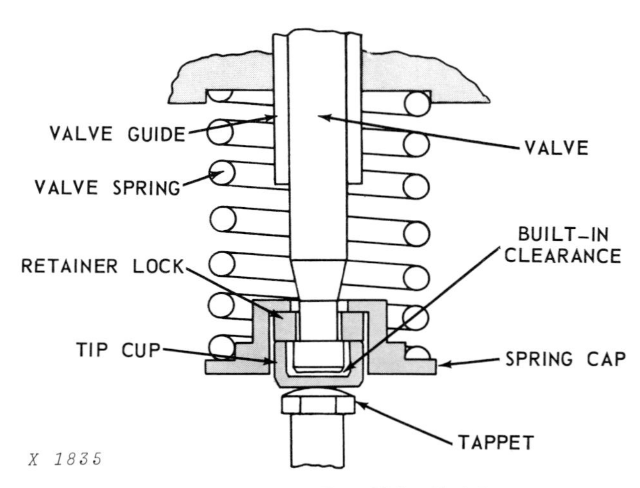

Fig. 40 — Release-Type Valve Rotator

POSITIVE-TYPE VALVE ROTATORS

The positive-type rotator is self-contained and installed in the same way as ordinary valve spring retainers. Rotation occurs when the valve moves and depends on the difference in spring loads between the valve open and closed positions.

Fig. 41 shows this rotator in operation.

When the valve is closed, the spring washer is located at point 1 and at point 2.

As the valve starts to open, the extra spring load causes the washer to flatten, transferring the load from point 2 to point 3. This forces the balls to roll down the inclined races.

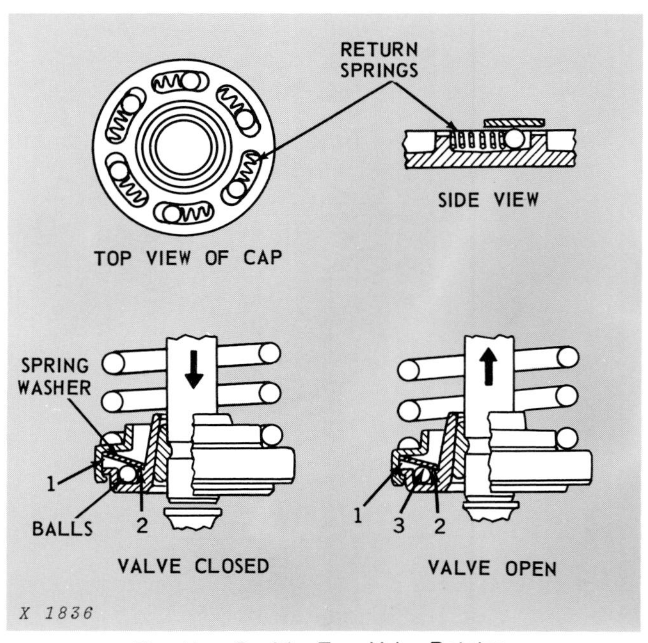

Fig. 41 — Positive-Type Valve Rotator

This rotates the whole assembly, which transmits the movement to the valve.

As the valve closes again, the spring washer is released from the balls. The balls then return to their closed position by action of the return springs (shown at top).

SERVICING VALVE ROTATORS

Positive-Type Rotators (Rotocap)

There is no maintenance for these valve rotators. However, when valves are replaced or reground, replace the valve rotators.

Check rotation of the rotator by looking for carbon build-up on the valve or by inspecting the seat. (Rotation can also be checked by watching the valve with the engine idling.)

The valve rotator may not rotate when held in the hand, but this does not mean it won't operate in the engine.

On valves which use rotators, inspect the valve stem, valve stem cap, and the keepers or locks for wear.

Wear at these points results in false tappet settings and rapid increases in tappet clearance.

Release-Type Rotators

Check the release-type rotator for the clearance built into the valve tip cup. This built-in clearance should be maintained. Wear is uniform up to the wear limit, after which the wear rapidly increases and may cause excessive valve clearance.

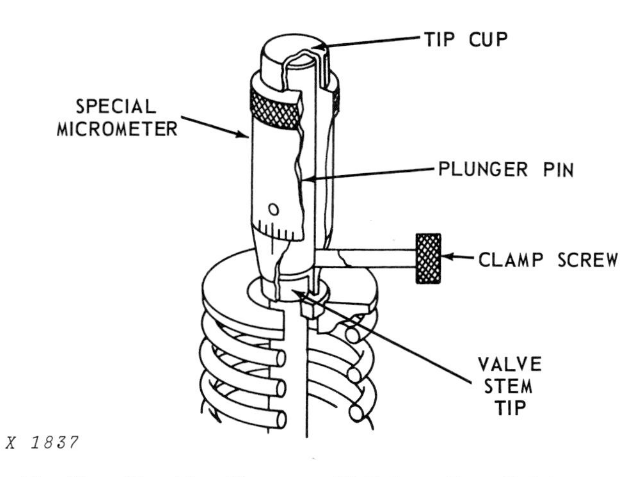

Fig. 42 — Checking Clearance Of Release-Type Rotator Using Micrometer

To check clearance, use a special micrometer as in Fig. 42 and do the following:

1. Remove tip cup from valve end.

2. Place micrometer on end of valve stem and set at the "0" mark. Hold the plunger pin firmly on top of the valve stem tip and tighten the clamp screw.

3. Place the tip cup against the plunger pin and check the clearance or lack of clearance which is transferred to the micrometer spindle barrel.

4. Readings to the right of "0" indicate too much clearance and readings to the left of "0" indicate too little clearance.

Another means of checking valve tip clearance is the Plastigage method. A plastic strip is inserted between the tip cup and the valve stem tip. With the locks held firmly against the shoulder on the valve, the cup is pressed in place. The width of the plastic strip is then checked.

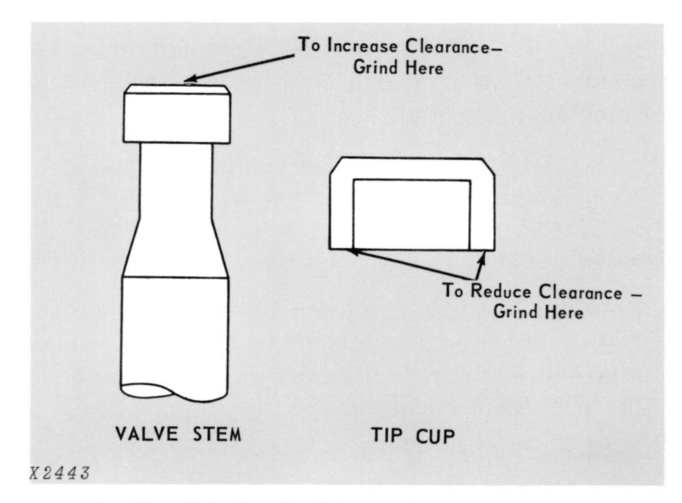

Fig. 43 — Grinding To Maintain Clearance Between Valve Stem And Cup

To reduce clearance, grind the open end of the tip cup (Fig. 43). Grind the valve stem tip to increase clearance.

CAMSHAFTS

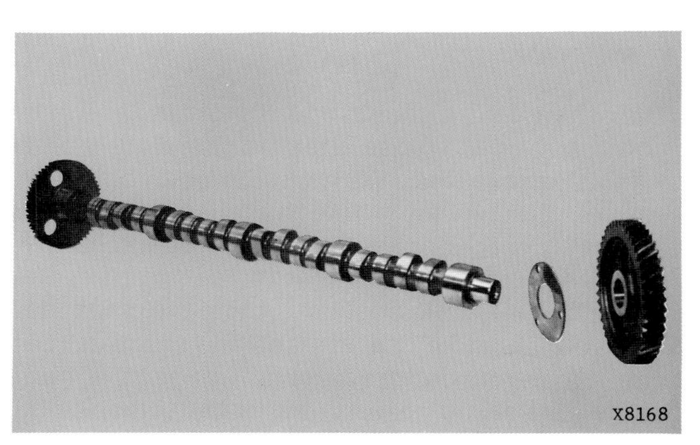

Fig. 44 — Camshaft, Thrust Plate, and Drive Gear

Camshafts for small and medium size engines are generally made of a one-piece casting or forging (Fig. 44). One intake and one exhaust cam is provided for each cylinder, along with bearing journals. The journal diameters are large to permit endwise removal of the shaft from its bore.

The camshaft is normally driven by gearing from the crankshaft.

How the cams are arranged on the shaft determines the firing order of the engine. The contour of each cam decides the time and rate of opening of each valve.

Camshafts are made of a low-carbon alloy steel with the cam and journal surfaces carburized before finish grinding. Some high-speed engines use alloy cast-iron camshafts with hardened cams and journals.

On some large engines the camshaft may have integral cams or it may have individual cams assembled on a shaft.

Overhead camshafts for car engines are supported on pedestals mounted on the cylinder head with removable caps, or else in a tunnel formed in the valve housing.

In some dual overhead camshafts for high-speed racing engines, one camshaft operates the two intake valves per cylinder while the other operates the two exhaust valves.

Camshafts may also drive oil pumps, fuel pumps, and distributors using extra lobes or a gear on the shaft.

CAMSHAFT TIMING

On 4-cycle engines, the camshaft turns at one-half the speed of the crankshaft, so that each valve is opened and closed once during two revolutions of the crankshaft.

The **exhaust valve** should open before the end of the power stroke and close after the completion of the exhaust stroke (Fig. 45). It should open before the end of the power stroke because, due to the angularity of the connecting rod and crankshaft, the pressure in the cylinder near the end of the power stroke has very little effect in turning the crankshaft. The valve should close after the end of the exhaust stroke to permit better scavenging due to the inertia of the outflowing gases in the exhaust manifold.

The **intake valve** should open before the end of the exhaust stroke and close after the end of the intake stroke (Fig. 45). The intake valve opens before the end of the exhaust stroke to take advantage of the inertia of the outflowing gases in the exhaust manifold. The valve should close after the end of the intake stroke to take advantage of the ramming effect, due to inertia, of the incoming air or air-fuel mixture, thereby more completely filling the cylinder.

To summarize:

- **Exhaust valve opens before end of power stroke, closes after end of exhaust stroke.**
- **Intake valve opens before end of exhaust stroke, closes after end of intake stroke.**

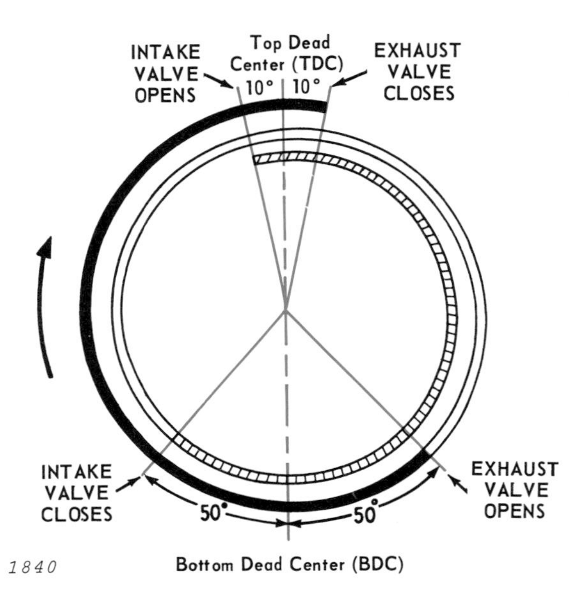

Fig. 45 — Typical Valve Timing (Two Engine Cycles Shown)

Fig. 45 illustrates the valve timing for a typical high-speed engine.

The **intake valve** opens at 10 degrees *before* top dead center (TDC). It stays open until 50 degrees *after* BDC as shown.

During this time the engine has taken in its full charge of air-fuel and the piston has started its compression stroke.

The fuel is then compressed and ignited, and the power stroke begins.

When the piston reaches a point 50 degrees before bottom dead center (BDC), the **exhaust valve** opens as shown and stays open until the piston has traveled to 10 degrees *beyond* TDC, when it closes.

Cam Profiles

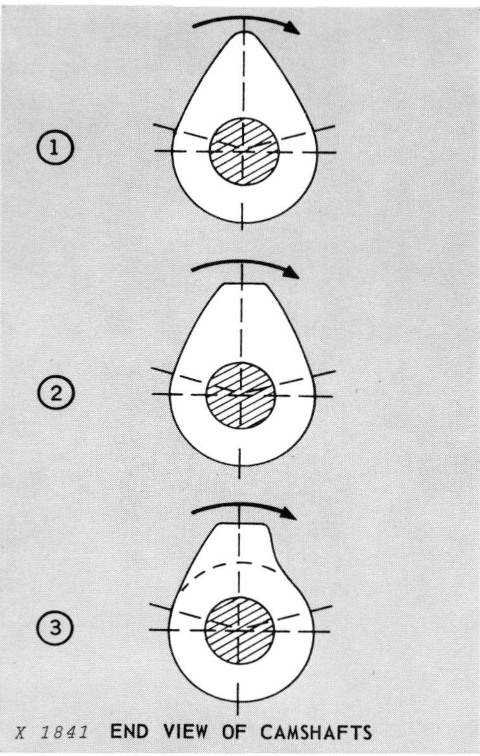

Fig. 46 — Cam Profiles

The **cam profile** is the opening ramp on the cam lobe which "takes up the slack" in the valve train. It also includes the nose which holds the valve open and the backside which allows the valve to close.

The slope of the cam face determines how fast the valve opens.

Cam No. 1 in Fig. 46 has curved faces which cause the valve to open faster at the start and to be held open wide until the end of the closing face is under the valve tappet.

Cam No. 2 provides a rapid opening and closing with a long period at wide open.

Cam No. 3 is used in high-speed engines (racing cars) to give the longest possible opening of the valve.

SERVICING CAMSHAFTS

Always check the camshaft when you recondition the engine.

In modern engines, the cam lobes have a difficult job to do. They must raise the valves, one by one, at the proper time and under pressure, by rapidly moving the tappets, push rods, and rocker arms at high engine speeds.

Inspection Of Bearing Journals And Bore

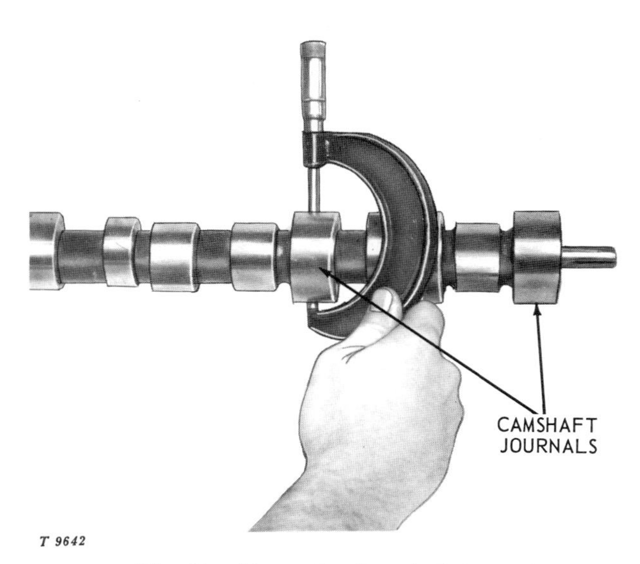

Fig. 47 — Measuring Camshaft Journals

Inspect camshaft journals for signs of wear or out-of-round condition.

Measure the camshaft journals with an outside micrometer as shown in Fig. 47.

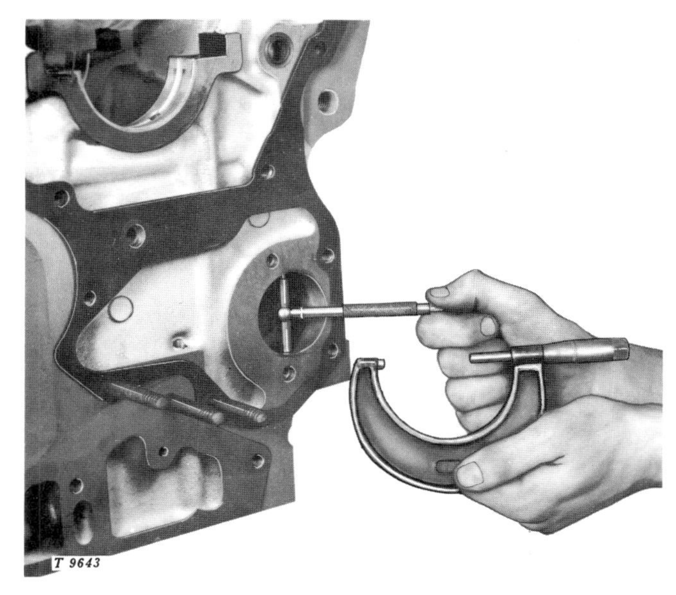

Fig. 48 — Measuring Camshaft Bores

Measure the camshaft bores or bearings with an inside micrometer (or telescoping gauge and outside micrometer) as shown in Fig. 48.

Compare the results with specifications.

Checking Cam Lobes

Using a micrometer, check the height of each cam lobe. Compare intakes to other intakes, and exhaust cams to their counterparts, then check all with specifications. Also inspect the cam faces on which the tappets slide for surface fatigue. The nose of the cam must be polished and flat. If rough or wavy, inspect the cam and its mating follower for possible service.

Regrinding and polishing the cams is not satisfactory on modern high-speed engines. Originally the cam surfaces are treated to provide an oil-absorbing surface for better lubrication. Most engine reconditioners cannot duplicate this process. Therefore, the camshaft must be replaced.

Measuring valve lift can give an indication of wear on cam lobes, cam followers, and push rods.

Selecting Bearings

Select the proper bearings to match the camshaft journals. Some manufacturers have undersized bearings for replacement parts as well as special installation tools.

Line up oil holes prior to assembly and after assembly, be sure to check the registry of these holes with those in the cylinder block with a wire.

CAM FOLLOWERS

Cam followers work off the lobes of the camshaft and drive the push rods to operate the valves (Fig. 49).

A cam follower is usually a piston-like part working in a vertical guide against the cam lobe. The follower is commonly hardened and ground to provide a long wearing surface face against the cam lobe.

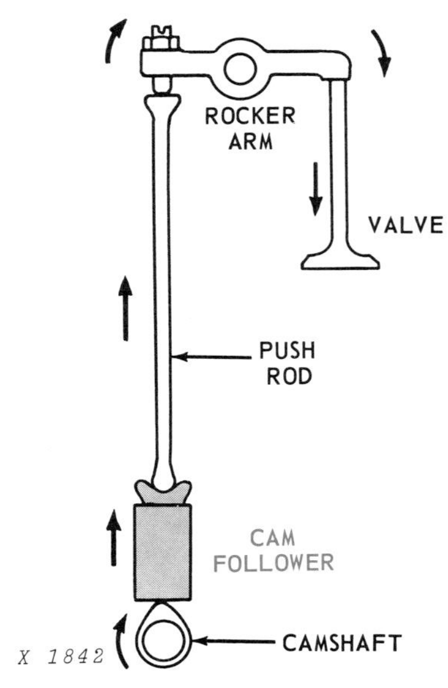

Fig. 49 — Cam Follower

Some cam followers have a roller on the driven end that works against the camshaft lobe. This reduces friction and wear on the follower.

NOTE: Cam followers may also be called "valve tappets."

HYDRAULIC VALVE LIFTERS

With hydraulic valve lifters, the valve clearances are always zero and valve operation is quiet.

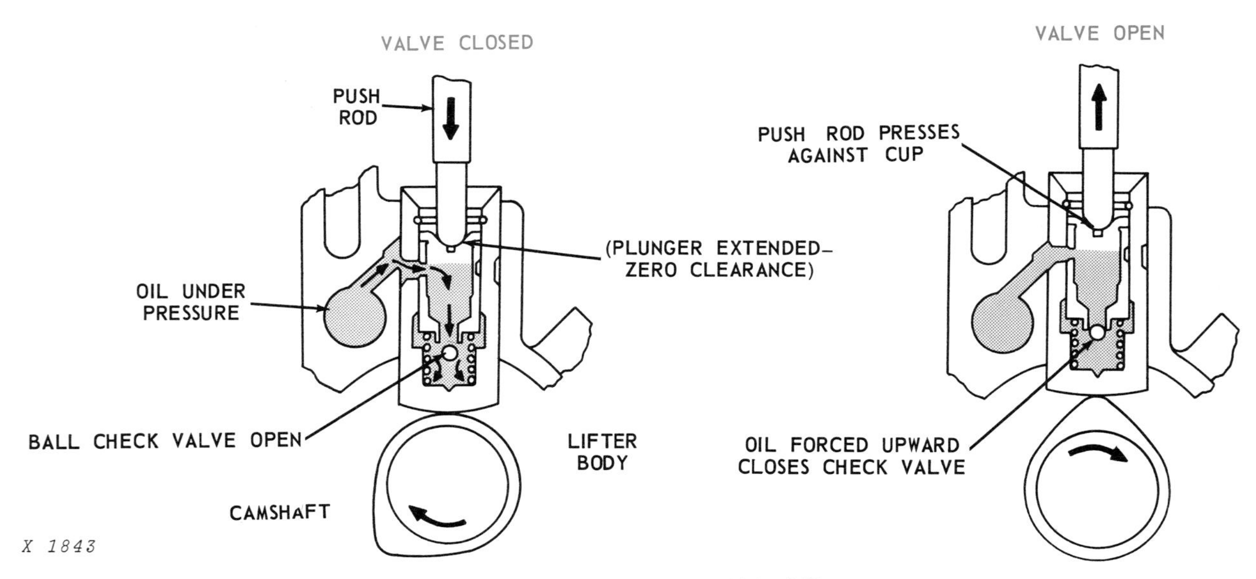

Fig. 50 — Hydraulic Valve Lifter

The lifter body has an opening through which engine oil under pressure is supplied to the pressure chamber in the lower part of the lifter (Fig. 50).

The oil unseats the check ball and fills the pressure chamber below the plunger. The small spring below the plunger keeps the lifter in contact with the cam when the valve is closed.

As the camshaft moves to open the valve, the pressure of the oil under the plunger increases and closes the check ball. The plunger and valve are lifted by the oil trapped above the check ball.

As the valve is closed, the pressure of the oil under the plunger decreases, and the valve ball opens.

SERVICING OF DAM FOLLOWERS

Proper inspection and service of cam followers is vital to good engine performance.

A change in valve clearance or valve lift can usually be detected by excessive noise at idle speeds.

If you suspect a change in clearance, remove the followers and inspect them for excessive wear.

Mark each one so it can be returned to the same bore.

After the followers are removed, clean all parts thoroughly with fuel oil or solvent and dry them with compressed air.

Look for signs of fatigue on each tappet. If the follower is scuffed or "dished", the camshaft may need replacing.

Examine follower bores to make sure they are clean, smooth, and free of score marks. Clean up any marks.

If the camshaft is replaced, also replace the cam followers.

On roller-type followers, the rollers must turn smoothly and freely on their pins and be free of flat spots or scuff marks.

If the rollers do not turn freely or have been scored or worn flat, examine the cam lobes on which they operate (see "Camshafts".)

Measure the maximum wear on roller-type followers. If clearances do not meet specifications, replace the cam roller and pin, which are usually serviced as a set.

When installing cam followers, lubricate the body and bore with engine oil so that the follower slides in easily. Also apply oil to the camshaft lobes.

VALVE PUSH RODS

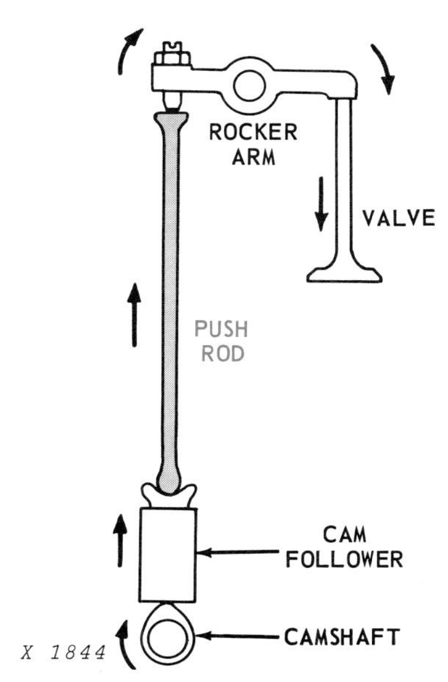

Fig. 51 — Valve Push Rod For Valve-In-Head Engine

The push rod, used in valve-in-head engines, is a steel rod which transmits camshaft motion to the rocker arm (Fig. 51).

Push rods are usually made from hollow steel pipe with solid steel ends.

The lower end of the push rod is usually formed into a half-round head to match a spherical seat in the tappet. The top end usually has a spherical socket to match the adjusting screw at the rocker arm.

SERVICING PUSH RODS

1. When removing push rods, identify each rod so it can be assembled with the same mating parts.

2. Inspect push rod ends for wear or damage.

3. Examine push rods for bent condition. Push rods should be replaced if they are bent more than specifications allow. Measure by placing the push rods on a flat plate. Rotate the push rod and measure the deflection with a dial indicator.

4. If a failed valve has been replaced, also inspect its mating rocker arm and push rod. Often the rocker arm or push rod is bent or otherwise damaged when the valve fails, even though this condition may not be noticeable.

VALVE SPRINGS

The main function of the valve spring is to close the valve and keep it closed and seated until opened by the camshaft.

Cylindrical springs are used by most engine manufacturers. In some cases, two springs are used for each valve to cut down spring vibration and valve flutter. Special spring dampers are sometimes used to reduce vibration and humming.

SERVICING VALVE SPRINGS

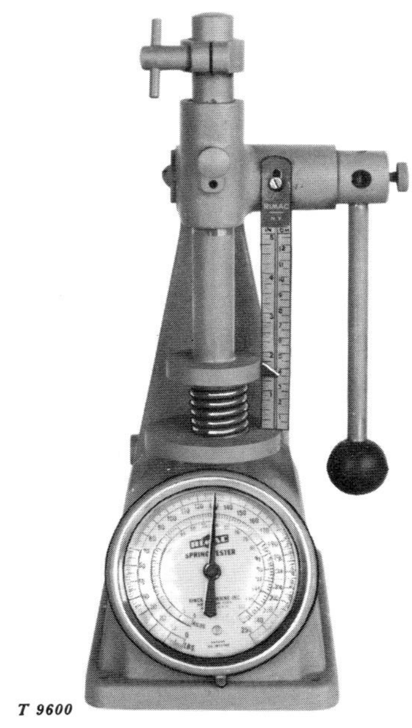

Fig. 52 — Checking Valve Spring In Tester

When **inspecting valve springs,** look for the following:

1. Wear on the casting where the springs rotate.

2. Wear on the spring caps. Spring caps can wear almost to knife edge.

3. Wear on ends of spring.

4. Lack of spring tension. Check against specifications using a spring tester (Fig. 52). A tension loss exceeding 10 percent of the specified load means that the spring needs replacing.

5. Also check for any pressure loss caused by valve grinding. As the valve seat is ground deeper, the spring operating height is longer, decreasing the spring load. Washers can sometimes be added to correct this, but be sure to allow for the washer when testing the spring.

6. Inspect for "cocked" springs. This places a greater than normal side thrust on the valve stem and results in excessive wear. Measure by placing the spring on a flat surface and using a square. Replace springs if they are cocked.

7. Don't be concerned if the valve springs have different free lengths. This can vary and the springs will still have the same length when compressed. Remember, the **force** *of the spring is the critical factor.*

VALVE ROCKER ARMS

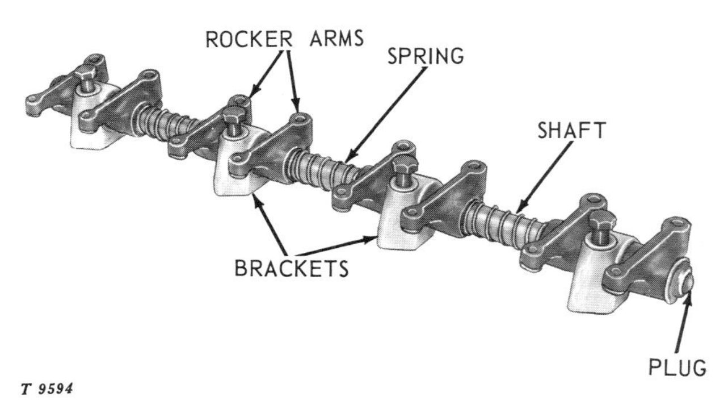

Fig. 53 — Valve Rocker Arm Assembly

In valve-in-head engines, the rocker arms transmit camshaft motion to the valves.

Most rocker arms are mounted on a single hollow shaft located at the top of the engine (Fig. 53).

When the push rods move up, the mating rocker arm is moved down, contacting its valve stem tip and opening the valve.

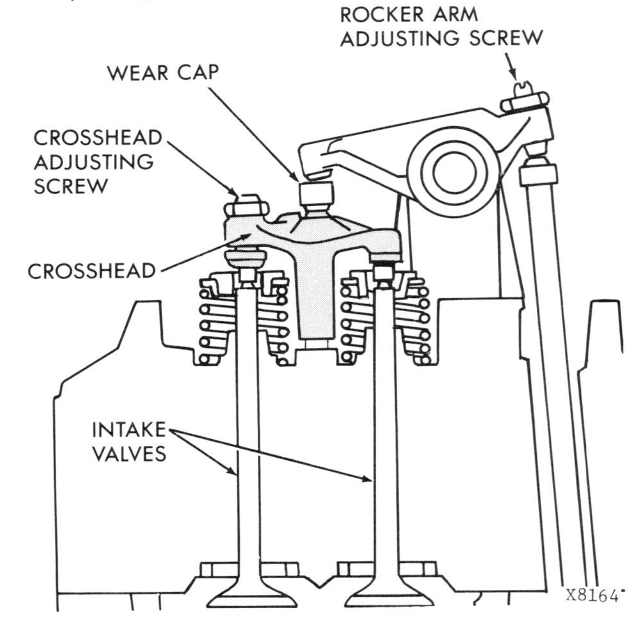

Fig. 54 — Rocker Arm Contacts Crosshead

On engines with two intake and two exhaust valves per cylinder, the rocker arm contacts a crosshead (Fig. 54) instead of the valve stem tip. The crosshead then contacts both intake or both exhaust valves and causes two valves to open simultaneously.

Lubricating oil is usually supplied through the drilled rocker arm shaft to each rocker arm.

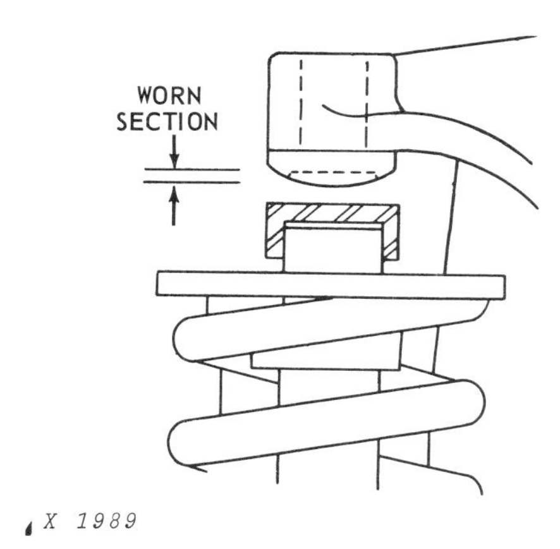

Fig. 55 — Worn Rocker Arm Tip

SERVICING ROCKER ARMS

1. Inspect the rocker arm shaft for scratches, scores, burrs, or excessive wear at points of rocker arm contact. Be sure that all oil holes are open and clean. If a valve has failed, examine the shaft for cracks.

2. Check for cups or concave wear on ends of rocker arms where they contact the valve tips (Fig. 55). Concave wear at this point makes it difficult to get the proper tappet adjustment. A worn rocker arm also creates side thrust on the valve stem and causes wear.

3. Replace or recondition rocker arms when noticeable grooving or wear occurs.

4. Check the rocker arm adjusting screw and nut for damage. Examine the spacer springs on shaft between rocker arm and be sure they are strong enough to exert a positive pressure on the arms.

5. Thoroughly clean holes in rocker arm mounting brackets. This is very important in some engines, because these holes feed oil to the rocker arm shaft.

6. When installing rocker arms, be sure to install the arms and springs in the same sequence as they were removed.

VALVE CLEARANCE

When valves are properly adjusted, there is a small clearance between the valve stem and the end of the rocker arm. (This clearance is sometimes referred to as "valve lash" or "tappet clearance".)

Valve clearance allows for the heat expansion of parts. Without clearance, expansion of the heated parts would cause the valves to stay partly open during operation.

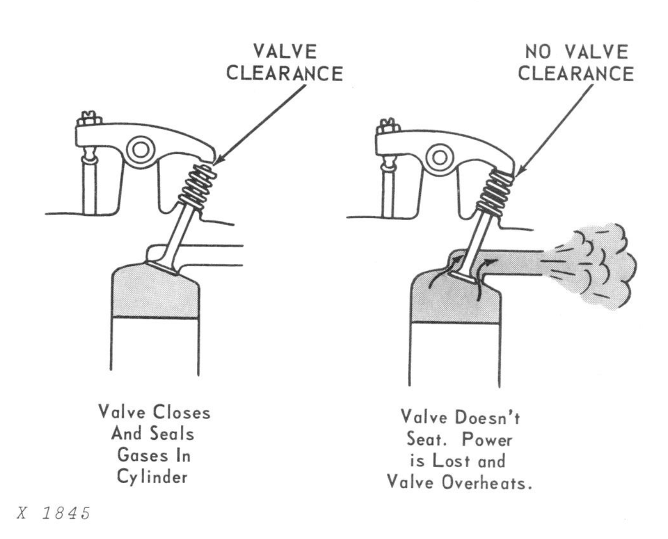

Fig. 56 — Problem Of Too Little Valve Clearance

This clearance is small, varying from approximately 0.006 inch (0.15 mm) to 0.030 inch (0.76 mm). Each engine manufacturer recommends a definite clearance for his engine model.

The valve clearance may vary depending on the engine model and whether the engine is hot or cold during adjustment (some engines run hotter than others).

Too little valve clearance throws the valves out of time. This causes valves to open too early and close too late. Also, valve stems may lengthen from heating and prevent valves from seating completely (Fig. 56). Hot combustion gases rushing past the valves cause overheating because the valves seat so briefly or so poorly that normal heat transfer into the cooling system does not have time to take place. This causes burned valves.

Too much valve clearance causes a lag in valve timing which throws the engine out of balance. The fuel-air mixture is late entering the cylinder during the intake stroke. The exhaust valve closes early and prevents waste gases from being completely removed.

The valves themselves also become damaged. When valve clearance is properly adjusted, the camshaft slows the speed of valve movement as it closes. But with too much clearance, the valves close with great impact, cracking or breaking the valve and scuffing the cam and follower.

HOW TO ADJUST VALVE CLEARANCE

Follow the manufacturer's procedure for correct valve adjustment. Here is a typical valve-in-head procedure:

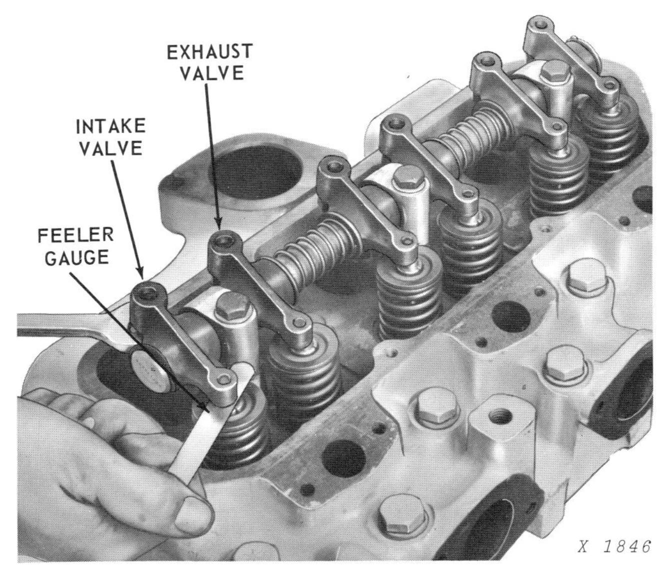

Fig. 57 — Adjusting Valve Clearance

1. Be sure the engine is at the recommended temperature before you begin valve adjustment.

2. Clean all dirt and oil from around valve cover and remove the cover.

3. Turn engine over until piston in the No. 1 or first cylinder is at top dead center (TDC) of its compression stroke. Most engines have timing marks on the flywheel or fan drive pulley to mark the "TDC" or other timing point.

A positive way to find when a piston is at "TDC" is to remove the spark plug or injection nozzle and hold your finger over the opening. On the compression stroke, air will be forced out against your finger until the piston reaches the "TDC" position.

4. With the No. 1 piston at "TDC", check the valve clearance using a feeler gauge (Fig. 57). Adjust clearance if necessary by turning valve adjusting screw up or down until clearance is to specifications.

5. Be sure to determine which are intake and which are exhaust valves because the clearances are usually different for the two.

6. Rotate the engine crankshaft in its firing order and adjust valve clearances when each piston reaches "TDC" of its compression stroke. Often two or three sets of valves may be set at a time with one rotation of the crankshaft. Refer to the engine Technical Manual for the correct method for each engine.

NOTE: On engines with two intake and two exhaust valves per cylinder, there are adjusting screws on both the crosshead and the rocker arm (Fig. 54). The crosshead adjustment is to insure both intake or both exhaust valves are actuated at the same time. The rocker arm adjustment sets valve clearance.

7. Install the valve cover, using a new gasket.

NOTE: Recheck the valve clearances after a "run-in" of the engine and a retightening of the cylinder head.

CYLINDER BLOCKS

The cylinder block is the main support for the other basic engine parts (Fig. 58). The block is usually a one-piece casting made from gray cast iron.

Blocks are cast with end walls and center webs to support the crankshaft and camshaft and have enlargements in their walls to make room for oil and coolant passages.

The cylinders are often cast into the block and may be bored into the block or machined to receive replaceable liners.

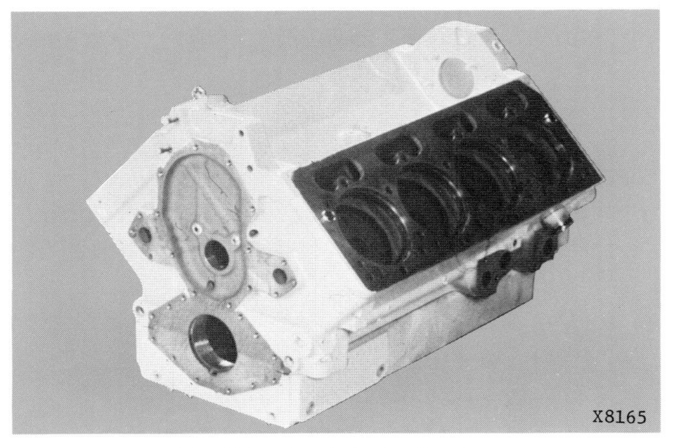

Fig. 58 — V-8 Cylinder Block

SERVICING CYLINDER BLOCKS

Strip the cylinder block as outlined in the engine Technical Manual.

After stripping, inspect the block for damage. If the block is still serviceable, prepare it for cleaning as follows:

Scrape all gasket material from the block. Then remove all oil gallery plugs and core hole plugs to allow the cleaning solution to contact the inside of the oil and water passages.

Cleaning Of Cast-Iron Cylinder Blocks

1. Remove grease by agitating the cylinder block in a hot bath of commercial heavy duty alkaline solution (Fig. 59).

2. Rinse the block in hot water or steam clean it to remove the alkaline solution.

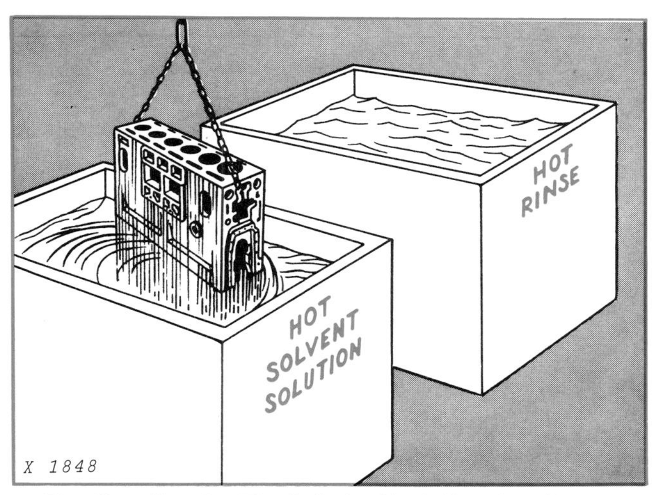

Fig. 59 — Cleaning The Cylinder Block (Cast-Iron Type)

3. If the water passages are heavily scaled, the block can only be cleaned using special equipment as follows:

a. Agitate the block in a bath of inhibited commercial pickling acid.

b. Allow the block to remain in the acid bath until the bubbling action stops (about 30 minutes).

c. Lift the block, drain it, and immerse it in the same acid solution for 10 minutes.

d. Repeat Step (c) until all scale is removed.

e. Rinse the block in clear hot water to remove the acid solution.

f. Neutralize the acid clinging to the casting by immersing in an alkaline bath.

g. Rinse the block in clean water or steam clean it.

4. Be sure that all water passages, oil galleries, and oil holes have been thoroughly cleaned.

Cleaning Of Aluminum Cylinder Blocks

1. Make a solution of one part emulsion-type soap and five parts of water. Using this solution in a steam cleaner, steam clean the cylinder block.

2. Rinse the block in clear water and dry it with compressed air.

3. If further cleaning is necessary:

a. Brush a commercial chlorinated solvent, suitable for aluminum, on the block. Allow the solvent to remain on the block for several hours.

b. Steam clean the block with the soap solution used in Step (1).

c. Rinse the block with clear water and dry it with compressed air.

4. After the cylinder block is thoroughly cleaned and dried, reinstall any core hole plugs. Coat the threads of the plugs with a sealer.

Testing The Cylinder Block For Cracks Or Leaks

Overheating of the engine may result in cracks between the water jackets and the oil passages.

The block can be pressure tested for cracks and leaks by the following method.

1. Seal the water openings in the block by using plates with gaskets. Drill and tap one of the plates to provide a connection for an air line.

2. Immerse the block for 15 minutes in a tank of hot water (180-200 °F) (82-93 °C).

3. Apply air pressure to the block and observe the water in the tank for bubbles, which mean cracks or leaks. If a large tank is not available, fill the block with water and apply air pressure.

4. Always replace a cracked cylinder block.

Inspecting The Cylinder Block

After cleaning and pressure testing, inspect the cylinder block.

1. Check all dowel pins, pipe plugs, expansion plugs, and studs for looseness, wear or damage and replace as necessary. Coat all parts with joint sealing compound before installing.

2. Inspect all machined surfaces and threaded holes in the block. Carefully remove any nicks or burrs from the machine surfaces with a file. Clean out tapped holes and clean up damaged threads.

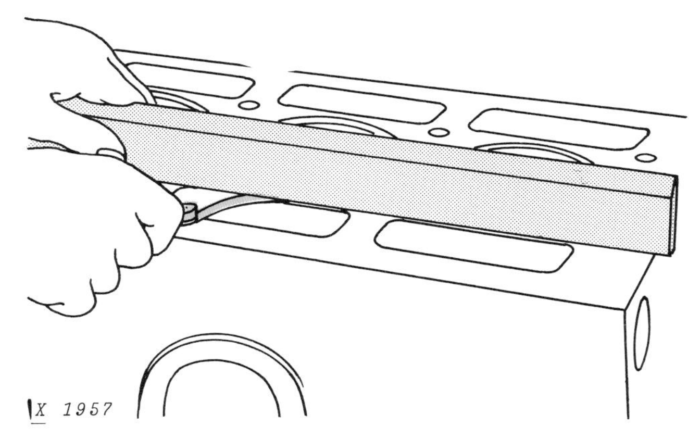

Fig. 60 — Checking Top Face Of Cylinder Block For Flatness

3. Check the top of the block for flatness with an accurate straight edge and a feeler gauge (Fig. 60). This is the critical area for sealing oil, water, and compression. If warped, replace or resurface the block as recommended in the engine Technical Manual.

4. Inspect the cylinder bores (for details, see "Cylinders" and "Cylinder Liners" later in this chapter).

5. After inspection, spray the machined surfaces with engine oil. If the cylinder block is to be stored for a long period, spray or dip the block in a rust-preventive solution. Cast iron will rust when exposed to the atmosphere unless it is greased or oiled.

CYLINDERS

The cylinder is basically a hollow tube in which the piston works.

There are two basic types of cylinders:

- **Cast-in-Block (Enbloc)**
- **Individual Castings (Liners or Sleeves)**

Let's see how they differ.

CAST-IN-BLOCK (ENBLOC)

Automotive engines generally feature the "enbloc" design. The cylinders are cast into the cylinder block so that the cylinders and block form a single unit.

INDIVIDUAL CYLINDER CASTINGS

The chief advantage of individual castings is that replacement is less expensive.

In larger engines, where each separate casting is bolted to a base, it is possible to build up an engine of almost any number of cylinders.

In high-speed engines, these individual castings are called liners or sleeves.

Cylinder liners are of two types:

- **Dry Liners**
- **Wet Liners**

DRY LINERS are sleeves which fit inside an already completed cylinder (Fig. 61). This liner is simply a wearing surface for the piston. They are not exposed to the engine coolant—so they are called "dry."

WET LINERS form not only the cylinder wall but also the inside of the water jacket (Fig. 61). When using wet liners, water seals must be provided for the liners.

Wet cylinder liners usually have a flange at the top of the liner to seat in a mating counter bore at the top of the cylinder block. Then when the cylinder head is installed, the liner is held firmly in place.

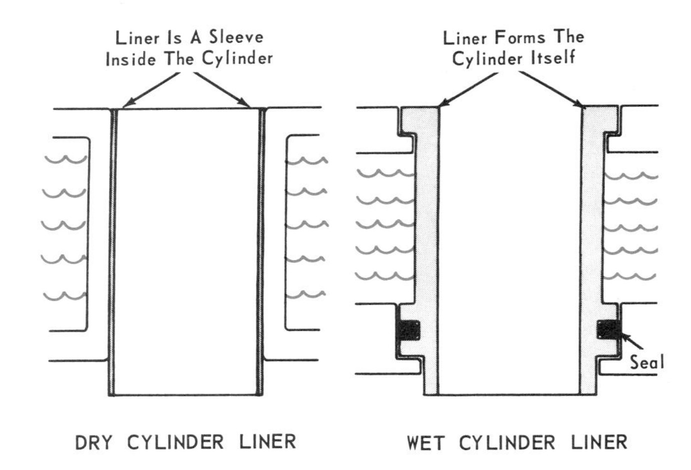

Fig. 61 — Types Of Cylinder Liners

SERVICING CYLINDERS

Because the servicing for integral cylinders and removable cylinder liners is identical except for removal and installation of the liners, the information in this section will serve for both types of cylinders.

Preliminary Inspection

As soon as the engine head or heads have been removed, inspect the condition of the cylinders in the area of ring travel. When the appearance is good and there are no scuff marks deep enough to require cylinder re-conditioning, check the cylinder for wear or out-of-roundness, described later in this chapter. These measurements help you decide whether to rebore or resleeve the engine.

Removing Ridge From Cylinders

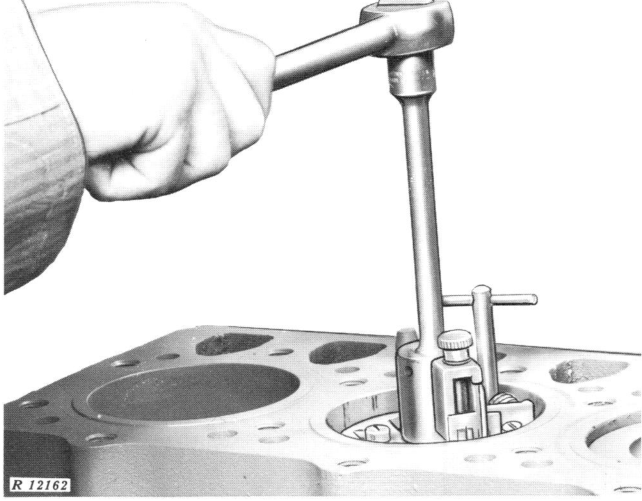

Fig. 62 — Removing Ridge From Cylinder

As the cylinder wears, a ridge is formed at the top of the piston ring travel zone. If this ridge gets too high, pistons can be damaged when they are removed.

Remove any ridges from cylinders using a ridge reamer (Fig. 62).

Do not cut down into the ring travel zone when removing the cylinder ridge.

It is possible to cut so deeply into the cylinder wall or so far down into the ring travel that reboring or replacement of the engine block or liner is necessary.

Most worn cylinders are out-of-round when cold. Regardless of how uneven the wear, blend the cut made with the ridge reamer so that the area where the worn surface meets the reamed surface is as smooth as possible.

Removing Cylinder Liners

Fig. 63 — Removing Cylinder Liner

After long service, liners collect carbon and rust or "freeze" in place. Details of removal will vary with the size and type of engine, but use these general tips:

1. Use pullers to remove liners (Fig. 63). They may be the conventional screw-threaded type or impact pullers.

2. In stubborn cases, other aids must be used. Removing scales from the water jacket, with inhibited acid compounds, may do the trick. Solvents and penetrating oil may help. Heat may also be used—putting steam or hot water into the water jackets while filling the liner with cold water.

NOTE: Measure cylinder liners for wear before removing them from the block. See "Measuring Cylinders for Taper and Out-of-Roundness".

Cleaning The Cylinder Area

Clean cylinders as instructed under "Cylinder Blocks" earlier in this chapter. Use a wire brush and solvents to clean out hardened carbon and gum deposits.

Thoroughly clean the counterbores in the cylinder block with a scraper. Deposits can unseat the liner, distort it, and possibly break the flange.

On wet liners, thoroughly clean the lower sealing ring surface in the block to prevent coolant leakage past the sealing rings.

Measuring Cylinders For Taper And Out-of-Roundness

Measure cylinder bores as follows:

1. Measure the bore parallel to the crankshaft at the top end of the ring travel zone.

2. Measure the bore in the same position at the bottom end of the ring travel zone.

3. Measure the bore at right angles to the crankshaft at the top end of the ring travel zone.

4. Measure the bore in the same position at the bottom end of the ring travel zone.

Compare the measurements (1) and (3) to find the *out-of-round wear* at the top end of the bore.

Compare the measurements (2) and (4) to find the *out-of-round wear* at the bottom end of the bore.

Fig. 64 — Measuring Cylinder Bore

Litho in U.S.A.

Compare the results of measurements (1), (2), (3), and (4) to find out whether or not the bore has worn tapered.

To measure, use a cylinder dial gauge, an inside micrometer, or a telescope gauge and outside micrometer (Fig. 64).

How Much Cylinder Wear?

The amount of cylinder taper and out-of-roundness that can be tolerated without reboring or resleeving the engine depends on the design, the condition, and the type of service for which the engine is used. There is no general rule for all engines.

Generally, whenever cylinder taper or out-of-roundness exceeds 0.005 inch (0.127 mm), the engine should be rebored or resleeved.

However, follow the engine manufacturer's rules for the exact wear limits.

When a cylinder bore is worn beyond the wear limit, the piston generally is also worn beyond its limits. Therefore, *normally replace the piston when the cylinder is replaced or reconditioned.* Refer to "Pistons" later in this chapter.

Reboring The Cylinder

Cylinder liners for smaller engines are generally priced for replacement when they have reached their wear limits.

Liners for larger engines and all integral cylinders are commonly rebored.

They are sized to the smallest standard oversize diameter at which they will clean up. Oversize pistons must then be fitted to provide the correct piston-to-liner clearance. The final finish should then be obtained by honing.

A number of boring bars are available which can produce a good finish (ready for honing) in the cylinders. Boring should be done only by competent servicemen who are careful in their work.

When **reboring** cylinders, take these precautions:

1. Grind the cutting tool properly before using it.

2. Be sure the top of the engine block is free of all deposits and irregularities.

3. Clean the base of the boring bar before the bar is set up. Otherwise the boring bar will tilt and the cylinder will be bored crooked.

4. Make an initial rough cut, followed by a finish cut. Then hone the cylinder to the exact size.

When the boring bar is operated wrong, it will produce a rough cylinder surface that may not clean up even when honed. The result: a noisy engine at least during break-in, and faster ring wear.

Honing Cylinders After Reboring

Use the recommended grit size for the honing stone to produce the specified finish. Too smooth a finish can retard piston ring seating, while too rough a finish will wear out the rings.

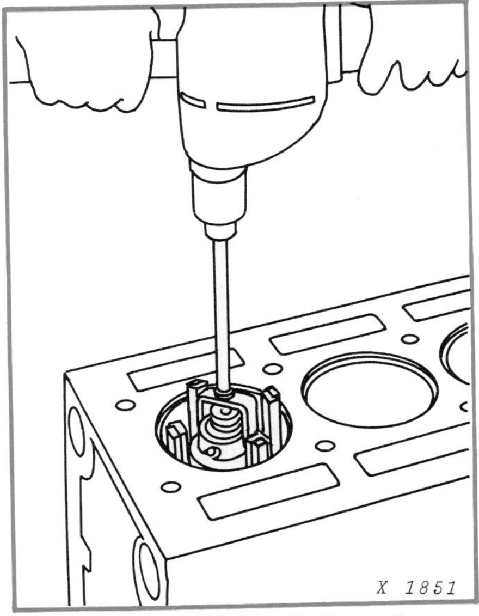

Fig. 65 — Honing The Cylinder Bore

Hone the cylinders as follows:

1. Clean the stones frequently with a wire brush to prevent "stone loading."

2. Follow the hone manufacturer's instructions for the use of oil or kerosene on the stones. **Do not use cutting agents with a dry hone.**

3. Insert the hone in the bore (Fig. 65) and adjust the stones snugly to the narrowest section. When correctly adjusted, the hone will not shake in the bore, but will drag freely up and down when the hone is not running.

4. Start the hone and "feel out" the bore for high spots, which cause an increased drag on the stones. Move the hone up and down the bore with short, overlapping strokes about one inch. Concentrate on the high spots in the first cut. As these are removed, the drag on the hone will become lighter and smoother. Feed the hone lightly to

avoid too much increase in the bore diameter. Some stones cut rapidly even under low pressure.

5. When the bore is fairly clean, remove the hone, inspect the stones, and measure the bore. Determine which spots should be honed most.

6. Moving the hone from the top to the bottom of the bore will not correct an out-of-round condition. Do not remain in one spot too long or the bore will become irregular. Where and how much to hone can be judged by feel. A heavy cut in a distorted bore produces a steady drag on the hone and makes it difficult to feel the high spots. Therefore, use a light cut with frequent stone adjustments.

7. Wash the cylinder block thoroughly after honing.

Deglazing Cylinder Liners

On smaller engines, cylinder liners are usually deglazed, but not rebored.

Deglazing gives a fine finish by removing any scuffs or scratch marks, but does not enlarge the cylinder diameter, so standard-size pistons may still be used.

The purpose for deglazing a crosshatch surface in the cylinder is to provide cavities for holding oil during piston ring break-in.

The reason for angled crosshatch pattern is to prevent the rings from catching in the grooves.

If the cylinder liners are wavy, scuffed, or scratched, **deglaze** them as follows:

1. Remove the pistons and rods. Cover the crankshaft with clean rags or damp paper to prevent the abrasives and dirt caused by deglazing from falling on the crankshaft. The cylinder liner may also be removed and placed in a fixture or scrap cylinder block if desired.

2. Swab the cylinder walls with a clean cloth or cotton string mop which has been dipped in clean, light engine oil.

3. Use a recommended deglazing tool. A brush-type tool with coated bristle tips often gives the best job of deglazing (Fig. 66).

4. Surface hone each cylinder for 10 to 12 complete strokes. The hone should be driven by a slow-speed drill. Move the hone up and down in the cylinder rapidly enough to obtain a crosshatch pattern as shown in Fig. 67. Use the proper grit deglazing cloth or honing stones to produce the recommended fine finish on the cylinder walls.

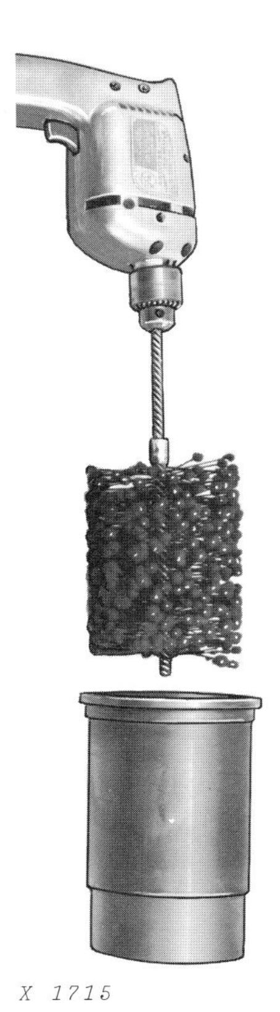

Fig. 66 — Cyliner Liner Deglazing Tool

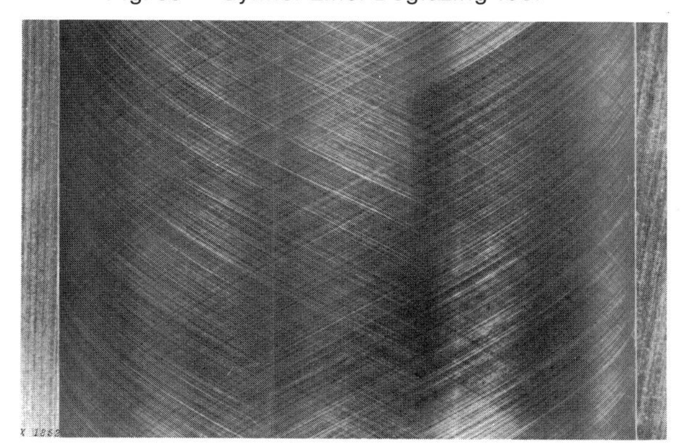

Fig. 67 — Deglazing Home Pattern

5. **Be sure to thoroughly clean the cylinder bores after honing.** If cylinders are not cleaned, hard abrasives remain in the engine. This rapidly wears rings, cylinder walls, and the bearing surfaces of all lubricated parts.

Wipe as much of the abrasive deposits from the cylinder wall as possible. Then clean with hot water and detergent and carefully wipe with a clean cloth (Fig. 68). Oil the cylinder immediately after cleaning.

One swabbing and wiping is not enough. Three operations are usually required—and more may be necessary. Keep on cleaning until a clean white rag shows no discloring when wiped through the cylinder bore.

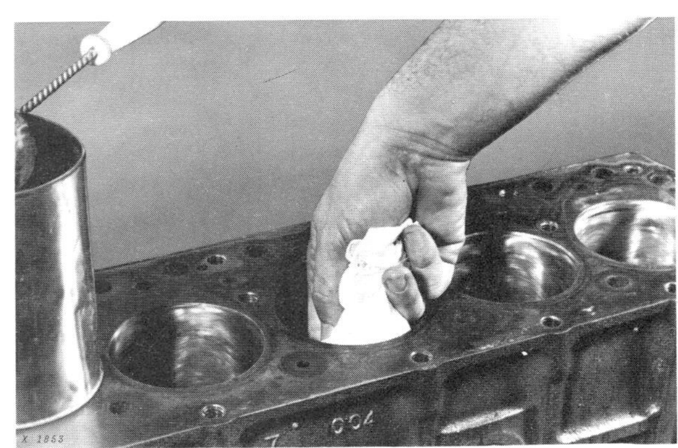

Fig. 68 — Thoroughly Clean Cylinders After Honing Or Deglazing

CAUTION: *Never use gasoline, kerosene, or commercial cleaners to clean cylinders after honing. These solvents will not remove abrasives from the walls.*

Installing Dry Cylinder Liners

Before installing dry-type cylinder sleeves, inspect and measure the cylinder bores.

If the bores are distorted, the block must be rebored to take oversized sleeves. Otherwise the cylinder sleeves may conform to the distorted bores, or air pockets will form between sleeves and block, causing hot spots that often result in a break-down of the rubbing surfaces.

Carefully clean all carbon and gum deposits from the cylinder bores as previously described.

The recommended fit of the sleeve in the bore may be either a loose fit or an interference fit. Always follow the recommendations when installing sleeves.

Installing Wet Cylinder Liners

1. *Clean all deposits from under flange of liner and mating bore in block. The liner must rest flat to prevent distortion.*

2. Check the liner bore in the block. If the flange surface is worn unevenly, remachine it.

3. *Clean the lower sealing surfaces in the block and liner to prevent coolant leaks when the liner is installed.*

4. If a new liner is being installed, install it **without seals** and check its height in relation to the top of the block (Fig. 69). Check the height in several places to be sure the liner is seated squarely. Compare the height to engine specifications. *Always pull the liner and install the seals before final assembly.*

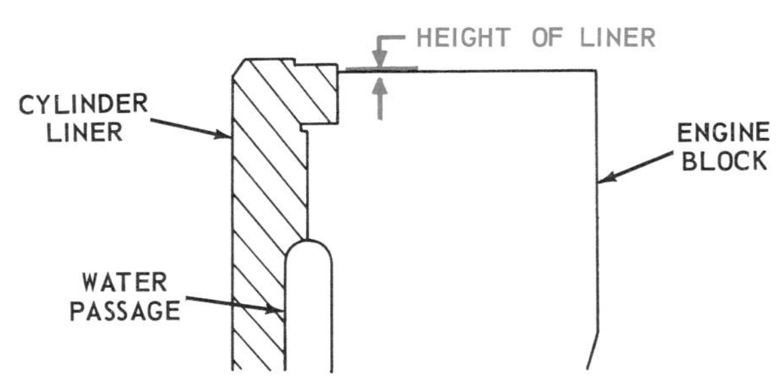

Fig. 69 — Checking Position Of Cylinder Liner In Block

5. If necessary, machine the block to square up the liner, or add shims (if recommended).

6. Place new seals on the liner. Lubricate both the seals and mating surfaces in the block. **Check to be sure the seals are not twisted or crimped.**

7. Work the liner gently into place as far as possible by hand. Finish seating the liner by placing a hard wooden block over it and tapping lightly with a hammer.

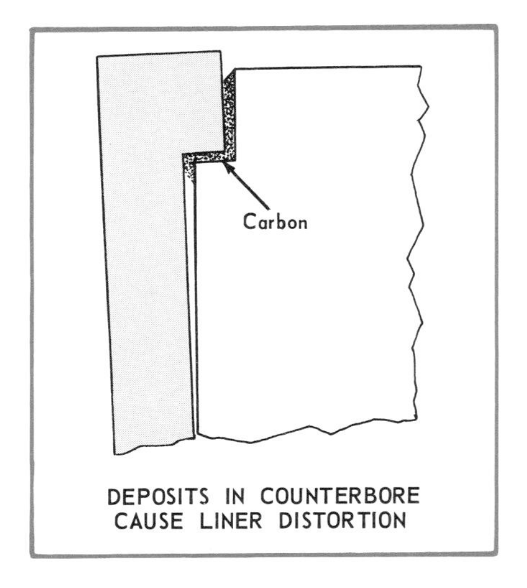

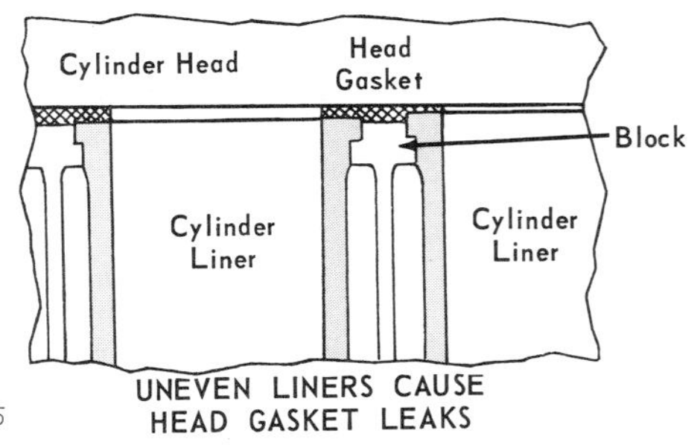

Fig. 70 — Problems With Cylinder Liners Of Different Heights

 Litho in U.S.A.

8. The liner may protrude above the block because of the uncompressed seals. But if in doubt, check for twisted seals or carbon deposits (Fig. 70). When the cylinder head is installed, the liners must seat evenly or it may become distorted or head gasket leaks may develop if liners are too low in the block.

9. Further check for correct liner installation by measuring for out-of-roundness in the liner bore. If a liner is distorted, check for twisted seals or deposits on the liner bore seats.

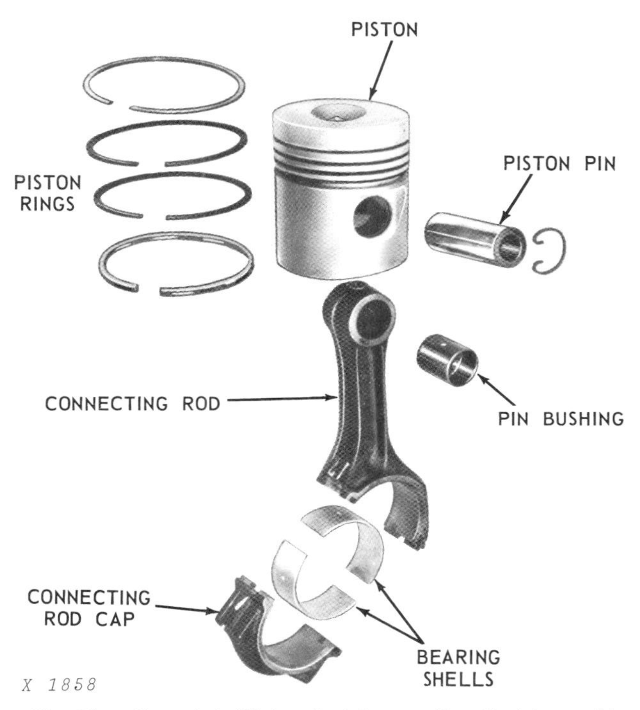

Fig. 71 — Complete Piston And Connecting Rod Assembly

Now let's turn to the parts within the cylinder—the piston with its rings (Fig. 71).

PISTONS

The piston does three jobs:

- **Receives the force of combustion**
- **Transmits this force to the crankshaft**
- **Carries the piston rings which seal and wipe the cylinder**

CONSTRUCTION OF THE PISTON

The piston must be:

1) **Precision-made**—to fit snugly in the cylinder, yet slide freely up and down.

2) **Rugged in Construction** — to withstand the forces of combustion and the rapid stop and start at the end of each stroke.

3) **Carefully Balanced and Weighed** — to overcome inertia and momentum at high speeds.

To be strong yet light, pistons are constructed of cast iron or aluminum alloys. Reinforcing ribs are used to keep the piston as light as possible.

PARTS OF THE PISTON

The main parts of the piston are:

- **Head**
- **Skirt**
- **Ring Grooves**
- **Lands**

These parts are shown on a typical piston in Fig. 72.

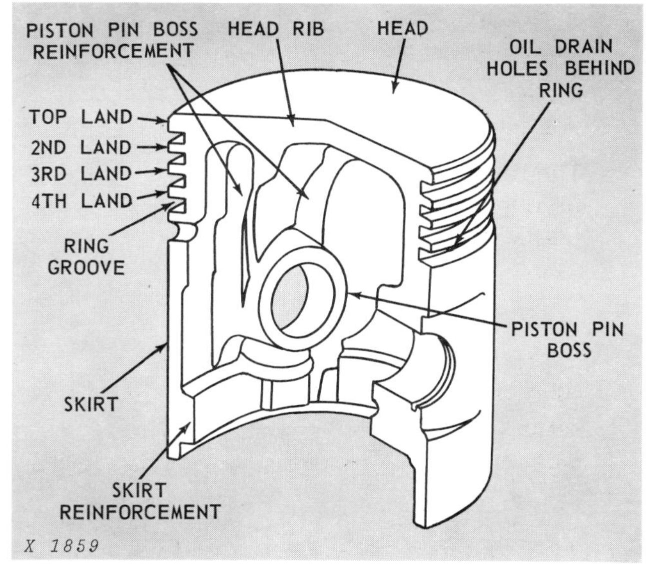

Fig. 72 — Piston

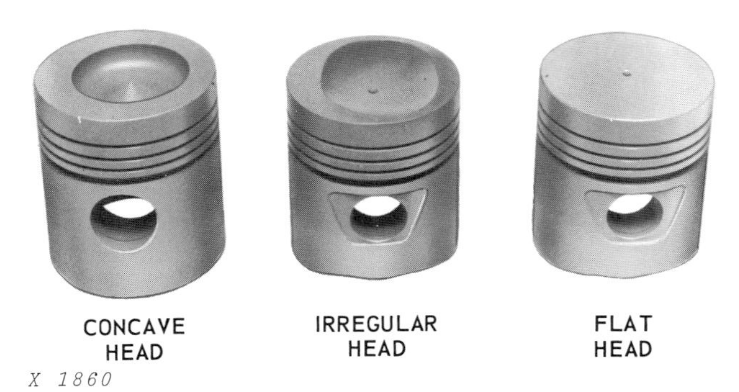

Fig. 73 — Pistons With Three Types Of Heads

The HEAD of the piston is the top surface where the combustion gases exert pressure. The piston head may be flat, concave, convex, or irregular (Fig. 73).

The different shapes of the piston head allow for more or less compression and swirling as needed for the different engines and fuels. Diesel pistons may also have a combustion chamber recessed in the head.

Ribs inside the piston head reinforce it and help to carry heat away from the head to the rings.

The SKIRT of the piston is the outside part below the ring grooves. The piston is kept in alignment by the skirt.

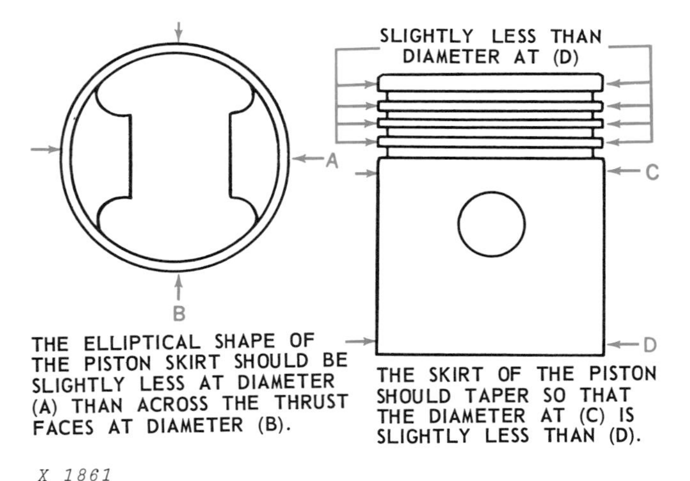

Fig. 74 — Cam-Ground Piston

The piston skirt is usually *cam-ground, tapered* and *elliptical* in shape (Fig. 74).

Fitting the piston to the cylinder is very critical because:

1) Metals expand when heated and,

2) Space must be provided for lubricants between the piston and cylinder wall.

The piston clearance depends upon the size and thickness of the piston and whether it is made of cast iron or aluminum (which expands more when heated). Also, the piston skirt runs much cooler than the top of the piston, so needs less clearance.

By making the piston skirt *elliptical* in shape, it will fit the cylinder during operation when it is hot. The narrower part is at the pin mounts (A), where the metal is thickest, while the wider part is where the metal is thinnest (B).

This fitting prevents noise or slap of the piston during warm-up. Then as the piston gets hot, it expands and becomes round.

Skirts of different shapes are used to allow for the heat and stress on the piston. However, most engines use the *trunk*-type skirt shown in Fig. 74.

The RING GROOVES are cut around the piston to accept the piston rings (see Fig. 72). They are shaped to the proper rings for good control of oil and blow-by. The lower groove has openings for oil collected by the oil control ring to flow back to the crankcase.

The piston LANDS are the areas between the ring grooves which hold and support the piston rings in their grooves (Fig. 72).

PISTON RINGS

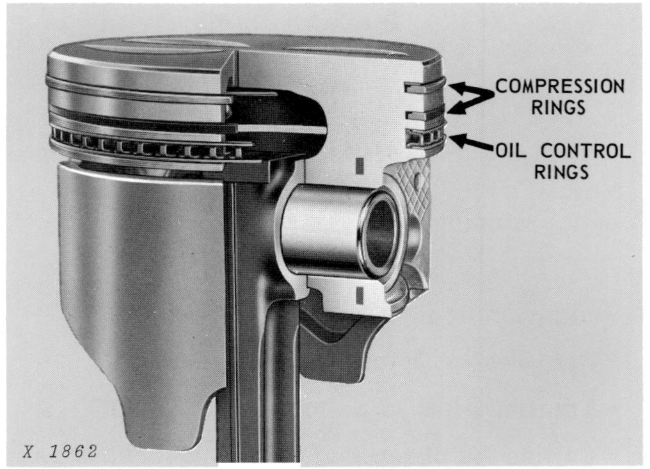

Fig. 75 — Piston Rings For A Typical Piston

Piston rings do three jobs:

1) Form a gas-tight seal between the piston and cylinder.

2) Help cool the piston by transferring heat.

3) Control lubrication between piston and cylinder wall.

Piston rings are of two types:

- **Compression Rings**
- **Oil Control Rings**

COMPRESSION RINGS (Fig. 75) prevent gases from leaking by the piston during the compression and power strokes. They seal by expanding out against the cylinder wall. The rings expand by their own tension and also by combustion pressure behind the rings during the power stroke.

Compression rings are split for easy assembly on the piston. The ends that are split do not form a perfect seal, so more than one ring is used.

If the cylinders are worn, ring expanders are sometimes used inside the rings for tighter sealing.

The OIL CONTROL RING is the bottom ring on the piston. Its job is to wipe the excess oil from the cylinder walls. This oil is fed through slots in the rings to holes in the piston groove, where it returns to the crankcase. For better oil control, spring expanders are often used under the oil control ring.

In some engines, oil rings are used both above and below the piston pin.

Piston rings are usually made of hardened cast iron and are often plated with metals such as chrome on their contact faces. The hard plating reduces the wear on the ring.

Types Of Piston Compression Rings

Various shapes of compression rings are used for pistons (Fig. 76). Many pistons use a combination of these rings.

Rectangular Ring—The common ring bears evenly against the cylinder wall along its whole face.

Taper-Faced Ring—assures that the lower outside edge will have positive contact with the cylinder. This gives quick seating and good wiping of oil on downstroke, but only fair control of blow-by. Seat very quickly.

Barrel-Faced Ring—used as a top ring on some pistons for better control of blow-by.

Inside Bevel Ring—Cutout along the edge of this ring allows it to twist in its ring groove and form a tighter seal during combustion. Used often as an intermediate ring.

Keystone Ring — helps prevent ring sticking by reducing deposits that fill ring grooves and freeze rings. An insert may be used in the groove as a seat for the ring (Fig. 76). Used mostly in heavy-duty diesel and aircraft engines.

Keystone rings may also be barrel-faced or taper-faced.

Joints For Compression Rings

Compression ring joints must be wide enough to prevent the ends from touching when the ring heats up and expands in operation.

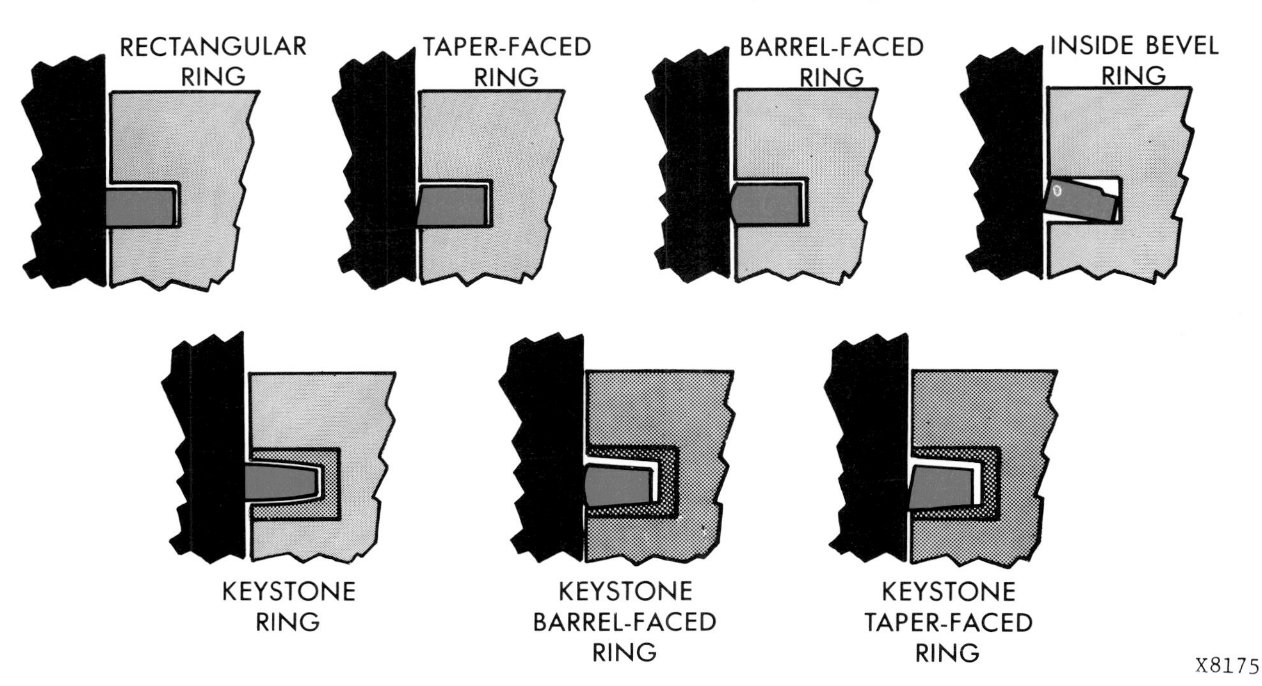

Fig. 76 — Compression Rings — Seven Major Types

Litho in U.S.A.

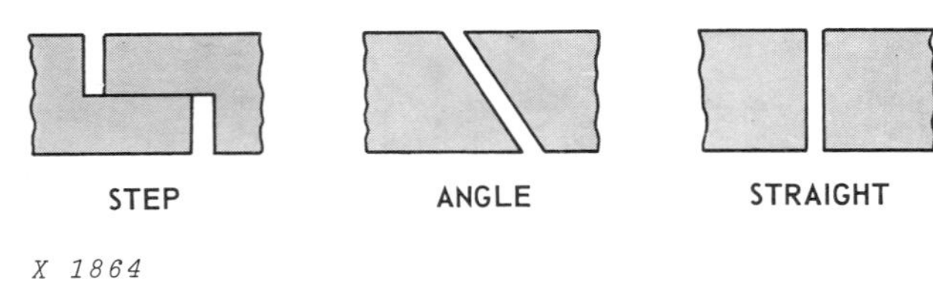

Fig. 77 — Joints For Compression Rings

The three common ring joints are shown in Fig. 77. The *step joint* gives better sealing, but the *straight joint* or *angled joint* is used most where joint leakage is not a great problem.

Types Of Oil Control Rings

Fig. 78 — Slotted Oil Control Ring

Today's engines throw more oil onto the cylinder walls than is needed for lubrication. (This extra oil absorbs heat.)

To control oil consumption and smoky exhaust, the oil control rings must wipe the excess oil from the cylinder walls and allow it to return to the crankcase.

Oil control rings have a groove in their outer face where oil is collected, fed on through the piston, and returned to the crankcase (Fig. 78).

The expander springs that are sometimes used under the oil control ring are made of a steel strip or coil.

DIAGNOSING PISTON FAILURES

Why do pistons fail? What do you look for? These are the questions that will be answered on the following pages.

Three Big Engine Problems

Three problems cause trouble in the piston area:

- **Oil Consumption and Blow-by**
- **Combustion Knock (Detonation)**
- **Preignition (Gasoline Engines)**

Let's look at each one of these problems and see what the causes and signs are.

Oil Consumption And Blow-By

The piston and rings are supposed to *seal* the cylinder against two things:

1) Excess oil passing up into the combustion chamber.

2) Gases or blow-by passing down to the crankcase.

Oil Consumption is the problem in Step 1 above. The piston rings must leave a film of oil on the cylinder walls or the parts will wear out too fast. Yet if the rings pass too much oil, the engine will consume too much oil.

Fig. 79 — Oil Consumption — How The Piston Rings Pump Oil Up Into The Cylinders, Where It is Burned

Slight oil consumption can be expected (Fig. 79). The piston rings must pass a small amount of oil for lubrication as the piston works.

One possible cause of excessive oil consumption results from the pumping action of worn piston rings as shown in Fig. 79. The condition has been exaggerated in the drawings for clarity.

However, the more the piston rings and cylinder wall are worn, the greater the oil consumption. (A lighter oil also causes more oil to be used.)

The problem is that engine overhauls are also expensive. The cost of overhaul must be balanced with the cost of oil consumed before deciding when to overhaul the engine.

There are also other ways that oil can reach the combustion chamber from the crankcase: worn valve guides, or leaking manifolds or turbochargers. Check all possible causes when diagnosing engine oil consumption and don't forget that external leaks also consume oil!

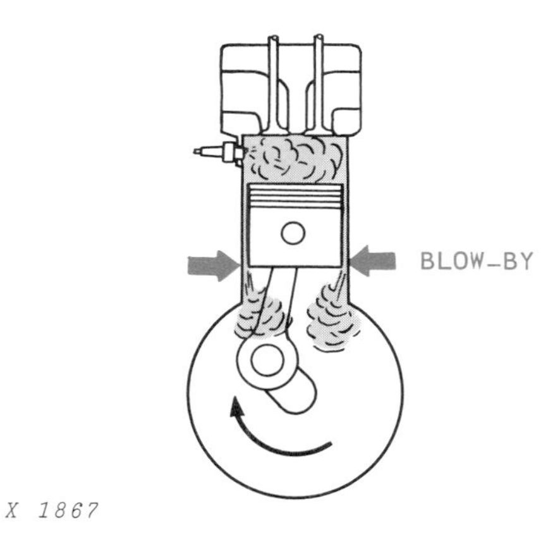

Fig. 80 — Blow-By Of Gases In Cylinder

Blow-by occurs when combustion gases get past the piston rings and into the crankcase (Fig. 80).

A slight amount of blow-by can be expected—this is why crankcases are usually ventilated.

However, severe blow-by can create real problems:

1. Pistons are overheated and expand, scoring the piston and cylinder wall. 2) Compression is lost, reducing power; 3) Crankcase oil is contaminated, causing wear.

CAUSES OF OIL CONSUMPTION AND BLOW-BY

1. *Piston rings installed wrong*
2. *Stuck rings or plugged oil ring*
3. *Top ring broken or top groove worn*
4. *Overall wear in piston, rings, and cylinder:*
   a. *Abrasive wear*
   b. *Scuffing and scoring*
   c. *Corrosive wear*
5. *Physical damage to pistons*

Let's look at each of these causes in detail.

PISTON RINGS INSTALLED WRONG

For a good seal between the piston and cylinder, the piston rings must conform to the cylinder wall and have plenty of tension.

If rings are installed upside down, are the wrong type or size, or are stretched or even broken by bad installation, more oil and gas vapor get by the piston.

For correct procedures, see "Installing Rings on Piston" later in this chapter.

STUCK RINGS OR PLUGGED OIL RING

Deposits caused by too much heat, unburned fuel, and excess lubricating oils, collect in the piston ring area. Ring failure usually occurs when these deposits **harden** and freeze the rings in their grooves.

Fig. 81 — Stuck Rings That Have Broken

When the rings are completely stuck, they often break (Fig. 81).

Fig. 82 — Plugged Oil Control Ring

Deposits in the top ring groove can cause sticking, scuffing, and scoring because they keep out oil and trap metal particles that wear off the piston.

Sludge deposits in the oil control ring can cause it to plug (Fig. 82). This means that oil control has been lost.

Other conditions that lead to **stuck or plugged rings** are:

- *Top groove failure*
- *Cylinder Liner distortion*
- *Combustion knock*
- *Preignition*
- *Overloading*
- *Cooling system failure*
- *Improper lube oil*
- *Cold engine operation (stop and go service)*

TOP RING BROKEN OR TOP GROOVE WORN

The top piston ring acts as both a compression and final oil control ring. It must form a seal between its sides and the ring groove, and between its face and the cylinder wall.

A poor seal will allow oil to by-pass the ring seal and be lost.

A poor seal will allow blow-by to contaminate the crankcase oil, forming sludge, which leads to ring sticking, clogging, and possibly scuffing.

The top ring and groove wears most since this area is exposed to the most heat, pressure, and abrasives and gets the least lubrication.

Fig. 83 shows *new* compression rings in *new* groove at the left. The sides of the groove are flat, parallel, and smooth. The ring also has the correct side clearance. Combustion gases in the power stroke force the ring down against the lower side of the groove. At the same time, the gases pass behind the ring and force it out against the cylinder wall. The result is a *good seal.*

At the right in Fig. 83, we see a *new* ring installed in a *worn* groove. The groove permits the ring to sag. This causes the upper corner of the ring face to contact the cylinder wall, resulting in oil being wiped up into the combustion chamber. The flat new ring cannot mate with the worn sides of the ring groove. The result is a bad seal at both the face and sides of the ring.

Top ring *groove* wear results in broken top rings or damaged pistons.

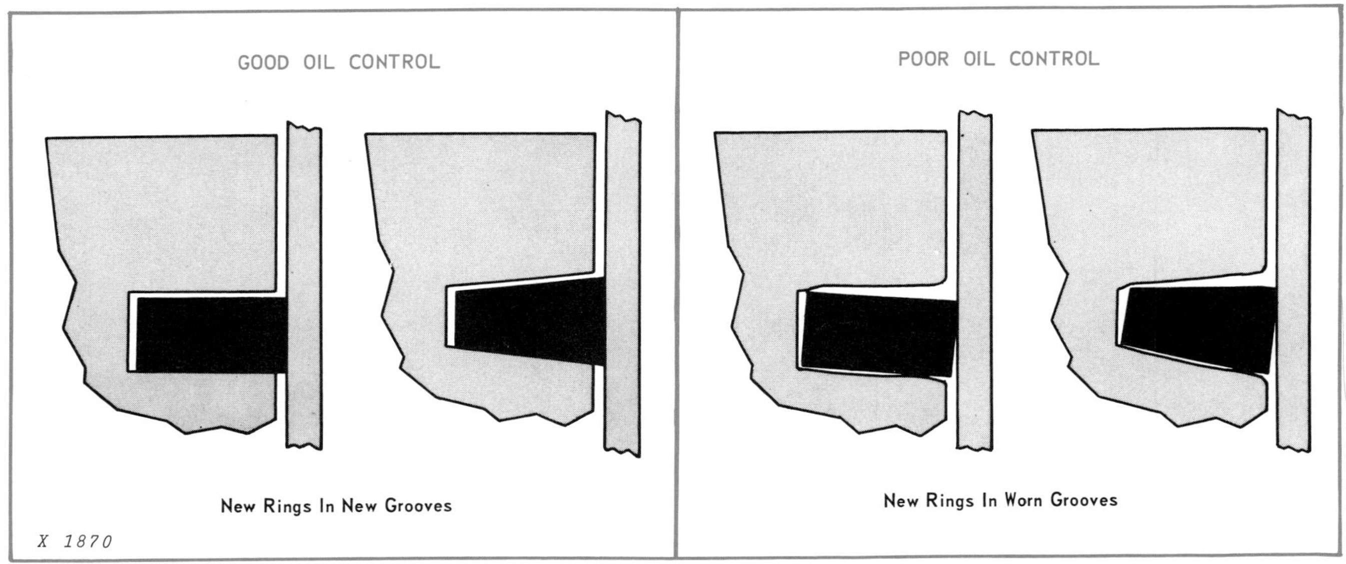

Fig. 83 — How A Worn Ring Groove Affects Oil Control

The following conditions can cause top groove failure:

- *Abrasives entering the engine through the intake system*
- *Combustion knock*
- *Preignition*
- *Installation of new ring in worn groove*
- *Assembly of new ring without ridge reaming the cylinder*
- *Use of wrong-size piston or ring, resulting in too much or too little side or end clearance*

OVERALL WEAR ON RINGS, PISTON, AND CYLINDER

1. Abrasive Wear

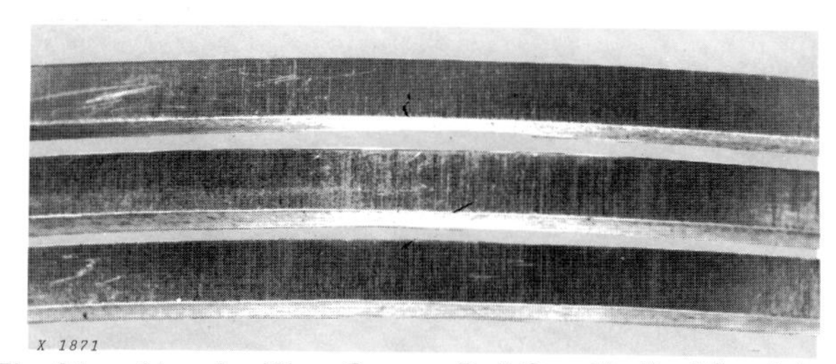

Fig. 84 — Abrasive Wear Causes Dull Gray Vertical Scratches On Ring Faces

When the ring faces are covered with dull gray, vertical scratches and have excessive ring gap (preferably checked in a new cylinder), the rings have been worn by abrasives. See Fig. 84.

Other indications of abrasives present in the engine are dull gray vertical scratches on piston skirts, scratched cylinder bore, high ridge at the top of the cylinder, loose piston fit, or badly scratched rod and main bearings.

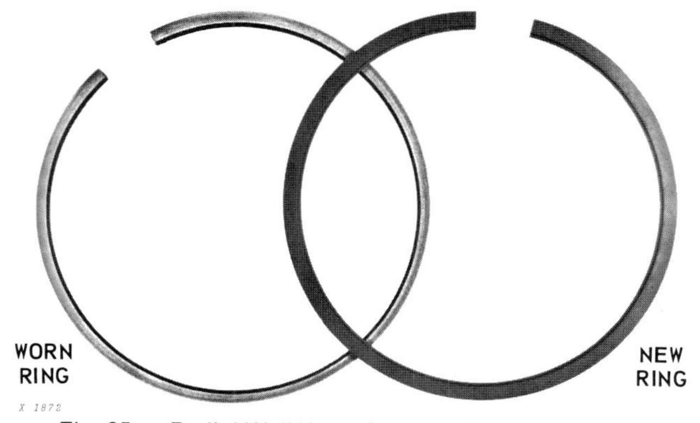

Fig. 85 — Radial Wall Wear On Worn Compression Ring

Badly worn compression rings can be identified by their reduced radial wall. Fig. 85 compares the reduced radial wall of the worn ring on the left with that of a new ring on the right.

We know what causes abrasive wear, but how do the abrasives enter the engine? Possibly from dirt left in the engine during the last engine overhaul. Through the air intake, crankcase breather, or fuel systems. Or from scuffing and scoring of parts as they wear.

2. Scuffing and Scoring

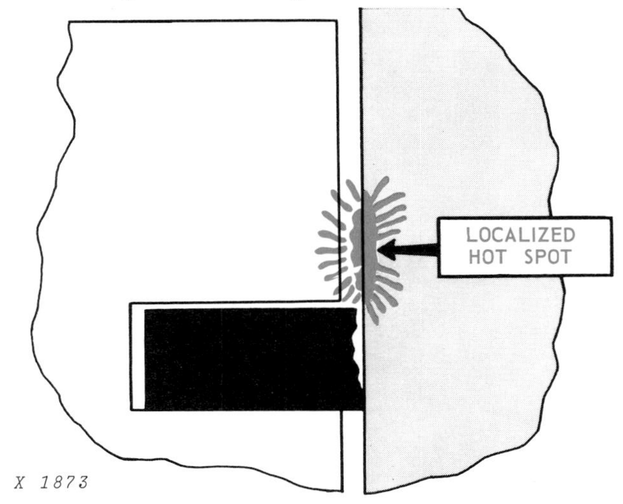

Fig. 86 — How Scuffing And Scoring Occur

Fig. 87 — Severe Piston Scoring

 Litho in U.S.A.

Scuffing is caused by too much heat. When two metal parts rub and the heat builds up to the melting point, a small deposit or "hot spot" of melted metal is pulled out and deposited on the cooler surface (Fig. 86).

Scuffing leaves discolored areas on the surface of rings, pistons, and cylinder walls. Under a magnifying glass, the metal in the center of these areas is burnished and smeared in the direction of motion of the part.

Scuffing starts as a very small surface disturbance, which may be difficult to see and identify. If they are not removed, scuffing spreads to adjacent areas and become noticeable and more severe, at which time it is called **scoring** (Fig. 87).

Fig. 88 — Corroded Piston Has Mottled, Grayish, Pitted Appearance

Any engine condition which heats rubbing parts to the melting point, or which prevents the transfer of heat from these surfaces, has an influence on scuffing.

The following is a list of possible causes of scuffing and scoring:

- *Improper warm-up (fast speeds or big loads too soon)*
- *Lubricating system not functioning*
- *Cooling system plugged*
- *Combustion knock and preignition*
- *Lugging or overloading*

3. Corrosive Wear

The third cause of overall wear on pistons and cylinders is corrosion.

Leaking coolant can corrode pistons. Cold engine operation or the wrong lubricating oils can also deposit chemicals in the crankcase which will corrode parts.

Severe corrosion will show up as a mottled, grayish pitted surface on pistons or cylinder walls (Fig. 88). The corrosion is caused by acids from the products of combustion.

Other corrosion may be harder to find. If excessive wear is found, and scuffing and scoring are eliminated as possible causes, suspect corrosive wear.

PHYSICAL DAMAGE TO PISTONS

Fig. 89 — Piston Damaged By Loss Of Pin Lock

An example of physical damage to the piston is shown in Fig. 89. This piston has lost its pin lock.

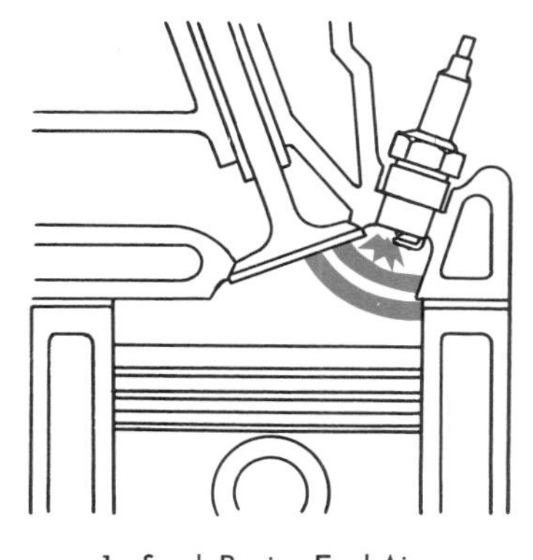

1. Spark Begins Fuel-Air Mixture Burning.

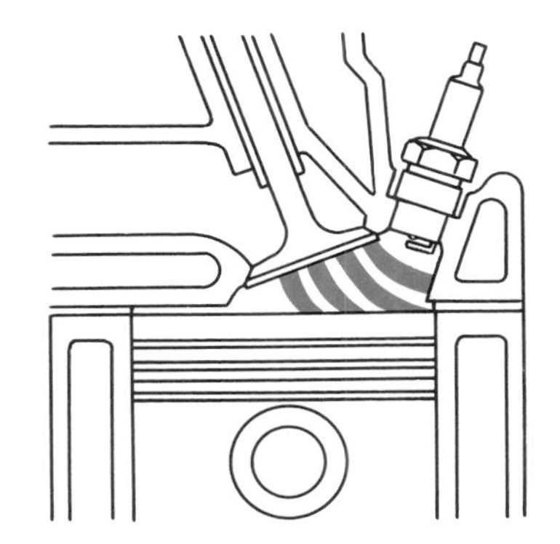

2. Flame Advances Smoothly, Compressing and Heating End-Charge.

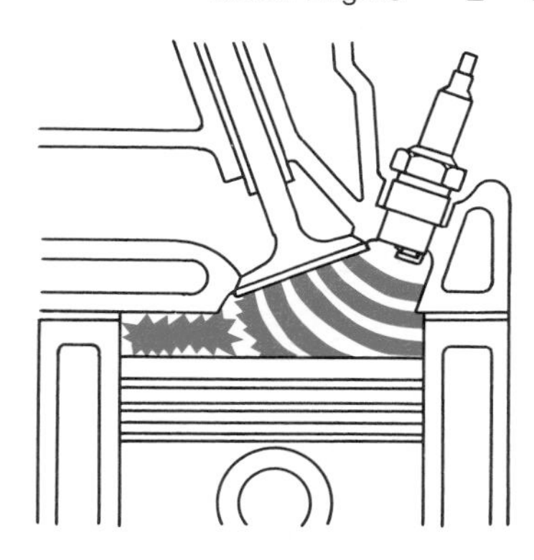

3. End-Charge Suddenly Ignites With Violence, Producing A Knock.

X 1877

Fig. 90 — How Combustion Knock Or Detonation Occurs

Physical damage to pistons can be caused by:

- *Connecting rod out of alignment.*
- *Crankshaft has too much endplay.*
- *Crankshaft journal has too much taper.*
- *Cylinder bore out of alignment.*
- *Piston pin locks installed wrong.*
- *Piston pin locks are faulty.*
- *Ring groove scratched while trying to clean out carbon.*
- *Piston handled carelessly or dropped.*

Combustion Knock (Detonation)

Fig. 91 — Combustion Knock Damaged This Piston And Broke The Top Ring

Knocking occurs when combustion of fuel in the cylinder is too early, too rapid, or uneven (Fig. 90).

The result is a "knock" which can burn the piston, wear out the top groove, or cause the ring to break or stick (Fig. 91).

CAUSES OF COMBUSTION KNOCK (DETONATION)

1. *Lean fuel mixtures*
2. *Fuel octane too low*
3. *Ignition timing advanced too much*
4. *Lugging the engine or overfueling*
5. *Cooling system not working*

Preignition (Gasoline Engines)

Preignition is when the fuel ignites before the spark occurs.

As a result, part of the fuel burns while the piston is still coming up on its compression stroke.

The burning fuel is compressed and overheated by the piston and by further combustion. The heat can get so high that engine parts are melted.

The piston in Fig. 92 was damaged by the heat of preignition.

Fig. 92 — Preignition Burned A Hole Through The Head Of This Piston

CAUSES OF PREIGNITION

1. *Carbon deposits that remain incandescent and ignite fuel early*
2. *Valve operating too hot because of excessive guide clearance or bad seats.*
3. *Hot spots caused by damaged rings.*
4. *Spark plugs of wrong heat range.*
5. *Spark plug loose (means "hotter" plug).*

Summary: Diagnosing Piston Failures

A premature failure of the engine must have a cause.

That cause must be found and corrected if the engine is to run properly again.

SERVICING PISTONS

Removing Piston Rings

Fig. 93 — Remove Piston Rings

Clamp the connecting rod in a vise to prevent damaging or burring of the piston (Fig. 93). Remove the old rings, using a ring expander.

Never reinstall piston rings once they have been removed. Discard the old rings and replace with new ones.

If the pistons are to be removed from the rods, remove the piston pin locks (if used) and place the pistons in hot water. This will expand them so the pins can be pushed out easily. *Never force pins either in or out of cold pistons.*

Service Hint: Use an electric fry pan and set it at low temperature to heat pistons.

Cleaning Pistons

Careful cleaning of pistons is most important, especially the ring grooves.

Do not scrape with a groove cleaner or a broken piston ring. Hard scraping will scratch the fine surfaces of the piston.

Two methods of cleaning pistons are recommended:

1) Chemical Soaking

2) Glass Bead Cleaning

CHEMICAL SOAKING OF PISTONS

A recommended piston cleaner will soften the carbon on pistons without a lot of effort or any damage. The carbon and residue can then be easily removed with a pressure spray rinse. Follow these steps:

1. Mix the cleaner solution and heat it as recommended.

2. Before soaking, use a solvent to remove oil film from the pistons (otherwise the cleaning solution will be weakened).

3. Soak the pistons in the cleaning solution for the specified period.

4. Soak for a second period if needed. A very light scraping may be required in some cases.

5. After soaking, drain and spray rinse the pistons with water and air.

Caution: *Be careful with the cleaning solution. Store and handle with the recommended safety.*

Fig. 94 — Piston Ring Groove Cleaning Machine

GLASS BEAD CLEANING OF PISTONS

This method of cleaning has proven successful if the proper machine is used. The pistons are blasted with glass beads in the machine shown in Fig. 94. Take the following steps:

1. Carefully remove old rings from the pistons.

2. Wash pistons in solvent to remove grease and oil. Use a stiff brush—not a wire brush—to help get them clean.

3. Dry the pistons using compressed air.

4. Clean the pistons in the glass bead cleaning machine (Fig. 94). Use the proper size of beads and the correct pressure.

5. Be careful not to hold the bead blast on one area too long or the metal may be eroded. Avoid this by keeping the blast moving.

6. Hold the nozzle away from the surface to be cleaned. The distance will vary depending on the recommended pressure.

PRECAUTIONS WHEN CLEANING PISTONS

1. Be careful to avoid scratching the sides of the piston ring groove.

2. Never use a wire brush to clean pistons.

3. Be sure the piston ring grooves are thoroughly cleaned. Excessive deposits in ring grooves can force the rings out, causing scuffing and scoring.

Inspecting Pistons

After cleaning, examine the piston for score marks, damaged ring grooves, or signs of overheating. Always replace a piston that has been severely scored, overheated, or burned.

Carefully inspect the pistons for cracks in the head or skirt areas and for bent or broken lands. Fatigue failures will often show up as cracks in the area of the pin boss.

The magnetic particle inspection methods outlined under "Crankshaft Inspection" may be used in locating cracks in cast-iron pistons.

Damaged pistons must be replaced.

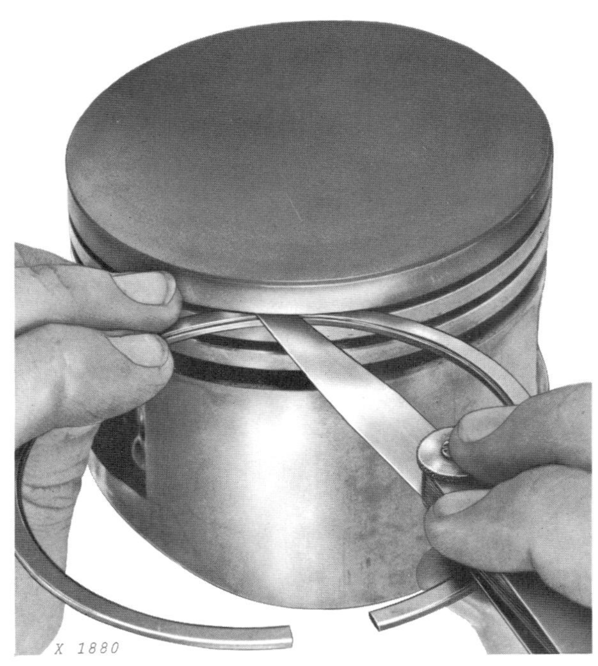
Fig. 95 — Measuring Ring Clearance With A Feeler Gauge

Checking Ring Grooves For Wear

Pistons should be replaced or ring grooves reconditioned whenever the ring grooves are worn excessively. Suspect that wear is excessive when side clearance is over 0.008 inch (0.200 mm). (Different wear limits apply to many two-cycle and large-bore diesels. Refer to the engine manufacturer's recommendations for each application.)

Check the piston grooves at several points because the grooves wear unevenly. The top groove receives the most wear, but check all of them.

Check for ring groove wear by installing a *new* ring in the groove. Then insert a feeler gauge between the upper surface of the new ring and the land to check the clearance (Fig. 95).

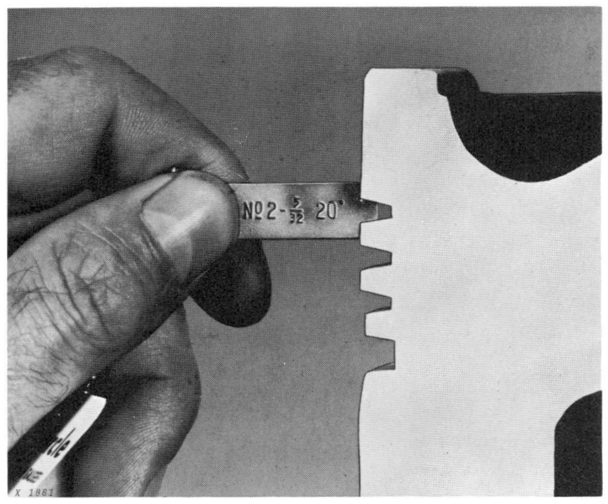
Fig. 96 — Checking Keystone Ring Groove For Wear

Keystone groove wear may also be checked using a special wear gauge (Fig. 96).

Many ring manufacturers provide special gauges for checking ring groove wear. These gauges are available for checking both rectangular and Keystone grooves.

Replace or regroove all pistons which have excessively worn grooves. Regroove with a ring groove cutting tool (only if recommended).

Piston Ring Groove Inserts

Many pistons for heavy-duty service have a steel or alloy insert cast into the top groove. This retards side wear on the groove and ring and lengthens the life of the pistons and rings.

Remachining Of Piston Ring Grooves

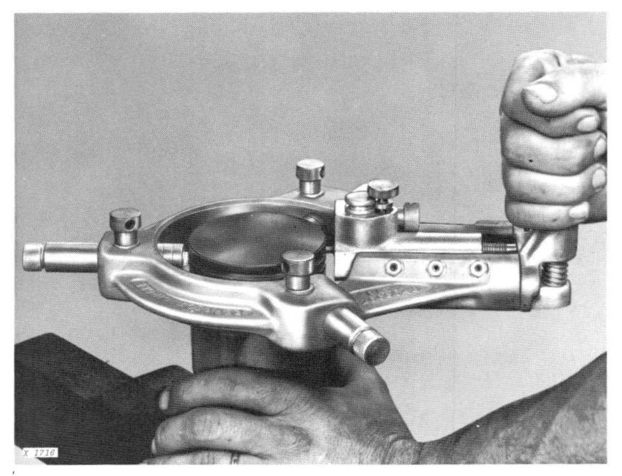
Fig. 97 — Remachining A Piston Ring Groove

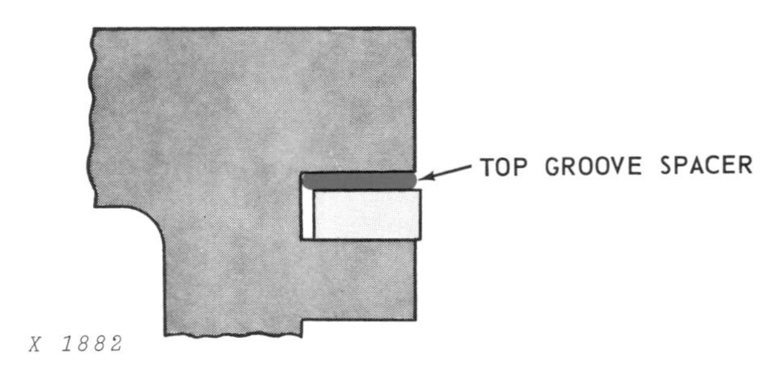

Fig. 98 — Groove Spacer Above Piston Ring

Worn piston ring grooves on some pistons can be remachined with a ring groove cutting tool (Fig. 97).

If recommended, machine these grooves for a new standard ring with a flat steel spacer above the ring.

Heat-treated steel spacers must be installed above the ring to compensate for the metal removed by regrooving (Fig. 98). The spacers are flat and supply a bearing surface for the full depth of the groove. This assures a good seal and heat transfer between ring and groove. Top groove spacers are easily installed without special tools.

Checking Piston Wear

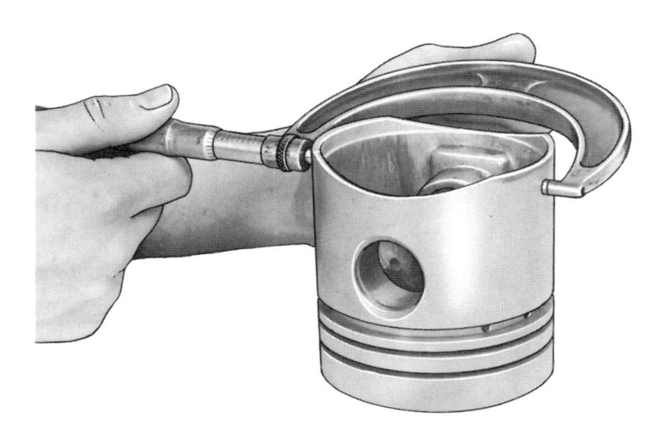

T 9616

Fig. 99 — Measuring The Piston For Wear

To check for piston wear, measure the diameter of the piston skirt across the thrust faces (at right angles to the piston pin bore). Take a reading at both the top and bottom of the skirt (Fig. 99).

Compare these measurements with new dimensions given in the engine Technical Manual. The difference between these two measurements determines the piston wear.

Checking Piston-To-Cylinder Clearance

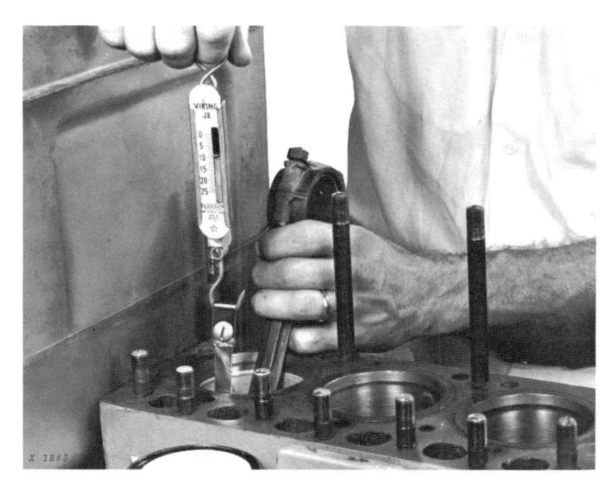

Fig. 100 — Determining Piston Clearance With Feeler Ribbon And Scale

Since most pistons wear with use and may have excessive clearance, be sure to check the piston clearance at every overhaul.

Pistons should be replaced if their clearance exceeds the specifications. (Some automotive pistons can be resized by knurling.)

Measure piston clearance as follows:

1. Measure the cylinder diameter at right angles to the crankshaft in the lower or least-worn area of the cylinder. Use a cylinder dial gauge, an inside micrometer, or a telescope gauge with outside micrometer.

2. Measure the diameter of the piston across the thrust faces with an outside micrometer (see instructions above).

3. The difference between these two measurements is the piston clearance.

Piston clearance may also be determined with a feeler ribbon and spring scale as follows:

1. Place the feeler ribbon in the cylinder along one of the thrust sides. (Several thicknesses of feeler ribbon are available.)

2. Turn the piston over and insert it in the cylinder with a thrust face centered against the ribbon (Fig. 100).

3. Push the piston down into the lower part of the cylinder. The feeler ribbon must be centered between a thrust side of the cylinder and thrust face of the piston or an incorrect measurement will result.

4. Pull out the feeler ribbon from between the cylinder and piston, using a 5- to 10-pound (10- to 20-kg) pull on the scales.

5. The thickness of feeler ribbon that can be removed with this pull is the piston clearance.

Knurling Pistons (Automotive Engines)

Knurling increases the diameter of a piston by precisely displacing metal in the two thrust faces. Many automotive pistons are resized in this way (Fig. 101).

The metal of the piston skirt is upset by a special knurling tool as shown. The displaced metal is forced upward between the teeth of the tool, producing a checked surface and increasing the diameter of the piston.

NOTE: Knurling of pistons is not done on engines for farm and industrial machines.

 Litho in U.S.A.

Fig. 101 — Knurling A Piston (For Automotive Engine)

Installing Rings On Piston

First be sure the piston is cleaned thoroughly. *Check the ring grooves for any deposits.*

Rings must be installed with the top side upward to provide good oil control. Refer to engine Technical Manual for directions on installing different ring types properly. Also, each new ring set usually contains an instruction sheet.

Fig. 102 — Installing Piston Rings Using A Ring Expander

Use a ring expander to prevent twisting or stretching the rings during installation (Fig. 102).

Be extremely careful not to twist or expand the rings too much as this will permanently distort them and reduce their performance. A ring expanding tool helps to avoid this danger.

Normally, ring ends may be located anywhere around the piston, since compression and oil rings are not pinned and usually can rotate around the piston. However the ring ends should be staggered, and locating of the rings may be specified by the manufacturer.

Installing Pistons

NOTE: Before installing pistons, complete all the piston pin, connecting rod, and bearing services which follow in this chapter.

Fig. 103 — Lubricate The Piston And Rings Before Installing

Thoroughly lubricate each piston and its rings with engine oil (Fig. 103) just prior to installation on cylinder. This lubricates the pistons, rings, and cylinders during engine cranking and until the "oil throw-off" from the connecting rod journals is adequate.

Several hundred revolutions may be necessary in a dry engine before the lubrication system supplies oil to all the moving parts.

Use a ring compressor to compress the rings while installing the piston (Fig. 104). Engine manufacturers usually recommend a specific type of compressor for this purpose.

Apply a gentle downward pressure on the piston to compress the rings into the cylinder. If the piston sticks, compress the rings again and check the cylinder for a partially removed ridge.

Do not pound on the piston head, as this may damage the piston or rings.

Refer to the engine manufacturer's instructions on how to position the piston in the cylinder.

PISTON PINS

The **piston pin** (or wrist pin) connects the piston to the connecting rod.

Piston pins can be fastened in three ways (Fig. 105).

- **Fixed Pin—pin fixed to piston, moves in rod**

Fig. 104 — Installing Pistons

- **Semi-Floating Pin—pin fixed to rod, moves in piston**
- **Full-Floating Pin—pin moves in both rod and piston**

Most piston pins are "full-floating" with bearings in both the rod and the piston. These pins must be held within the piston bore to prevent contact with the cylinder. Note the pin locks at the right in Fig. 105.

In high-speed engines most of the floating pins bear directly on the piston material. This is true with both aluminum-alloy and cast iron pistons.

SERVICING OF PISTON PINS

Piston pins should be checked for looseness, etching, scoring, and excessive damage or wear. Replace them is damaged.

Normally, piston pins need not be replaced except when installing new pistons. This avoids the possibility of noisy pins when the job is done.

Always determine the type of pin and rod assembly from the engine Technical Manual before attempting to drive the pin from the piston. This will save time and prevent piston damage.

Precision Pin Fits

The piston pin must take impact loads under all conditions without wearing out or failing. To do this, the pin bearing must have a full bearing surface and still be able to oscillate without any drag.

The requirements for a *precision pin fit* are:

1. A perfectly round pin hole, free from high spots and chatter marks.

2. A straight pin hole, free from taper, waviness and bellmouthing.

3. A perfectly aligned pin hole between bosses, free from bind and deflection.

4. A correct surface finish to sustain and support an adequate oil film.

5. A good oil clearance, as recommended for each type of piston and rod.

Checking The Pin Fit

Without a precision pin hole gauge, it is impossible to check a pin fit accurately.

However, these three simple tests will show whether or not you should install new bushings:

1. Check the pin for out-of-roundness or looseness. Clamp the pin in a pin vise and rotate the

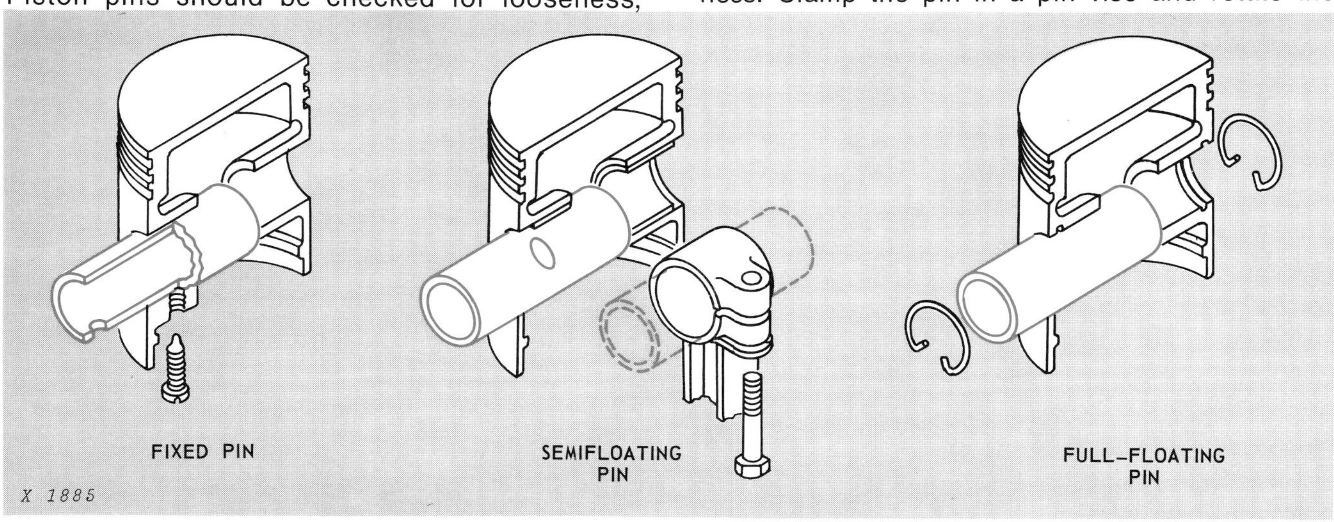

Fig. 105 — Pistons Pins — Three Ways Of Fastening

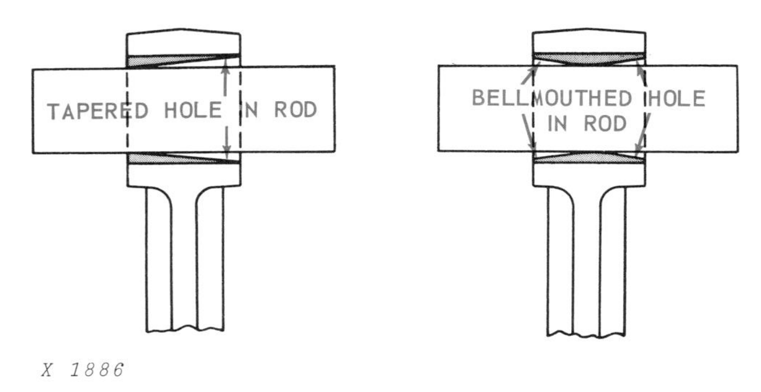

Fig. 106 — Bad Pin Fits In Connecting Rod

rod back and forth on the pin several times. Then remove the rod and examine the shiny contact spots. A good pin fit will show pin contact over the entire surface of the bushing.

2. Check the bore for taper or bellmouthing (Fig. 106). Insert the pin from each end of the bushing. If it is free on one end but tight on the opposite end, the pin hole is tapered. If it enters easily from either end but becomes tight in the center, the hole is bellmouthed. A good pin fit must have parallel surfaces.

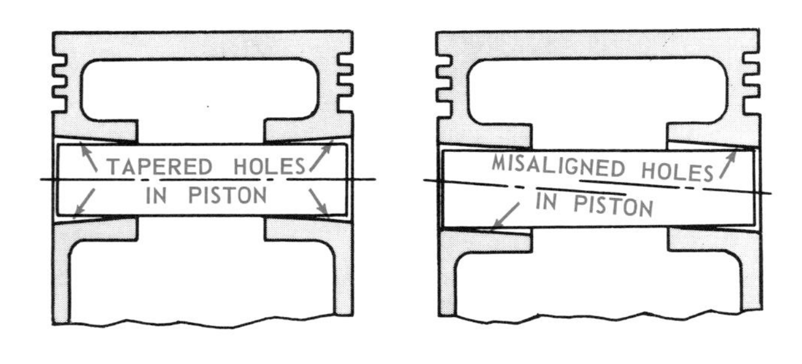

Fig. 107 — Bad Pin Fits In Piston Bosses

3. Check for taper or misalignment between the piston pin holes (Fig. 107). Check each pin hole in the piston separately with the pin for equal size and for taper. Both holes should be straight and of equal size.

If pin holes are not tapered, push pin through toward second piston boss. Pin should enter second boss without a "click" and without forcing or binding. A good pin bearing should also have an even drag through both pin holes.

Also make these checks after installing new bushings to be sure the new pin fit is accurate.

Know the exact clearance between pin and pin holes for each piston and pin design to get the precision fit required for today's high-compression engines.

CONNECTING RODS

Connecting rods must be light and yet strong enough to transmit the thrust of the piston to the crankshaft.

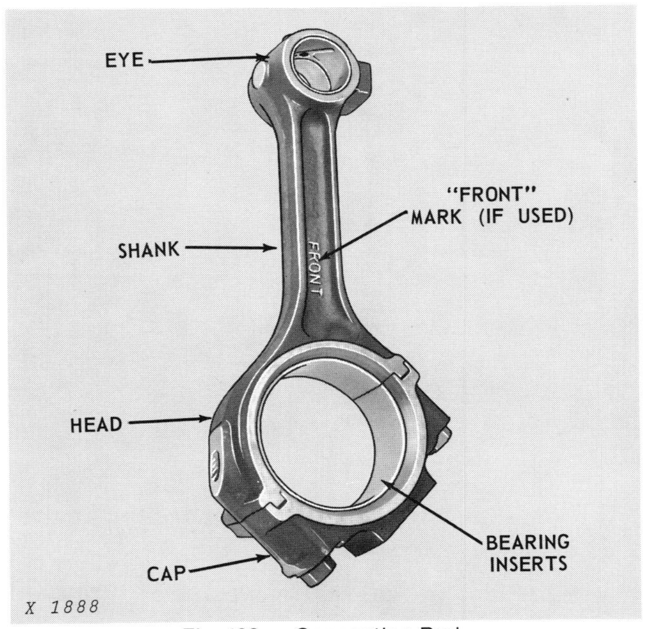

Fig. 108 — Connecting Rod

The connecting rod (Fig. 108) has a **shank** with a small **eye** at one end and a large **head** at the other. The eye forms a bearing for the piston pin, while the head forms a bearing for the crankshaft. The head end of the rod is split and has a bearing **cap** which is bolted on. Two split bearing halves are inserted into the head and cap.

Most connecting rods are forged in one piece and the cap is then cut off and fitted to the head. Some connecting rods are designed with a taper at the eye end (Fig. 109). This provides for added bearing surface in the heavily loaded areas of the piston and connecting rod during the power stroke (arrows).

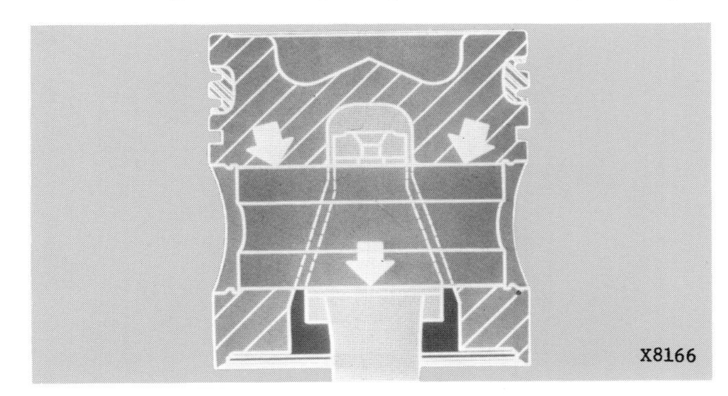

Fig. 109 — Connecting Rod With Tapered Eye

SERVICING OF CONNECTING RODS

Bent Or Twisted Rods

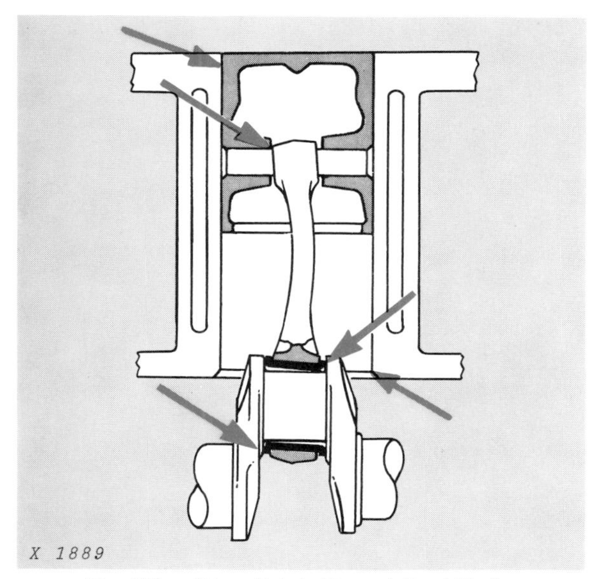

Fig. 110 — Wear Points From A Bent Rod

Generally, connecting rods do not become bent or twisted as the result of normal operation. Both bending and twisting usually result from poor machining or mishandling when reconditioning the engine.

A twisted connecting rod not only places false loads on the connecting rod bearings, but also on the piston, sometimes leading to piston scuffing.

Fig. 110 shows the points of wear from a *bent* connecting rod.

Rod Alignment

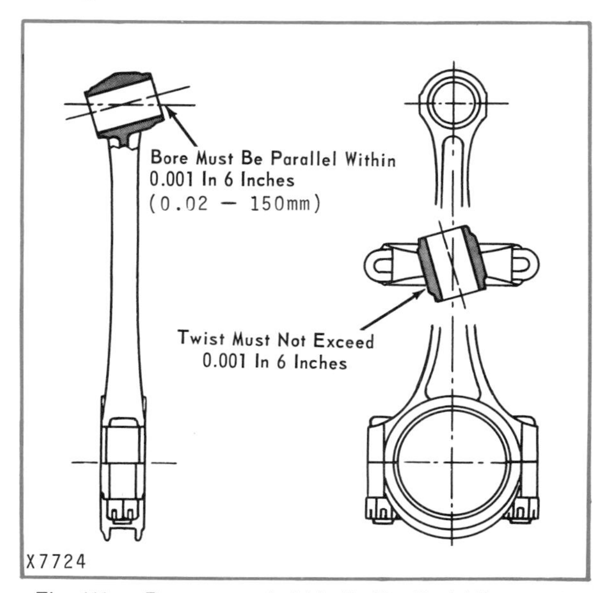

Fig. 111 — Recommended Limits For Rod Alignment

The rod must be aligned within close limits (Fig. 111).

The crankshaft bearing bore and the piston pin bushing bore should be parallel to each other within 0.001 inch in six inches (0.025 mm in 152 mm).

New or reconditioned connecting rods usually have enough material in the piston pin bushing end to permit fitting of a standard or oversize piston pin by the removal of stock.

The connecting rods from an engine are normally reused, after checking, by pressing out the old bushing and inserting a new one.

Before a new bushing is installed, be sure to chamfer the pin end to remove the sharp edge that receives the bushing. Otherwise, the pin's sharp edge removes stock from the bushing and the bushing is then loose in the bore.

Piston And Rod Alignment

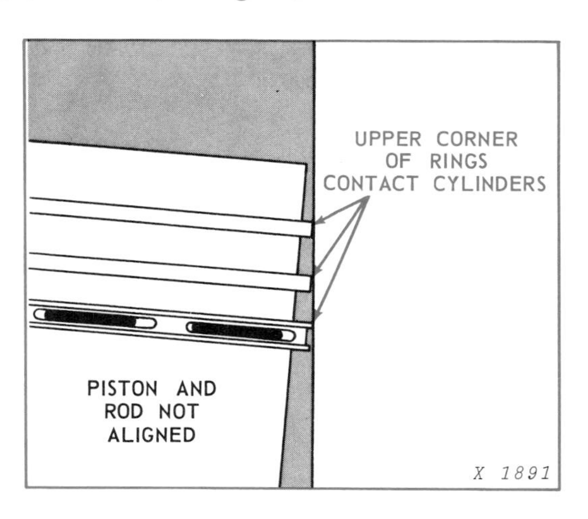

Fig. 112 — If Piston And Rod Are Not Aligned, Engine Uses More Oil

The piston must work squarely in its bore or the rings don't seal (Fig. 112). Misaligned rods not only cause oil consumption and blow-by, they also impose big loads on the rod bearings, the pistons, and the cylinder walls.

To check piston and rod alignment, use a fixture such as the one shown in Fig. 113.

Corrections are usually made by twisting or bending the rod with a notched bar.

Heavy rods seldom remain aligned after this operation because the rod is not permanently set and the rod soon returns to its warped condition. It is better to replace these rods if bent.

 Litho in U.S.A.

Fig. 113 — Checking Piston And Rod Alignment

Bearing Clearance

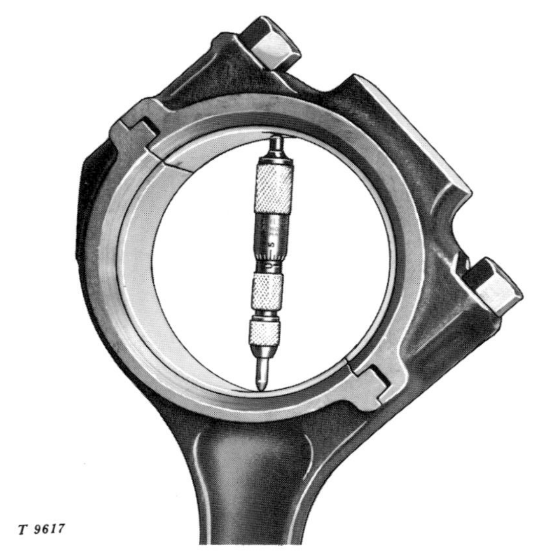

Fig. 114 — Measuring Connecting Rod Bearing Clearance

Install connecting rod caps with bearing inserts in place and tighten to the specified torque. Measure the inside diameter of the connecting rod bearings at several places (Fig. 114). Also measure the outside diameter of the crankshaft connecting rod journals at several places (Fig. 115). Compare the two measurements to find bearing clearance.

Measuring Bearing Clearance Using Plastigage

Bearing clearance can also be measured with the crankshaft in place by using "Plastigage," a plastic thread which "crushes" to the exact clearance. While this method will give the bearing clearance, it will not tell you whether the wear is on the bearing or on the crankshaft journal.

Fig. 115 — Measuring Crankshaft Rod Journals

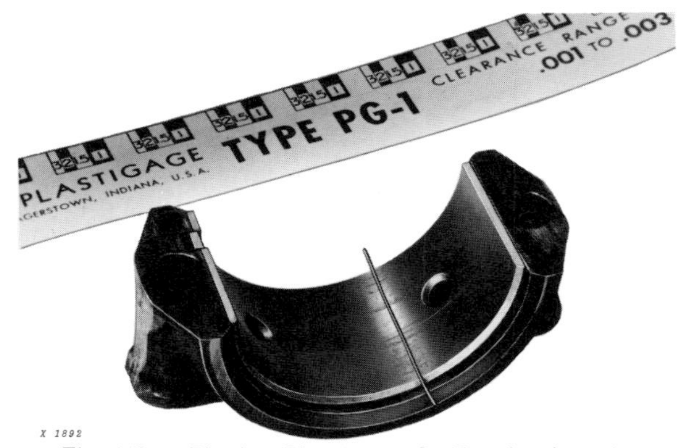

Fig. 116 — Placing Plastigage On Bearing Insert

1. Place a piece of Plastigage the full length of the bearing insert about 1/4 inch (6 mm) off center (Fig. 116).

2. Install the bearing cap and tighten the bolts to the specified torque.

3. Remove the bearing cap. The flattened Plastigage will be found adhering either to the bearing insert or the crankshaft.

4. Compare the width of the flattened Plastigage at its widest point with the graduations on the envelope (Fig. 117).

The number within the matching graduation on the envelope indicates the total clearance in thousandths of an inch.

Approximate taper may be seen when one end of the flattened Plastigage is wider than the other.

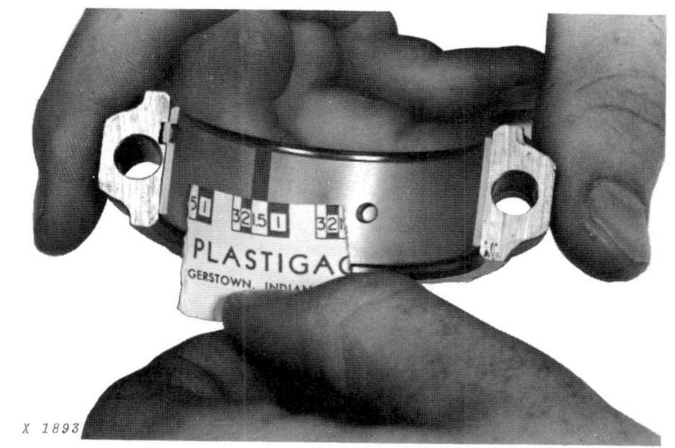

Fig. 117 — Determining Bearing Clearance With Plastigage

Measure each end of the flattened Plastigage. The approximate taper is the difference between the two readings.

ASSEMBLY OF RODS

Install each connecting rod and piston in the cylinder bore from which it was removed.

Some connecting rods are offset with the center of the bearings to one side from the center line of the shank. This makes the engine as compact as possible without skimping on the bearing surface. Be sure that offset rods face the proper way before you install them.

CRANKSHAFT

The crankshaft converts the up-and-down motion of the pistons into rotary motion. It ties together the reactions of all the pistons into one rotary force that drives the machine.

The crankshaft is forged or cast from a heat-treated steel alloy for extra strength. It is usually made in one piece (Fig. 118).

PARTS OF THE CRANKSHAFT

The main parts of the crankshaft are:

- **Journals—bearing surfaces for support and for connecting rods**
- **Throws—offsets which help provide leverage to rotate the crankshaft**
- **Counterweights—balancing weight opposite the rod journals.**

Fig. 118 shows a typical crankshaft.

ARRANGEMENT OF CRANKSHAFT THROWS

The arrangement of the crankshaft throws affects:

1) The balance of the engine

2) Vibration from turning of the shaft

3) Loads on the main bearings

4) The firing order of the engine

The throws are placed so they counterbalance each other when the crankshaft is driven at great force and speed.

Fig. 119 shows the arrangement of throws on several crankshafts.

Four-Cylinder Engine Crankshafts—have either 3 or 5 main support bearings and 4 throws in one plane. As shown, the throws for No. 1 and No. 4 cylinders are 180 degrees from those for No. 2 and No. 3 cylinders.

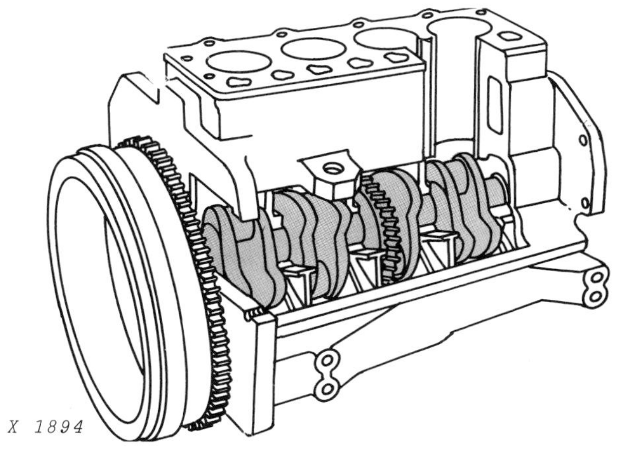

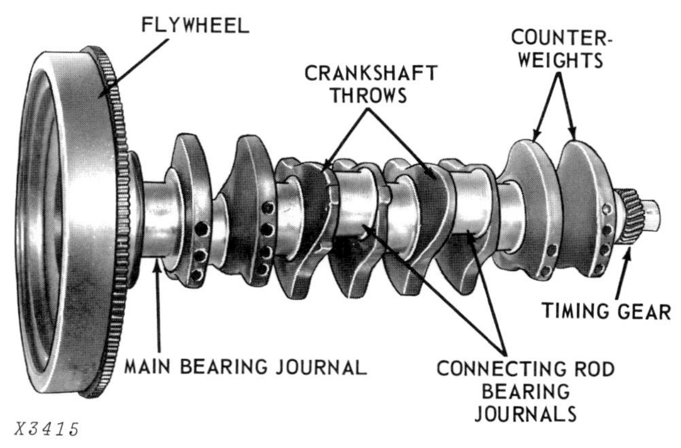

Fig. 118 — Crankshaft Of A Typical 4-Cylinder Engine

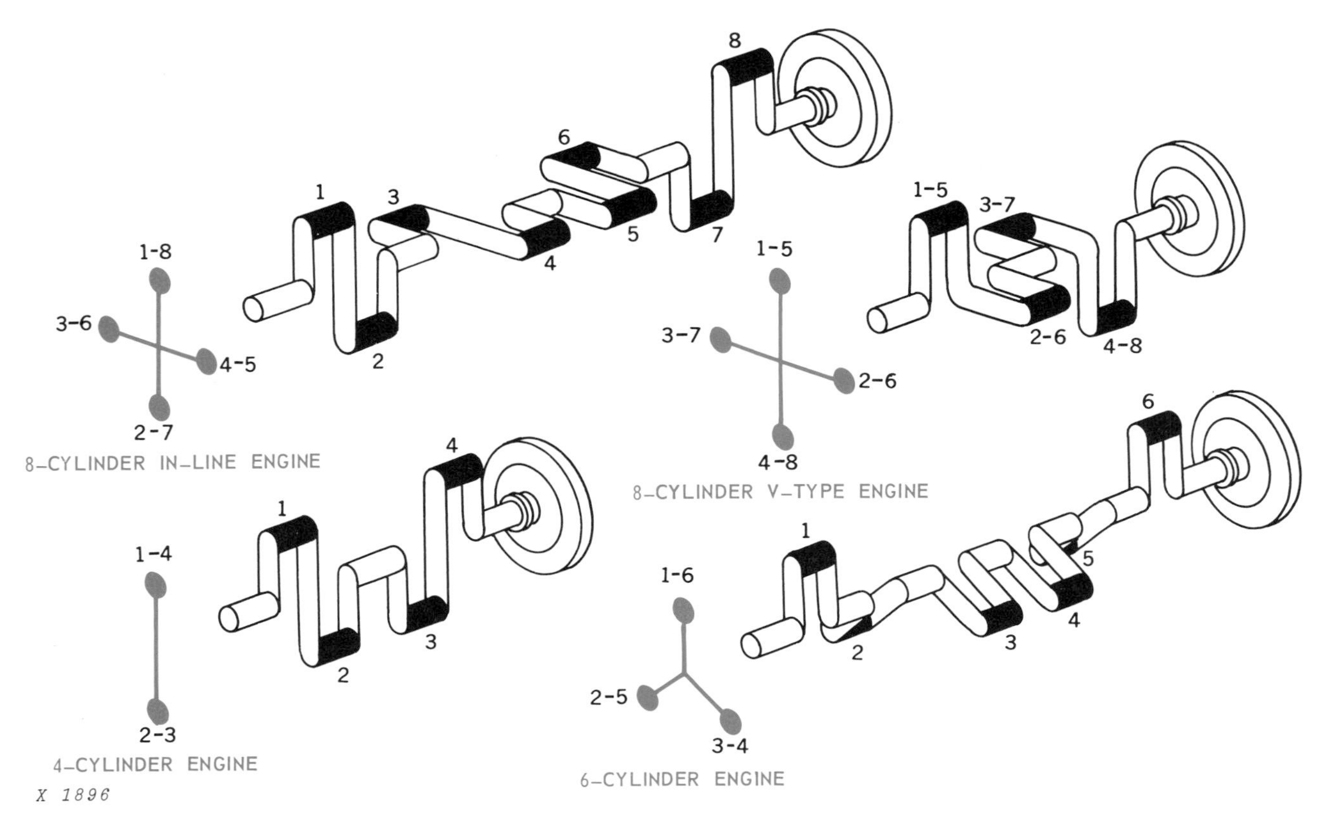

Fig. 119 — Crankshaft And Throw Arrangements

Six-Cylinder Engine Crankshafts—have each of 3 pairs of throws arranged 120 degrees apart. They may have as many as 7 main bearings—one at each end and one between each pair of throws.

Eight-Cylinder Engine Crankshafts — have different throw arrangements depending upon whether the engine is a V-8 or in-line model (Fig. 119). The crankshafts of V-8 engines are similar to 4-cylinder types or they have the 4 throws fixed 90 degrees from each other, as shown, for better balance and smoother operation. V-8 engines usually have *two* connecting rods side-by-side fastened to one throw.

BALANCING THE CRANKSHAFT

The power impulses of the engine tend to set up torsional vibrations in the crankshaft.

These vibrations must be controlled or the crankshaft might break at high speeds.

To balance the force of the pistons and connecting rods, *counterweights* are placed opposite the rod journals (Fig. 118).

The weight of the flywheel and the use of vibration dampeners also helps to stabilize the turning crankshaft (see later in this chapter).

LUBRICATING THE CRANKSHAFT

Most engines have pressure oil lubricating the crankshaft (Fig. 120).

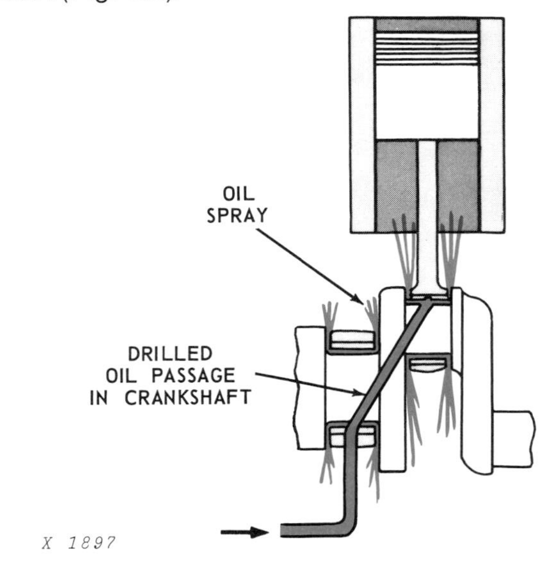

Fig. 120 — How The Crankshaft Is Lubricated

Oil holes are drilled through the crankshaft journals to match the holes leading in from the block. This oils the main and rod bearings as shown, while the excess oil sprays out to help lubricate the pistons and cylinder.

BEARING JOURNALS

Main and connecting rod journals of most crankshafts are induction hardened. They are also ground and polished to the exact bearing size.

Chrome plating of the journals is not very common, although it is one way to reclaim old journals.

SERVICING OF CRANKSHAFTS

Inspecting The Crankshaft

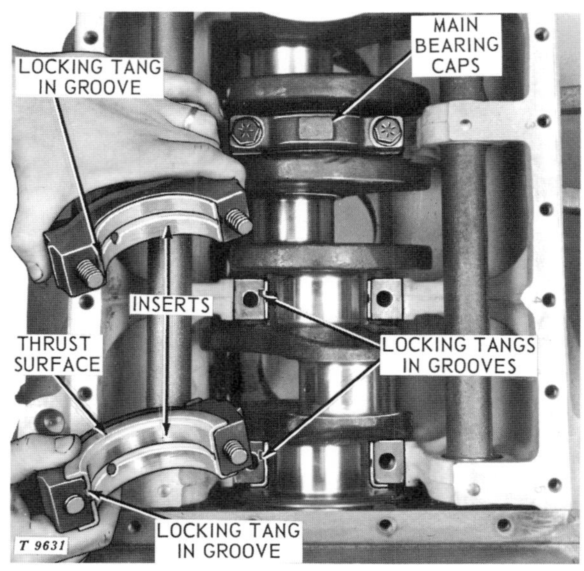

Fig. 121 — Removing Main Bearing Caps From Crankshaft

Bearing trouble is usually related to crankshaft wear. Whenever the main or connecting rod bearings are inspected (see "Bearings" in this chapter), the crankshaft should also be inspected.

PRELIMINARY INSPECTION OF BEARINGS

1. Remove the bearing caps one at a time (Fig. 121). Examine the crankshaft for scoring, ridging, overheating, cracks, or abnormal wear.

2. Identify the main bearing and connecting rod caps so that they can be installed in the same order they were removed.

3. Remove the bearing inserts from the main bearing bore in block and from the main bearing caps. Examine the inserts for evidence of scoring, wear, or "flaking out" of bearing material. Also look for a worn spot on the bearing which means a particle has lodged behind the bearing.

Fig. 122 — Checking Crankshaft Alignment With A Dial Indicator

4. *If one main bearing insert needs replacing, always replace both bearing inserts.*

5. Check clearance and condition of all main bearing inserts at this time. Wear on the damaged insert may be caused by another being out of specifications.

6. If any other inserts are within specifications but show excessive wear, replace them.

7. *Install new main bearing inserts at every major overhaul.*

COMPLETE INSPECTION OF CRANKSHAFT

When a crankshaft has been removed for reconditioning, inspect it as follows:

1. Clean the crankshaft with fuel oil and dry it with compressed air. Clean and blow out all the oil passages thoroughly.

2. Check the alignment of the crankshaft. Support it on its front and rear journals in V-blocks or a lathe and check the alignment at the center

Fig. 123 — Measuring Crankshaft Journals With A Micrometer

or intermediate journals using a dial indicator (Fig. 122). Protect the crankshaft journals from scratches while turning in the V-blocks by laying a strip of paper in the V's.

The center or intermediate journals should not vary more than the average oil clearance allowable in the total indicator reading. Refer to the engine specifiations for the allowable oil clearances.

To test either the front or rear main journals, move one of the V-blocks to the center or to an intermediate journal.

3. Measure all of the main connecting rod bearing journals (Fig. 123). Measure at several places around the journal to find the smallest diameter, in case the journals have worn out-of-round. Refer to the engine Technical Manual for wear limits.

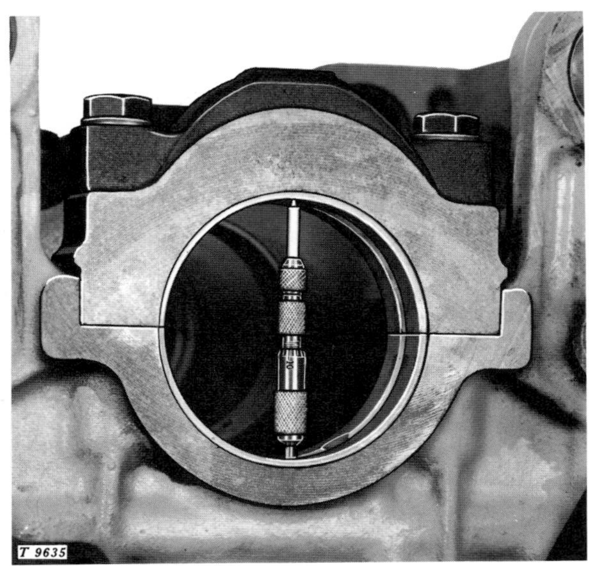

Fig. 124 — Measuring The Main Bearing Clearance

4. Measure the main bearing clearance. With the crankshaft out of the block, install the main bearing caps with bearing inserts in place and tighten to specifications. Using an inside micrometer, measure the inside diameter of the main bearings (Fig. 124). Compare the reading with the outside diameter of crankshaft main journals (see Fig. 123). Calculate this difference between the two readings to determine the bearing clearance.

Main bearing clearance can also be measured with the crankshaft in place by the use of "Plastigage." See earlier in this chapter under "Connecting Rods" for details.

5. Used crankshafts may have some ridging caused by the oil groove in the upper bearing (Fig. 125). If this ridge is not removed before new bearings are

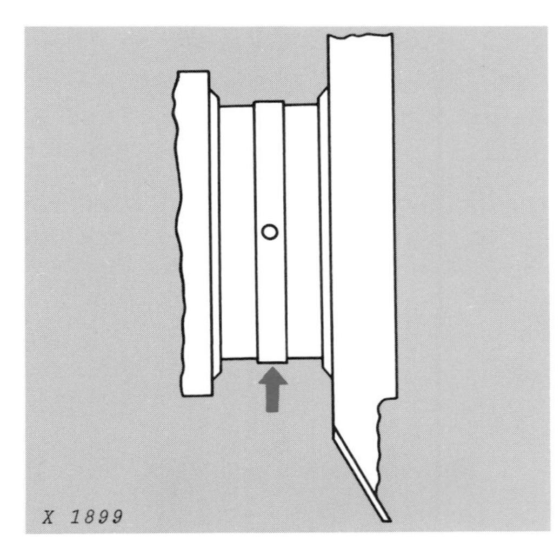

Fig. 125 — Badly Ridged Crankshaft Journal

installed, pressure on the bearings will be too high during operation.

Low ridges can be removed by working crocus cloth (wet with fuel oil) around the journal. Rotate the crankshaft frequently to eliminate an out-of-round condition. If the ridges are greater than 0.0005 inch (0.01 mm), first use 120-grit emery cloth to clean up the ridge, then use 240-grit emery cloth for finishing. Finally, use wet crocus cloth for polishing.

If the ridges are greater than 0.001 inch (0.025 mm), the crankshaft may have to be reground.

6. Check the surfaces of the crankshaft for cracks. Several methods of finding minute cracks not visible to the eye are outlined below:

Magnetic Particle Test—The part is magnetized and then covered with a fine magnetic powder or solution. Flaws, such as cracks, form a small local magnet which cause the magnetic particles in the powder or solution to gather, marking the crack. *The crankshaft must be demagnetized after this test.*

Fluorescent Magnetic Particle Test—This method is similar to the magnetic particle method, but is more sensitive since its employs magnetic particles which are fluorescent and glow under ultraviolet light. Very fine cracks that may be missed under the first test, especially on discolored or dark surfaces, will be disclosed under the light.

Fluorescent Penetrant Test—This is a method which may be used on *nonmagnetic* materials such as stainless steel, aluminum, and plastics. A highly fluorescent liquid penetrant is applied to the part. Then, the excess penetrant is wiped off and the part is dried. A developing powder is then applied which helps to draw the penetrant out of the flaws by capillary action. Inspection is carried out under an ultraviolet light.

A majority of the cracks revealed by the above tests are *normal and harmless* and only a few will damage the part. *Remember that interpreting the results is the most important step.*

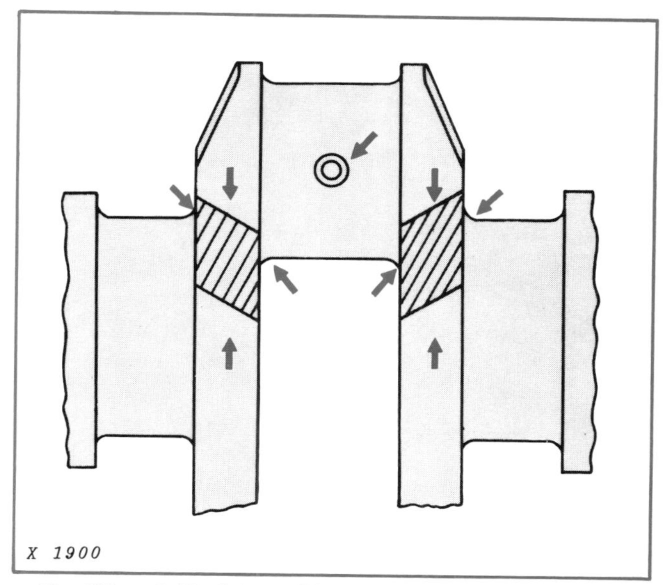

Fig. 126 — Critical Areas Of Load Stress In A Crankshaft

Crankshaft failures are rare; when one cracks or breaks completely, be sure to make a thorough inspection for the causes. Unless these causes are discovered and corrected, there may be a repeat of the failure.

Two types of loads are imposed on a crankshaft:

1) Bending forces

2) Twisting, torsional forces

The design of the shaft is such that these forces produce almost no stress over most of the surface. Certain critical areas, however, sustain most of the load. See Fig. 126.

BENDING FATIGUE FAILURES result from crankshaft bending, which takes place once per revolution.

The crankshaft is supported between each of the cylinders by a main bearing, and the force of combustion on the piston is divided between the adjacent bearings. Abnormal bending stress, particularly in the crank fillet, may result from misalignment of the main bearing bores, improperly fitted bearings, failed bearings, a loose or broken bearing cap, or an unbalanced fan pulley.

Bending failures start at the crank fillet and progress throughout the crank, sometimes extending into the journal fillet.

If the main bearings are replaced due to one or more badly damaged bearings, make a careful inspection to determine if any cracks have started in the crankshaft. These cracks are most likely to occur on either side of the damaged bearing.

TORSIONAL FATIGUE FAILURES result from twisting vibrations occurring at high frequency.

A combination of abnormal speed and load conditions may cause the twisting forces to set up a vibration which imposes high stresses at the locations shown in Fig. 126.

In addition, these stresses occur at the crankshaft journal oil holes near the flywheel end of the shaft.

Torsional stresses may produce a fracture in either the crankshaft rod journal or the main journal. Failures of crankshaft journals are usually at the fillet at a 45 degree angle to the axis of the shaft.

Causes of torsional failures are: a loose, damaged, or defective vibration damper, a loose flywheel, or improper or additional fan pulleys or couplings. Other causes may be overspeeding the engine or resetting the governor.

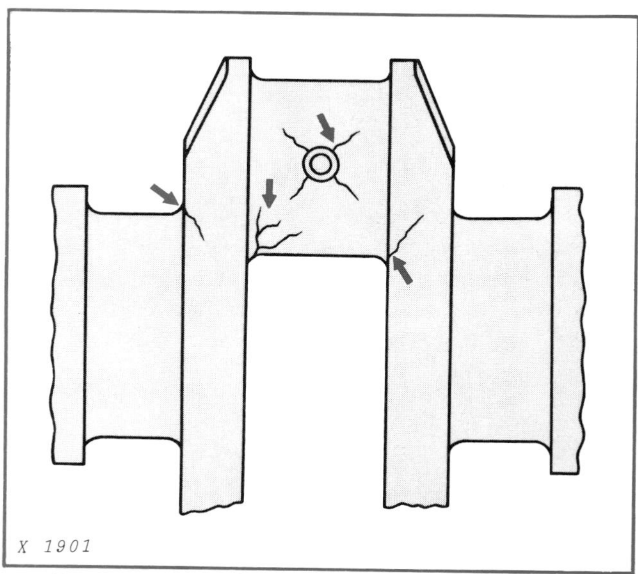

Fig. 127 — Crankshaft Fatigue Cracks That Require Replacement Of The Crankshaft

To repeat: *Most small cracks found in inspection of the crankshaft are harmless.*

Two types to look for are circumferential fillet cracks at the critical areas, and 45 degree cracks (with axis of shaft) starting from either the critical fillet locations or the crankshaft journal holes as shown in Fig. 127. These cracks require replacement of the crankshaft.

7. Check the crankshaft thrust surfaces for evidence of excessive wear or roughness. In many instances, only slight grinding or "dressing up" of the thrust surfaces is necessary. In these cases, new standard thrust washers will probably hold the end thrust clearance within the specified limits (if oversized thrust bearings are used).

8. Inspect the crankshaft keyways for evidence of cracks or wear, and replace the shaft if necessary.

9. Carefully inspect the crankshaft in the area of the rear oil seal contact surface for evidence of rough or grooved conditions. Any marring of this surface will cause oil leaks.

Reconditioning Crankshaft Journals

Two methods are used to recondition the rod and main journals:

1. *Grinding the journals by removing material from the surface.*

2. *Rebuilding the journals by adding material to the surface.*

REGRINDING THE CRANKSHAFT

1. Before grinding the crankshaft, make a careful check for cracks which start at an oil hole and follow the journal surface at an angle of 45 degrees to axis. Any crankshaft with such cracks must be replaced—regrinding only increases the effect of stress.

Also, when a shaft is inspected by the magnetic particle method, minute cracks may be found beneath the surface. These are not harmful *provided* the regrinding does not bring them out onto the surface.

2. Measure the crankshaft journals at A and B (Fig. 128) and compare with the diameters required for various undersize bearings (obtained from engine or bearing manufacturer). This will determine the size to which the crankshaft journals must be reground.

In addition to the standard main and connecting

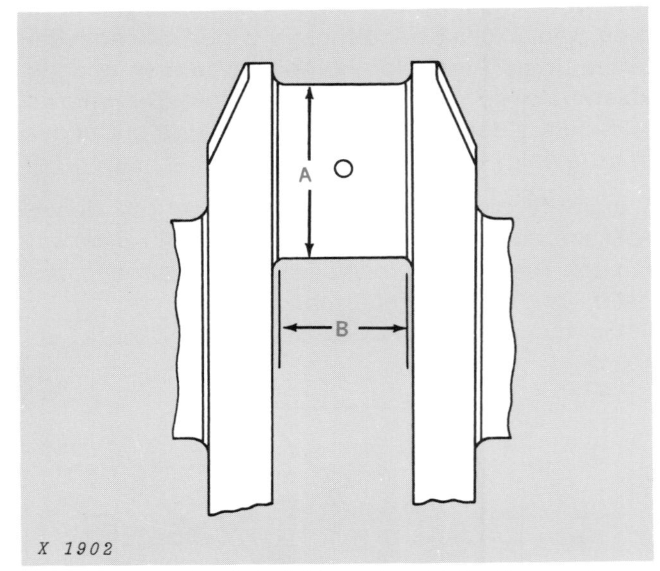

Fig. 128 — Crankshaft Dimensions To Be Measured Before Regrinding

rod bearings, undersize 0.002, 0.010, 0.020 and 0.030-inch (0.05, 0.25, 0.50, 0.75 mm) bearings are usually available.

It is not advisable to regrind the average crankshaft below 0.030 inch (0.75 mm). This applies to small highspeed engines, but there are some large heavy-duty engines that permit regrinding to sizes below this. As a guide, check to see if undersized bearings are available below 0.030 inch (0.75 mm).

REBUILDING CRANKSHAFTS

Rebuilding crankshafts is adding material to the surface of the crankpins and journals.

This is a specialty job and should not be attempted unless the equipment is available and the job is recommended by the engine manufacturer.

Several methods of rebuilding include:

1. *Chromium plating*

2. *Electro-welding*

3. *Metal Spraying*

4. *Gas Welding*

Installing The Crankshaft

1. Install new bearing inserts in the cylinder block and place the bearing caps in the same location from which they were removed. Usually the inserts and bores have locking devices which align to insure proper fit and avoid turning (see Fig. 129). Be sure that these locks align and that the oil holes in the inserts line up with oil passages in the cylinder block.

2. Apply a few drops of clean engine oil to the bearing and spread it over the bearing surface. Carefully position the crankshaft on the bearings. Be sure to hold the crankshaft parallel to the bore and gently lower it into position. This must be done with extreme care because the thrust bearing surface can be easily damaged.

3. Install the remaining bearing halves and caps, making sure the caps are installed on the mains from which they were removed by referring to identification marks made during removal. Loosely install the cap screws in the bearing caps until they are finger tight.

4. Before tightening the caps, align the thrust bearing or washers as follows: Tap the crankshaft to the rear to line up the front flanges. Then tap the crankshaft to the front to line up the rear flanges.

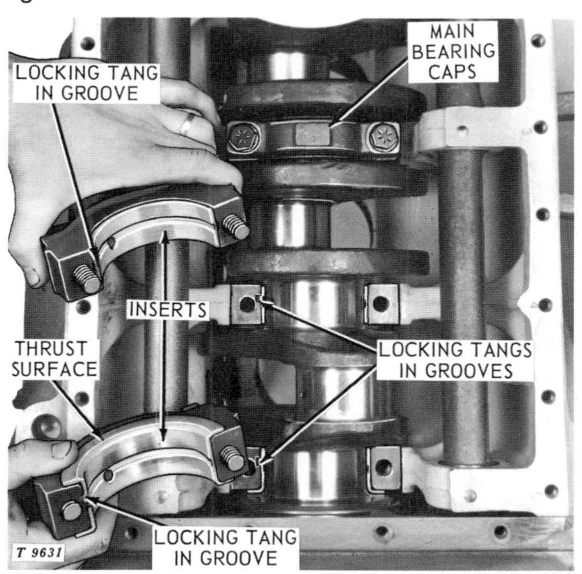

Fig. 129 — Installing A Typical Crankshaft

5. Now tighten the bearing cap screws to the specified torque, starting with the center cap and working alternately towards both ends of the block.

6. If the bearings have been installed properly, the crankshaft will turn freely after all the main bearing caps are drawn down to the specified torque.

Checking Crankshaft Endplay

The crankshaft must have a certain amount of endplay to get the proper thrust during operation and to avoid excessive wobble and wear.

The main bearing which absorbs the thrust usually has a double flange (see Fig. 137). Separate thrust washers are sometimes used for the same purpose.

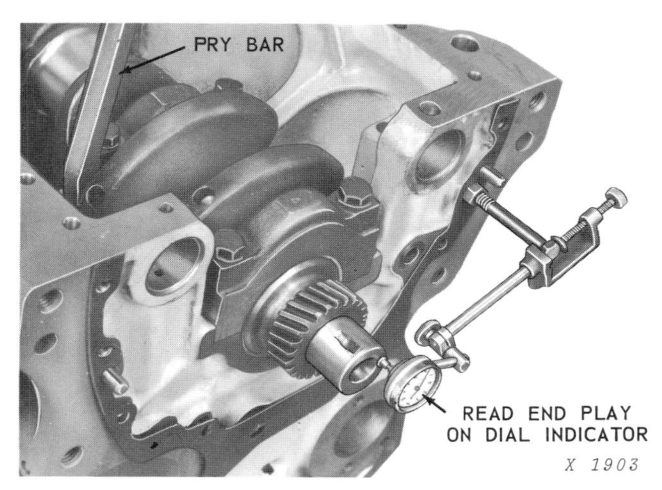

Fig. 130 — Checking Crankshaft Endplay

To check the crankshaft endplay, force the crankshaft toward the dial indicator as shown in Fig. 130. Keep a constant pressure on the tool and set the dial indicator on zero. Then, force the crankshaft in the opposite direction and note the amount of endplay shown on the indicator. Refer to the engine specifications for the correct endplay.

Too little endplay can be the result of a misaligned thrust bearing, or a burr or dirt on the inner face of the bearing flange or thrust washer.

Too much endplay means that the thrust surfaces are worn and need replacement.

ENGINE BALANCERS

Just as the crankshaft must be balanced, so the whole engine must be balanced. This is a harder job because some parts rotate while others reciprocate—or move up and down.

UNBALANCING FORCES IN THE ENGINE

Two unbalancing forces act on the working engine:

- **Centrifugal Force—rotary outward force at the crankshaft**
- **Inertia Force—up-and-down force at the pistons.**

CENTRIFUGAL FORCE at the CRANKSHAFT is created by the evolving weight of the heavy crankshaft throws (Fig. 131). Many engines can use counterweights on the crankshaft to balance this force and reduce the bending force on the crankshaft and main bearings. Such a crankshaft is said to be *statically and dynamically balanced.*

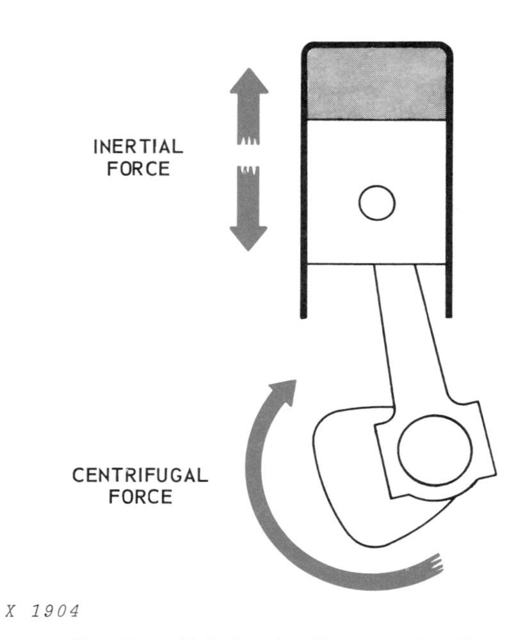

Fig. 131 — Unbalancing Forces In The Engine

INERTIAL FORCE at the PISTONS is greatest at the top and bottom of each stroke where the piston must reverse itself at high speeds (Fig. 131). Other inertial forces are created by the motions of the connecting rod and the piston as they "pedal" at high speeds. Counterweight on the crankshaft can also balance these forces in some cases, but not all.

BALANCING THE ENGINE

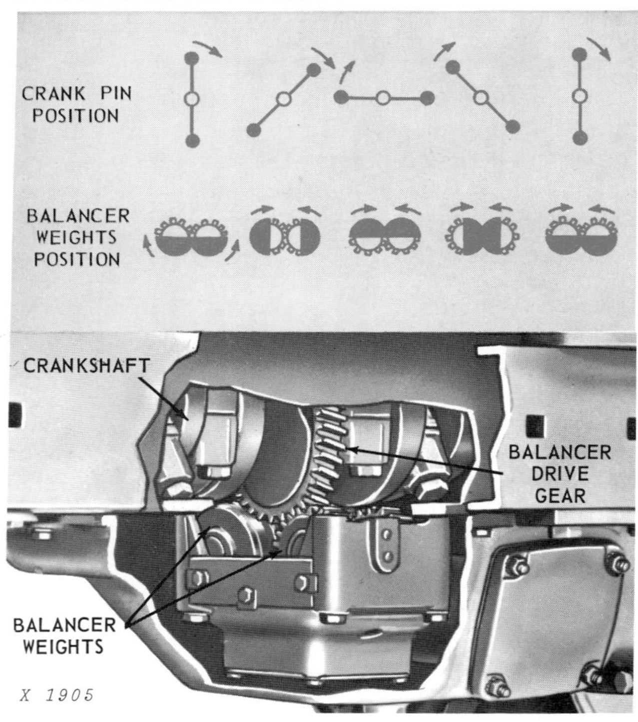

Fig. 132 — Engine Balancer On A Four-Cylinder Four-Cycle Engine

In four-cycle engines with more than four cylinders, inertial forces can be balanced by correct arrangement of the crankshaft throws.

With four cylinders or less, however, the crankshaft can be balanced as a whole but the individual cranks are difficult to balance. To reduce the vibrations on high-speed engines, balance weights or balancing shafts are used.

Fig. 132 is a diagram of how balancing weights are able to cancel the inertial forces in a four-cylinder, four-cycle engine.

The balancer shown is driven by the crankshaft and has two rotating gears with counterweights.

The balancer weights are timed to the crankshaft so that their position (shown in dark color) counteracts the forces on the cranks. These forces are caused by the different speeds of the pistons at the tops and bottoms of their strokes.

For example, when the crank is pulled up, the weights are down. Result: the upward force of the piston is cancelled by the downward centrifugal force of the weights.

Other positions of the crankshaft and weights are also shown in Fig. 132.

ANOTHER FORCE—TORSION VIBRATION

As the cylinders fire, the pressure "twists" the crankshaft, producing *torsional vibrations* in the crankshaft. We saw earlier that the crankshaft throws and the flywheel are weighted and arranged to reduce these forces. However, some

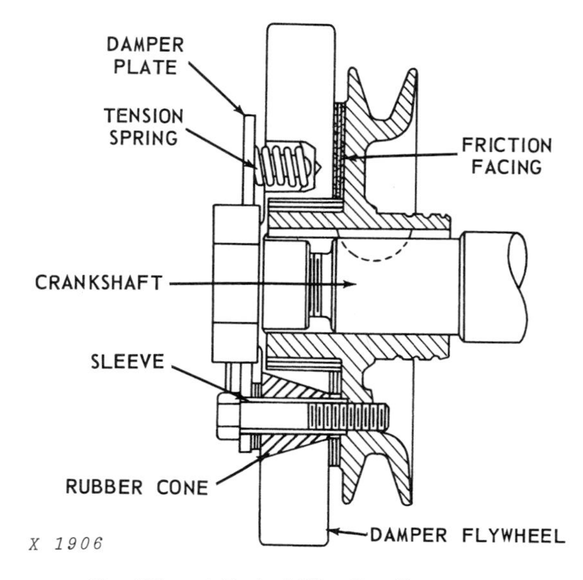

Fig. 133 — A Typical Vibration Dampener

engines are operated under extra stress—sudden loads, varying speeds, and the like.

A **vibration dampener** can be mounted on the free end of the crankshaft to control the vibrations.

Most vibration dampeners resemble a miniature flywheel (Fig. 133). A friction facing is mounted between the hub face and a small damper flywheel. The damper flywheel is mounted on the hub face with bolts that go through rubber cones in the flywheel. These cones permit the damper to move slightly on the end of the crankshaft. This reduces the effects of torsional vibration in the crankshaft.

Several other types of vibration dampeners are used which employ springs, rubber bonding, "floating" action of fluids, and loose pins. However, the principle is the same as described above.

MAIN BEARINGS

All the major wear and load points in an engine use *bushings* or *bearings* to reduce friction (Fig. 134).

Let's define the two words:

- **Bushing—small full-round sleeve, pressed in, for lighter loads or slower speeds.**
- **Bearing—full-round or halves, for heavier loads and higher speeds.**

BUSHINGS are used at the piston end of the connecting rod, at the rocker arms, the oil pump, etc.

BEARINGS are used at the crankshaft main journals, at the connecting rod journals, at the camshaft, etc.

Actually, the words "bushing" and "bearing" are synonymous when talking in general about a wear or friction point in an engine.

In the remainder of this section we will use the word "bearing" for all applications.

WHAT BEARINGS ARE MADE OF

The bearing material depends upon the expected wear and stress. Generally the bearing has a steel backing with one of three linings:

1) Tin- or lead-base babbitt

2) Copper or aluminum alloys

3) Multi-layer bearings in copper or aluminum alloys and silver combinations

Normally these materials are applied to the steel backings as a thin layer which is 0.013 to 0.025 inch (0.33 to 0.64 mm) in thickness in small bearings and somewhat thicker in large bearings. Bearings having one material deposited on the backing are referred to as "bimetal" bearings, while those having extra overlay coatings are called "trimetal" bearings.

WHAT THE BEARING MUST DO

Each type of bearing lining must have these qualities:

1. **Conformability** — the ability to creep or flow slightly so that the shaft and bearing will conform to each other.

2. **Embeddability**—the ability to let small dirt particles embed themselves to avoid scratching the shaft.

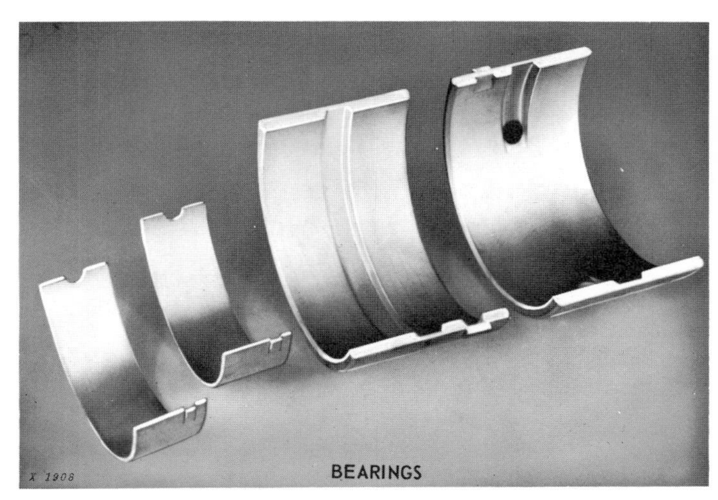

Fig. 134 — Bushings And Bearings For Typical Engine

3. **Seizure Resistance**—smooth surface action, since it is difficult to avoid some metal-to-metal contact during starting or when the oil film becomes thin during operation.

4. **Corrosion Resistance**—the ability of the bearing material to resist chemical corrosion.

5. **Temperature Strength**—the ability of a bearing material to carry its load at high operating temperatures.

Other items such as Wear Rate, Cost, Thermal Conductivity, and the Ability to Form a Good Bond with the backing material must also be considered.

BEARING LOCKS

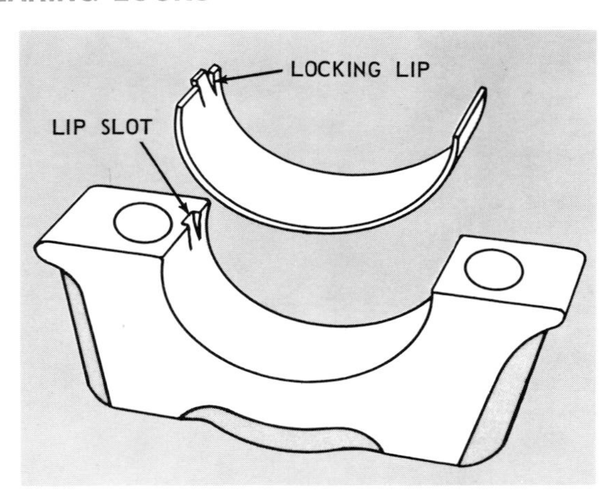

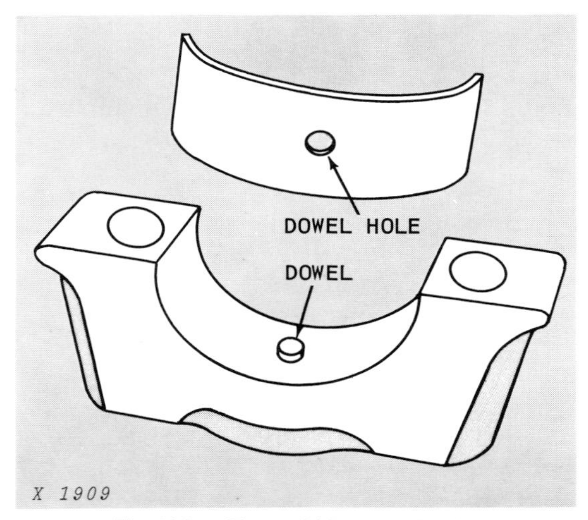

Fig. 135 — Types Of Bearing Locks

Unless it is designed for full-floating operation, the bearing insert must be locked in place to prevent rotation with the shaft (Fig. 135).

This is usually done by means of a locking lug on each half of the insert, which fits in a notch in both the housing and bearing cap. The lug on one insert prevents rotation in one direction while the other lug works in the opposite direction.

In some thick-walled bearings, a dowel may be used in the cap or housing to hold the bearing.

BEARING CRUSH

A *crush fit* provides a close contact between a bearing insert and its housing and cap. When the cap is tightened down fully, the bearing is crushed into place.

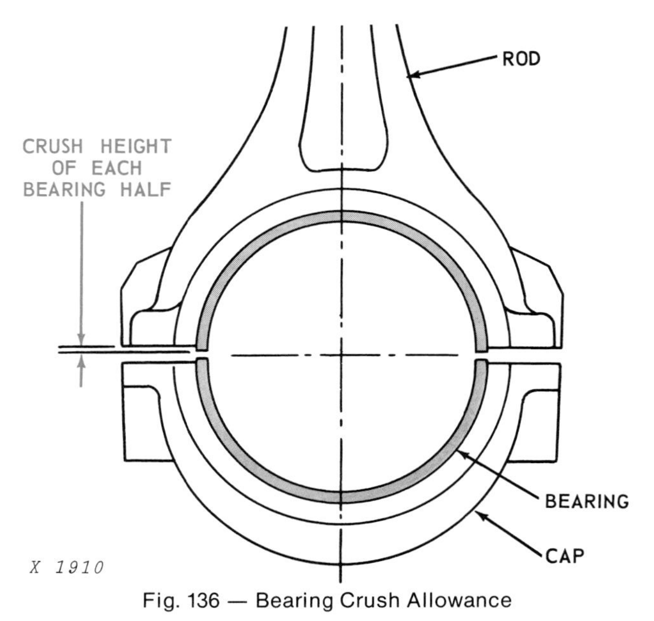

Fig. 136 — Bearing Crush Allowance

Fig. 136 shows the allowable crush height of a bearing insert half.

This tight fit aids the bearing lock in preventing movement of the bearing in the housing and provides full support.

BEARING SPREAD

Most main and connecting rod bearing halves are purposely manufactured with *spread* for a tighter fit. This is an extra distance across the parting faces of the bearing half in excess of the actual diameter of the housing bore.

BEARING OIL GROOVES

In most engines the lubricating oil is supplied to the main bearings through the cap or housing. Part of this oil must pass through a hole in the main bearing journals and then through the cranks to the connecting rod journals.

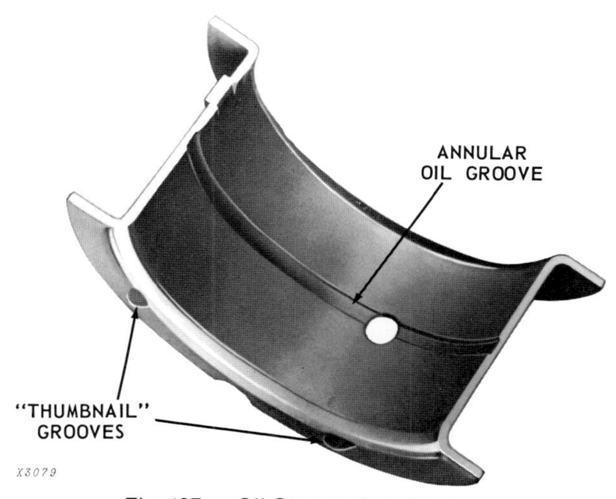

Fig. 137 — Oil Grooves In A Bearing

Some of this oil must get to a passage in the connecting rod and on to the piston pin and piston. Oil grooves must be provided for this oil flow (Fig. 137). In many cases either partial or complete circumferential grooves are provided in the bearing surface. In other cases at least part of the grooving is in the cap or housing outside of the bearing insert.

Distributing grooves are often machined into each bearing half along the parting faces to help distribute the oil the full length of the bearing so that the shaft can pick up the oil and carry it on to the load-supporting area. The term "mud pocket" is sometimes used to describe these grooves because the grooves will temporarily trap wear particles or dirt.

"Thumbnail" grooves are often placed in select locations on flange bearing faces to help distribute the oil evenly over the thrust surfaces (Fig. 137).

CRANKSHAFT THRUST BEARINGS

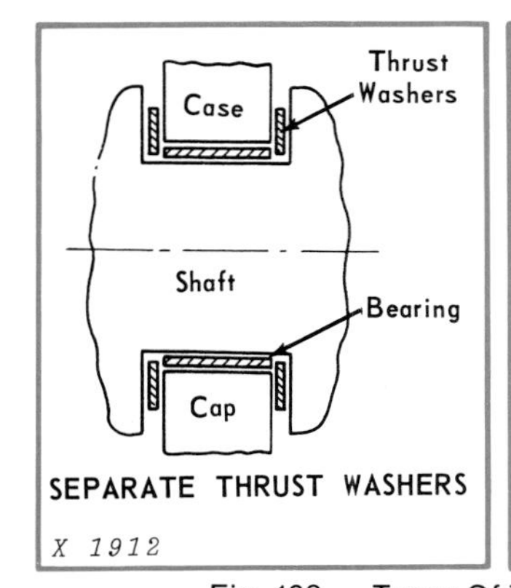

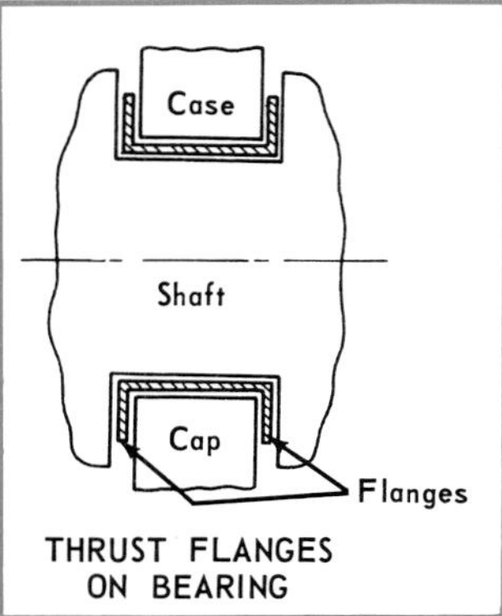

Fig. 138 — Types Of Thrust Bearings

The crankshaft is usually held in position axially at one of the main bearings near the flywheel end. This may be done by flanges which are part of the bearing inserts or separate flat thrust washers on each side of one main bearing (Fig. 138). Thrust flanges must be on both halves of the bearing.

SERVICING OF BEARINGS

Analyzing Bearing Failures

Failure of a bearing in service can be recognized in most cases by the following signs:

1) A drop in the lubricating oil pressure

2) Excessive oil consumption

3) Engine noise—rhythmic knock

Wear will vary widely with different engines and operations. For example, continued overloads or frequent shutdowns. "Normal" wear rate is affected by an "abnormal" operation, even starting the engine.

Major Causes Of Bearing Failure

When bearings fail, find the cause and correct it before installing new inserts. Otherwise, another failure can be expected in a very short time.

One bearing manufacturer found the causes of bearing failures ran like this:

Dirt	*42.90%*
Lack of Lubrication	*15.30%*
Improper Assembly	*13.40%*
Misalignment	*9.80%*
Overloading	*8.70%*
Corrosion	*4.50%*
Undetermined and Other Causes	*5.40%*

Diagnosing Bearing Failures

To locate the cause of an engine bearing failure, take the following steps:

1. Remove the oil pan from the engine after draining.

2. While removing the bearings, mark each one for its correct location. Mark it lightly with a soft lead pencil (Fig. 139). A code can be used: "1 U" for No. 1 cylinder — upper half of bearing, "1 L" for the lower half, etc.

NOTE: Any bearings showing signs of damage or excessive wear should be discarded and new bearings installed.

3. Clean all of the bearing halves removed in a suitable solvent to remove sludge and oil. Blow or wipe them dry.

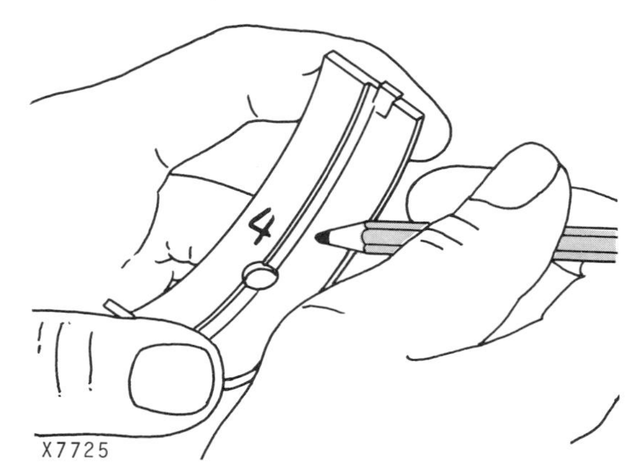

Fig. 139 — Mark The Bearing To Identify Its Position In The Engine During Removal

4. Observe the sludge or settlings in the engine oil pan. Sometimes sand or metal particles as well as disintegrated bearing linings can be found and identified.

5. Place the cleaned bearings out on a flat surface in the order of position. Make two groups: Connecting rods farther away, main bearing nearer. The bearing surfaces should face up, with 1U on the upper left, 1L below it, and 2U next to 1U, and so on.

6. Keep an open mind; that is, avoid reaching a conclusion at this point.

7. Examine the connecting rod bearing operating surfaces, noting the distress areas, the rubbing areas, and so on.

8. Examine the main bearings in the same way. Closely inspect the mains that match the connecting rod bearings that are most distressed.

9. Turn all the bearings over to show the bearing backs. Examine the fit of the bearings into their housing by noting the transfer pattern on each half. Inspect the backs for interference by dirt particles, and check for a damaged fit of the locking lips into the recesses.

10. Turn the bearings over once more. After noting the condition of all bearings—overheated condition, poorly fitted, poorly installed, dirty, and so on—then make a conclusion.

11. Repair the engine as necessary to eliminate the cause of bearing failure. Then replace the bearings with new, approved sets.

Damage Revealed By Bearings

Here are some examples of bearing damage plus the causes and remedies.

Oil starvation was the cause of damage to the bearing shown in Fig. 140. Lack of oil can occur immediately after overhaul. This is when priming of the engine's lubricating system is vital to assure initial lubrication.

After break-in, other things can happen. Both local and general oil starvation can result from external leaks and mechanical supply failures. Blocked oil suction screen, oil pump failure, oil passages plugged or leaking, failed pressure relief valve springs, or badly worn bearings can stop the circulation of lubricating oil.

A mislocated oil hole will also cut off the oil supply to a bearing, causing rapid failure. Always check to be sure that the oil hole in the bearing is in line with the oil supply hole in the connecting rod or crankcase.

Also, the oil supply may become diluted by seepage of fuel into the crankcase from a defective fuel pump. This will reduce the oil's film strength and score the bearings.

Fig. 140 — Oil Starvation Caused This Damage

Corrosion from acid formation in the oil is seen by a finely pitted surface and large areas of deterioration (Fig. 141).

Corrosion occurs when oil temperature goes above 300°F (150°C) and when excessive blow-by occurs. It is also aggravated by stop-and-go operation, which causes condensation in the crankcase.

Fig. 141 — Corrosion From Acid Formation In Oil

Fig. 142 — Damage From Dirt Embedded In Bearing

Prevent corrosion by changing oil at correct intervals and by selecting oil of the proper quality and classification for the machine and type of service.

Follow the manufacturer's recommendations.

Dirt can embed in the soft bearing material (Fig. 142). This causes wear and decreases the life of both the bearing and its journal.

Prevent this by cleaning the engine thoroughly during bearing installation and by proper maintenance of both air and oil filters.

Bent connecting rods can cause concentrated wear on the bearings (Fig. 143). This angular loading from a bent rod will cause the excessive wear shown on one edge of the upper bearing insert and the opposite edge of the lower bearing insert. When this wear pattern is found, check the connecting rod for poor alignment.

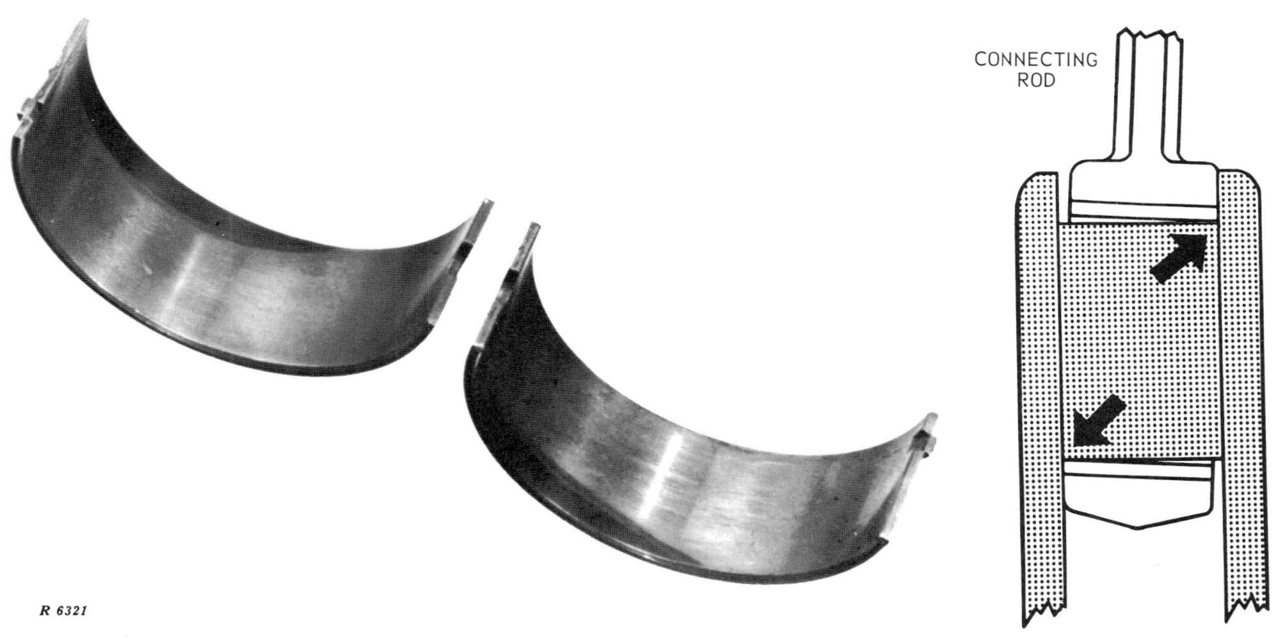

Fig. 143 — Excessive Wear Caused By A Bent Connecting Rod

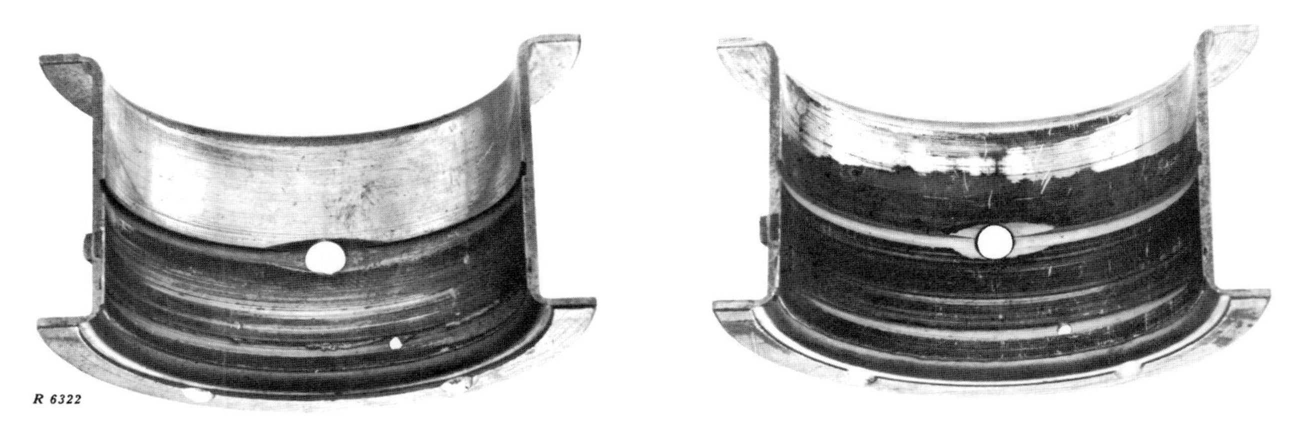

Fig. 144 — Wear On One Edge Of Bearing Caused By Tapered Journals

Tapered journals allow areas of excessive clearance between the journal and bearing, which distributes more wear on one edge of the bearing (Fig. 144). This wear is further concentrated by the force of the piston on the insert carrying the greater load.

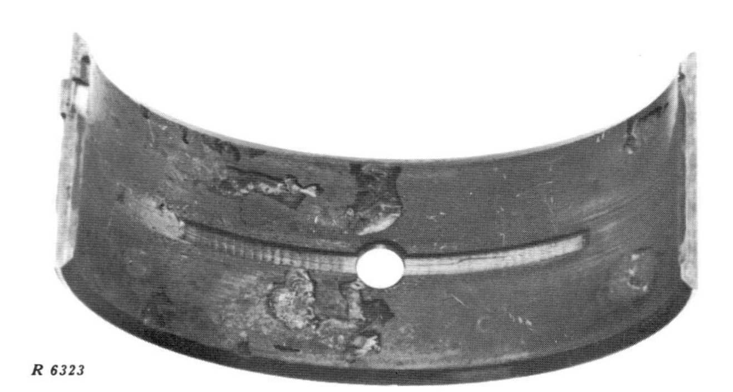

Fig. 145 — Bearing Fatigue Caused By Overloading And Heat

Overheating from overloads on the engine causes a metal fatigue which breaks away and voids the surface of the bearing (Fig. 145).

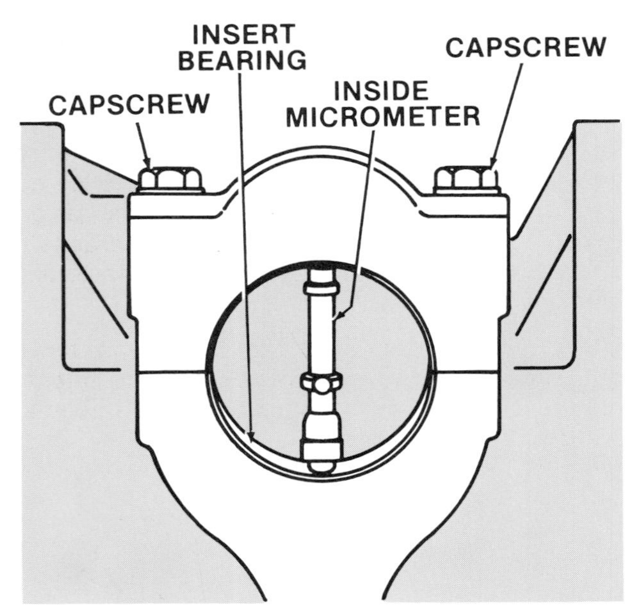

Fig. 146 — Checking Bearing For Wear

To measure bearing wear, assemble the bearing without the crankshaft. Properly torque the capscrews. Use an inside micrometer to measure the bearing surface (Fig. 146). Also check for wear on the matching crankshaft journals.

FLYWHEEL

The flywheel does three things for the engine:

- **Stores energy for momentum between power strokes**

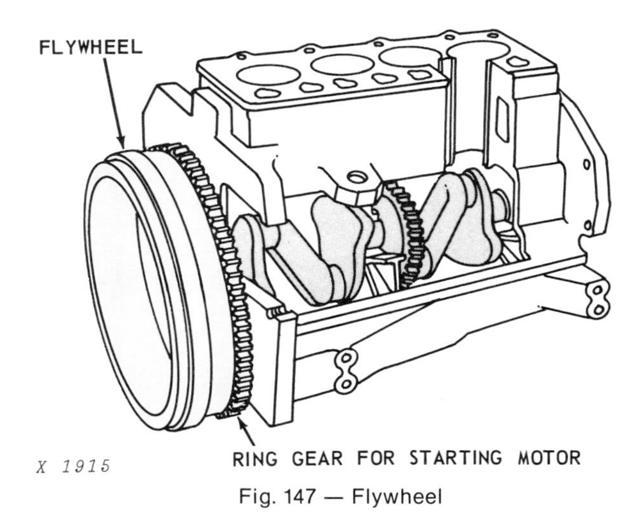

Fig. 147 — Flywheel

- **Smooths out speed of crankshaft**
- **Transmits power to the machine**

Mounted on the rear of the crankshaft (Fig. 147), the heavy flywheel is a stabilizer for the whole engine.

In a 4-cycle engine, the flywheel must be heavy enough to turn the engine during the exhaust, intake, and compression strokes. At the same time, it must transmit power to the driven machine.

The more cylinders in the engine, the less need for a flywheel. This is because the power impulses are closer together and the engine has more momentum of its own.

A lighter flywheel is needed for an engine operated at variable speeds where faster acceleration is required.

Flywheels are generally made of heavy cast iron or steel and are fastened to the crankshaft by dowel pins and bolts.

Flywheels may also have two other jobs:

1) Provide a drive from the starting motor via the ring gear.

2) Serve as a facing for the engine clutch.

SERVICING THE FLYWHEEL

The flywheel is rugged in construction and has no moving parts.

It does place a heavy load on the engine rear main bearing, which should be checked carefully at each overhaul.

The flywheel must also be removed when servicing many parts of the engine.

Inspecting The Flywheel

After removal, check the clutch contact face of the flywheel for scoring, overheating, or cracks. If scored, most flywheels can be refaced. However, do not remove too much material from the flywheel and keep the surface perfectly flat all around.

The ring gear for the starting motor is generally shrunk in place on the flywheel rim. If it becomes damaged, remove and replace it.

Removing Ring Gear From Flywheel

Note whether the ring gear teeth are chamfered. The replacement gear must be installed so that the chamfer on the teeth faces the same direction. Then remove the ring gear as follows:

1. Support the flywheel, crankshaft side down, on a solid flat surface or hardwood block which is slightly smaller than the inside diameter of the ring gear.

2. Drive the ring gear off the flywheel, with a suitable drift and hammer. Work around the circumference of the ring gear to avoid binding the gear on the flywheel.

Installing Ring Gear On Flywheel

1. Support the flywheel, ring gear side up, on a solid flat surface.

2. Rest the ring gear on a flat metal surface and heat the ring gear uniformly with an acetylene torch, keeping the torch moving around the gear to avoid hot spots.

IMPORTANT: Never overheat the gear because this may destroy the original heat treatment.

If the engine manufacturer specifies a certain heat range, heat crayons which give the temperature may be obtained from most tool vendors.

3. Use a pair of tongs to place the gear on the flywheel with the chamfer, if any, facing the same direction as on the gear just removed.

4. Tap the gear into place against the shoulder on the flywheel. If the gear cannot be tapped into place readily, remove it and apply additional heat, but DO NOT OVERHEAT.

CLUTCH ASSEMBLY

For information on clutch assemblies refer to FOS Power Trains.

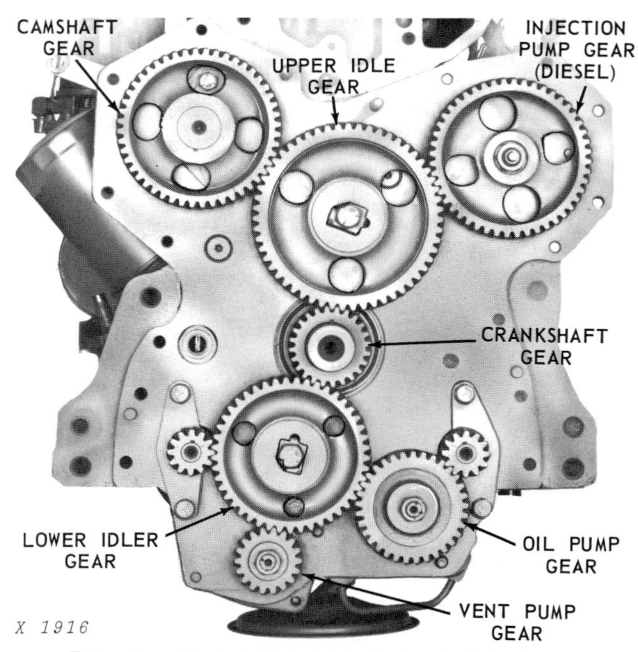

Fig. 148 — Timing Gear Train On Typical Engine

TIMING DRIVES

The crankshaft is the "hub" around which other parts of the engine can be timed and driven (Fig. 148). This is done by the meshing of gears as shown.

The camshaft runs at one-half engine speed, so a 2-to-1 reduction of gears is used. An idler gear transmits the rotation to the large camshaft gear (which turns half as fast as the smaller crankshaft gear). A chain drive can also be used to turn the camshaft.

Other accessories that can be driven by the crankshaft are fuel pumps, oil pumps, injection pumps (diesel), ventilator pumps, and water pumps.

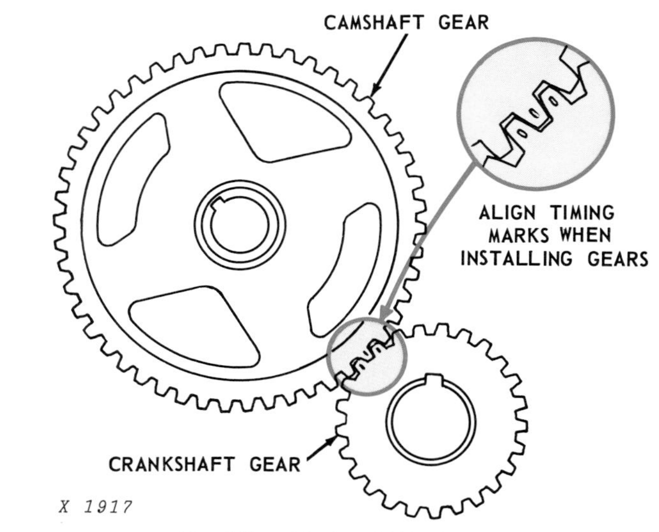

Fig. 149 — Timing Of Engine Gears

TIMING THE GEAR TRAIN

Many gears in the engine gear train must be timed to each other.

For example, the camshaft operates the valves and must be synchronized with the pistons and crankshaft.

Timing of these gears is done by matching timing marks when the gears are installed (Fig. 149). A timing tool may also be required for timing the gears on some engines. Check the engine Technical Manual for timing instructions.

GEAR TRAIN BACKLASH

Backlash is the amount of "play" between two gears in mesh. As gear teeth wear, more backlash occurs.

Check the engine Technical Manual for the allowable backlash between the engine timing gears.

ENGINE CLUTCHES

For details on engine clutches, refer to FOS Power Trains.

ENGINE BREAK-IN

Modern engines need less break-in than the old ones did. The reason is that design, workmanship, fuels, and lubricants are all better today.

Short but careful break-in periods are the rule on new or overhauled engines.

It is a mistake to "baby" the engine and then suddenly put it under full load. This breaks in the engine for light loads only.

In all cases, follow the recommendations in the engine Technical Manual.

The steps below are widely recommended.

BREAK-IN STEPS

1. Adjust valve tappets, carburetor or injection pump, and engine timing as accurately as you can before starting the engine.

2. Run the engine at half-throttle for a short period until the engine coolant is warmed up to normal.

3. Check for good oil pressure in the engine. Look for oil or coolant leaks.

4. Operate the engine at the specified speed and load for the short time recommended to help seat the head gasket, etc.

5. Shut down the engine and:

a) Retighten the cylinder head

b) Recheck valve tappet clearances

c) Recheck engine timing

6. Operate the engine at normal loads for the first 100 hours (or as recommended). Avoid light loads and excessive idling. Never "lug down" the engine. Check the crankcase oil level more often during this time (special break-in oil may be recommended).

7. At the end of the break-in period, service the engine as specified. This may include changing the oil and replacing the filter.

TEST YOURSELF

QUESTIONS

1. What is one sequence for tightening cylinder head bolts?

2. What does "knurling" a valve guide do?

3. Why are valve rotators used on some valves?

4. The camshaft turns at (1/4 - 1/2 - 2/3) the speed of the engine crankshaft.

5. Valve clearance is normally adjusted with the piston at ________ of its ________ stroke.

6. Explain the terms "wet" and "dry" cylinder liners.

7. Where is a cylinder normally worn the most by its piston? a. Top inch of ring travel. b. Bottom inch of ring travel. c. Center of ring travel.

8. True or false? "Measure the cylinder liner as soon as it is removed from the engine block."

9. Why are piston skirts not straight up-and-down?

10. Match the three items at left below with the correct item on the right.

a. Blow-By	1. Gasoline igniting before spark occurs
b. Knock(Detonation)	2. Leaking of gases past the pistons
c. Preignition	3. Too-rapid combustion of fuel

11. True or false? "When removing piston rings, replace only the damaged ones."

12. Why mark the position of each connecting rod and piston when removing them from the engine?

13. Why are crankshafts so heavy?

14. What engine part gives momentum between power impulses?

15. True or false? "For an engine operated at variable speeds, a lighter flywheel is needed."

16. When breaking in an engine, what three adjustments should be rechecked after a brief run-in?

(Answers in back of text.)

Litho in U.S.A.

GASOLINE FUEL SYSTEMS / CHAPTER 3

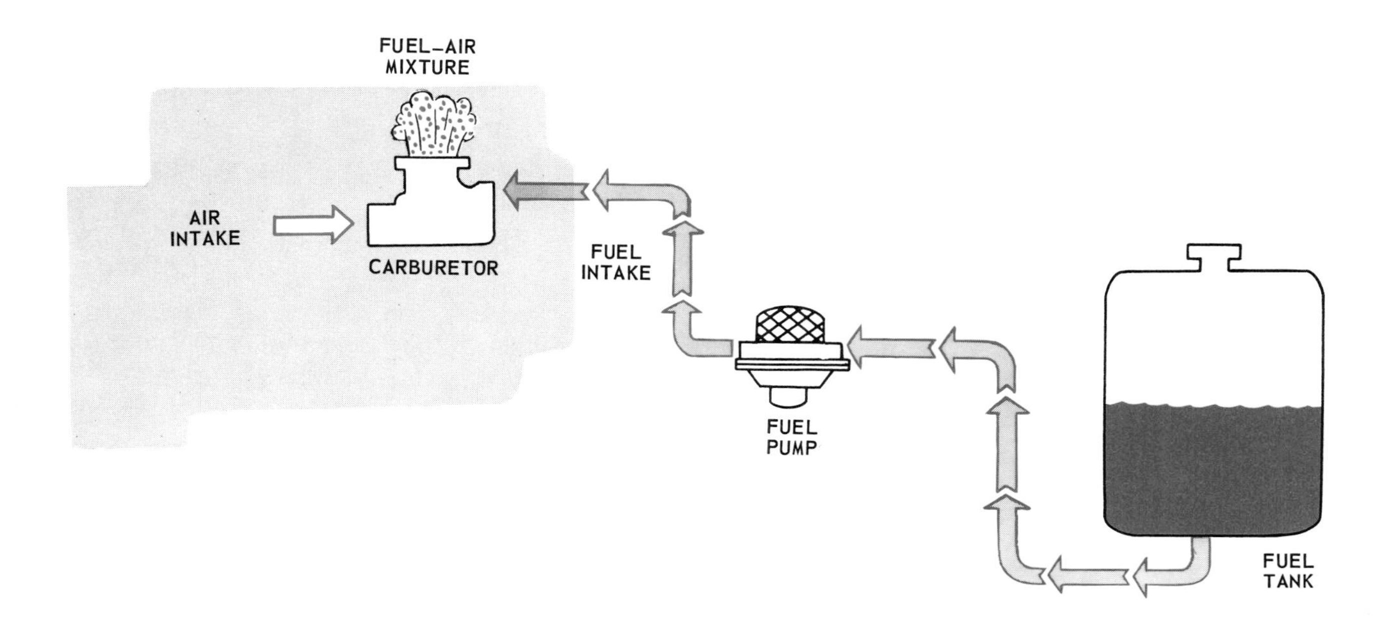

Fig. 1 — Gasoline Fuel System

INTRODUCTION

The gasoline fuel system (Fig. 1) supplies a combustible mixture of fuel and air to power the engine.

The gasoline fuel system has three basic parts:

- **Fuel Tank**
- **Fuel Pump**
- **Carburetor**

The FUEL TANK stores the gasoline for the engine.

The FUEL PUMP moves the fuel from the tank to the carburetor. This is an optional feature needed with force-feed supply systems. The **fuel pump** draws the gasoline through a fuel line from the tank and forces it to the *float chamber* of the carburetor, where it is stopped.

The CARBURETOR atomizes fuel and mixes fuel and air in the proper ratio. Filtered air is forced in at one end and vapor out at the other end (Fig. 2).

The pressure differential is created when air flows through the narrow neck, called the *venturi*. Air flow moves faster through a restriction and this lowers the air pressure.

At the same time, the engine creates a partial vacuum on intake stroke and this causes vapor to press into the combustion chamber of the cylinder as shown in Fig. 2.

The fuel is forced into the airstream from the *nozzle* which projects into the tube at the venturi. As low-pressure air rushes by, small drops of fuel are forced out and mixed with the air.

The fuel-air mixture must pass the *throttle valve* which opens or closes to let the right volume of fuel-air mixture into the engine. This is controlled by the operator at the throttle as he sets the engine speed.

The *choke valve* also controls the supply of fuel to the engine. When starting the engine in cold weather, for example, it can be partly closed, forming a restriction. This restriction causes more

 Litho in U.S.A.

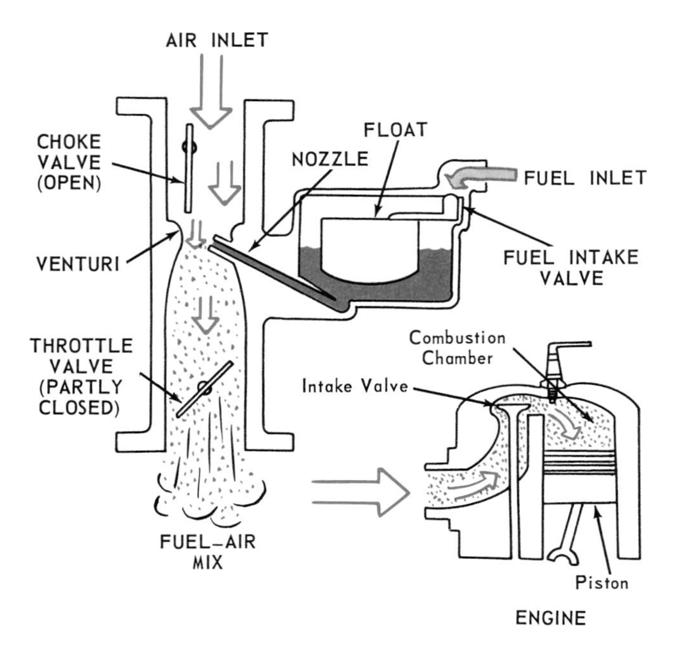

X 1648

Fig. 2 — Basic Carburetor at Part Throttle or Low Power

fuel and less air to be drawn into the combustion chambers. This results in a richer mixture in the cylinders for the harder job of starting.

FUEL SUPPLY SYSTEMS

Fuel is supplied to the gasoline fuel system in two major ways:

- **Gravity-Feed System**
- **Force-Feed System**

GRAVITY-FEED SYSTEM

The **gravity-feed system** (Fig. 3) has a fuel tank placed *above* the carburetor; fuel lines; fuel filter; and the gravity-fed carburetor.

A float attached to a valve allows fuel to enter the carburetor at the same rate at which the engine is consuming it. This system maintains a uniform level in the carburetor regardless of the amount of fuel in the tank.

FORCE-FEED SYSTEM

The **force-feed system** allows the fuel tank to be located at a level *below* the carburetor as shown in Fig. 3.

However, a fuel pump is required to raise the fuel from the tank to the carburetor. Fuel pump operation is described later in this chapter.

FUEL TANK

The location of the **fuel tank** varies on each machine. The tank is usually made of sheet metal and attaches to the frame.

Most systems have the fuel line attached at or near the bottom of the tank and usually have a filter screen at the fuel line connection.

A stand pipe is commonly used on tractor tanks. It provides a fuel outlet which projects above the bottom of the tank to avoid drawing out sediment.

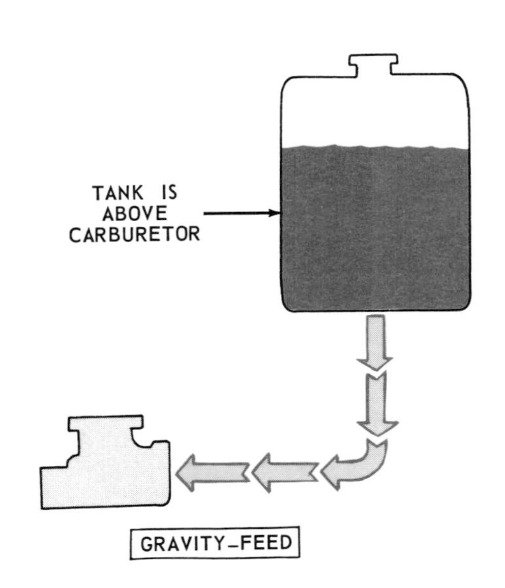

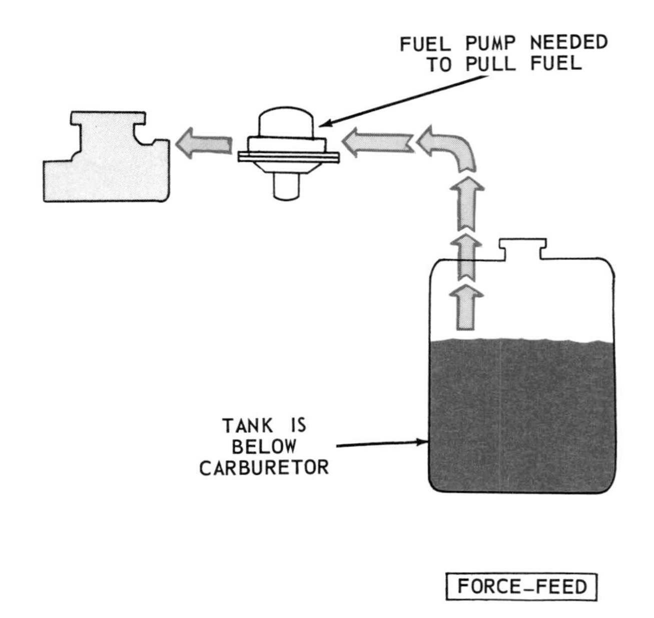

X 1650

Fig. 3 — Fuel Supply Systems — Two Types

Litho in U.S.A.

X8160

Fig. 4 — Fuel Tank On Modern Machine

A drain cock at the bottom of the tank allows water and sediment to be drained off periodically.

A shut-off valve may be used to close the fuel outlet before removing the tank.

To ventilate the fuel tank, a vent mechanism may be built into the filler cap or as a separate opening near the top. The vent allows air to replace the fuel as it is drawn out and prevents restriction of the fuel flow or a vacuum in the tank.

INSPECTION AND REPAIR OF FUEL TANKS

Cleaning

Flush the tank for 15 minutes with hot water. Run in at the bottom and allow it to overflow at the top.

Steam the tank for 30 minutes. Force in live steam at top of tank and allow it to escape through bottom. If live steam is not available, again flush the tank with boiling water continuously for 30 minutes and dry thoroughly with compressed air.

CAUTION: Cleaning and repairing a fuel tank is very dangerous. Never permit live sparks, smoking, or fire of any nature in the vicinity.

Inspecting for Leaks

Refer to the section in Chapter 5 titled "Inspecting Tank for Leaks."

Repair Of Fuel Tanks

Refer to the section in Chapter 5 titled "Repair of Fuel Tanks." Be sure to observe all the precautions on filling the tank with water and venting it before welding.

Fuel Tank Caps

The fuel tank cap must do three jobs:

1) Seal out dust and dirt.

2) Keep fuel from splashing out of the tank.

3) Allow air to enter the tank to force fuel out (unless the tank has other ventilation).

Be sure the gasket on the cap seals the tank and that the vents are open so the tank can breathe.

If the fuel filler cap is vented, *always replace it with a vented cap.* Otherwise, the tank may collapse when a vacuum is created as fuel flows out.

FUEL LINES

Fuel lines, usually made of steel tubing, transfer fuel from one location to another. Recently, polyethylene lines have become popular where temperatures permit their safe use.

Maintenance involves watching for leaks from too loose connections or damaged fittings, and checking for bends or dents that might restrict fuel flow.

FUEL GAUGES

Refer to Chapter 5 of this manual for information on fuel gauges.

 Litho in U.S.A.

FUEL PUMPS

The more basic fuel systems depend on gravity or air pressure to get fuel from the tank to the carburetor.

However, for many years, the fuel pump has been widely used on cars, trucks, buses, tractors, stationary, marine and aircraft engines.

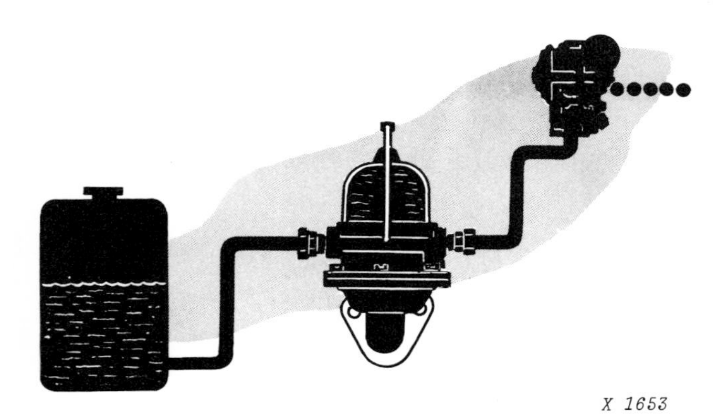

X 1653

Fig. 5 — The Fuel Pump Draws Fuel From The Tank To The Carburetor

The fuel pump automatically supplies the needed fuel from the tank to the carburetor (Fig. 5).

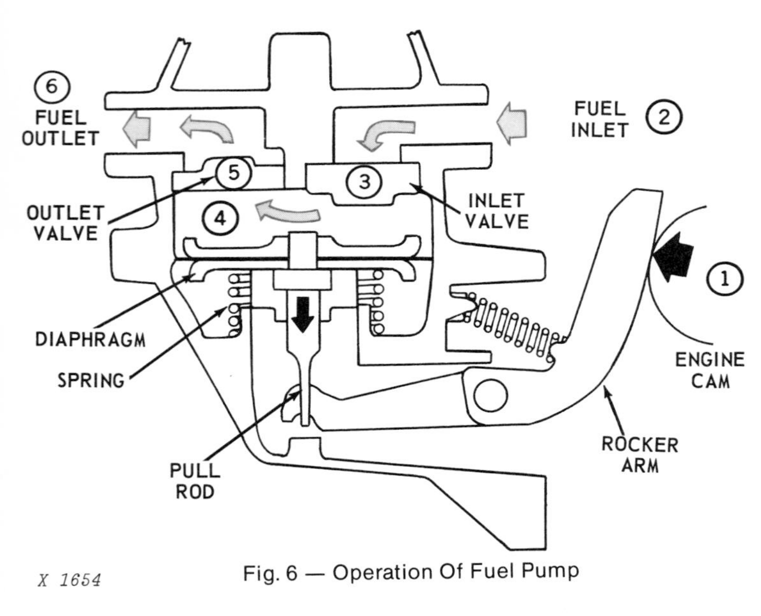

X 1654 Fig. 6 — Operation Of Fuel Pump

The fuel pump operates as shown in Fig. 6.

Power is applied to the pump rocker arm at (1) by an eccentric on the engine camshaft. As the camshaft rotates, the eccentric causes the rocker arm to rock back and forth. The inner end of the rocker arm is linked to a flexible diaphragm located between the upper and lower pump housing. As the rocker arm rocks, it pulls the diaphragm down, then releases it. A spring located under the diaphragm forces it back up. Thus the diaphragm moves up and down as the rocker arm rocks.

When the diaphragm is pulled down, a low vacuum or low pressure area is created above the diaphragm. This causes atmospheric pressure in the fuel tank to force fuel into the pump at (2). The inlet valve (3) opens to admit fuel into the center chamber (4).

When the diaphragm is released, the spring forces it back up, causing pressure in the area above the diaphragm. This pressure closes the inlet valve and opens the outlet valve (5), forcing fuel from the pump through the outlet (6) to the carburetor.

If the needle valve in the float bowl of the carburetor closes the inlet so that no fuel can enter the carburetor, the fuel pump can no longer deliver fuel.

In this case, the rocker arm continues to rock but the diaphragm remains at its lower limit of travel so the spring cannot force the diaphragm up. Normal operation of the pump resumes as soon as the needle valve in the float bowl opens the inlet valve, allowing the spring to force the diaphragm up.

COMBINATION PUMP (Fuel and Vacuum)

These pumps contain not only a fuel pump but also a vacuum pump. Both fuel and vacuum sections of the combination pump are actuated by a single rocker arm. It has a pair of valves and a spring-loaded diaphragm. However, it pumps air instead of fuel, thus creating a vacuum for operating an accessory such as a windshield wiper or a vacuum brake.

ELECTRIC FUEL PUMP

This type of fuel pump is used where a mechanical drive is not practical. It contains flexible metal bellows operated by an electromagnet (Fig. 7). When the electromagnet is connected to the battery (by turning on the ignition switch), it pulls down the armature which extends the bellows. This action creates a vacuum in the bellows and fuel from the fuel tank enters the bellows through the inlet valve.

When the armature reaches the lower limit of travel, it opens a set of contact points. This disconnects the electromagnet from the battery and thus allows the spring to push the armature upward and collapse the bellows. This in turn forces the fuel from the bellows through the outlet valve to the carburetor.

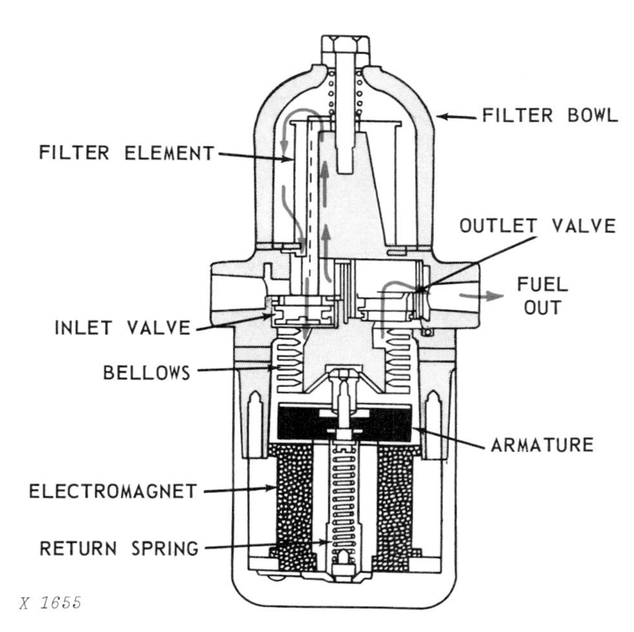

Fig. 7 — Cutaway View Of Electric Fuel Pump

When the armature reaches the upper limit of its travel, the contact points are closed. This energizes the electromagnet, causing it to again pull the armature down, starting the cycle again.

SERVICING FUEL PUMPS

Testing The Fuel Pump

A fuel pump analyzer can be used to test the fuel pump for delivery and pressure. Usually, however, the pump is visually checked for defects.

For a visual test, disconnect the pump-to-filter line at the filter.

Set the throttle so that the engine will not start and turn the engine over several times.

If fuel spurts from the line, the pump is operating properly.

If little or no fuel flows, check the following items:

- *Primer lever left in upward position*
- *Leaking sediment bowl gasket*
- *Plugged screen inside sediment bowl*
- *Loose or damaged connections*
- *Clogged fuel lines*
- *Loose cover screws on the pump*

If the problem is not within these areas, repair or replace the pump.

Repairing The Fuel Pump

Disassemble the pump as outlined in the machine Technical Manual.

Inspect the pump parts as follows:

1) Look for diaphragm punctures or leaks. Check the slot in the diaphragm pull rod for wear.

2) Examine the cover and body assembly for cracked or warped gasket surfaces.

3) Examine the valve and cage assemblies for worn valves or broken springs.

4) Check the diaphragm and rocker arm spring for sufficient tension.

5) Inspect the rocker arm link and pin for worn holes and other wear or damage.

6) Inspect the filter screen for punctures and clogging.

Reassembly of Fuel Pump

Reassemble the pump as instructed in the Technical Manual. Replace all defective parts with new parts from the repair kit, or replace the fuel pump.

FUEL FILTERS

Contamination of fuel is a major cause of excessive engine wear and failures.

Some engines have a separate filter bowl to clean the fuel before it enters the fuel pump. The purpose of the separate filter is to trap water and any foreign objects that may contaminate the fuel system. These filters should be checked and drained periodically. Most of them have a drain plug which can be loosened; the fuel pump primer lever can be actuated until all deposits are drained out.

If the filter has a sediment bowl and screen, remove and clean them periodically.

CARBURETORS

Engines will not run on "liquid" gasoline. The gasoline must be vaporized and mixed with air for all types of conditions. For example:

- **Cold Or Hot Starting**
- **Idling**
- **Part Throttle**
- **Acceleration**
- **High Speed Operation**

 Litho in U.S.A.

By mixing fuel with air for each of these conditions, the **carburetor** regulates the combustion and so the power of the engine.

To get the right fuel-air mix, the carburetor must "atomize" the fuel and mix the fine particles of fuel with air.

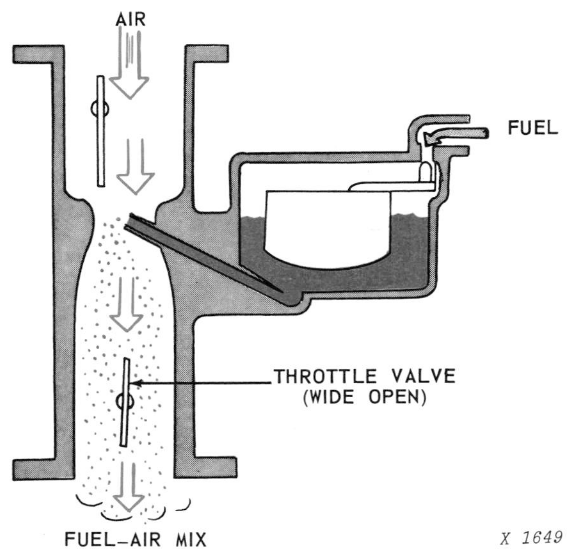

Fig. 8 — Basic Carburetor

Fig. 10 —

Atomizing is done by adding air to the liquid fuel as it moves through the carburetor passages and then spraying this fuel air mixture through nozzles or jets into a stream of moving air flowing into the engine's intake manifold. See Fig. 8.

The fuel in the fuel-air mixture is then vaporized before it enters the combustion chamber of the engine.

The various speed and load conditions demand a different volume of air for the mixture. The fuel-air ratio must be kept within flammable limits to permit combustion.

The modern gasoline engine works best when about 15 parts of air are mixed with 1 part fuel.

The primary job of the carburetor is to produce this ratio or fuel-air mix for any operating condition.

THEORY OF PRESSURE DIFFERENTIAL

Since the carburetor operates by pressure differentials, let's look at these terms:

- **Vacuum**
- **Atmospheric Pressure**
- **Venturi Principle**

VACUUM

Absolute vacuum is any closed area completely free of air or atmospheric pressure. We can better understand the carburetor by calling any pressure less than atmopheric a *vacuum* or low pressure area.

ATMOSPHERIC PRESSURE

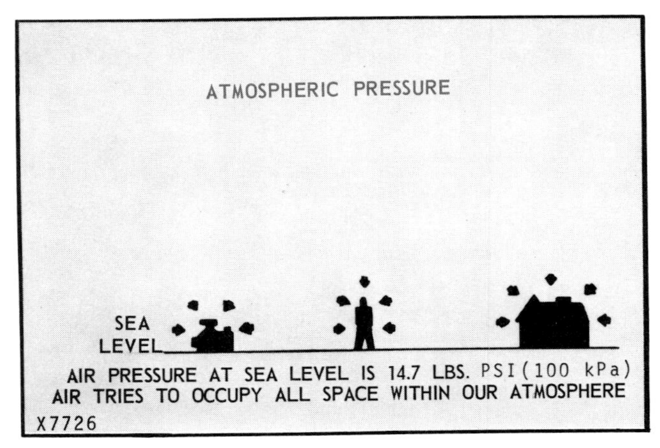

Fig. 9 — Atmospheric Pressure

Atmospheric pressure is the weight or pressure of the air around us (Fig. 9). Air pressure at sea level is 14.7 pounds per square inch (100 kPa). This air tries constantly to occupy all space within our atmosphere.

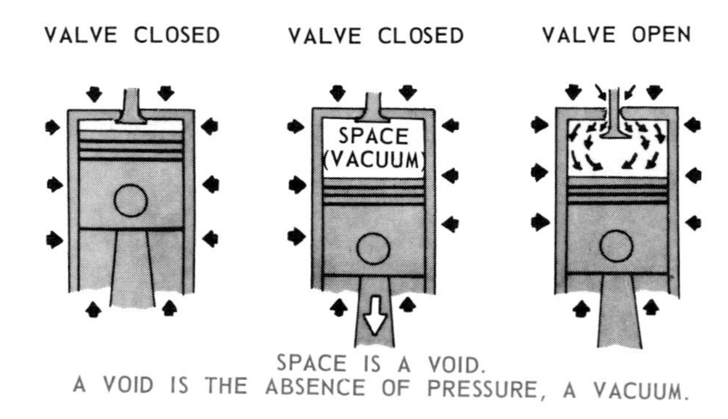

Fig. 10 — The Engine Creates a Vacuum

Piston movement in an engine creates a vacuum or low pressure area. Atmospheric pressure forces air to flow into this vacuum. See Fig. 10.

VENTURI PRINCIPLE

Pressure differentials are basic to how a carburetor works. A **venturi** is used in the throat or bore of all carburetors to create this pressure differential. The

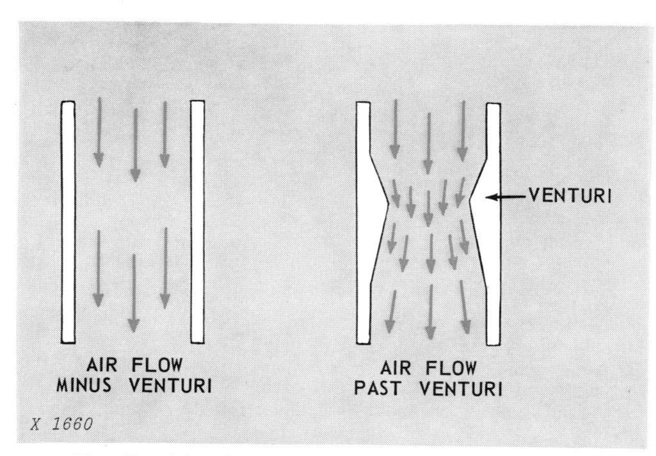

Fig. 11 — The Venturi Is A Restriction Of Air Flow Which Lowers Air Pressure

faster the air moves, the lower the air pressure at the venturi. This low pressure is the basic force by which a carburetor works (Fig. 11).

BASIC TYPES OF CARBURETORS

There are three basic types of carburetors:

- **Natural draft**
- **Updraft**
- **Downdraft**

In all three, there is a fuel supply in the *fuel bowl,* a passage or *air tube* through the carburetor for the stream of air going to the engine, and a *nozzle* connecting the bowl to the air tube. The *venturi* is also a feature of all three, creating the pressure drop by which the carburetor works.

Uses of The Basic Carburetors

The *natural draft* carburetor (Fig. 12) uses a cross-draft to supply the air flow and is used on engines where there is little space on top or where the atomized fuel in the mixture is vaporized by heat from the water in the engine water jacket.

The *updraft* carburetor (Fig. 13) can be placed low on the side of the engine and supplied fuel by gravity feed. However, the fuel must then be lifted up into the engine. This means that air velocities must be high, which can only be attained by using small passages in the carburetor and manifold.

These carburetors are best adapted for use on most modern farm and industrial machines.

The *downdraft* carburetor (Fig. 14) allows for larger volumes, since the fuel will reach the engine even though the air velocity is low. These carburetors can be used when high speeds and high power outputs are required.

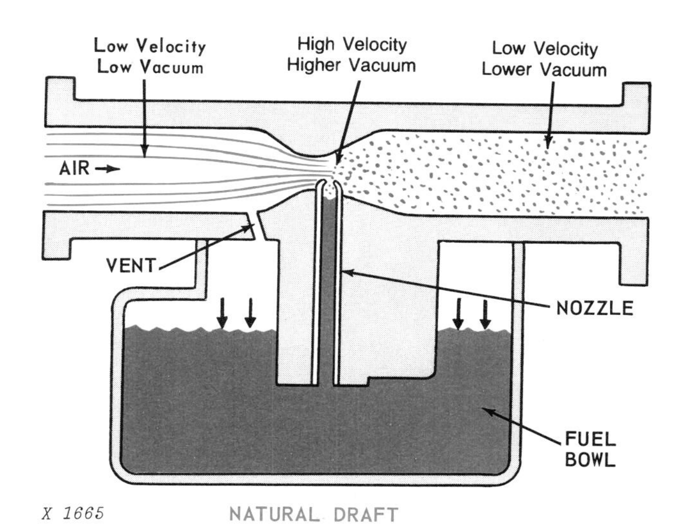

Fig. 12 — Natural Draft Carburetor

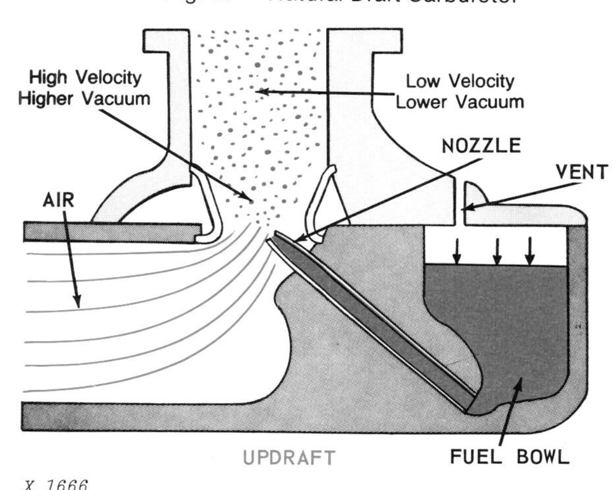

Fig. 13 — Updraft Carburetor

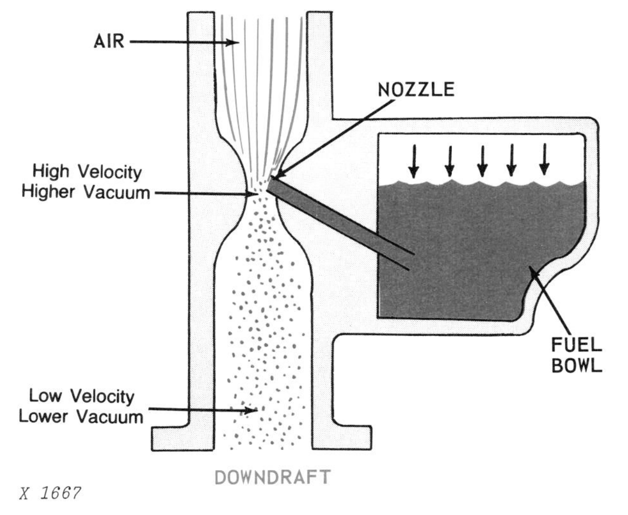

Fig. 14 — Downdraft Carburetor

Litho in U.S.A.

FUEL SUPPLY SYSTEMS

All carburetors require a source of fuel and a means of maintaining the proper level in the fuel bowl.

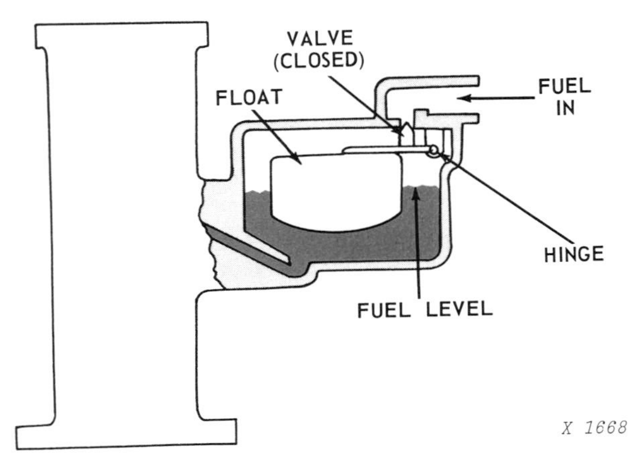

Fig. 15 — Fuel Supply System for Updraft Carburetor

Fuel enters the bowl under pressure (either gravity or pump) through a valve which is controlled by the float. As fuel rises in the bowl the float rises with it. When the correct fuel level is reached, the float shuts off the fuel supply by forcing the valve against its seat (Fig. 15).

As fuel is used by the carburetor the float lowers with the fuel supply, allowing the valve to move from its seat and admit more fuel.

CHOKE SYSTEMS

The choke system provides an extremely rich fuel mixture during starting, particularly in cold weather.

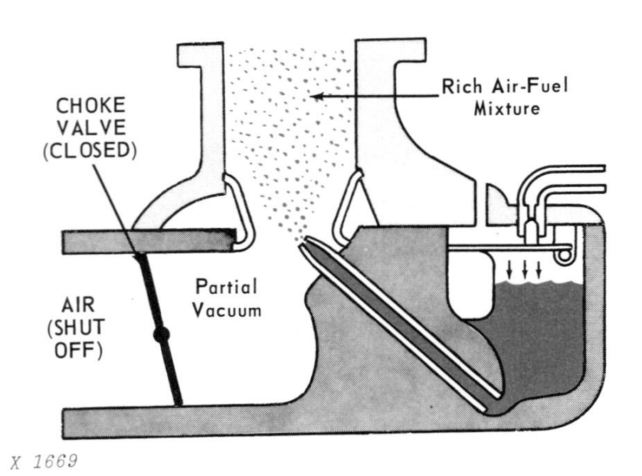

Fig. 16 — Choke System for Updraft Carburetor

To provide this rich mixture, a disk or choke valve is located in the air intake side of the carburetor tube (Fig. 16).

When the choke valve is closed, the vacuum created inside the intake manifold extends beyond the nozzle to the choke valve. Because of this vacuum, atmospheric pressure in the fuel bowl can force more fuel into the carburetor, resulting in a richer fuel-air mixture as shown.

Automatic Chokes

The operation of an automatic choke depends primarily on the unwinding of a thermostatic coil spring as heat is applied. Fig. 17 shows a typical model.

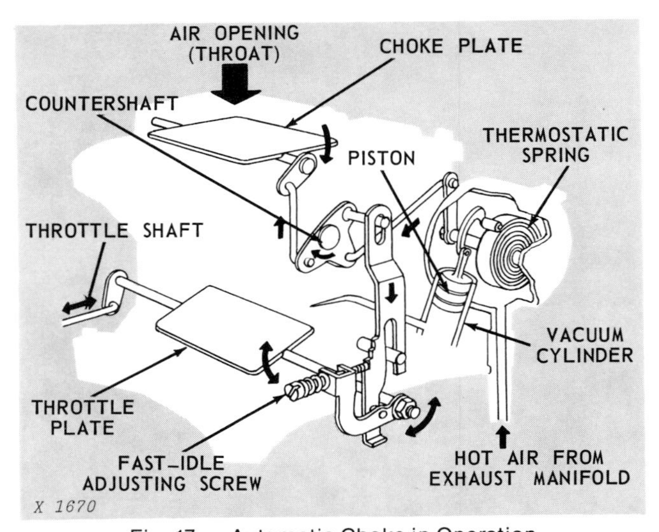

Fig. 17 — Automatic Choke in Operation

When the engine is cold, the thermostatic spring holds the choke valve closed.

When the engine starts, the intake manifold vacuum acts on the vacuum piston and partially opens the choke valve.

Hot air from the exhaust manifold heats the thermostatic spring. This heat expands the spring, allowing the choke valve to open wide.

THROTTLE SYSTEMS

The **throttle system** regulates the amount of fuel-air mixture entering the engine cylinders (Fig. 18).

This is required for two reasons:

1) To vary the engine speed.

2) To keep a uniform engine speed under varying loads.

On many engines, the throttle valve is connected by a linkage to a governor, which in turn is connected to a speed control lever. When the speed control lever is set to operate the engine at a given speed, the governor will maintain that speed (unless the engine is overloaded).

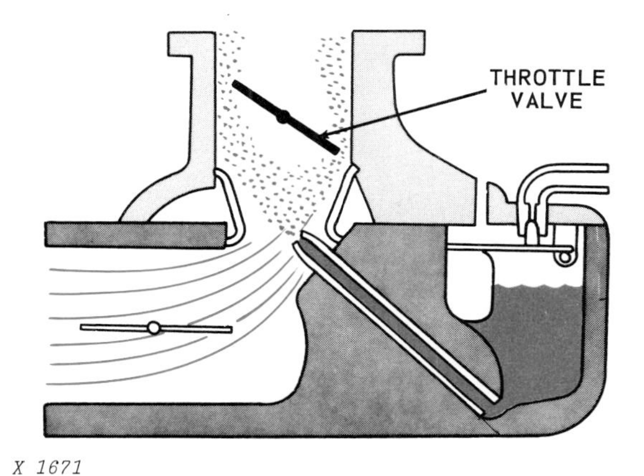

Fig. 18 — Throttle System for Updraft Carburetor

If the load is increased, the governor will automatically open the throttle valve wider, permitting more fuel-air mixture to enter the engine, thus maintaining a uniform speed.

Refer to Chapter 9 of this manual for details on "Governing Systems."

LOAD SYSTEMS

The load system delivers the proper fuel-air mixture to the engine in all ranges of speed and load above idling.

Fig. 19 illustrates a load system for an updraft carburetor. The amount of fuel entering the nozzle is regulated by a **load adjusting needle.**

In many carburetors, a **fixed jet or orifice** allows the proper amount of fuel for maximum power and economy to enter the nozzle.

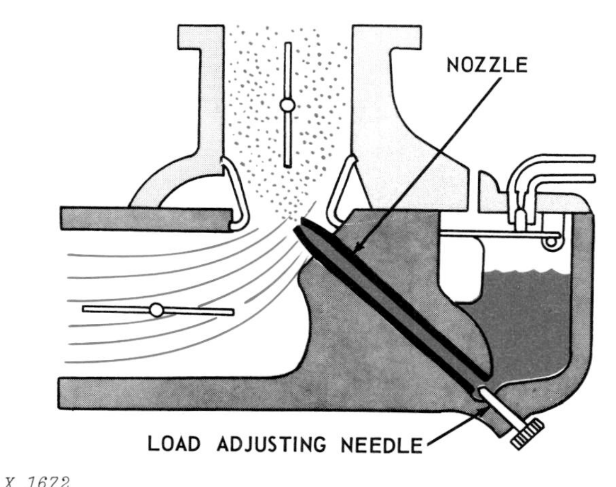

Fig. 19 — Load System for Updraft Carburetor

ACCELERATING SYSTEMS

Whenever the throttle is opened quickly to give extra power for a sudden load, extra fuel is required for a momentarily richer fuel-air mixture.

This extra fuel can be supplied in two ways:

1) An acceleration pump for rapid acceleration. This can be operated either by the throttle or by a vacuum.

2) An acceleration well or similar device for less rapid acceleration.

Acceleration Pump

The acceleration pump is often a simple piston-type pump (Fig. 20) which delivers fuel into the air-stream at the start of acceleration.

The pump is actuated either by the throttle or by vacuum. The throttle actuated pump is shown in Fig. 20.

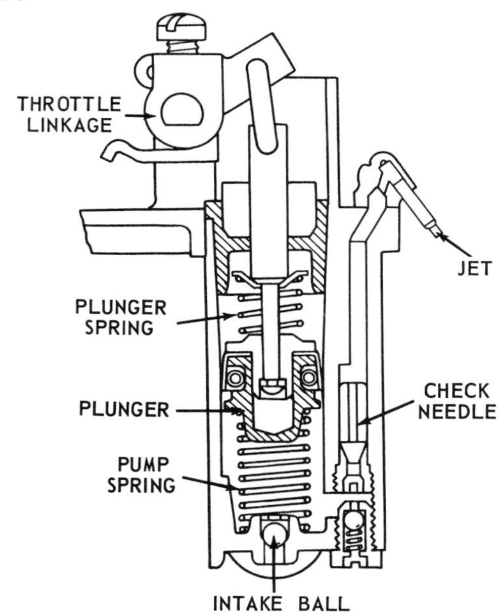

Fig. 20 — Acceleration Pump (Operated By Throttle)

When the throttle is opened, the plunger descends under the pressure of the plunger spring and fuel is forced out through the accelerating jet. When the throttle is fully depressed and at rest, the spring forces the plunger to the bottom of the well, thus continuing the flow of the accelerating charge.

Acceleration Well

When the engine is idling, fuel rises inside the load nozzle and passes through holes in the side of the nozzle into a chamber or well surrounding the nozzle.

When the throttle is suddenly opened, the fuel stored in the accelerating well (Fig. 21) pushes through the holes in the side of the nozzle without being metered by the adjusting needle and combines with the normal flow in the nozzle.

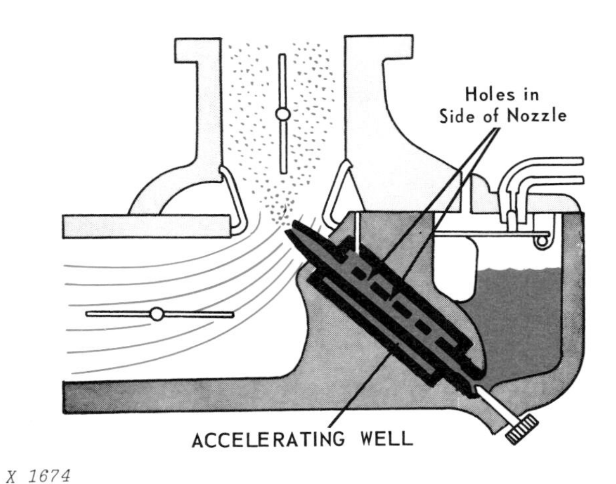

Fig. 21 — Acceleration Well On Updraft Carburetor

The two quantities of fuel enter the airstream, making a much richer fuel-air mixture to satisfy the sudden need for more power.

As the fuel supply drops in the accelerating well, the holes that are uncovered become *air* bleeds.

The accelerating well then remains drained of fuel until the throttle returns to one-fourth load or to fast-idle, at which time it refills.

IDLING SYSTEMS

When the throttle valve is closed, the idling system supplies just enough fuel-air mixture to keep the engine running at slow idle.

The idling system is different for updraft and downdraft carburetors.

Idling System for Updraft Carburetors

In the updraft carburetor shown in Fig. 22, air for the idle system goes around the venturi and enters the fuel system through a drilled passage (see arrows). The idle adjusting needle regulates the amount of air used.

Fuel is forced through drilled passages from the accelerating well to be mixed with air and delivered to the engine through the primary idle orifice. The operation of the primary and secondary orifices is the same as in the natural draft carburetor.

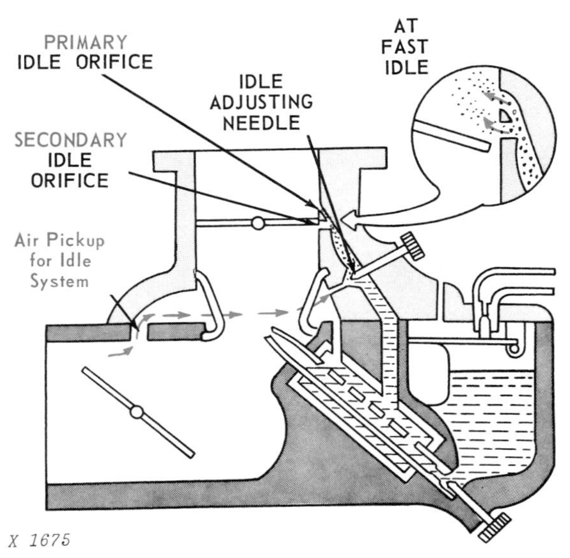

Fig. 22 — Idling System for Updraft Carburetor

Idling System for Downdraft Carburetors

In the downdraft carburetor (not shown), air for the idle system enters a small passage above the venturi. Atmospheric pressure pushes air and fuel through the passages. Then they mix and flow past the tapered point of an idle adjusting screw. The fuel-air mixture is very rich but leans out somewhat as it mixes with the small amount of air that gets past the closed throttle valve. The idle adjusting screw regulates the amount of fuel-air mix used.

The operation of the primary and secondary orifices is the same as for the natural draft and updraft carburetors.

ECONOMIZER SYSTEMS

The purpose of the economizer system is to retard the flow of fuel to the engine at part throttle when the full capacity of the nozzle is not required. The basic operation is the same for all three types of carburetors.

The passage providing air for the bowl vent is extended to a point near the throttle valve. When the throttle valve is partially open, the passage is on the engine side of the throttle. Action of the engine draws air through the passage, reducing the air pressure in the bowl.

When the bowl pressure is reduced, the **difference** between the pressure in the venturi and on the fuel in the bowl is also reduced. This retards the flow of fuel out the nozzle.

In the updraft carburetor (Fig. 23), an economizer jet in the passage regulates the air pressure at part throttle.

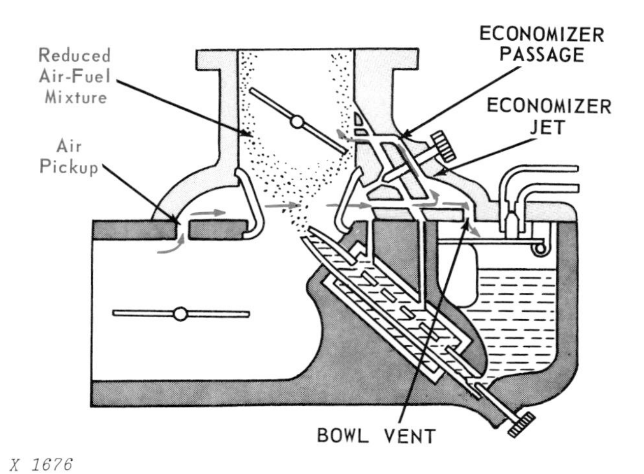

Fig. 23 — Economizer System for Updraft Carburetor

In some updraft carburetors, the economizer passage is brought around the throttle shaft so that when the throttle valve is closed or partially opened, a slot in the shaft permits air to enter the engine.

As the throttle valve is opened, the shaft rotates until its slot closes the passage completely at full throttle and no air can be drawn out of the fuel bowl.

SUMMARY: OPERATION OF CARBURETORS

The chart below gives a summary of the job of each major system in the carburetor.

OPERATION OF CARBURETOR SYSTEMS

System	*Function*
1. **Float**	Controls flow of fuel from tank to carburetor to maintain a constant level of fuel in carburetor.
2. **Choke**	Creates a rich fuel mixture for starting a cold engine.
3. **Throttle**	Allows engine to accelerate smoothly for idle to high speeds.
4. **Load**	Allows engine to operate at its maximum power output.
5. **Idle**	Allows engine to operate economically at idle speeds when power is not needed.

SERVICING OF CARBURETORS

Often, the carburetor is the first component to be blamed for bad engine operation.

An experienced serviceman will first check out other faults:

1) Faulty ignition.

2) Low engine compression.

3) Faulty supply of fuel or air to carburetor.

4) Worn or badly adjusted governor linkage.

5) A faulty carburetor—checked last of all.

Proper carburetion can only be obtained if the pistons, rings, valves, gaskets, manifolds, camshaft, combustion chambers, air cleaner, fuel pump, governor, and ignition system are in good condition and are functioning properly.

Fig. 24 shows some of the things which can affect good carburetion.

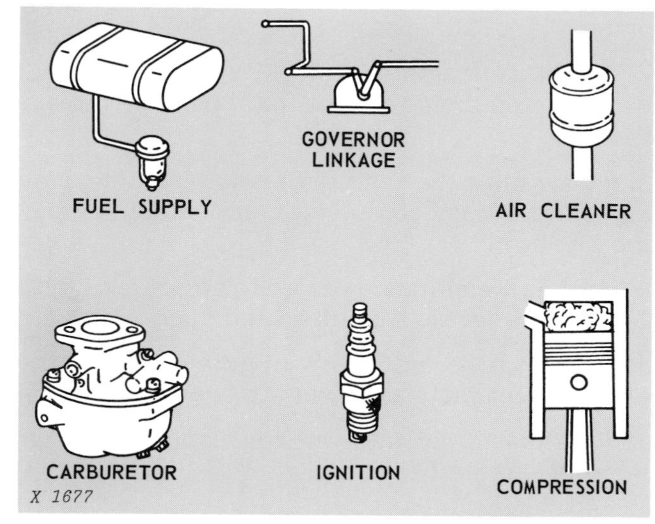

Fig. 24 — Things Which Affect Carburetion

MAINTENANCE TIPS AND PRECAUTIONS FOR CARBURETORS

Below are listed some items which will help you in servicing carburetors:

1. Always **service** the carburetor at these times:

a. After engine valve grinding or major engine overhaul

b. Every year or at the beginning of each season on seasonal machines. At this time, clean the carburetor, repack the shaft, check the bearings, and replace the seals and gaskets.

2. Always **adjust** the carburetor at the following times:

a. During engine tune-up

b. After major overhaul of the engine

c. Whenever the carburetor has been removed for service

d. Anytime the engine idles badly or requires speed adjustment

3. Repair kits are provided for many carburetors. When repairing, be sure to use ALL of the new parts in the kit.

4. Clean all parts thoroughly when repairing the carburetor. Use a carburetor cleaning solution for removing varnish-like deposits from all metal parts and rinse them in solvent.

5. Never use small wires or drills to clean out jets or orifices. This may enlarge or burr the precision bores and upset the performance of the carburetor.

6. Never use compressed air to clean a completely assembled carburetor. To do so may cause the metal float to collapse.

7. To test a metal float for leaks, immerse it in hot water. If air bubbles escape from the float, replace it.

8. Always drain the carburetor before any long storage period. Many carburetors have a drain plug in the fuel bowl.

9. Most carburetors are vented to avoid vapor lock. Be sure the vent is kept clean and open.

10. Never turn adjusting needles too tightly against their seats as you may damage them.

11. Use special tool kits, when available, to recondition the carburetor.

12. Be sure to tighten the screws which hold the throttle disk in place.

13. Always check the height of the float when assembling the carburetor (Fig. 25). See the machine Technical Manual for the correct height. If the float is badly bent or warped, replace it.

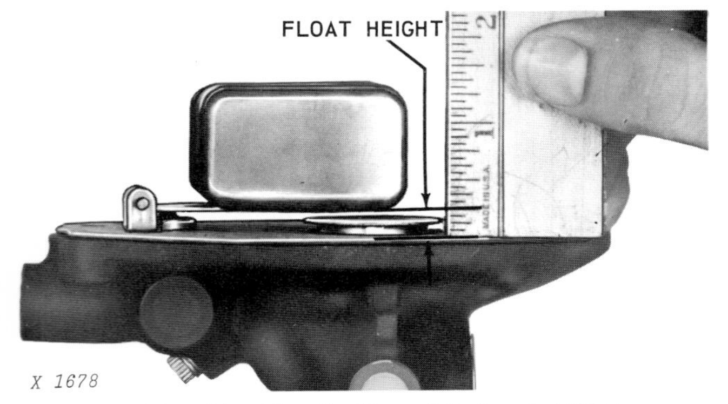

Fig. 25 — Checking Position of Carburetor Float

ADJUSTING THE CARBURETOR

Keep the carburetor adjusted for smooth, economical operation of the engine.

However, remember that engine speed adjustments and ignition timing both affect and are affected by carburetor settings. Don't change one without checking the other!

Most carburetors have three basic adjustments:

- **Idle Speed Adjustment**
- **Idle Fuel Adjustment**
- **Full-Load Fuel Adjustment**

Adjust the carburetor as follows:

1. Warm up the engine fully.

2. Set the *idle speed* (Fig. 26) with the engine throttle closed. Adjust the screw to get the slowest engine idling without stalling or "roughening" the engine. See the Technical Manual for the recommended idle rpm.

3. Adjust the *idle fuel* mixture for smooth engine idling. Turn the screw in until the engine begins to stall . . . then turn out the screw slightly until the engine idles smoothly.

4. If necessary, readjust the idle speed (step 2) after getting the correct idle fuel adjustment.

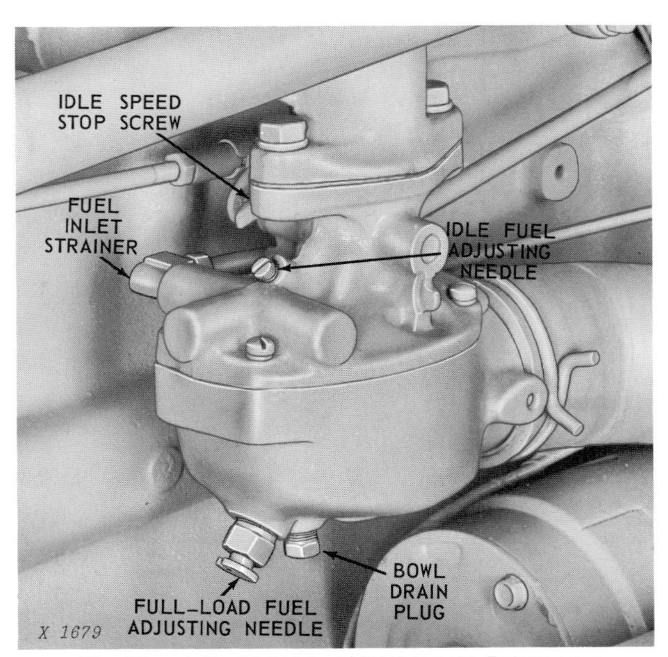

Fig. 26 — Adjustment and Maintenance Points on a Typical Carburetor

5. Adjust the *full-load fuel* mixture last for good fuel economy under load. For safety, adjust while the machine is under load—but is stationary. Adjust at fast idle by turning in the load screw until the engine starts to lose power . . . then back it out a little until the engine picks up speed and runs smoothly.

6. Check the operation of the speed control or governor mechanisms. If they are erratic or overspeed, adjust the linkage. See Chapter 9, "Governing Systems".

CARBURETOR TROUBLE SHOOTING CHARTS

While the basic causes of carburetor trouble will vary, the common difficulties and their respective causes are outlined below.

POOR PERFORMANCE

Possible causes of **poor performance** generally result from too lean a fuel-air mixture. If the carburetor is correctly adjusted, a lean mixture and poor performance may result from the following:

a. Air leaks at carburetor or intake manifold

b. Clogged engine air cleaner

c. Clogged fuel lines

d. Defective fuel pump

e. Low fuel level

f. Clogged fuel screen

g. Dirt in carburetor jets and passages

h. Worn or inoperative accelerating pump

i. Load needle valve, economizer or jet not operating

j. Damaged or wrong-size main metering jet

k. Worn idle needle valve and seat

l. Loose jets in carburetor

m. Defective gaskets in carburetor

n. Clogged mufflers, ignition and poor compression

o. Worn throttle shaft or bushings.

POOR IDLING

Poor idling is usually caused by a defective ignition system, leaking engine valves, or uneven engine compression. In the carburetor, check the following:

a. Incorrect adjustment of idle needle valve

b. Incorrect float level

c. Sticking float needle valve

d. Defective gaskets between carburetor and manifold

e. Defective gaskets in carburetor

f. Loose carburetor to manifold nuts

g. Loose manifold to cylinder block nuts

h. Idle discharge holes partly clogged

i. Loose jets in carburetor

j. Vacuum leaks which are partly compensated for by a rich idle adjustment

k. Worn main metering jet

l. Restricted or clogged air cleaner

HARD STARTING

In addition to the fuel system troubles listed below, **hard starting** may be caused by use of engine oil that is too heavy, defective ignition system, low compression, weak starting battery, defective starting motor, or excessive engine friction.

a. Incorrect choke adjustment

b. Defective choke

c. Incorrect float level

d. Incorrect fuel pump pressure

e. Sticking fuel inlet needle

f. Improper starting

POOR ACCELERATION

Poor acceleration may be caused by defective ignition system, excessive engine friction, lack of compression, and incorrect carburetion.

In the case of the carburetor, check the following:

a. Incorrect fuel level

b. Accelerator pump incorrectly adjusted

c. Accelerator pump not operating

d. Corroded or bad seat on accelerator bypass jet

e. Accelerator pump leather hard or worn

f. Clogged accelerator jets or passages

g. Defective ball checks in accelerator system

CARBURETOR FLOODS

The usual causes of **carburetor flooding** are:

a. Dirt on float valve seat

b. Sticking float valve

c. Leaking float

d. Fuel level too high

e. Defective gaskets in carburetor

f. Excessive fuel pump pressure

EXCESSIVE FUEL CONSUMPTION

There are many causes of **excessive fuel consumption** other than defective carburetion. Among the causes are: poor engine compression, excessive engine friction, clogged mufflers, defective ignition. In the carburetor and fuel system, check the following:

a. Poor adjustment of idle mixture

b. Wrong setting of load adjusting needle

c. Fuel leaks in carburetor or lines

d. Clogged air cleaner

e. Defective fuel economizer

f. Defective carburetor gaskets

g. Excessive fuel pressure

h. Sticking fuel inlet needle

TEST YOURSELF

QUESTIONS

1. What are the three basic components of a gasoline fuel system?

2. What is the purpose of a fuel pump?

3. What is the primary function of a carburetor?

4. Is the following statement true or false? "The greater the air velocity, the greater the air pressure."

5. Name the three basic types of carburetors.

6. What is the purpose of the choke in a carburetor?

7. What is the purpose of an acceleration pump in a carburetor?

8. Write the correct statement for question No. 5.

(Answers in back of text.)

 Litho in U.S.A.

LP-GAS FUEL SYSTEMS / CHAPTER 4

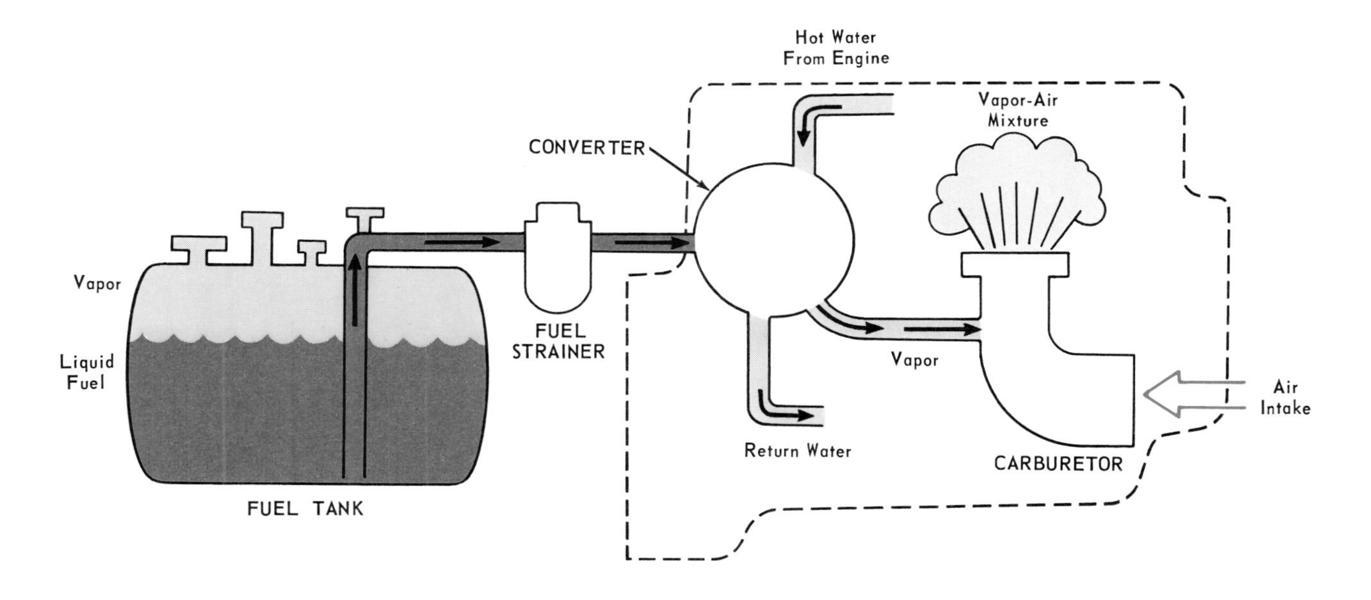

Fig. 1 — LP-Gas Fuel System

Liquefied petroleum (LP) gas vaporizes very easily. In fact it normally remains a liquid only when under pressure. Therefore it must be kept in strong heavy tanks.

LP-gas is vaporized *before* it reaches the carburetor, while gasoline remains a liquid until this point. This is the basic difference between the two systems.

HOW FUEL IS WITHDRAWN

To withdraw LP-gas fuel from the tank, two methods are used:

1) Liquid withdrawal

2) Vapor withdrawal

These two systems are compared in Chapter 1.

Most all modern systems use the *liquid withdrawal* method. The fuel is drawn from the tank under pressure as a liquid and then vaporized.

The *vapor withdrawal* method is normally used only for starting the engine. Here the vapor is drawn off the top of the tank and used as starting fuel. This is done because in a cold engine there is not enough heat to convert liquid fuel to vapor. Once the engine is started and warmed up, it can be switched to *liquid withdrawal* again.

PARTS OF LP-GAS FUEL SYSTEM

The LP-gas fuel system (Fig. 1) has four basic parts:

- **Pressurized Fuel Tank**
- **Fuel Strainer**
- **Converter**
- **Carburetor**

The PRESSURIZED FUEL TANK stores the liquid fuel under pressure. A space for vapor is left at the top of the tank.

 Litho in U.S.A.

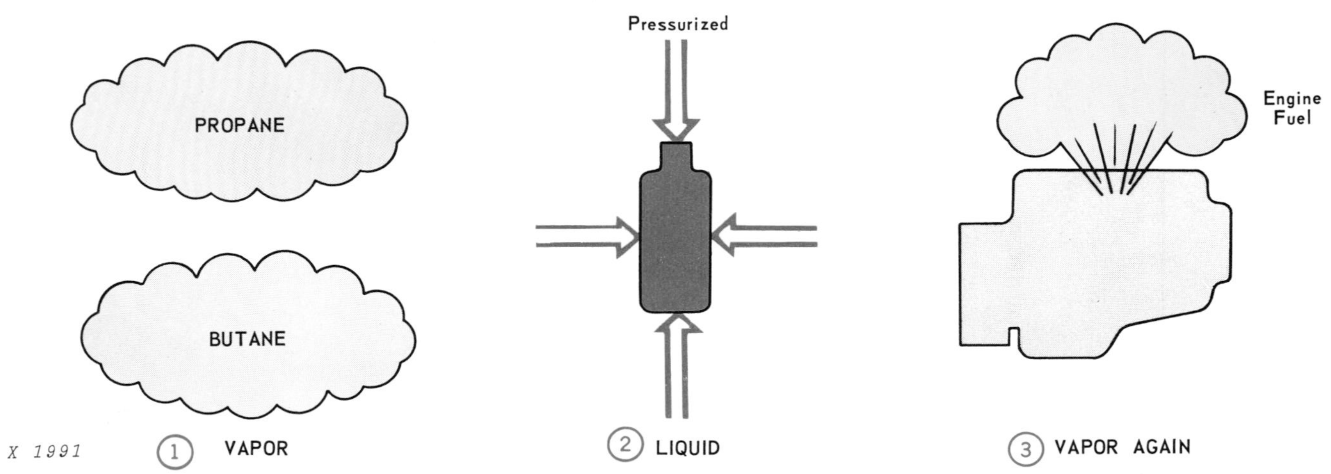

Fig. 2 — How LP-Gas Becomes Fuel For The Engine

The FUEL STRAINER cleans the liquid fuel. It normally has a solenoid which permits flow only when the engine ignition is turned on.

The CONVERTER changes the liquid fuel to vapor by warming it and then drops the pressure of the vapor.

The CARBURETOR mixes the fuel vapor with air in the proper ratio for the engine.

FACTS ABOUT LP-GAS FUEL

LP-gas is:

1. *A by-product of gasoline manufacture.*
2. *Also obtained from natural gas.*
3. *Made up mainly of propane and butane.*
4. *A vapor unless compressed or cooled.*
5. *Liquefied by compressing many gallons of vapor into one gallon of liquid.*
6. *Easier to handle and store as a liquid.*
7. *Expansive when heated (due to more vaporization).*
8. *Stored in strong tanks with pressure relief valves.*
9. *Converted to vapor again on way to engine.*

FUEL COMBUSTION

LP-gas burns slower than gasoline because it ignites at a higher temperature.

For this reason, the spark is often advanced farther on LP-gas engines.

More voltage at the spark plugs may be needed for LP-gas engines than for gasoline.

To solve this, "colder" plugs or smaller spark plug gaps may be recommended in the engine Technical Manual.

LP-gas engines do not require as much heat at the intake manifold as gasoline models. This is because LP-gas will vaporize at lower temperatures. The result: Less heat is wasted and more heat goes into engine power.

SAFETY IN HANDLING LP-GAS

LP-gas is not hazardous if handled safely.

However, there are special methods which must be used for LP-gas.

Be sure you understand LP-gas equipment before you handle it.

Many states or local areas have regulations for handling LP-gas. Check this with your local and state governments, or LP-gas distributor.

SAFETY RULES FOR LP-GAS

1. Remember first of all that **LP-gas is always under pressure.** It wants to get out of the tank, lines, and other components. They MUST be leak-proof.

2. Never loosen a line or fitting without first reducing the pressure.

3. Natural LP-gas cannot be smelled or seen. It is odorless and colorless—and therefore **dangerous.** However, commercial gases are generally odorized to help in detecting them.

4. LP-gas is heavier than air and will settle in low spots. These pockets are explosive until the leak is stopped and the area is ventilated.

5. Stay outdoors away from buildings when transferring fuel from storage to fuel tank. Some fuel always escapes when hoses are disconnected.

6. Never smoke or use any flame near the fuel system while filling or servicing it.

7. Never fill an LP-gas fuel tank more than 80% full.

8. LP-gas exhaust smoke can fool you—it doesn't smell bad. But it still contains some carbon monoxide. Be sure to ventilate the area before you start or operate an LP-gas engine.

9. When stopping the engine, always close both the vapor and liquid withdrawal valves. Leave the valves closed until the engine is to be started again.

10. Before taking LP-gas equipment indoors, check the state or local regulations. For your own safety, follow the rules.

11. LP-gas expands when heated. If the machine must be taken indoors for service in cold weather, be sure the tank is as near empty as possible. Otherwise, the fuel in the tank may expand until the safety valve opens.

12. Ventilation is a must when LP-gas equipment is indoors. Open doors and windows as wide as possible. Or use blowers or fans placed near the floor where vapors might collect if a leak developed. Protect against sparks or fires by using explosion-proof motors, switches, and conduits.

FUEL TANK, VALVES, AND GAUGES

The LP-gas fuel tank must be strong and heavy to withstand the pressure of its fuel.

The fuel tank is never filled completely full of liquid fuel because room must be left for vapors and expansion.

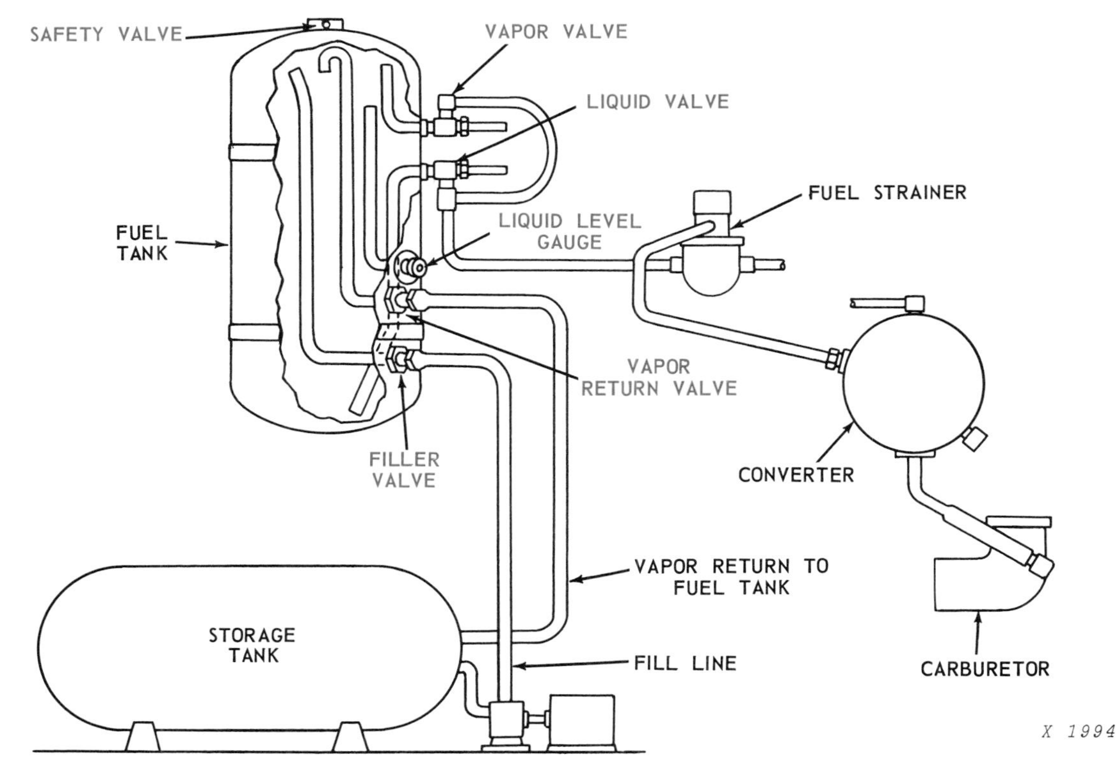

Fig. 3 — Valves And Gauges In LP-Gas Fuel System

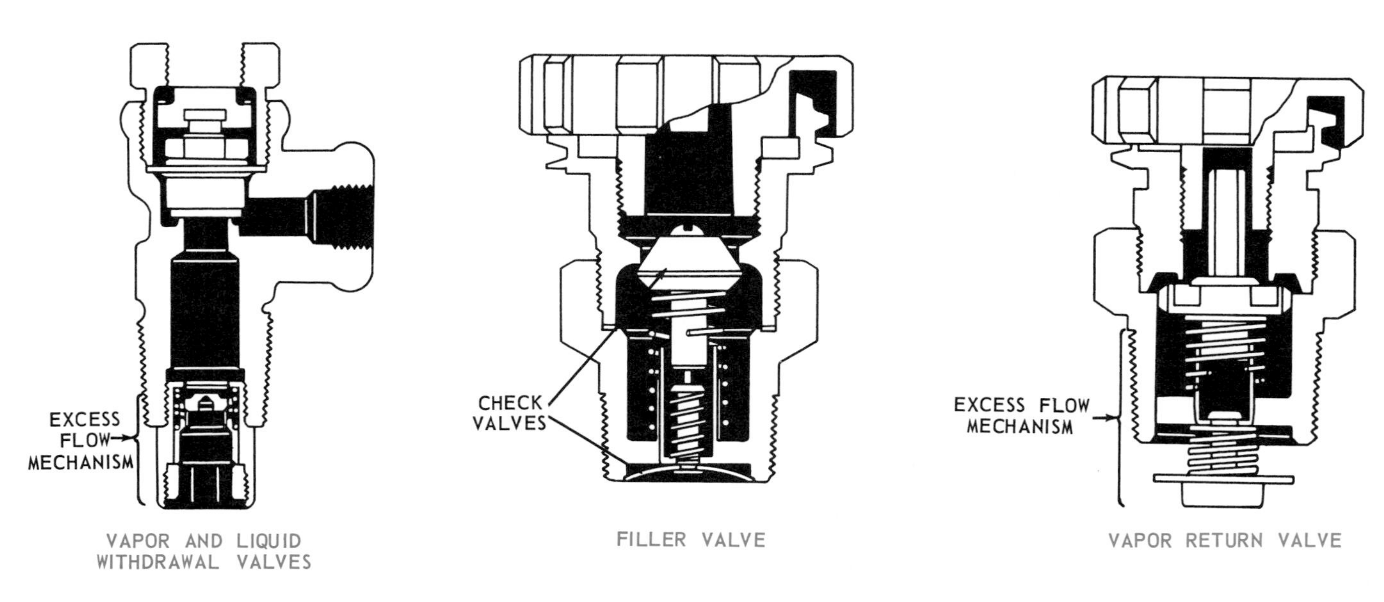

Fig. 4 — LP-Gas Fuel Tank Valves

The fuel tank also has these valves and gauges:

- **Liquid Withdrawal Valve**
- **Vapor Withdrawal Valve**
- **Safety Relief Valve**
- **Liquid Level Gauge**
- **Filler Valve**
- **Vapor Return Valve**
- **Fuel Gauge**

LIQUID WITHDRAWAL VALVE pulls liquid fuel from the bottom of the tank for normal engine operation.

VAPOR WITHDRAWAL VALVE pulls vapor from the top of the tank for starting the engine.

Both of these valves have safety devices called excess flow mechanisms (see at left in Fig. 4). The valves automatically shut off when too much fuel is flowing out. If a leak develops in the system, the valve instantly closes and permits only a small amount of gas to flow.

SAFETY RELIEF VALVE will open if pressure in the tank becomes too high.

LIQUID LEVEL GAUGE tells when the tank is 80% full during filling of the tank. This assures that space is left for vapors. By opening the gauge, the level can be checked. Before the 80% fill, vapor will issue from the gauge. When the 80% fill is reached, a spray or liquid will issue. Caution must be taken, however, not to leave the gauge open during filling.

FILLER VALVE (Fig. 4) has a double check valve which prevents fuel from escaping when the filler hose is disconnected or accidentally breaks.

VAPOR RETURN VALVE (Fig. 4) permits vapor to return to the storage tank as the fuel tank is filled. This equalizes the pressures and makes it easier to fill the tank. An excess flow mechanism (shown) automatically closes if flow through the vapor return line is too much. This protects against escape of gas if the return line is broken or detached.

FUEL GAUGE tells the level of fuel in the tank. (For details, see below.)

FUEL GAUGES

Two types of fuel gauges are used for LP-gas fuel tanks:

- **Magnetic**
- **Electrical**

MAGNETIC fuel gauges (Fig. 5) use a float in the tank attached by a pivoting shaft to a magnet. As the float rises and falls with the fuel level, it rotates the magnet, which operates another magnet on the gauge. This moves the point to indicate the amount of fuel on the face of the gauge.

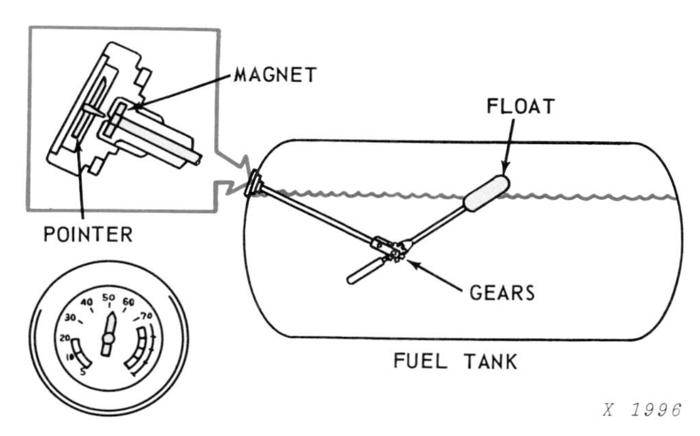

Fig. 5 — Magnetic Fuel Gauge

ELECTRICAL fuel gauges are similar to gasoline gauges. Details are given in Chapter 5. A sender located on the fuel tank signals a receiver (Fig. 6) on the machine instrument panel.

Fig. 6 — Electrical Fuel Gauge (Receiver Shown)

AUXILIARY FUEL CONNECTION

This connection allows a portable fuel tank to be attached for moving the machine without filling the main fuel tank. The connection is usually made at the fuel strainer.

SERVICING FUEL TANK, VALVES, AND GAUGES

Be sure that you are familiar with all rules and regulations before servicing LP-gas equipment.

Two rules are important here:

1) never service the fuel tank except to remove and install valves or gauges.

2) Never service the valves or gauges except to remove and install them or to replace caps or dust covers.

Replacing Parts

If a valve or gauge is damaged or worn and must be replaced, first empty all fuel from the tank to remove the pressure. If possible, *do this outdoors by running the engine until all fuel is exhausted.* If not possible, consult your local LP-gas distributor for the approved method in your locality for emptying the tank.

When replacing a valve, use a small amount of Permatex or similar sealer on the threads. Install the valve and tighten securely. Test for leaks by using soapy water. **Never use an open flame.**

For other information, refer to the engine Technical Manual.

FUEL STRAINER

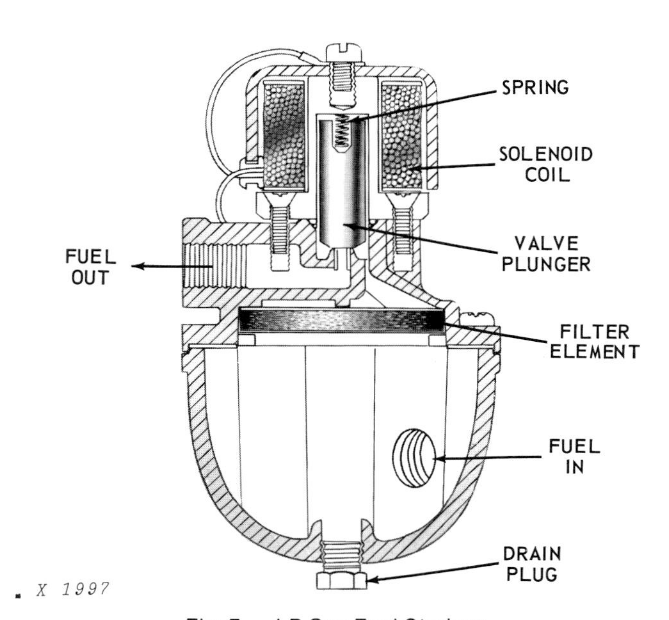

Fig. 7 — LP-Gas Fuel Strainer

The fuel strainer (Fig. 7) is located between the fuel tank and the converter.

The fuel strainer has two functions:

1) Strains fuel to clean it.

2) Shuts off fuel when system is not operating.

The SHUT-OFF is an electrical solenoid coil (Fig. 7). When the engine ignition is turned on, the solenoid magnetizes the valve plunger and opens it, allowing fuel to flow. When the ignition is turned off or fails, the valve closes automatically. Pressure of fuel holds it tightly on its seat.

SERVICING THE FUEL STRAINER

If the fuel strainer frosts up, its filter element is probably clogged and needs cleaning. *Before attempting to clean the strainer, make sure both withdrawal valves are closed, the engine is cold, and the lines and strainer are emptied of gas.*

 Litho in U.S.A.

To clean the strainer, first remove the drain plug (Fig. 7) and drain out foreign matter. Follow instructions on blowing out the strainer.

To clean the strainer parts, see the engine Technical Manual.

If the solenoid fails, it may show up as follows: When the ignition is turned on, gas fails to flow to the converter. In this case, the valve plunger may be stuck or restricted.

CONVERTER

The converter does two things:

- **Heats the liquid fuel to vaporize it**
- **Reduces pressure of the vapor**

The liquid fuel must be converted to vapor before the engine can use it. The pressure must also be cut so the carburetor can handle it.

The converter has two chambers separated by a wall (See Fig. 8).

The *heat exchanger chamber* is divided into two spiral passages—one for fuel, the other for engine coolant.

The *low-pressure chamber* contains fuel vapor under low pressure.

HEAT EXCHANGER CHAMBER

Liquid fuel under pressure enters the converter through the *high-pressure valve* (Fig. 8) and passes into the spirals of the heat exchanger where it rapidly changes to a vapor.

As the fuel vaporizes, pressure builds up and pushes on the flexible *diaphragm* (Fig. 8). This diaphragm is connected to the *high-pressure valve* as shown. When pressure reaches a preset point, it overcomes the spring pressure and closes the valve. This seals the fuel vapor in the heat exchanger.

LOW-PRESSURE CHAMBER

The second chamber of the converter is connected to the heat exchanger by a *low-pressure valve* (Fig. 9). This valve is held closed until opened by its *lever.*

The lever is actuated by a large flexible diaphragm as shown.

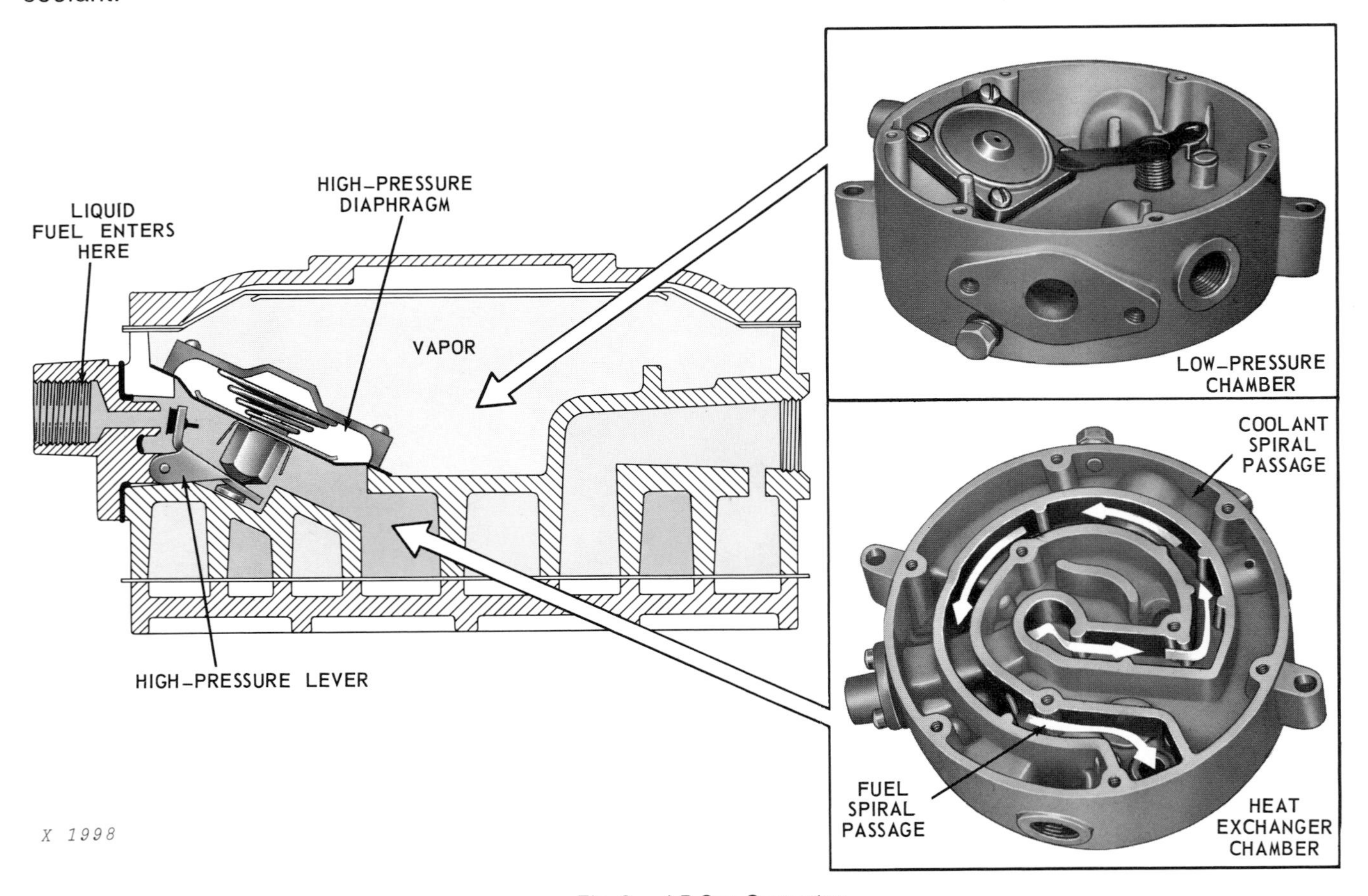

Fig. 8 — LP-Gas Converter

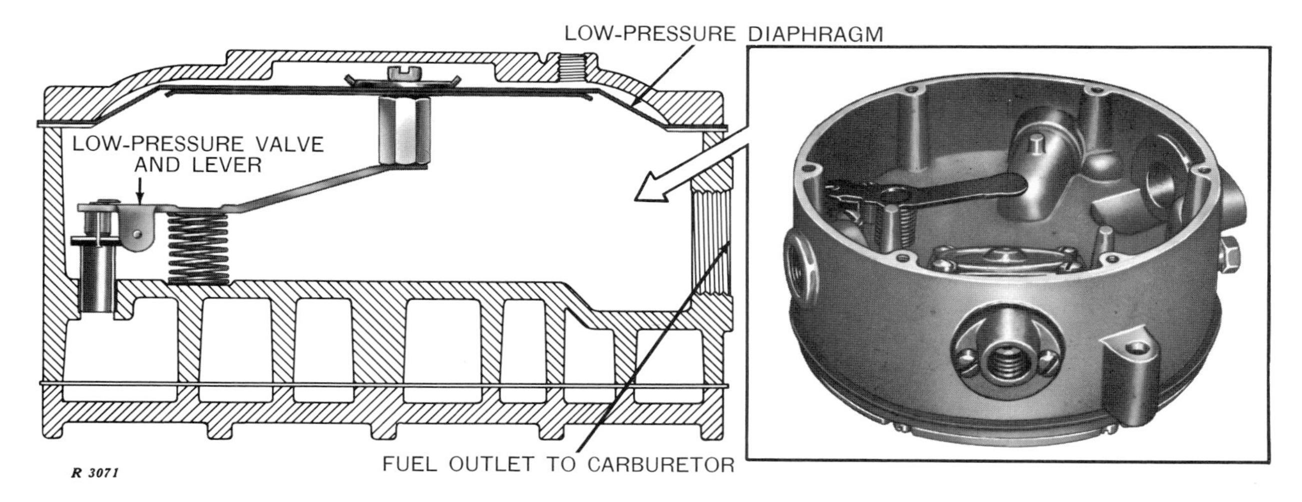

Fig. 9 — Low-Pressure Chamber of Converter

Operation is as follows:

When the engine piston is on the intake stroke, it draws gas (through the carburetor) from the low-pressure chamber, reducing the pressure beneath the large diaphragm to slightly below atmospheric pressure.

Since the other side of the diaphragm is vented to atmospheric pressure, the diaphragm is pushed down. As it moves downward, it actuates the low-pressure valve lever, opening the low-pressure valve and permitting gas from the heat exchanger to enter the low-pressure chamber. From here the gas goes to the carburetor and then into the engine cylinder.

When the engine intake valve closes and the "demand" for gas momentarily ceases, pressure on both sides of the large diaphragm becomes equal, and the spring again closes the low-pressure valve.

Since some gas has been drawn from the heat exchanger, pressure in the chamber is reduced, permitting the high-pressure valve spring to push the high-pressure diaphragm down and open the high-pressure valve.

Liquid fuel again flows into the heat exchanger, where it vaporizes, and again builds pressure, closing the valve.

PRECAUTION AGAINST FREEZING OF CONVERTER

If the converter should ever freeze up, an expansion device is built into the back cover (Fig. 10).

When the coolant freezes, it expands out into spirals in the back cover, pushing the flexible gasket with it. This avoids damage to the converter.

A drain plug at the bottom of the converter allows coolant to be drained out to prevent freezing. If the cooling system is drained, also drain the converter. Also disconnect the water inlet and outlet lines.

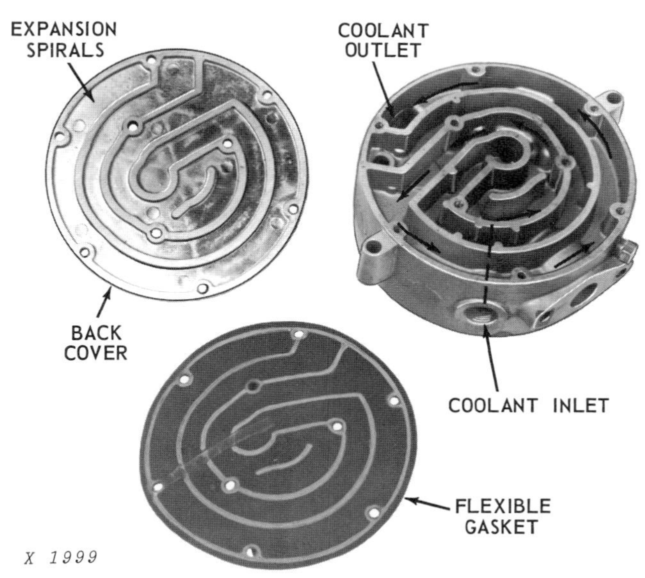

Fig. 10 — Expansion Device To Protect Converter If It Freezes Up

PROBLEMS OF ENGINE AIR PULSATIONS AFFECTING CONVERTER

In some four-cylinder engines, the air pulsations from the cylinders will affect the low-pressure diaphragm in the converter.

If the diaphragm works in unison with the engine intake, the flow of fuel vapor from the converter will be erratic.

A regulator valve (Fig. 11) can be used to prevent this. The valve in effect helps the diaphragm to "breathe" normally.

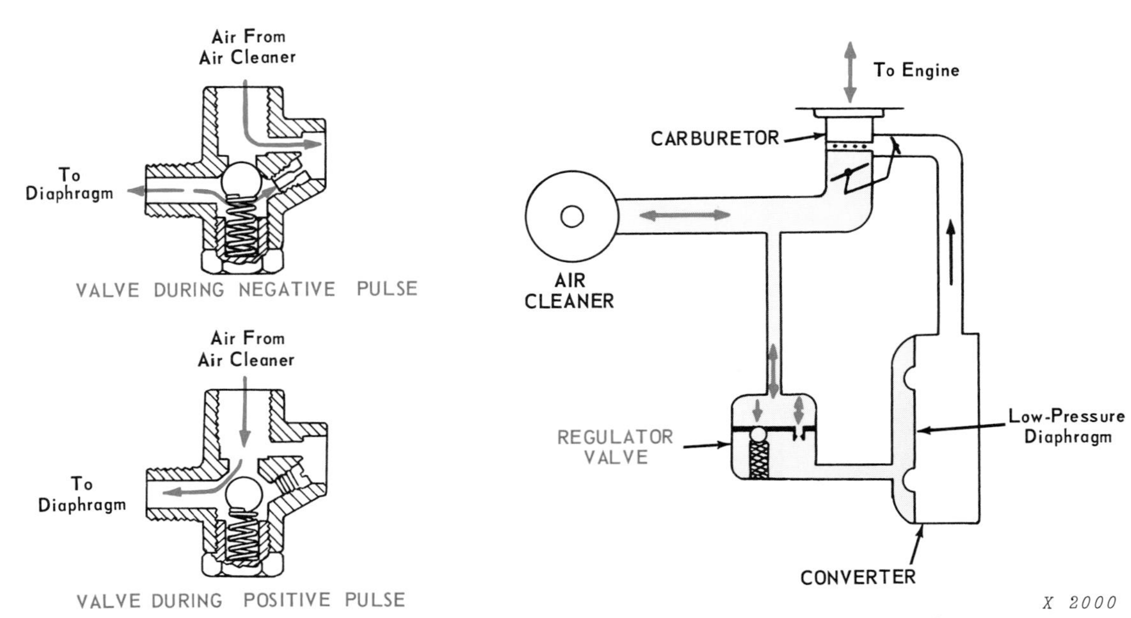

Fig. 11 — Regulator Valve For Stabilizing Converter Against Engine Air Pulsations

During positive pulses, the valve moves to direct pressure to the atmospheric side of the diaphragm as shown. This overcomes the pulse from inside the diaphragm, keeping the low-pressure valve open for normal operation.

During negative pulses, the diaphragm is allowed to breathe at air inlet pressure through the orifice in the regulator valve as shown.

The regulator valve ball is operated by air pressure changes.

SERVICING THE CONVERTER

The converter is constructed so that it will seldom cause trouble, providing clean fuel is used in the tank and clean, soft water in the cooling system.

If the converter frosts up when the engine is cold, it is probably due to a leaking high-pressure valve caused by dirt under the valve seat.

If frosting occurs when the engine is hot it is probably due to poor coolant circulation through the heat exchanger or a restriction in the high-pressure valve.

For details on converter service, see the engine Technical Manual.

CARBURETORS

The LP-gas carburetor does much the same job as gasoline models (Chapter 3). The main difference is that when fuel enters the LP-gas model, it is a *vapor,* while in the gasoline it is a *liquid.* Figure 13 shows an updraft model, but downdraft models are also used.

The **metering valve** varies the amount of fuel vapor entering the carburetor.

The **spray bar** takes the fuel vapor and mixes it with air flow coming in from the bottom in Fig. 12.

A **load adjusting screw** (see Fig. 13) meters the fuel vapor at the spray bar at full throttle for the best engine power.

The **throttle disk** controls the incoming air and works with the metering valve to get the best fuel-air mixture for each load and speed.

OPERATION

Both the metering valve and throttle disk are operated by the engine throttle (Fig. 12).

At low idle speeds, they close down to limit the fuel-air mixture fed to the engine (see Fig. 12).

At fast idle speeds, the valve and disk open up to feed more fuel-air to the engine (see Fig. 12).

Spray Bar Operation

Incoming air flows through the carburetor bore and hits a restriction—the spray bar. This deflects the air and accelerates it, creating a pressure drop (similar to the venturi in gasoline carburetors).

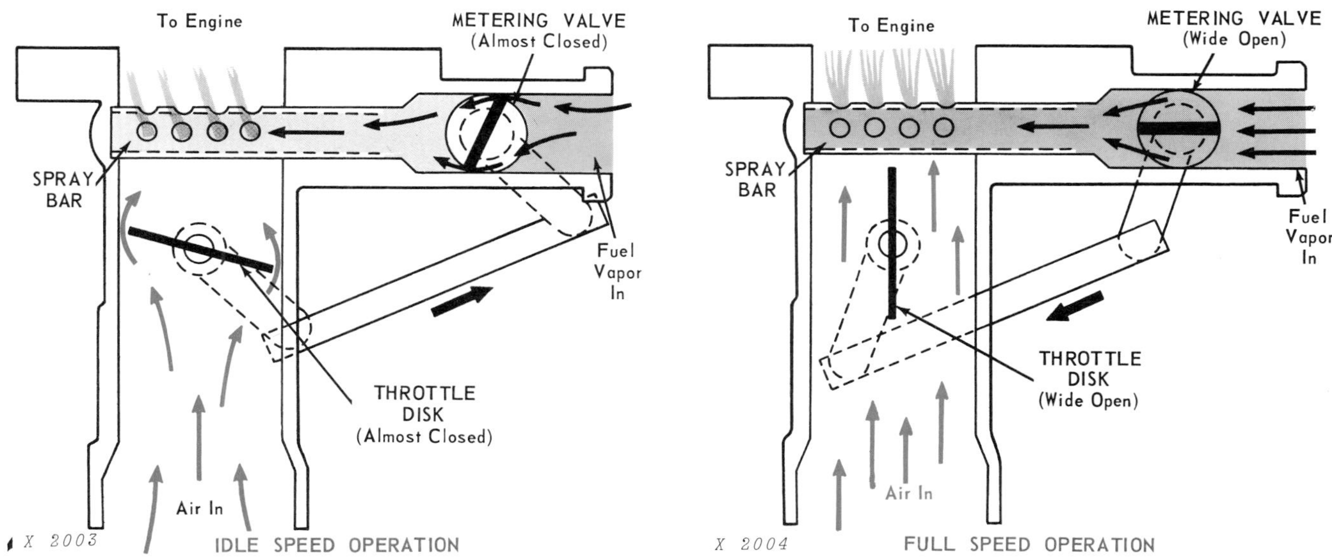

Fig. 12 — Metering Valve And Throttle Disk Operate Together

CARBURETOR MAINTENANCE

Refer to Chapter 3, page 3-11, for a list of "Maintenance Tips and Precautions for Carburetors."

LP-gas carburetors do not use floats, but most of the other rules apply.

CARBURETOR ADJUSTMENTS

The carburetor must be adjusted for smooth, economical operation of the engine.

However, remember that engine speed adjustments and ignition timing both affect and are affected by carburetor adjustments. Don't change one without checking the other!

Most LP-gas carburetors have these basic adjustments:

- **Idle Speed Adjustment**

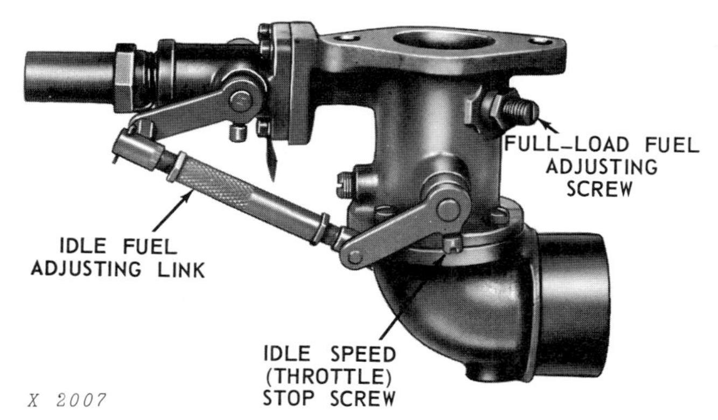

Fig. 13 — Adjusting Points On Typical LP-Gas Carburetor

- **Idle Fuel Adjustment**
- **Full-Load Fuel Adjustment**

These adjusting points are shown on the typical carburetor in Fig. 13.

Adjusting The Carburetor

For details, see the engine Technical Manual.

OTHER FACTORS IN USING LP-GAS

Air Cleaners

Too-heavy oil in the air cleaner can be a problem with LP-gas engines. The result is a restriction in air intake which overchokes the carburetor and "floods" the engine.

Crankcase Oil

LP-gas is know to burn "cleaner" than other fuel but this has led to the notion that the engine oil will last much longer.

This is not true, however. The old oil gets thicker with use regardless of how dirty it is, resulting in lack of lubrication and loss of power from the friction. Also, the additives may wear out before the oil itself.

Always follow the operator's manual on when to change the engine oil and filters.

LP-GAS TROUBLE SHOOTING CHARTS

Refer to the engine Technical Manual for all repairs suggested below.

HARD STARTING

1. Improperly blended fuel—See fuel dealer.

2. Over-priming—See operator's manual.

3. Improper starting—See operator's manual.

4. Fuel strainer solenoid failed—Check electrical circuit.

5. Air vent on converter plugged—Open it.

6. Defective low-pressure diaphragm in converter —Replace diaphragm.

7. Carburetor metering valve or linkage binding—Repair or replace.

LOSS OF POWER

1. Throttle not opening fully—Check governor and throttle linkage.

2. Air vent on converter restricted—Open it.

3. Clogged fuel strainer—Clean the strainer as recommended.

CAUTION: Do not clean the strainer while the engine is running or when near an open flame.

4. Liquid fuel outlet on tank clogged—Clean it.

5. Excess-flow valves (vapor and liquid) closed, indicated by frosting at withdrawal valve—Reset by closing valves and opening slowly.

6. Fuel strainer frosted, meaning restriction due to dirt, or water, or to stuck valves—Clean, or replace valves.

7. Lean fuel mixture caused by restricted or altered fuel hose or pipes—Replace with new approved parts.

8. Converter high-pressure valve sticking—Remove valve assembly. Clean, and reinstall. Be sure valve is free.

9. Low-pressure valve in converter restricted—Check for free flow of fuel as recommended. Clean valve as necessary.

10. Defective or ruptured diaphragm—Replace it.

11. Carburetor badly adjusted or metering valve binding—Adjust carburetor, free up or replace metering valve.

12. Air leak between carburetor throttle body and air horn—Replace gasket or tighten screws.

POOR ECONOMY (Also See "Power Loss")

1. Tank not properly filled—Fill as recommended in operator's manual.

2. Improper carburetor adjustment—See operator's manual.

3. Wrong fuel—See fuel dealer.

4. Metering valve worn or damaged—Replace with a new valve.

FREEZE-UP OF CONVERTER

Freeze-ups during normal operation are usually due to **lack of water circulation** through the converter. Causes may be:

1. Low coolant level in the radiator—Fill radiator

2. Defective water pump—Repair it.

3. Thermostat removed—Install thermostat.

4. Water circulation reversed—Inlet water pipe connected at wrong point on converter.

5. Restriction in coolant pipes or converter—Clean or replace.

6. Operating engine on liquid fuel before coolant is warm—Warm up engine on vapor.

FROST ON FUEL STRAINER OR WITHDRAWAL VALVES

Frost on **fuel strainer** *is caused by:*

1. Dirt, water, or other restrictions—Clean the strainer.

2. Excess flow mechanism in withdrawal valve closed—Clean or replace valve.

Frost on **withdrawal valves** *is caused by:*

1. Excess flow mechanism in valve closed—Reset by closing valve and opening slowly.

2. Water in fuel tank (if this is cause, ice will form on liquid withdrawal valve and frost will be present between valve and strainer)—Empty strainer hose of gas and flush out with antifreeze solution as recommended.

ROUGH IDLING

1. Improper carburetor throttle rod or drag link adjustment—Readjust.

2. Defective carburetor-to-manifold gasket—Tighten cap screws or replace gasket.

3. Converter-to-carburetor fuel hose leaking—Replace it.

4. Dirty or wrong type of spark plugs—Clean or replace.

5. Metering valve in carburetor binding—Remove and clean or replace.

POOR IDLE SPEED ACCELERATION

1. Diaphragm punctured—Replace it.

2. Lean fuel mixture due to restriction in fuel inlet hose—Replace hose.

3. Wrong carburetor idle adjustment—Adjust as recommended.

LACK OF FUEL AT CARBURETOR

NOTE: To check, press primer (if used) at converter with throttle in wide open position. An audible hiss should be heard if fuel is flowing.

1. No fuel supply—Fill tank if necessary.

2. Liquid or vapor withdrawal valve closed—Open valve.

3. Fuel lines kinked, bent, or restricted—Check fuel lines.

4. Broken or loose electrical wiring connections on fuel strainer—Repair.

5. Fuel strainer clogged—Clean it. See operator's manual.

ENGINE STOPS WHEN SPEED CONTROL LEVER IS MOVED TO SLOW IDLE

1. Slow idle speed incorrectly adjusted—See operator's manual.

2. Converter-to-carburetor hose leaking—Replace.

3. Carburetor gasket leaking—Replace, and tighten cap screws securely.

4. Converter back cover gasket leaking—Replace gasket.

OVERHEATING

1. Too-lean fuel mixture—See operator's manual for carburetor adjustment.

2. Clogged radiator—Clean.

3. Defective Water pump—Repair.

4. Overloaded engine—Reduce load.

5. Incorrect ignition timing—Check timing.

6. Slipping fan belt—Adjust.

FUEL SEEPING INTO WATER PASSAGES IN CONVERTER

The transfer of fuel into the water passages in the converter is caused by failure of the back cover gasket.

To check, remove radiator cap, turn ignition switch on, and watch for bubbles or water blowing out of the radiator filler neck. If so, replace the gasket.

TEST YOURSELF

QUESTIONS

True or False?

1. "In both gasoline and LP-gas engines the fuel is a vapor when it reaches the carburetor." ________

2. "LP-gas is liquefied by putting it under high pressure." ________

3. "Frosting of LP-gas components is usually caused by a restriction." ________

4. "Always fill the LP-gas fuel tank completely full." ________

5. "Change the ignition timing when converting an engine from gasoline to LP-gas." ________

6. "Keep the doors and windows closed in winter to prevent freezing of LP-gas equipment." ________

7. "The converter changes liquid fuel to vapor and raises its pressure." ________

(Answers in back of text.)

Litho in U.S.A.

 Litho in U.S.A.

DIESEL FUEL SYSTEMS / CHAPTER 5

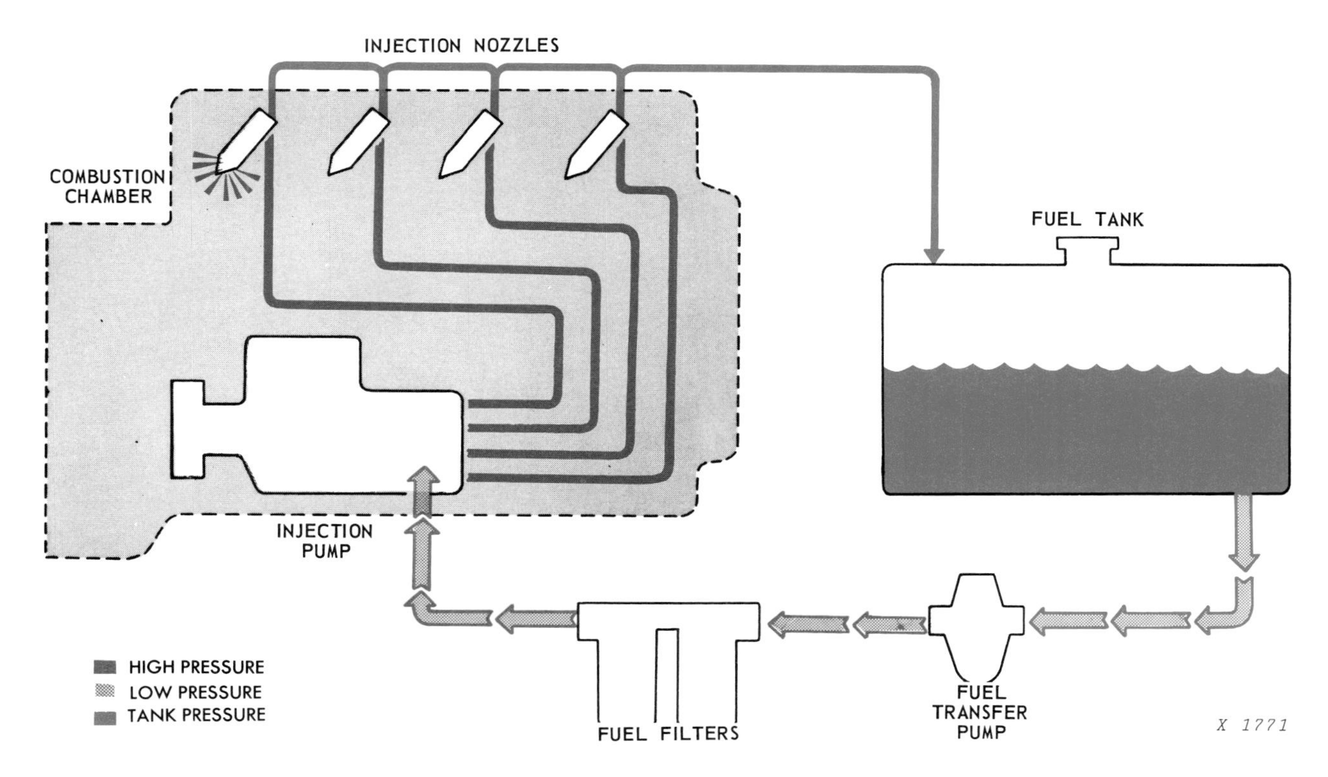

Fig. 1 — Diesel Fuel System (Distributor Type Shown)

INTRODUCTION

The prime job of the diesel fuel system is to inject a precise amount of atomized and pressurized fuel into each engine cylinder at the proper time.

Combustion in the diesel engine occurs when this charge of fuel is mixed with hot compressed air. No electrical spark is used (as in the gasoline engine).

The major parts of the diesel fuel system are:

- **Fuel Tank—stores fuel**
- **Fuel Transfer Pump—pushes fuel through filters to injection pump**
- **Fuel Filters—clean the fuel**
- **Injection Pump—times, measures, and delivers fuel under pressure to cylinders**
- **Injection Nozzles—atomize and spray fuel into cylinders**

Fig. 1 shows these parts in a basic system.

OPERATION

In operation, fuel flows by gravity pressure from the FUEL TANK to the Transfer Pump (Fig. 1).

The TRANSFER PUMP pushes the fuel through the FILTERS, where it is cleaned.

The fuel is then pushed on to the INJECTION PUMP where it is put under high pressure and delivered to each Injection Nozzle in turn.

The INJECTION NOZZLES atomize the fuel and spray it into the Combustion Chamber of each cylinder.

Later we will tell in detail how each part of the system works.

 Litho in U.S.A.

WHAT FUEL INJECTION MUST DO

The diesel fuel injection system must:

1. *Supply the correct quantity of fuel.*
2. *Time the fuel delivery.*
3. *Control the delivery rate.*
4. *Break up or atomize the fuel.*
5. *Distribute fuel evenly through the cylinder.*

Let's take a closer look at these requirements.

Correct Fuel Quantity—The fuel system must supply the exact amount of fuel to each cylinder, each and every time.

Fuel Delivery Timing—Fuel delivered too early or too late during the power stroke causes a loss of power. Fuel must be injected into the cylinder at the instant maximum power can be realized.

Delivery Rate—Smooth operation from each cylinder depends on the length of time it takes to inject the fuel. The higher the engine speed the faster the fuel must be injected.

Fuel Atomization—The fuel must be thoroughly mixed with the air for complete combustion. For this reason the fuel must be broken up into fine particles.

Fuel Distribution—The fuel must be spread evenly in the cylinder to unite with all the available oxygen. This makes the engine run smoothly and develop maximum power.

FUEL TANK

There are many different sizes and shapes of fuel tanks. Each size and shape is designed for a definite requirement. Here are some of them:

Capacity—The tank must be capable of storing enough fuel to operate the engine for a reasonable length of time.

Size—The size, of course, depends on the capacity needed. It also depends on the space on the machine available for the tank.

Shape—The shape of the tank can be altered to conform to the space available on the machine. For example, it can be either tall or short, round or square. This may allow for more fuel capacity in a restricted space.

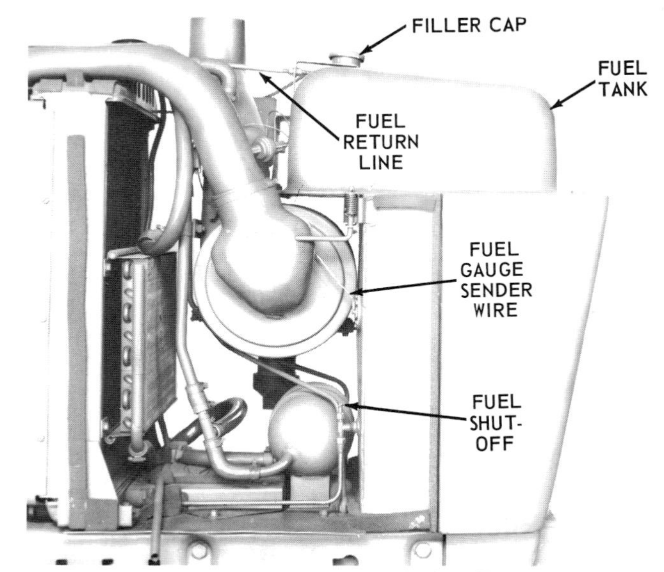

Fig. 2 — Fuel Tank On A Modern Tractor

Venting—The tank must be closed to prevent dirt from entering, yet it must also be vented to allow air to enter, replacing the fuel used.

Outlets—Three tank openings are also required—one to fill, one to discharge and one for draining. Sometimes another opening is used for leak-off fuel returned from the injection system.

FUEL TANK SERVICE

Leaks, condensation, and dirt are about the only problems of the fuel tank.

CAUTION: Cleaning or repairing a fuel tank can be dangerous. Sparks, smoking, or an open flame is hazardous in the area where the tank is cleaned or repaired.

Removing The Tank

The tank should be removed from the machine for repairing or checking for leaks.

Cleaning The Tank

After the tank is drained, clean it with water or steam.

1) Flush the tank with hot water until all fuels and vapors are removed.

2) Dry the tank with compressed air.

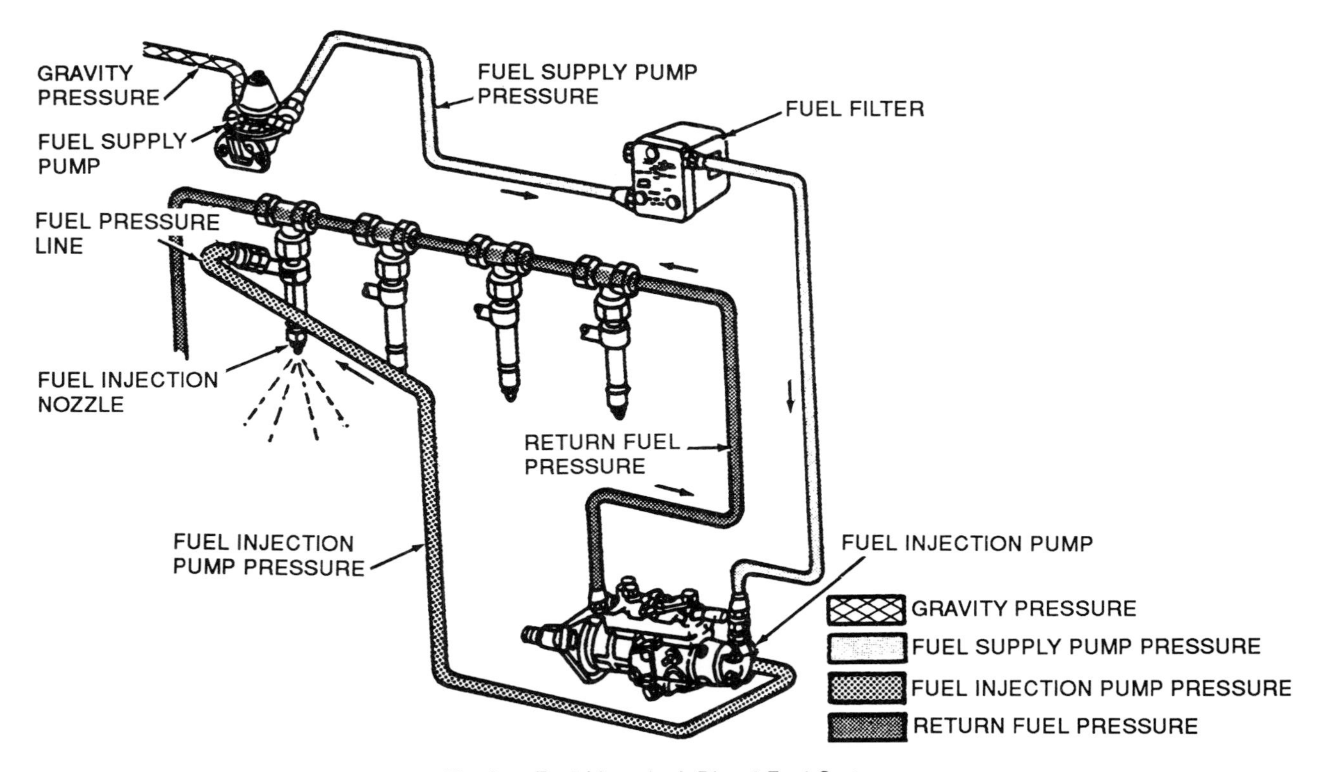

Fig. 3 — Fuel Lines In A Diesel Fuel System

Inspecting Tank for Leaks

Use one of two methods to find leaks in the fuel tank:

1) Wet method

2) Air pressure method

WET METHOD

Plug fuel outlet tightly. Dry entire outer surface of tank thoroughly with compressed air and a clean, dry rag. Place tank so that all surfaces may be easily seen, such as setting it on top of blocks. Then fill tank with water. Insert end of air hose in filler neck and apply approximately 3 pounds (20 kPa) air pressure against water. Examine tank surfaces for moist spots where water may have been forced through.

AIR PRESSURE METHOD

Plug filler neck and attach an air hose to fuel outlet. Submerge fuel tank in clean water and apply approximately 3 pounds (20 kPa) of air pressure. Draw a ring around each spot on fuel tank where bubbles appear. These bubbles indicate leaks in the tank that need repairing.

Repair Of Fuel Tanks

SOLDERING

Inspect and repair the tank as soon as possible after it has been cleaned. When soldering, the soldering iron should **not** be red hot. A red-hot iron can ignite any explosive mixture remaining in the tank.

WELDING

Tanks can be welded without danger, **if** precautions are taken.

Weld the tank as follows:

1. To prevent pockets of fuel, plug the tank outlet and *fill the tank full of water.*

2. Leave the filler cap off to allow for expansion of steam. (If the tank must be turned, weld a pipe to an old cap and bend up the pipe while installing the cap.)

3. After welding, retest the tank for leaks.

In summary: *Always fill tank with water and ventilate it before welding.*

Litho in U.S.A.

FUEL GAUGES

The **electric-type** fuel gauge is widely used today. There are two types:

- **Balancing coil gauge**
- **Thermostatic gauge**

Each type has a tank unit and a dash unit (Fig. 4).

BALANCING COIL FUEL GAUGES

The tank unit in this gauge has a sliding contact that slides back and forth on a resistance as the float moves up and down in the fuel tank. This changes the amount of electrical resistance the tank unit sends to the dash unit. Thus, as the tank empties, the float drops and the sliding contact moves to reduce the resistance. See at left in Fig. 4.

The dash unit has two coils as shown. When the ignition switch is turned on, current from the battery flows through the two coils, producing a magnetic pattern. This acts on the armature to which the indicator is attached.

When the resistance of the tank unit is high (tank filled and float up), then the current flowing through the E (empty) coil also flows through the F (full) coil. This pulls the armature to the right so the pointer indicates F (full) on the dial.

As the tank begins to empty, the resistance of the tank unit drops, causing more of the current flowing through the empty coil to pass through the tank unit. This weakens the magnetic field in the "full" coil, causing the "empty" coil to pull the armature toward it. Thus the pointer swings around toward the E (empty) side of the dial.

THERMOSTATIC FUEL GAUGES

This type of gauge has a pair of thermostat blades, each with a heating coil. The coils are connected in series through the ignition switch to the battery. See at right in Fig. 4.

The tank unit also has a float that actuates a cam. As the cam moves, it puts more or less flex on the tank thermostat blade. When the tank is full, the float is up and the cam bends hard on the blade. Then, when the ignition switch is turned on, current flows through the heater coils.

When the tank blade is hot enough, it bends further so that the contacts separate. Then, as the blade cools, the contacts close together. Again the blade is heated and the points reopen. This action continues as long as the ignition switch is on.

Meanwhile, the blade in the dash unit is heated and bends a like amount. Movement of this blade is carried through linkage to the pointer, which then moves to indicate on the "full" side of the dial.

However, if the tank is nearly empty, the float is down and the cam bends the tank thermostat blade only a little.

As a result, only a small amount of heating is enough to bend the blade further and open the contacts. Thus, the dash unit bends only a little and the pointer is toward the "empty" side.

NOTE: Refer to the FOS "Electrical Systems" Manual for diagnosing and repair of these fuel gauges.

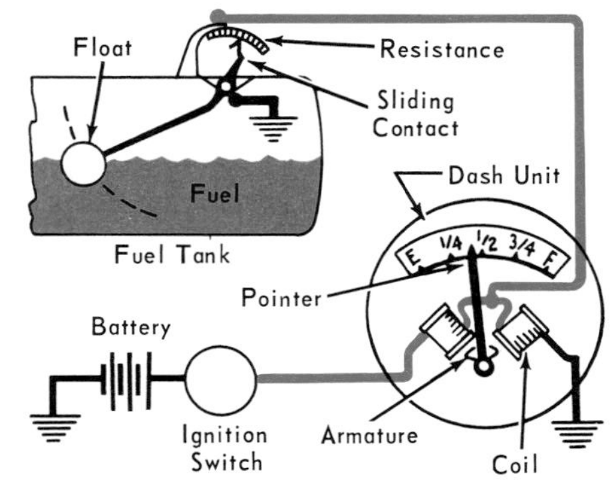

BALANCING COIL GAUGE

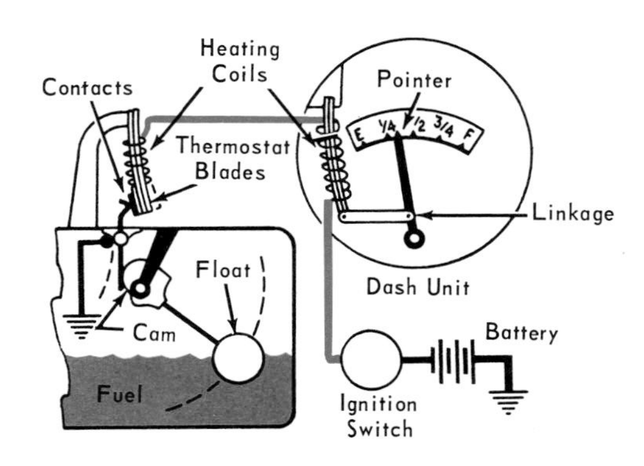

THERMOSTATIC GAUGE

Fig. 4 — Fuel Gauges — Two Types

FUEL LINES

The types of diesel fuel lines are:

Heavy-Weight Lines—for very high-pressure fuel between the injection pump and the nozzles (Fig. 3).

Medium-Weight Lines—for light or medium fuel pressures between the tank and injection pump.

Light-Weight Lines—for little or no fuel pressure of leak-off fuel from the nozzles to the tank or pump.

The heavy fuel injection lines should all be approximately the same length for proper injection timing in most systems. Note in Fig. 3 how some of the injection lines are shaped; this is done so that all the lines are of the same length.

SERVICING FUEL LINES

Periodically inspect the fuel lines for loose connections, breaks, or flaws. Any leak will pinpoint these failures.

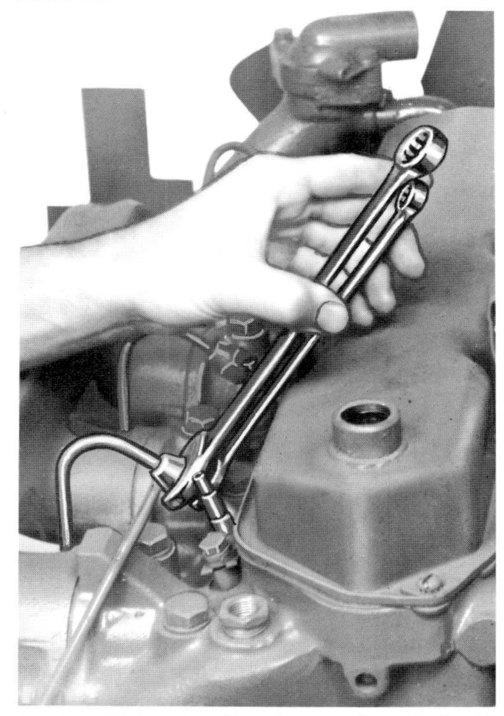

Fig. 5 — Tightening Line Connections To Avoid Bending The Lines (Final Tightening Shown)

Keep the line connectors tight—but not too tight. Tighten until snug, but don't strip the threads. To avoid bending the lines, use only one hand with two wrenches as shown in Fig. 5 for final tightening.

When replacing a fuel line, be sure to use an identical line in size, shape, and length. The **inside** diameter is very critical in injection lines.

FUEL TRANSFER PUMPS

Simple fuel systems use gravity or air pressure to get fuel from the tank to the injection pump.

On modern high-speed diesel engines, a fuel transfer pump is normally used. This pump, driven by the engine, supplies fuel automatically to the diesel system. The pump often has a hand primer lever for bleeding air from the system.

OPERATION, TESTING, AND SERVICING

Refer to Chapter 3 for details on fuel transfer pumps. Also consult the machine Technical Manual for complete procedures and specifications.

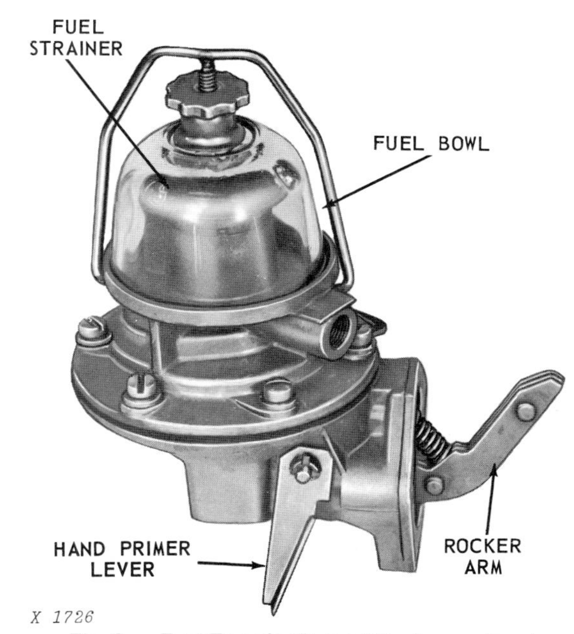

Fig. 6 — Fuel Transfer Pump (Diaphragm Type)

FUEL FILTERS

Fuel filtration is very important in diesel operation:

- **Diesel fuels tend to be impure**
- **Injection parts are precision-made**

As a result, diesel fuel must be filtered not once, but several times in most systems.

A typical system (Fig. 7) might have three stages of progressive filters:

1. Filter Screen At Tank Or Transfer Pump—removes large particles.

2. Primary Filter—removes most small particles.

3. Secondary Filter—removes tiny particles.

Some systems also have a "final" filter as a "watch dog" for the system.

 Litho in U.S.A.

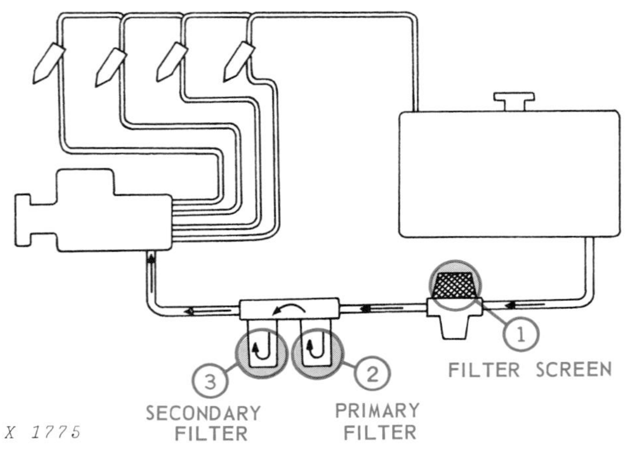

Fig. 7 — Diesel Fuel Must Be Filtered Thoroughly

Most diesel filters have a water trap where water and heavy sediment can settle and be drained.

TYPES OF FILTERS

Filtration removes suspended matter from the fluid. Some filters will also remove soluble impurities.

Filtration can be done in three ways:

- **Straining**
- **Absorption**
- **Magnetic Separation**

STRAINING is a mechanical way of filtering. It uses a screen which blocks and traps particles larger than the openings. The screen may be of wire mesh for coarse filtering or of paper or cloth for finer filtering.

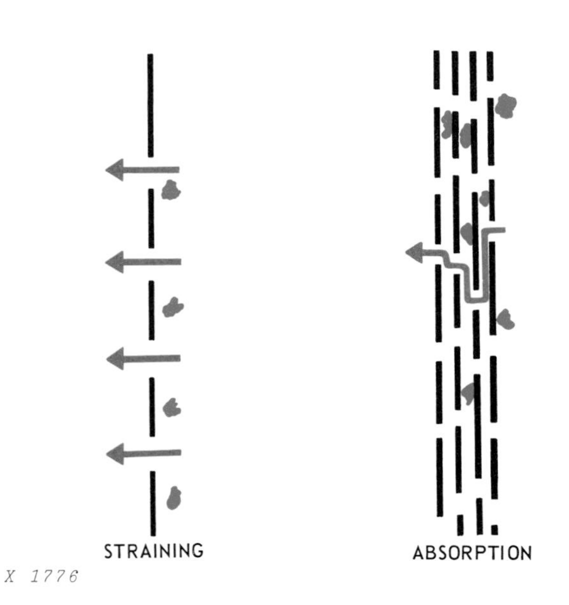

Fig. 8 — Types of Fuel Filtering

ABSORPTION is a way of trapping solid particles and some moisture by getting them to stick to the filter media—cotton waste, cellulose, woven yarn, or felt.

MAGNETIC SEPARATION—is a method of removing water from the fuel. By treating a paper filter with chemicals, water droplets can be formed and separated when they drip into a water trap. (The filter also removes solid particles by one of the other methods of filtration.)

FUEL FLOWS IN DUAL FILTERS

Dual filters may operate in two ways:

- **Series**
- **Parallel**

In SERIES FILTERS, *all* of the fuel goes through *one* filter, then through the *other.*

In PARALLEL FILTERS, part of the fuel goes through *each* filter.

Advantages of Series and Parallel Filters

Series filters clean the fuel better because the second filter can pick up dirt missed by the first one.

However, *parallel filters* can move a larger volume of fuel through faster.

SERVICING FILTERS

While the fuel filters help to clean the fuel, they are not meant to clean up dirty fuel from a bad fuel supply.

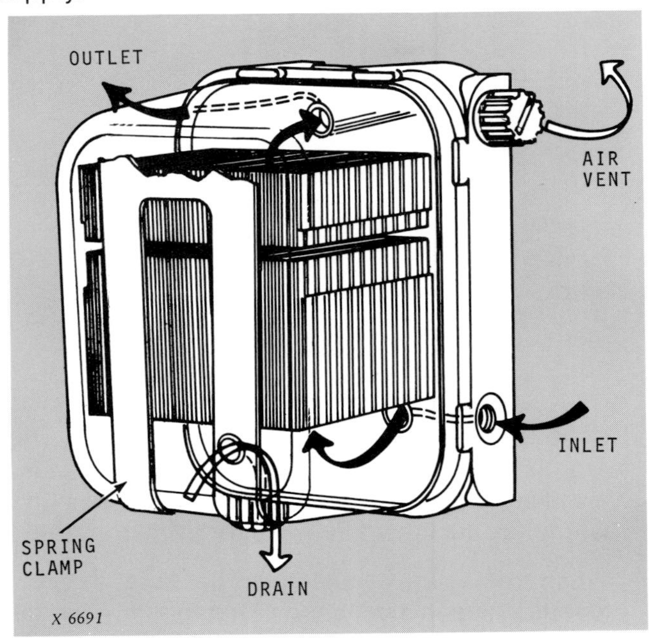

Fig. 9 — Flow of Fuel through Filter

During normal operation, the filter elements should be changed at regular intervals as recommended in the operator's manual.

Change the filter elements more often when operating in unusual conditions such as extreme dust or dirt.

On series filters, normally change the first-stage more frequently than the second-stage filter.

Check the water traps under the filter at frequent intervals. Drain out the water and sediment which collects there.

FUEL INJECTION SYSTEMS

The pump-and-nozzle combination which finally delivers the fuel is the heart of the diesel fuel system.

Let's repeat what we said earlier about what fuel injection must do:

1. *Supply the correct quantity of fuel.*
2. *Time the fuel delivery.*
3. *Control the rate of delivery.*
4. *Atomize the fuel.*
5. *Distribute fuel in the combustion chamber.*

These jobs are all handled by the pump-nozzle team.

But in the various systems, the jobs are handled in different ways.

Because the engine operates at variable loads and speeds, the pump-nozzle should be able to vary the quantity, timing, and rate of fuel delivery.

Let's look at the types of injection systems and see how well they are able to perform these tasks.

TYPES OF INJECTION SYSTEMS

Some early diesel engines used compressed air to blow fuel into the cylinders. But today's high-speed engines use a form of *mechanical* or *solid fuel injection.*

There are three major types:

- **Common Rail System**
- **Accumulator System**
- **Jerk Pump System**

The COMMON RAIL is an older system and keeps constant fuel pressure in a header that serves the nozzles. The pump and/or the nozzles are mechanically actuated by revolving cams. The quantity of fuel injected depends completely on how long the valve stays open.

In ACCUMULATOR SYSTEMS, the quantity of fuel injected can be varied regardless of pump speed by using an adjustable spring or a hydraulic accumulator to control the fuel pressures.

In JERK PUMP SYSTEMS, the pump times, meters, and forces the fuel at high pressures through the nozzles. The quantity and duration of fuel injected at various speeds are controlled better with this system. For this reason, the jerk pump system is widely used on modern engines.

Let's look at each system in more detail.

COMMON RAIL SYSTEM

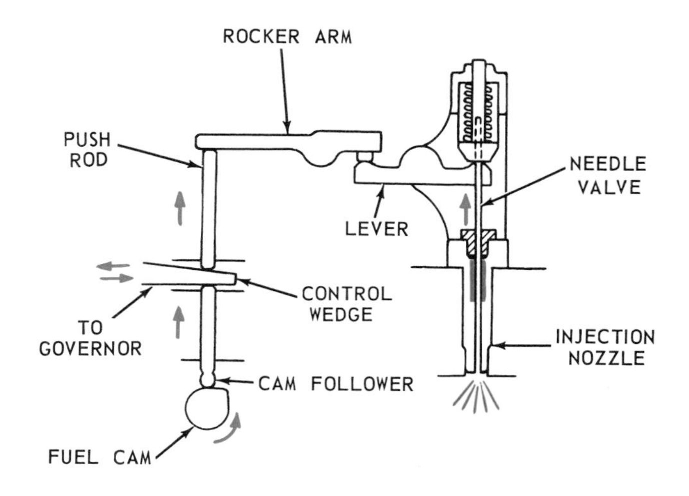

Fig. 10 — Common Rail Injection System

This system has a high-pressure, constant-stroke and constant-delivery pump which discharges into a common rail, or header, to which each injection nozzle is connected by tubing.

A spring-loaded bypass valve on the header maintains a constant pressure in the system, returning all excess fuel to the fuel tank.

The injection nozzles are operated mechanically, and the amount of fuel injected into the cylinder at each power stroke is controlled by the lift of the needle valve.

The operation of the injection system is shown in Fig. 10.

The fuel cam lifts the push rod; this motion is transmitted to the needle valve through the rocker arm and intermediate lever; the fuel area above the needle valve seat is connected at all times with the fuel header.

When the needle valve is lifted from its seat, the fuel is admitted to the combustion space through the small holes drilled in the injector tip, below the valve seat. Passing through these tiny holes, the fuel is divided into small streams which are broken up or atomized.

The amount of fuel injected is controlled by a control wedge (Fig. 10) which changes the lash of the fuel valve. When the wedge is pushed to the right, the valve lash is decreased. The motion of the cam follower will then be transmitted earlier to the push rod, the fuel needle will be opened earlier, closed later, and its lift will be slightly greater. Therefore, more fuel will be admitted per cycle.

When the wedge is pulled out, to the left, the valve lash is increased. The needle valve will be lifted later and closed earlier, and less fuel will be admitted.

The position of the control wedge is changed either by the governor, by hand, or by both.

The fuel injection pressure is adjusted to suit the operating conditions by changing the spring pressure in a bypass valve. Fuel injection pressures from 3200 psi (22 000 kPa) to about 5000 psi (34 000 kPa) are used at rated load and speed, depending upon the type of engine.

To reduce the pressure fluctuations in the system, due to the intermittent fuel discharge from the pumps and withdrawals by the fuel valves, the volume of the fuel in the system is increased by attaching to the fuel header an additional fuel container called the *accumulator* which has a relatively large capacity.

The common rail system is not suitable for high-speed, small-bore engines, because it is difficult to control accurately the small quantities of fuel injected into the cylinders at each power stroke.

ACCUMULATOR SYSTEM

The quantity of fuel injected per stroke can be varied regardless of pump speed in the accumulator system.

Regulating is by spring pressure or by hydraulic pressure (accumulator).

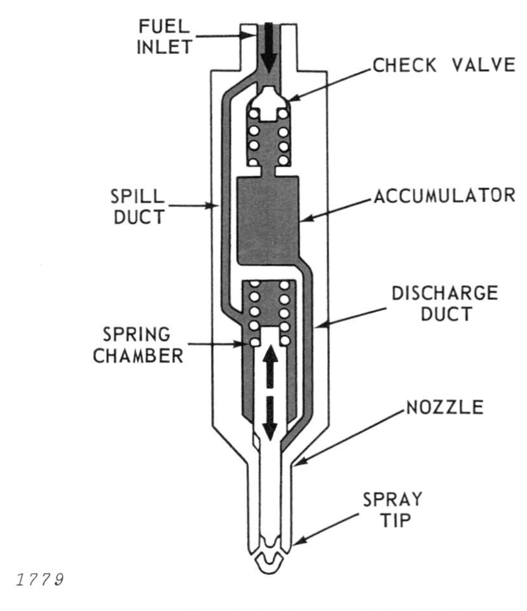

Fig. 11 — Nozzle in Accumulator System

A cam-driven pump sends fuel through the nozzle check valve into the accumulator as well as through the spill duct into the spring chamber (Fig. 11).

When the pump is too full and starts to bypass fuel, the nozzle check valve closes (as inlet pressure drops).

Fuel in the spring chamber is then vented through the spill duct back to the pump, while fuel now trapped in the accumulator passes through the discharge duct to the nozzle.

Since accumulator pressure is higher than the nozzle opening pressure, the nozzle valve lifts and injects fuel.

Injection stops when the accumulator pressure drops to the nozzle closing pressure.

The maximum injection pressure depends upon the accumulator volume and the quantity of fuel metered to it by the pump. Therefore, it is *independent* of the pump speed and nozzle orifices.

In summary, **the quantity of fuel injected per stroke can be varied in the accumulator system.**

JERK PUMP SYSTEM

In this system, both the quantity of fuel injected and its duration can be controlled better. Most all modern diesel systems are of this type.

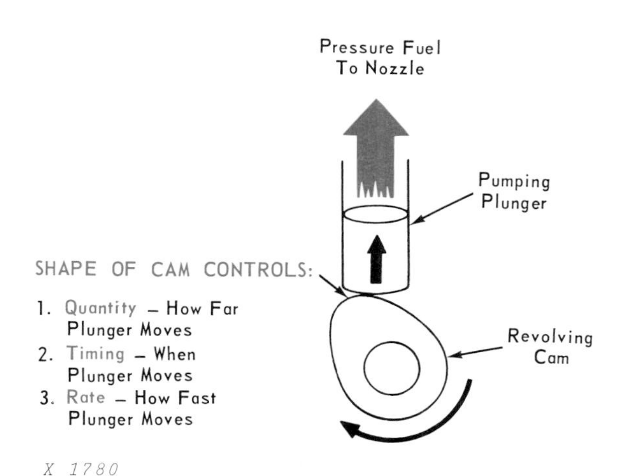

Fig. 12 — Plunger And Cam Principle Of Jerk Pumps

Jerk pumps operate on the **plunger and cam** principle (Fig. 12). As the cam revolves, it moves the plunger up and down, pumping fuel out to the nozzle.

The shape of the cam controls three things:

1. *Fuel Quantity—how far the plunger moves.*
2. *Timing—when the plunger moves.*
3. *Rate of Delivery—how fast the plunger moves.*

For example, if the revolving cam in Fig. 12 had a longer lobe, it would move the plunger farther, pumping a larger **quantity** of fuel.

If the cam lobe were thicker, it would move the plunger sooner, changing the **timing** of the plunger stroke.

If the cam lobe were thinner, the plunger would raise and lower faster, changing the **rate** of fuel delivery.

By adjusting these three factors—*fuel quantity, timing, and rate of delivery*—modern injection pumps can serve the engine better than the other systems we've discussed.

MODERN INJECTION PUMPS

As we said, modern injection pumps are almost all jerk pumps which use the plunger and cam method of fuel injection.

The injection pump-nozzle team can work in many ways. In a four-cylinder engine (Fig. 13), there are four major ways of injecting fuel:

- **Individual pump and nozzle for each cylinder**
- **Pumps in common housing, nozzles for each cylinder (in-line type)**
- **Combined pump and nozzle for each cylinder (unit injector type)**
- **One pump serving nozzles for several cylinders (distributor type)**

The unit injector pump is common on larger engines, while the in-line and distributor types are widely used, especially on off-the-road farm and industrial machines.

We will use the distributor and the in-line pumps in our examples of how an injection pump works—first the *distributor* type.

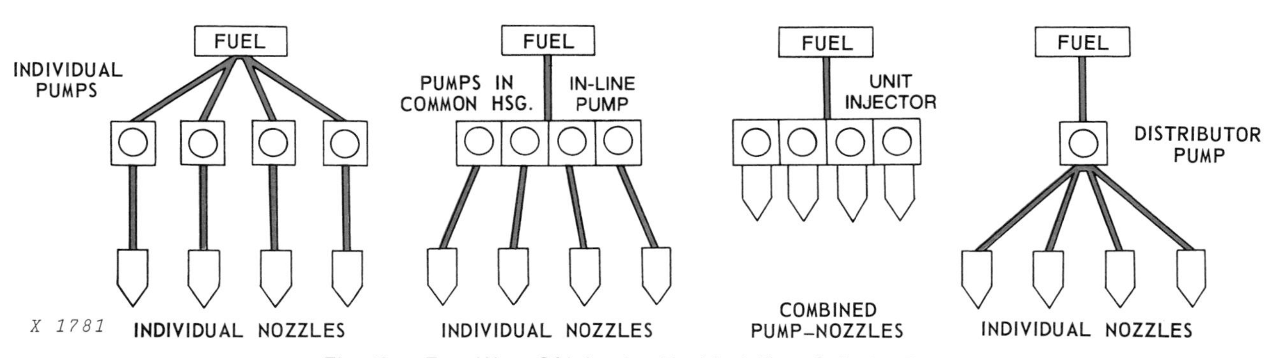

Fig. 13 — Four Ways Of Injecting Fuel In A Four-Cylinder Engine

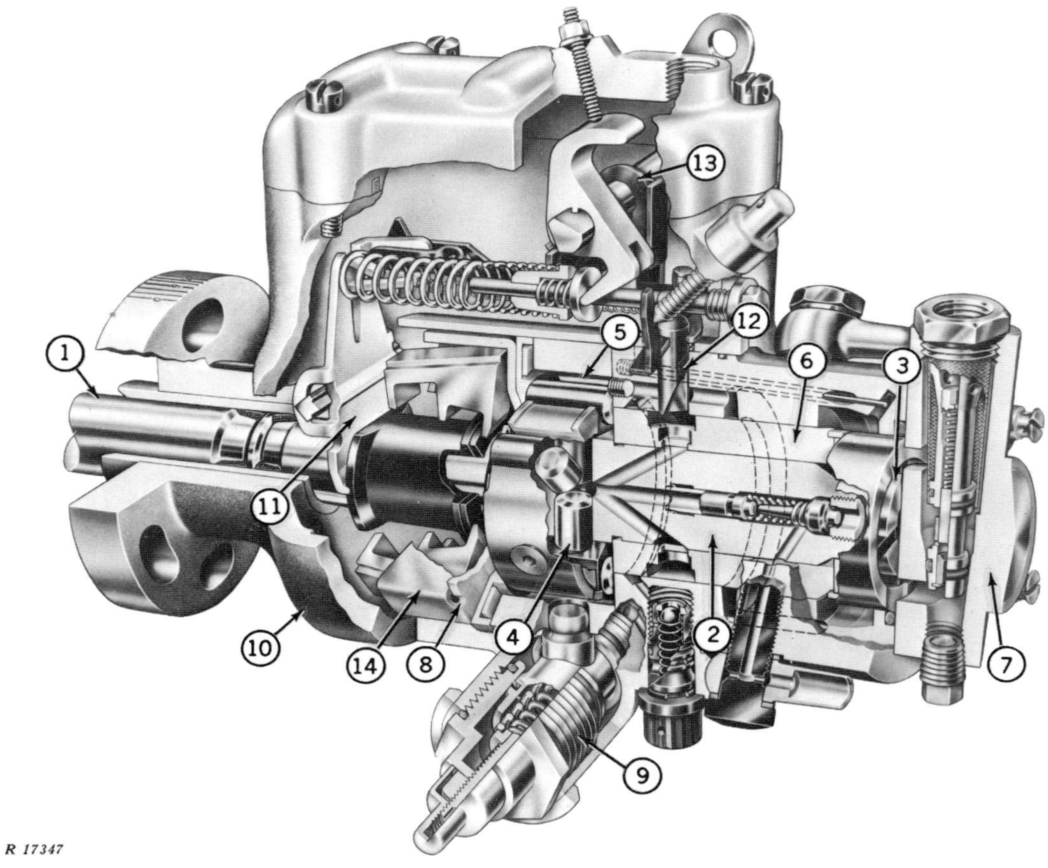

1. *Drive Shaft*
2. *Distributor Rotor*
3. *Transfer Pump*
4. *Pumping Plungers*
5. *Internal Cam Ring*
6. *Hydraulic Head*
7. *End Plate*
8. *Governor*
9. *Automatic Advance*
10. *Housing*
11. *Governor Arm*
12. *Metering Valve*
13. *Shut-Off Lever*
14. *Governor Weight Retainers*

Fig. 14 — Distributor-Type Injection Pump

DISTRIBUTOR INJECTION PUMP

A *distributor* injection system normally uses one pump to distribute fuel to all cylinders.

A typical pump is shown in Fig. 14.

MAIN PARTS OF THE PUMP

Here are the main parts of the pump as shown in Fig. 14:

1. *Drive Shaft*
2. *Distributor Rotor*
3. *Transfer Pump*
4. *Pumping Plungers*
5. *Internal Cam Ring*
6. *Hydraulic Head*
7. *End Plate*
8. *Governor*
9. *Automatic Advance*
10. *Housing*

ROTATING PARTS

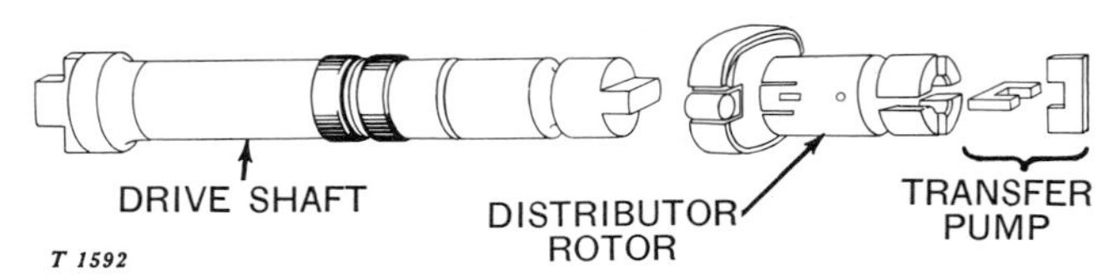

Fig. 15 — Main Rotating Parts Of Pump

The main rotating parts are shown in Fig. 15. They are: drive shaft, distributor rotor, and transfer pump. All rotate on a common axis.

HOW IT WORKS

In Fig. 14 above, the drive shaft (1) engages the distributor rotor (2) in the hydraulic head (6). The drive end of the rotor has two cylinder bores, each containing two plungers (4).

The plungers are actuated toward each other simultaneously to pump fuel by an internal cam ring (5). The cam ring has as many lobes as there are cylinders to be served. (However, a three-cylinder engine uses a pump with six cam rings.)

The transfer or supply pump (3, Fig. 14) in the opposite end of the rotor from the pumping cylinders, is of the positive displacement, vane type and is covered by the end plate (7).

The distributor rotor (2) has two angled inlet passages for charging and an axial bore to serve all outlets.

The hydraulic head (6) contains the bore in which the rotor revolves, the metering valve bore, the charging ports and the head outlets. These outlets are connected through appropriate fuel line fittings to the injection lines which lead to the nozzles.

Covering the transfer pump, on the outer end of the hydraulic head, is the end plate (7). This assembly houses the fuel inlet connection, fuel strainer and transfer pump pressure regulating valve.

The pump contains its own mechanical governor (8), capable of close speed regulation.

The action of the weights in their retainer (14) is transmitted through a sleeve to the governor arm (11) and through a positive linkage to the metering valve (12).

The metering valve is closed to shut off fuel through a solid linkage by an independently operated shut-off lever (13).

FUEL FLOW

To understand the operation, let's trace the fuel through the pump during a complete pump cycle as shown in Fig. 16.

Fuel is drawn from the supply tank into the pump through the inlet strainer (1) by the vane type fuel transfer pump (2).

Since transfer pump displacement greatly exceeds the injection requirements, a large percentage of fuel is bypassed through the regulating valve (3) back to the inlet side. The flow of this positive displacement pump increases with speed and the regulating valve is designed so transfer pump pressure also increases with speed.

Fuel, under transfer pump pressure, is forced through the drilled passage (4) in the hydraulic head into the annulus (5). It then flows around the annulus to the top of the sleeve and through a connecting passage (6) to the metering valve (7).

The radial position of the metering valve, controlled by the governor, regulates the flow of fuel into the charging ring (8) which incorporates the charging ports.

As the rotor revolves, the two inlet passages (9) register with two charging ports in the hydraulic head, allowing fuel to flow into the pumping cylinders.

With further rotation, the inlet passages move out of registry and the single discharge port is opened. The rollers (10) contact the cam lobes forcing the plungers together. Fuel trapped between the plungers is then delivered to the nozzle.

Lubrication of the pump is a part of the design. As fuel at transfer pump pressure reaches the charging ports, slots on the rotor shank allow fuel and any trapped air to flow into the pump housing cavity.

In addition, an air bleed in the hydraulic head connects the outlet side of the transfer pump with the pump housing cavity. This allows air and some fuel to be bled back to the fuel tank via the return line. The fuel thus bypassed helps lubricate the internal parts.

SMOKE REDUCTION FOR TURBOCHARGED SYSTEMS

An aneroid (not shown) is usually mounted on the fuel pump governor housing. It controls the flow of fuel into the distributor type or in-line type injection pump. It is a fuel flow inhibitor that is regulated by either mechanical or hydraulic pressure (see page 5-26 for more aneroid information).

When air is not sufficient in the manifold during rapid acceleration of turbocharged engines, the aneroid limits the fuel supply. This prevents a too rich fuel mix in the cylinders (due to insufficient air supply from the intake turbocharger) thereby reducing smoke. Aneroids also help reduce smoke density during the initial startup.

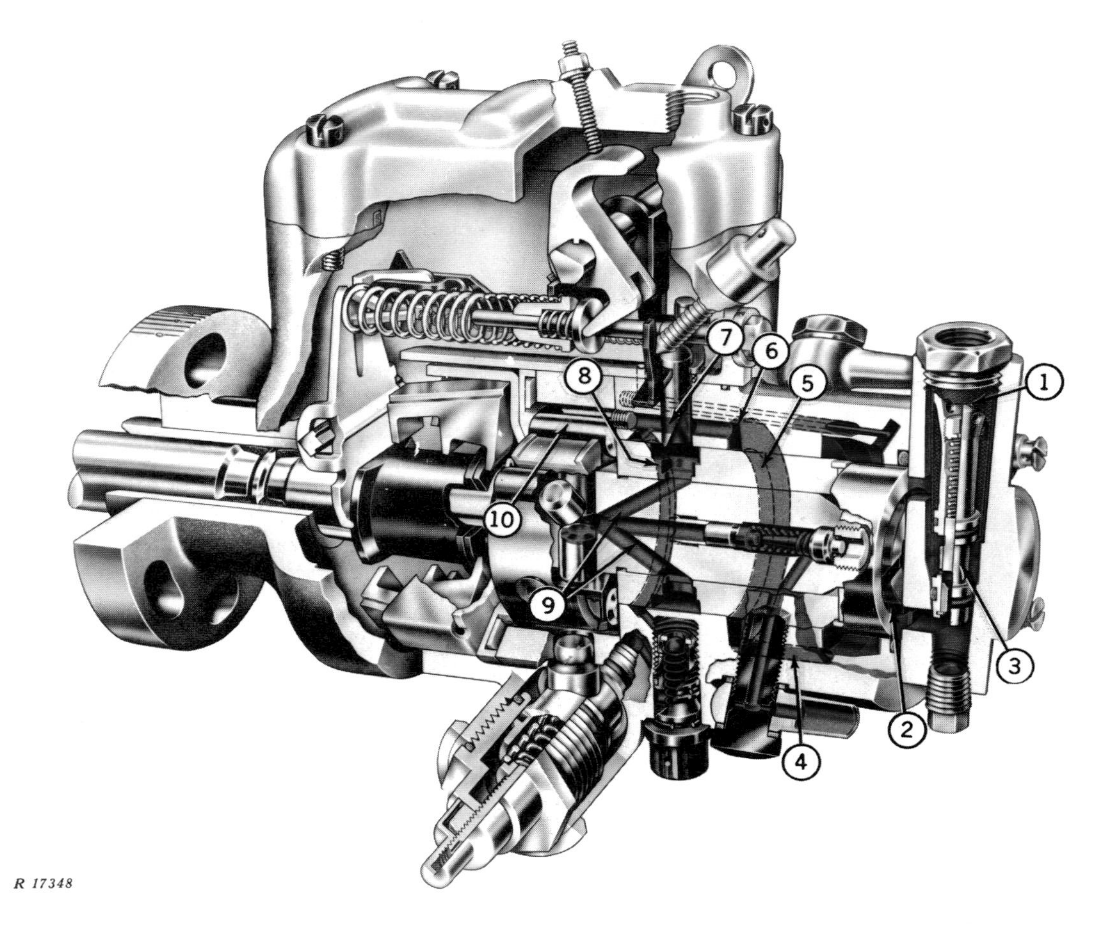

Fig. 16 — Fuel Flow

CHARGING AND DISCHARGING

Figs. 17 and 18 show the fuel flow during the charging and discharging cycles.

Charging Cycle

As the rotor revolves (Fig. 17) the angled inlet passages in the rotor register with opposite charging ports of the charging ring. Fuel under pressure from the transfer pump and controlled by the opening of the metering valve flows into the pumping cylinders, forcing all plungers apart.

The plungers move outward a distance proportionate to the amount of fuel required for injection on the following stroke. If only a small quantity of fuel is admitted into the pumping cylinders, as at idling, the plungers move out very little. Maximum plunger travel and, consequently, maximum fuel delivery is limited by adjusting leaf springs which contact the edge of the roller shoes. Only when the engine is operating at full load will the plungers move to the most outward position.

Note in Fig. 17 that the angled inlet passages in the rotor are in registry with the ports in the charging ring, but the rotor discharge port is not in registry with a head outlet. Note also that the rollers are between the cam lobes. Compare their relative positions in Figs. 17 and 18.

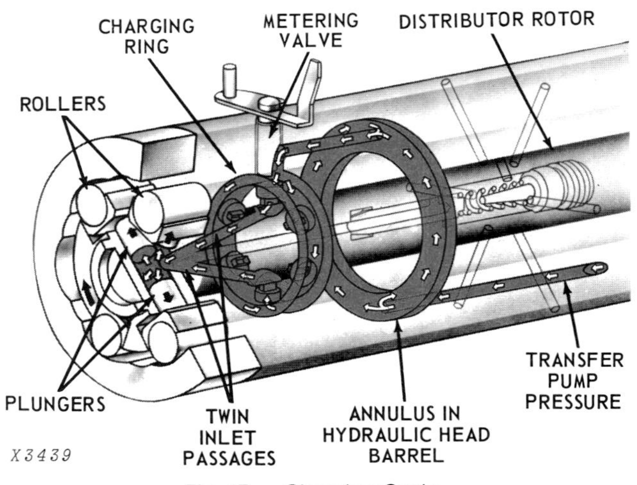

Fig. 17 — Charging Cycle

 Litho in U.S.A.

Discharge Cycle

As the rotor continues to revolve (Fig. 18), the inlet passages move out of registry with the charging ports.

For a brief interval the fuel is trapped until the rotor discharge passage registers with one of the head outlets.

As this registration takes place, both sets of rollers contact the cam lobes and are forced together.

During this stroke the fuel trapped between the plungers is forced through the axial passage of the rotor and flows through the rotor discharge passage to the injection line.

Delivery to the line continues until the rollers pass the highest point on the cam lobe and are allowed to move outward.

The pressure in the axial passage is then relieved, allowing the injection nozzle to close.

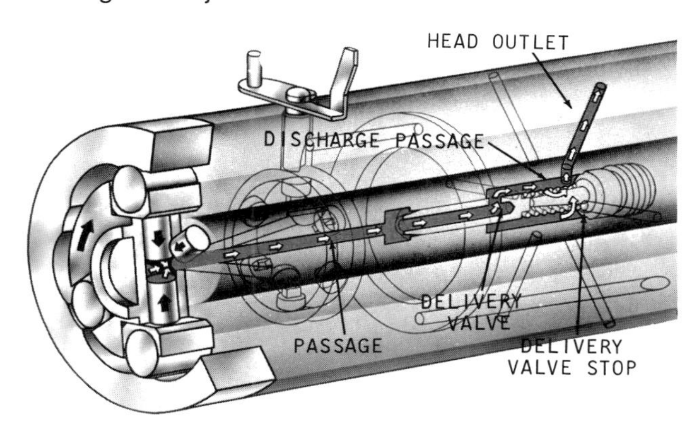

Fig. 18 — Discharge Cycle

DELIVERY VALVE

Controlled line retraction, the major function of the delivery valve, is done by rapidly decreasing the injection line pressure after injection to a predetermined value lower than that of the nozzle closing pressure.

This reduction in pressure causes the nozzle valve to return rapidly to its seat, achieving sharp delivery cut-off and preventing dribble of fuel into the engine combustion chamber.

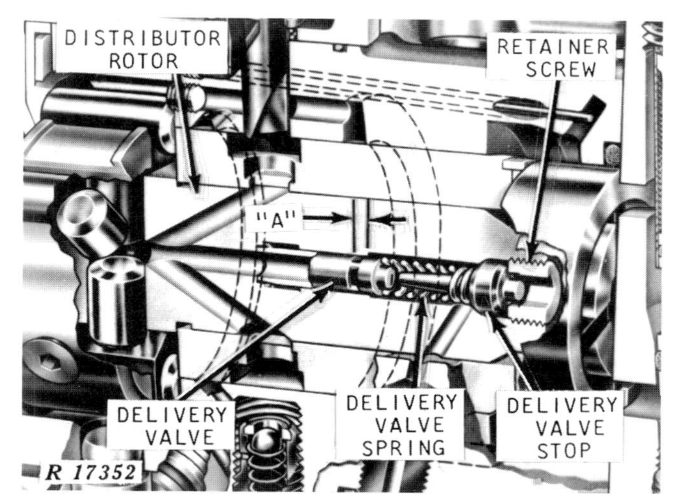

Fig. 19 — Delivery Valve

The delivery valve (Fig. 19) is located in the center of the distributor rotor. It requires no seat—only a shoulder to limit travel.

Since the same delivery valve performs the function of retraction for each line, the retracted amount will not vary from cylinder to cylinder. This results in a smooth-running engine at all loads and speeds.

When injection starts, fuel pressure moves the delivery valve slightly out of its bore and adds the volume of its displacement section "A" to the enlarged cavity of the rotor occupied by the delivery valve spring. This displaces a similar volume of fuel in the spring cavity before delivery through the valve ports starts.

At the end of injection, the pressure on the plunger side of the delivery valve is quickly reduced by allowing the cam rollers to drop into a retraction step on the cam lobes. Cam retraction volume is slightly more than delivery valve retraction volume.

As the valve returns to its closed position, its displacement (section "A") is removed from the spring cavity and, since the rotor discharge port is still partly in registry, fuel rushes back out of the injection line to fill the volume left by the retreating delivery valve.

After this, the rotor ports close completely and the remaining injection line pressure is trapped.

Litho in U.S.A.

RETURN OIL CIRCUIT

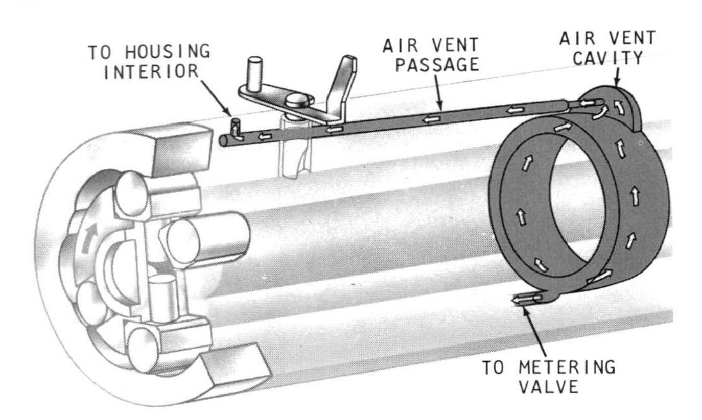

Fig. 20 — Return Oil Circuit

Fuel under transfer pump pressure is discharged into a cavity in the hydraulic head (Fig. 20).

The upper half of this cavity connects with a vent passage. Its volume is restricted by a wire to prevent undue pressure loss.

The vent passage is located behind the metering valve bore and connects with a short vertical passage entering the governor linkage compartment.

Should air enter the transfer pump because of suction-side leaks, it immediately passes to the air vent cavity and then to the vent passage as shown.

Air and a small quantity of fuel then flow from the housing to the fuel tank via the return line.

END PLATE OPERATION

The *end plate* (Fig. 21) has three basic functions:

1. Provides fuel inlet passages and houses pressure regulating valve.

2. Covers the transfer pump.

3. Absorbs end thrust of drive and governor.

During hand priming (Fig. 21), the fuel flows into the inlet side of the transfer pump through the port "A." Priming of the rest of the system is accomplished when the pump is rotated by the engine starting motor or on the test bench.

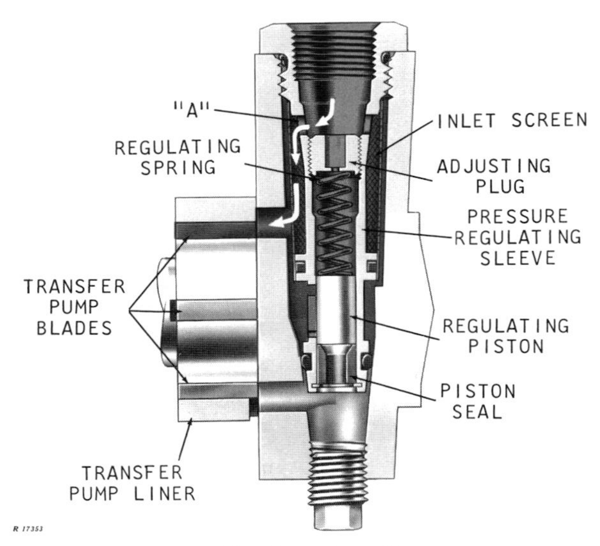

Fig. 21 — End Plate

Fig. 22 (below) shows the operation of the pressure regulating valve while the pump is running.

Fuel pressure from the discharge side of the transfer pump forces the piston up the sleeve against the regulating spring. As pressure increases, the regulating spring is compressed slightly until the lower edge of the regulating piston starts to uncover port "B." Since the pressure on the piston is opposed by the regulating spring, the delivery

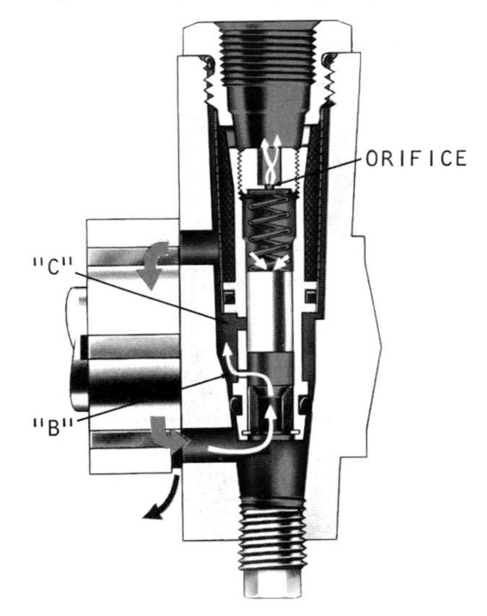

Fig. 22 — Pressure Regulating Valve

pressure of the transfer pump is controlled by the spring rate and size and number of regulating ports.

A high-pressure relief port "C" in the sleeve, above the regulating port, prevents high transfer pump pressures if the engine or pump is accidentally overspeeded.

VISCOSITY COMPENSATION

The distributor pump must work equally well with different fuels and varying temperatures which affect fuel viscosity. A feature of the regulating device offsets pressure changes caused by viscosity differences. Located in the bottom of the spring adjusting plug is a thin plate containing a sharp-edged orifice (Fig. 22). This orifice allows fuel leakage by the piston to return to the inlet side of the pump.

Flow through such a short orifice is virtually unaffected by viscosity changes. For this reason, the fuel pressure on the spring side of the piston will vary with viscosity changes. The pressure exerted on top of the piston is determined by the flow admitted past designed clearance of the piston in the sleeve.

With cold or heavy fuels, very little leakage occurs past the piston and flow through the adjusting plug is a function of the orifice size. Downward pressure on the piston is thus very little.

With hot (or light) fuels, leakage past the piston increases. Pressure in the spring cavity increases also, since flow through the short orifice remains the same as with cold fuel. Thus, downward pressure, assisting the regulating spring, positions the piston so less regulating port area is uncovered below it. Pressure is thus controlled and may actually over-compensate to offset other leakages in the pump, which increase with the thinner fuels.

CENTRIFUGAL GOVERNOR

In the centrifugal *governor* (Fig. 23), the movement of the flyweights against the governor thrust sleeve rotates the metering valve.

This rotation varies the registry of the metering valve opening with the passage from the transfer pump, thus controlling the flow to the engine.

This type of governor derives its energy from flyweights pivoting on their outer edge in the retainer. Centrifugal force tips them outward, moving the governor thrust sleeve against the governor arm, which pivots on the knife edge of the pivot shaft and, through a simple positive linkage, rotates the metering valve.

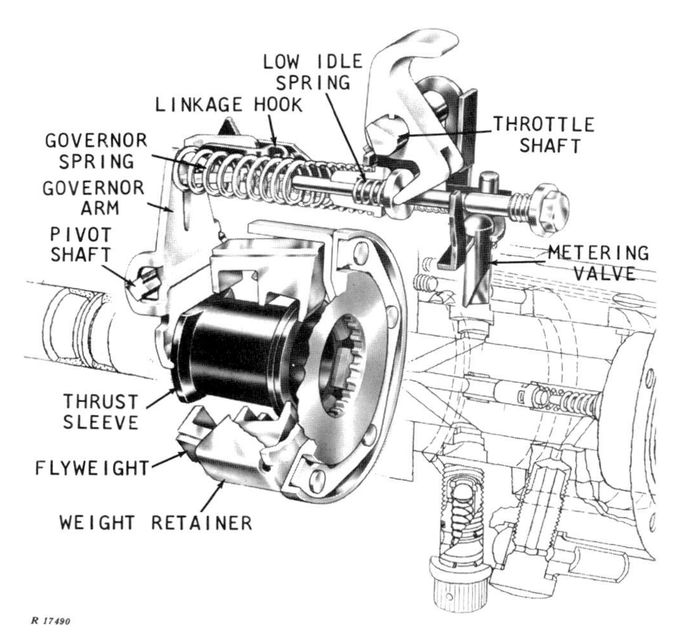

Fig. 23 — Centrifugal Governor

The force on the governor arm caused by the flyweights is balanced by the compression-type governor spring, which is manually controlled by the throttle shaft linkage in regulating engine speed.

A light idle spring is provided for more sensitive regulation at the low speed range.

The limits of throttle travel are set by adjusting screws for proper low idle and high idle positions.

A light tension spring takes up any slack in the linkage joints and also allows the stopping mechanism to close the metering valve without overcoming the governor spring force. Only a very light force is required to rotate the metering valve to the closed position.

AUTOMATIC LOAD ADVANCE

Pumps equipped with *automatic load advance* (Fig. 24) permit the use of a simple hydraulic servo-mechanism, powered by oil pressure from the transfer pump, to advance injection timing.

Transfer pump pressure operates the advance piston against spring pressure as required along a predetermined timing curve.

The purpose of the load advance device is to advance injection timing slightly as engine load decreases. This offsets the normal timing retardation that occurs at light loads.

Cam movement is induced by pressure developed at the transfer pump, admitted by the metering valve. Since the governor controls the metering valve position at all loads, it also controls the amount of fuel, under pressure, fed to the automatic advance device.

AUTOMATIC SPEED ADVANCE

Pumps equipped with *automatic speed advance* permit the use of a simple hydraulic servo-mechanism, powered by oil pressure from the transfer pump, to advance injection timing.

Transfer pump pressure operates the advance piston against spring pressure, as required, along a predetermined timing curve.

The purpose of the speed advance device, which responds to speed changes, is to advance the timing for the best engine power and combustion throughout the speed range.

Controlled movement of the cam in the pump housing (Fig. 24) is reduced and limited by the action of the piston of the automatic advance against the cam advance pin.

During cranking, the cam is in the retard position, since the force exerted by the advance spring is greater than that of transfer pump pressure. As the engine speed and transfer pump pressure increase, fuel under transfer pump pressure entering the advance housing behind the power piston moves the cam.

The amount of advance is limited by the length of the advance pistons. A ball check valve is provided to offset the normal tendency of the cam to return to the retard position during injection.

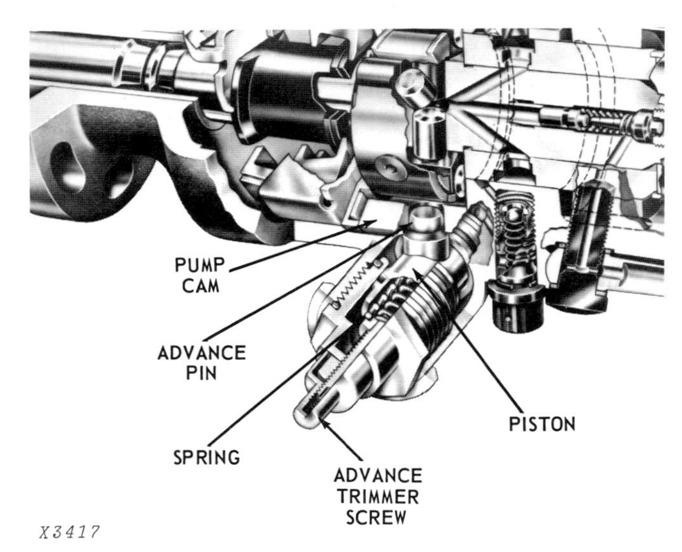

Fig. 24 — Automatic Advance Mechanism

Automatic Advance Fuel Circuit

Fuel, under transfer pump pressure, is forced through the drilled passage (1, Fig. 25) located in the hydraulic head to the annular ring (2).

Fuel then flows around and to the top of the annular ring, where it registers with the bore leading to the metering valve. The metering valve is designed to allow a quantity of fuel to flow into a second annular ring (3) which registers at the bottom with the bore of the advance clamp screw assembly.

As transfer pump pressure increases, the ball check (located in the clamp screw assembly) is lifted off its seat, allowing fuel to pass through the clamp screw into the passage located behind the power piston.

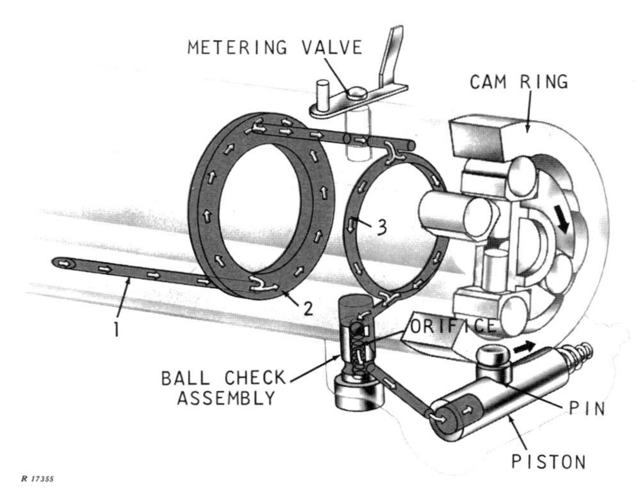

Fig. 25 — Automatic Advance Fuel Circuit

TORQUE CONTROL

Torque is commonly defined as the "lugging ability" of an engine.

Since torque increases with overload, a predetermined point at which maximum torque is desired may be selected for the engine. As engine speed decreases, the torque increases toward this preselected point. This desirable feature is called "torque back-up."

Three basic factors that affect torque reserve are: (1) metering valve opening area; (2) time allowed for charging; and (3) transfer pump pressure curve.

The only control between engines for purposes of establishing a desired torque curve is the transfer pump pressure curve, since the other factors are common to all engines.

Torque control works as follows:

When the engine is operating at high idle speed and no load, the quantity of fuel delivered is controlled only by governor action through the metering valve.

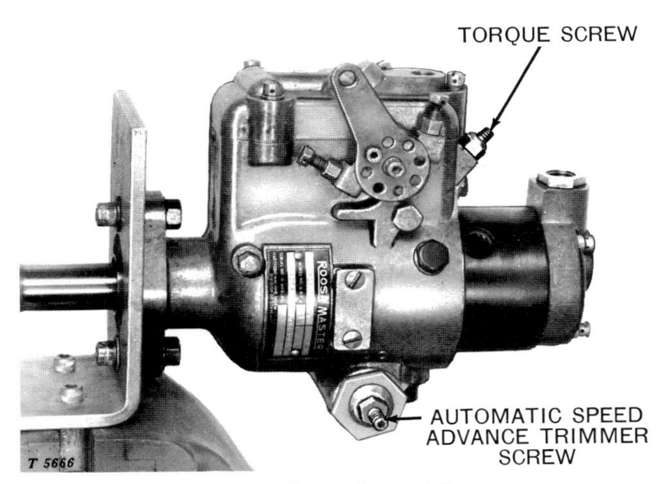

Fig. 26 — Torque Control Screw

NOTE: At this point, the torque screw (Fig. 26) and maximum fuel adjustment have no effect.

As load is applied, the quantity of fuel delivered remains dependent on governor action and metering valve position until full load governed speed is reached. At this point, further opening of the metering valve is prevented by its contact with the previously adjusted torque screw.

Thus, the amount of fuel delivered at full load governed speed is controlled by the torque screw.

Normally, the maximum fuel flow at full-load, full-speed operation is limited by the maximum roller-to-roller setting of the injection pump.

Where a torque screw is used, fuel flow is restricted at full load and full speed by limiting the metering valve opening. As additional load is applied to the engine, it slows down. The injection pump slows down as well, allowing more time between pumping strokes for fuel to flow past the metering valve.

At this slower operating speed, the pumping plungers are forced out to the maximum roller-to-roller setting and maximum fuel is delivered.

Torque adjustment can be done properly only during a dynamometer or bench test. Never attempt it on a unit in the field without means of determining actual fuel delivery.

An external screw (Fig. 26) is provided to adjust for proper engine torque reserve.

ELECTRICAL SHUT-OFF

The *electrical shut-off* device used on power units is an "energized-to-run" type. This device is housed within the governor control cover (Fig. 27).

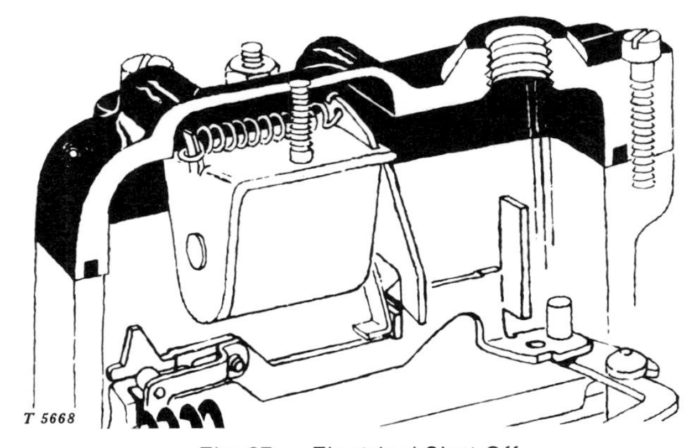

Fig. 27 — Electrical Shut-Off

De-energizing the coil allows the shutdown coil spring to release the armature. The lower end of the armature moves the governor linkage hook, rotating the metering valve to the closed position and cutting off the fuel.

1. Aneroid
2. Individual Pumping Element
3. Injection Line
4. Leak-Off Line
5. Pump Housing
6. Hand Primer
7. Sediment Bowl
8. Fuel Transfer Pump
9. Camshaft
10. Control Rack
11. Governor

Fig. 28 — In-Line Injection Pump

IN-LINE INJECTION PUMPS

The *in-line* injection pump uses an individual pump for each cylinder (Fig. 28).

The individual pumping elements are usually mounted together in "packs" or "in line" as shown.

In-line pumps give greater capacity for larger diesel engines.

MAIN PARTS OF THE PUMP

Main parts of the pump as shown in Fig. 28 are: individual pumping element (2), pump housing (5), fuel transfer pump (8), camshaft (9), control rack (10), and governor (11).

The typical pump shown is a single-acting plunger-type driven by the engine.

FUEL FLOW

Fuel flows from the fuel tank to the fuel transfer pump (8). The transfer pump sends the fuel on at low pressure to fuel filters and then back to the injection pump.

Each pumping element then meters the fuel at high pressure to its engine cylinder.

A hand primer (6) is located on the fuel transfer pump. This primer can be operated by hand to pump fuel when bleeding the system or when the fuel lines are disconnected.

 Litho in U.S.A.

FUEL TRANSFER PUMP

The fuel transfer pump (Fig. 29) assures that fuel is supplied to the injection pump at all times. The fuel is pumped through the fuel filters on its way, assuring that only clean fuel reaches the injection pump.

The transfer pump is a single-acting, piston-type pump mounted on the side of the injection pump, and is driven by a cam on the injection pump camshaft. All fuel flows through a preliminary filter located in the transfer pump sediment bowl.

The hand primer is screwed into the transfer pump housing. The pump is operated by unscrewing the knurled knob and working the knob up and down. When the knob is pulled up, the pump cylinder is charged with fuel. When the knob is pushed down, this fuel is forced through the fuel filters and into the injection pump.

INJECTION PUMP FUEL FLOW

The single-acting, plunger-type injection pump has an engine-driven camshaft which rotates at one-half engine speed. Roller cam followers, riding on the camshaft lobes, operate the plungers to supply high-pressure fuel through the delivery valves to the injection nozzles.

A governor-operated control rack is connected to the control sleeves and plungers, to regulate the quantity of fuel delivered to the engine.

Engine lubricating oil is piped into the injection pump to provide splash lubrication to the working parts.

PUMPING ELEMENTS

The pumping elements (Fig. 29) are a plunger and barrel — one set to supply each engine cylinder. The plunger is precisely fitted to the barrel by lapping to provide clearance of about 0.0001-inch (0.002 mm). With such small clearance, perfect sealing results without special sealing rings. A pumping element is always replaced as a complete unit (barrel with matching plunger).

The plungers are operated at a constant stroke; that is, they move the same distance each time the cam actuates them.

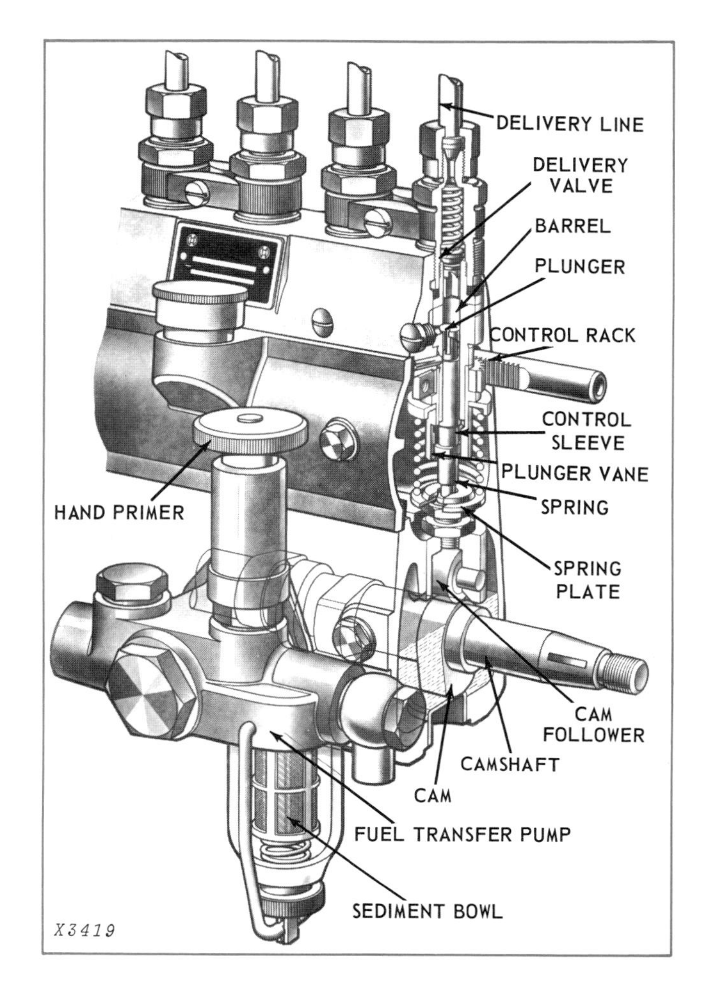

Fig. 29 — Pumping Elements

To vary the amount of fuel delivered per stroke for satisfying varying load demands, the upper part of the plunger is provided with a vertical channel extending from its top face to an annular groove, the top edge of which is milled in the form of a helix (called the control edge). On top of the plunger is a machined notch called the "retard" notch, which retards the injection timing for starting the engine.

The barrel has either a single or double control port, depending on the design. An annular groove inside the barrel routes any fuel leakage between the plunger and barrel through a hole in the barrel back to the fuel gallery. The top of the barrel is closed by a spring-loaded valve called the delivery valve.

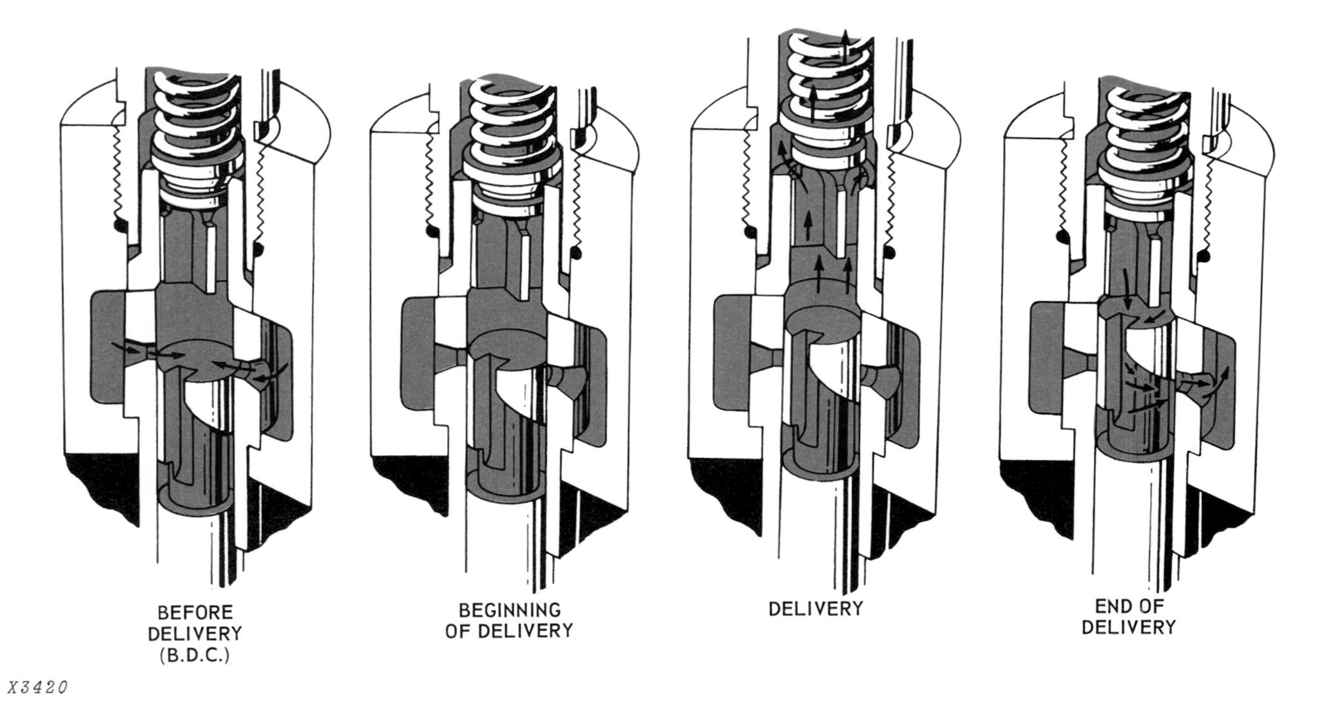

Fig. 30 — Plunger Operation (At Maximum Fuel Delivery)

OPERATION OF PUMPING ELEMENTS

When the plunger is at the bottom of its stroke (Fig. 30) fuel fills the space above the plunger, the vertical slot, and the cut-away area below the helix. The fuel flows into the area from the fuel gallery through ports in the barrel.

As the plunger moves upward, it closes the barrel ports, and discharges the fuel trapped in the pressure area through the delivery valve and line to the injection nozzle. Delivery of fuel stops as soon as the control edge of the helix uncovers the control port. Fuel then flows out through the vertical slot and annular groove back into the fuel gallery.

PLUNGER POSITIONS

No Fuel Delivery

No fuel delivery occurs when the plunger is rotated to a position where the vertical slot in the plunger aligns with the control port (Fig. 31). Since the vertical slot prevents the control port from being covered, pressure cannot built up. Hence, no fuel can be forced to the injection nozzles.

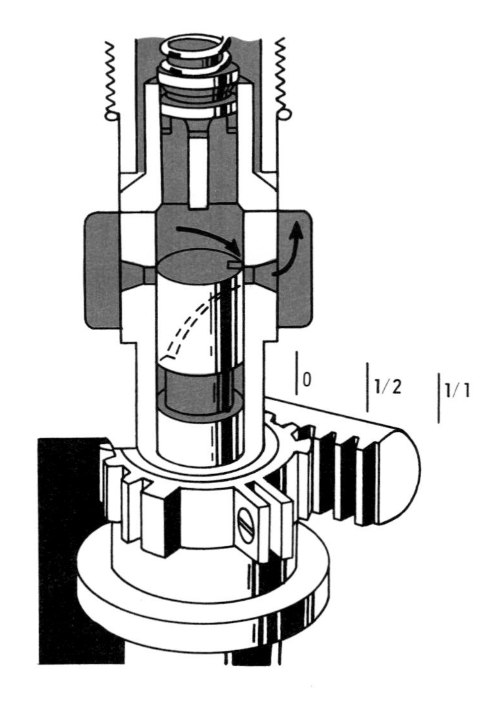

Fig. 31 — No Fuel Delivery

Partial Fuel Delivery

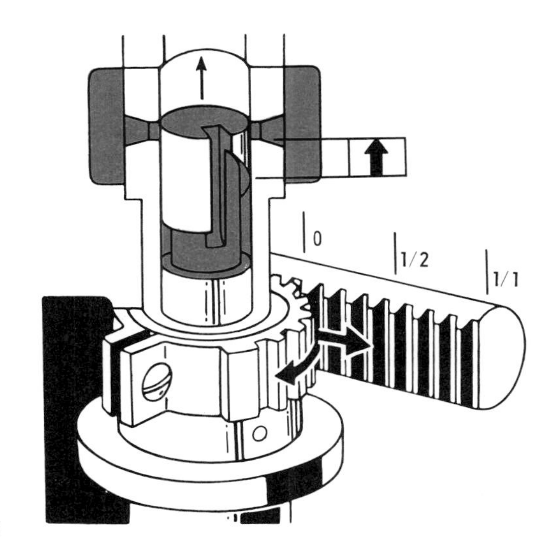

Fig. 32 — Partial Fuel Delivery

Partial fuel delivery occurs at any position of the plunger between no delivery and maximum delivery, depending on the position of the helix in relation to the control port (Fig. 32).

Maximum Fuel Delivery

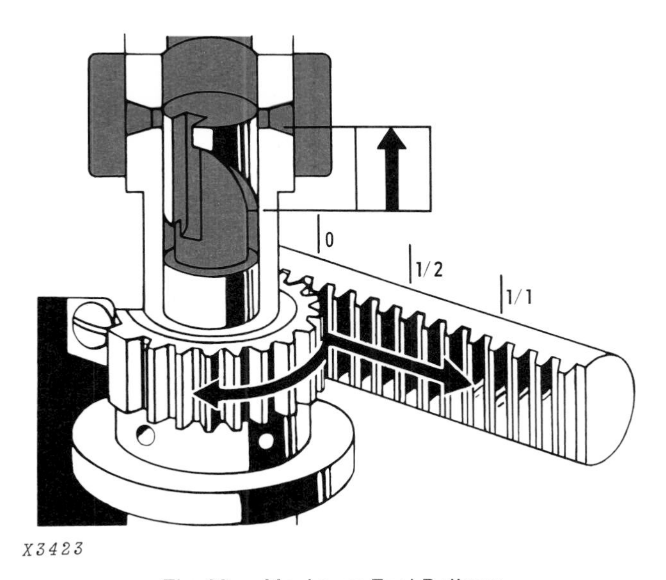

Fig. 33 — Maximum Fuel Delivery

Maximum fuel delivery (Fig. 33) occurs when the plunger is rotated to a position where the control port is covered by the plunger as it moves upward, for the greatest possible distance (effective stroke) permitted by the control rack.

Excess Fuel Delivery

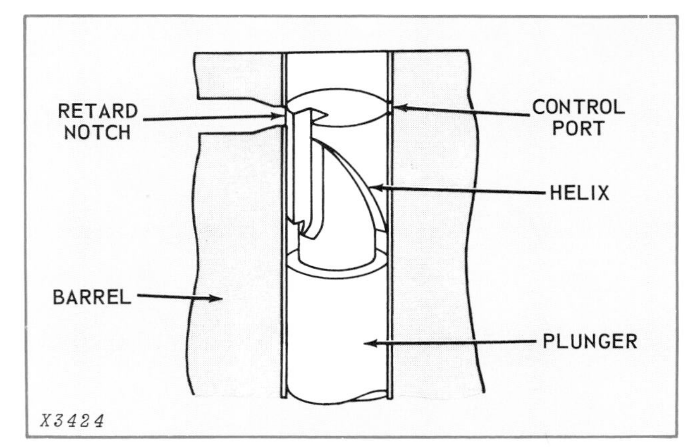

Fig. 34 — "Excess Fuel" Delivery and "Retard Notch" (Exaggerated)

For "excess fuel" delivery (Fig. 34), the plunger is rotated to a point where the extreme edge of the helix or annular groove provides the maximum effective pumping stroke permitted by design. This "excess fuel" position should not be confused with the maximum fuel position, since "excess fuel" is available for use only when starting the engine, while "maximum fuel" is used when the engine is running.

To obtain the "excess fuel" position, move the speed control lever to the slow idle position while the engine is stopped. Doing so allows the rack to move to its extreme forward position at cranking speed. This moves the plunger to a position to deliver "excess fuel" for starting.

Excess fuel aids in engine starting, especially when the engine is cold. As soon as the engine is started, the control rack moves the plungers from the "excess fuel" position to a fuel delivery position corresponding to the settling of the speed control lever.

Retard Notch

A "retard notch" (Fig. 34) on the upper face of the plunger also aids in starting a cold engine. This "retard notch" is aligned with the control port when starting the engine, which means that the plunger must move farther upward before the control port will be completely closed. This delay in port closing results in the desired timing retardation.

When the "retard notch" is aligned with the control port, the plunger helix is designed to be in the "excess fuel" position. In this way, both "excess fuel" and retardation of injection timing work together when starting the engine.

 Litho in U.S.A.

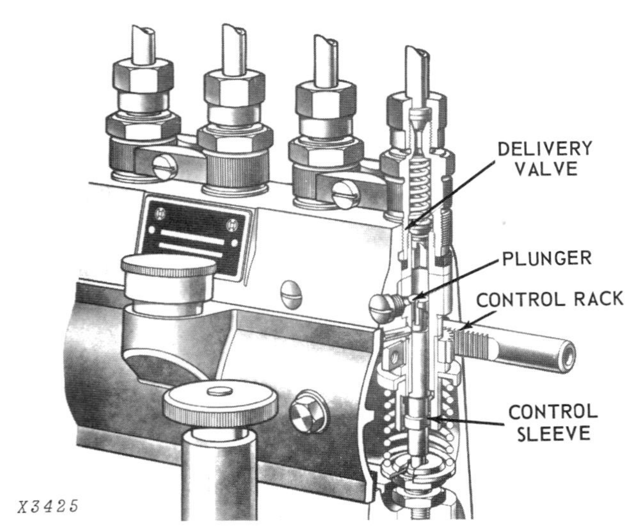

Fig. 35 — Control Rack, Sleeve, and Delivery Valve

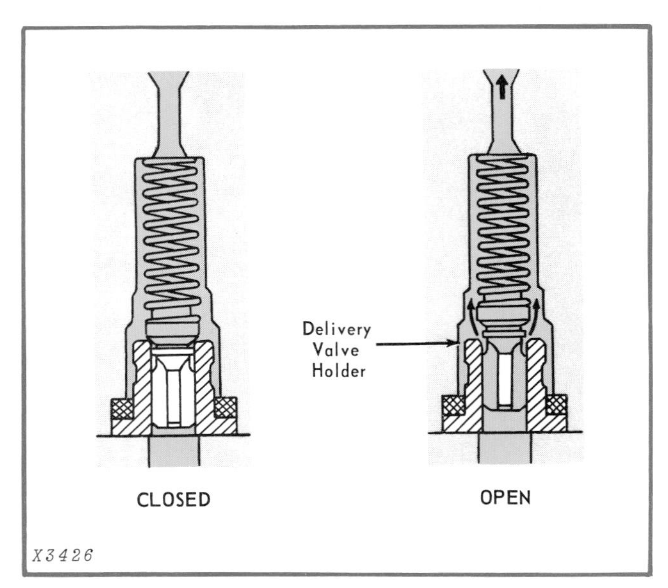

Fig. 36 — Delivery Valve

CONTROL RACK AND SLEEVE

The control rack (Fig. 35) is connected to the governor by linkage which moves the rack to regulate the speed of the engine.

The sleeve, which is actuated by the control rack, is fitted over the barrel and accepts the vanes on the plunger. This provides positive rotation of the plunger when the sleeve is rotated by the rack.

Some pumps have a clamped-on tooth segment which meshes with the teeth on the control rack.

Other pumps have a pin at the upper end of the control sleeve which engages a slot in the control rack.

Also affecting the control rack travel, is the starting fuel control shaft which permits the rack to move to the "excess fuel" position described earlier.

DELIVERY VALVE

The delivery valve (Fig. 36) is guided by its stem in the valve housing.

During the fuel delivery stroke, the valve is forced off its seat and fuel is forced along the longitudinal grooves over the valve face into the delivery line.

When the helix on the pump plunger uncovers the control port, the pressure drops suddenly in the pump barrel. The pressure in the delivery line and the valve spring force the valve back on its seat. When this happens, the movement of the valve relieves the pressure in the delivery line, thus preventing dribbling at the injection nozzle.

To do this, the delivery valve is constructed with a relief plunger, which fits into the valve holder. When the delivery stroke ends, and the valve starts to resume its seated position, the relief plunger will move into the bore of the valve holder, thus sealing the delivery line from the pressure chamber. After the relief plunger has entered the bore of the valve housing, the valve seats firmly. Now the space for fuel in the delivery line is increased by an amount equal to the volume of the relief plunger. The effect of this increase in volume is a sudden pressure drop in the delivery line, causing the injection nozzle valve to close instantly.

NOTE: For service instructions on in-line injection pumps, refer to the pump Technical Manual.

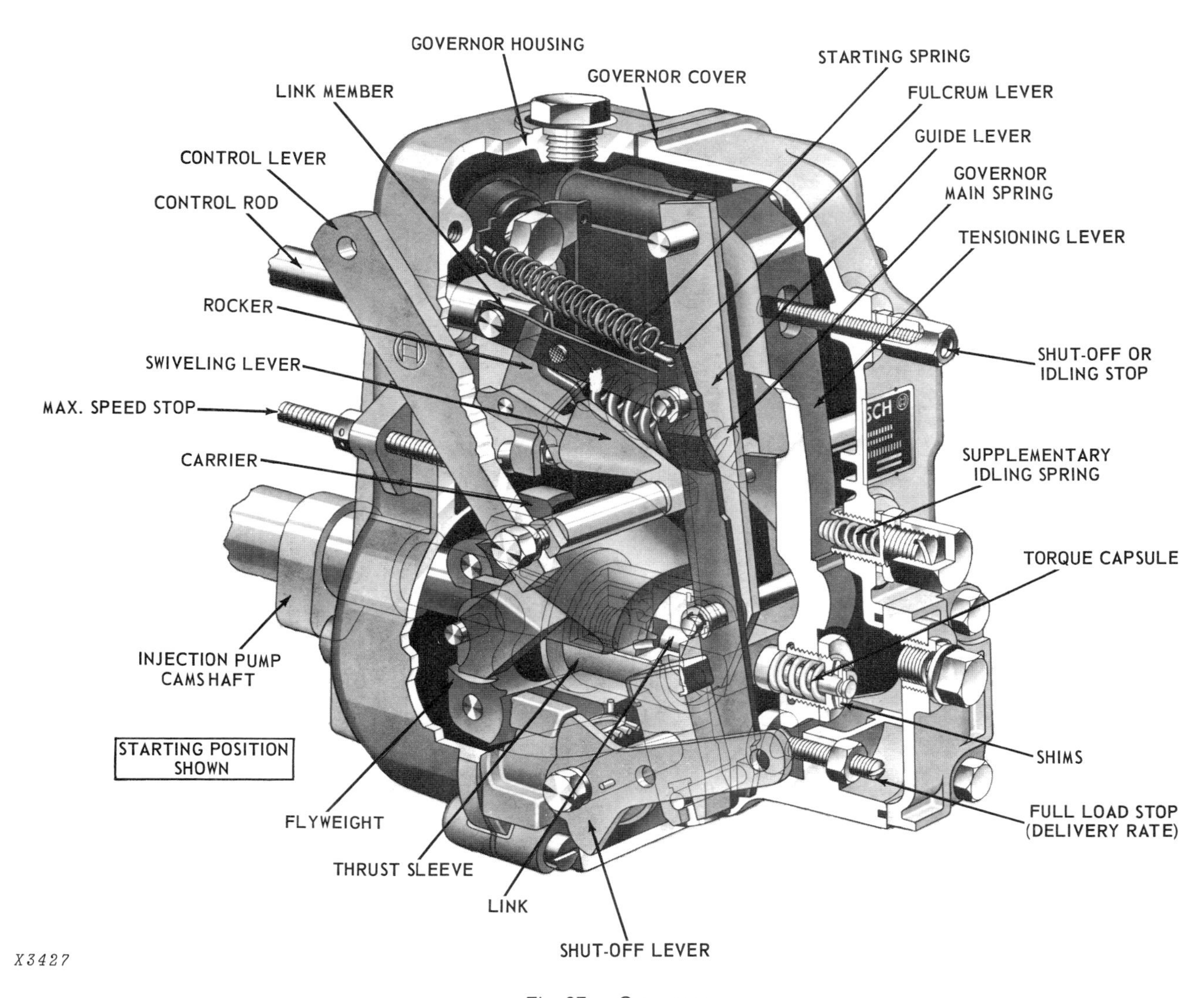

Fig. 37 — Governor

GOVERNOR

Description

A **governor** (Fig. 37) is a vital component of a diesel fuel system. It maintains a nearly constant engine speed, at any point between idling and maximum speed positions.

The desired speed is set by the operator, using the speed control lever. Then the governor, by operating the injection pump control racks, maintains a nearly constant speed by varying the amount of fuel supplied to the engine to satisfy varying load demands.

The governor illustrated in Fig. 37 is a centrifugal, variable-speed type. This governor is capable of maintaining a steady speed in the range between idling and maximum speed positions.

The governor is mounted at the rear of the injection pump housing. A governor flyweight assembly is mounted on the injection pump camshaft. The flyweight assembly force, acting through the governor linkage, moves the control rack to provide the desired speed regulation.

The governor is completely enclosed to permit splash lubrication of the working parts, using oil in common with the injection pump.

An operating lever shuts off fuel delivery to the engine by moving the control rack to the "stop" position.

OPERATION OF GOVERNOR

As engine speed increases, the governor flyweights move outward until the force created by the whirling weights equals the counterforce of the governor main spring (Fig. 37).

As engine speed decreases, centrifugal force diminishes and the weights swing inward.

Movement of the flyweights is transmitted to the injection pump control racks through the governor linkage to obtain fuel delivery corresponding to the desired speed setting.

Movement of the governor linkage under varying engine requirements are described on the following pages.

Starting the Engine

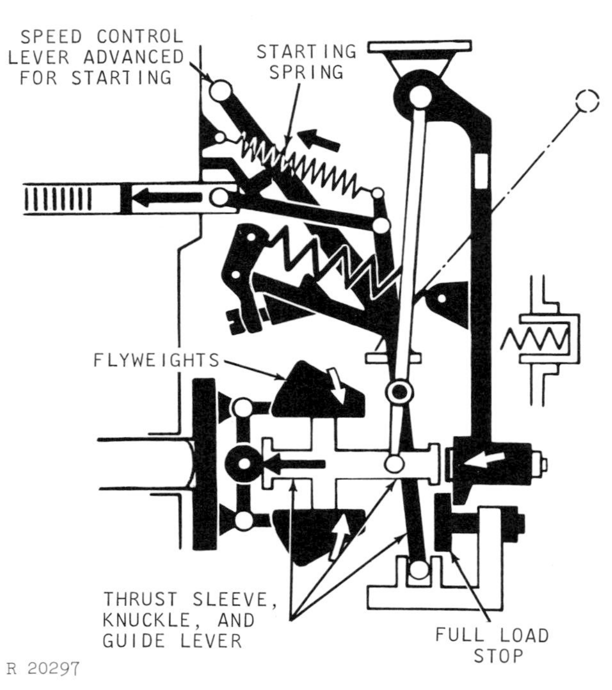

Fig. 38 — Governor Position When Starting

With the engine stopped, and the speed control lever in the slow idle position and then advanced slightly, the starting spring will pull the control rack to the "excess fuel" position (Fig. 38). At the same time the tensioning lever is moved up against the full-load stop, which also moves the guide lever, knuckle, and thrust sleeve forward. The flyweights then come to rest against the thrust sleeve (innermost position).

While the starter is cranking the engine, the injection pump begins supplying excess fuel to the engine.

Once the engine starts, the centrifugal force produced by the whirling flyweights overcomes the starting spring tension (even before idling speed is reached).

The engine speed increases until the centrifugal force of the flyweights and the governor main spring are balanced.

Engine Idling

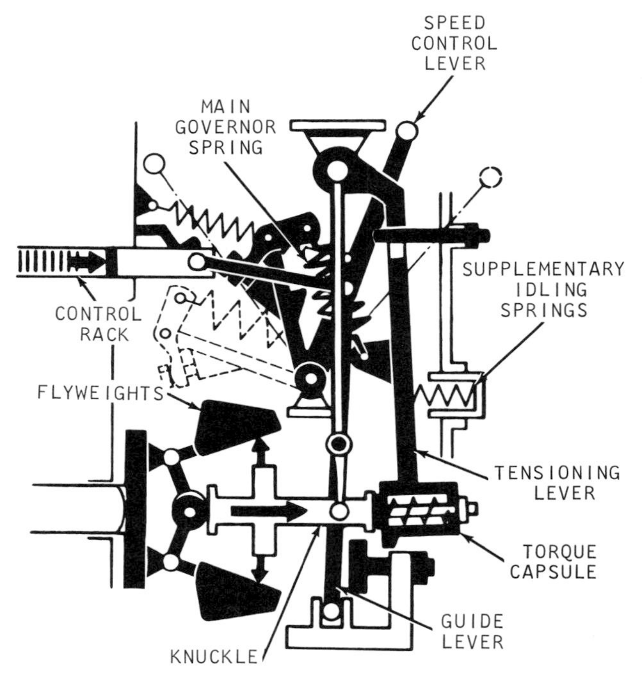

Fig. 39 — Governor Position with Engine at Slow Idle

When the engine is idling (Fig. 39), the governor starts to function automatically. At this speed the governor main spring is almost free of tension. Therefore, the governor main spring has only a slight effect on the governor linkage at idling speed. This means that the flyweights can swing outward (even at low speed) with very little resistance.

As the control lever and guide lever are moved, the fuel control rack is also moved, increasing the tension on the governor main spring.

Since the centrifugal force and spring tension are relatively low at idle speed, the torque capsule in the tensioning lever is only slightly compressed. The gap between the knuckle and the tensioning lever is therefore greater at low speed than at high speed, causing the tensioning lever to contact the supplementary idling spring, resulting in the desired speed regulation.

Engine at Medium Speed

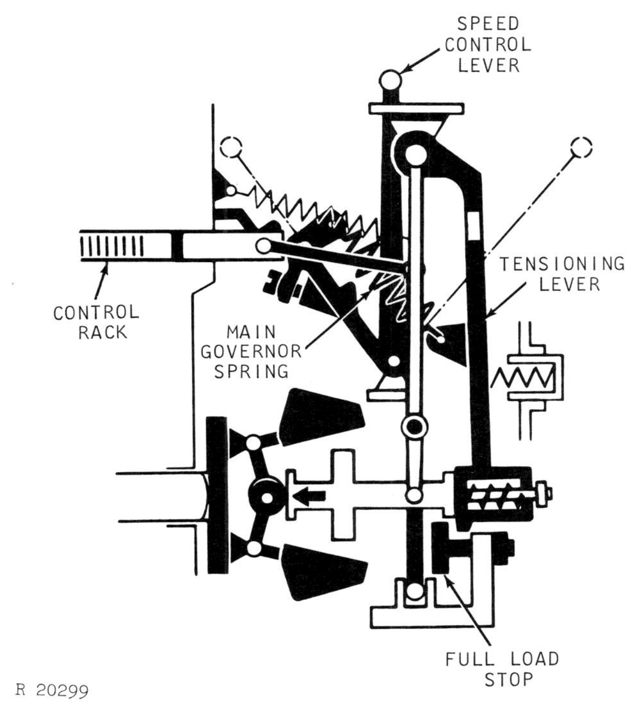

Fig. 40 — Governor Position with Engine at Medium Speed

Any movement of the speed control lever above idling will cause the control rack to move to the maximum fuel delivery position, with the tensioning lever moving to the full-load stop (Fig. 40).

The injection pump delivers more fuel to the engine, resulting in increased speed. As soon as the centrifugal force exceeds the force of the governor main spring (determined by the position of the speed control lever), the governor linkage moves the control rack to a position where the centrifugal force is just equal to the spring force. This results in a lower fuel delivery rate.

By this action, the governor will maintain a steady engine speed.

Engine at Maximum Speed

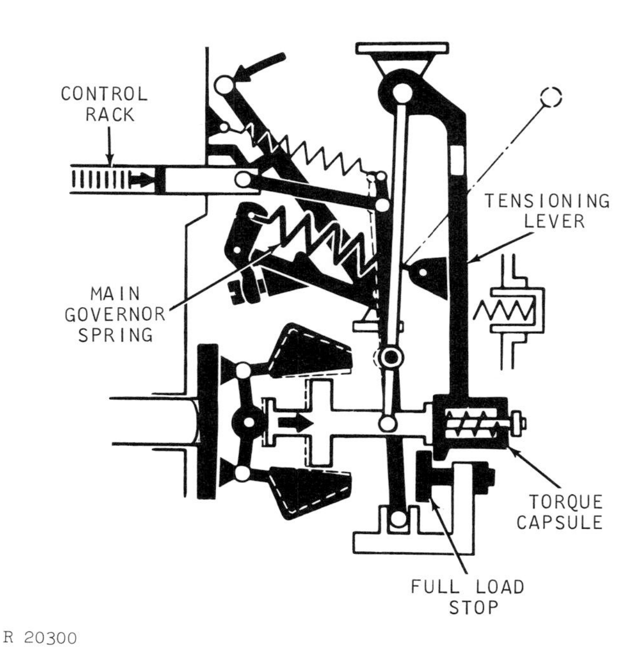

Fig. 41 — Governor Position with Engine at Maximum Speed

The governor operation at maximum rated speed is about the same at medium speed, except that the tension lever stretches the governor main spring to the fullest.

With the governor main spring fully tensioned, the tension lever is moved against the full-load stop with greater force. This means that the control rack is moved into the maximum fuel delivery position (Fig. 41).

The torque capsule is always compressed, once the tensioning lever is moved against the full-load stop. It will remain compressed until the engine speed is reduced enough to reduce the centrifugal force of the governor flyweights. This forces the fuel control rack into a position to provide adequate torque reserve.

Once governed full-load (maximum) speed is reached, governor response will regulate fuel delivery between full load and fast idle to handle varying loads so long as there is no overload.

Actuating the engine shut-off lever (Figs. 42 and 43) moves the stop device which, in turn, moves the control rack to shut off the fuel supply to the engine. This movement takes place independently of the flyweight and speed control lever positions.

Stopping The Engine

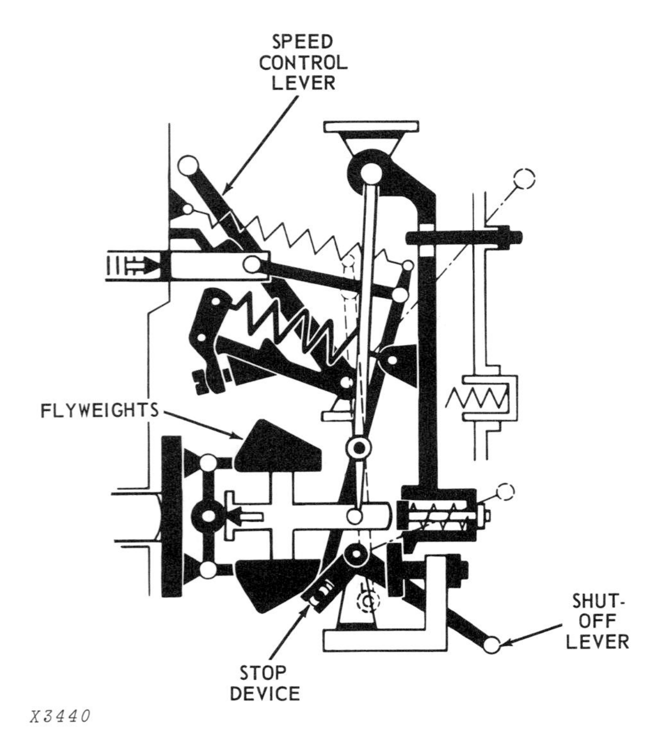

Fig. 42 — Governor Position When Stopping the Engine (With Stop Device)

The supporting lever of the stop device (Fig. 43) is coupled to the shaft and shut-off lever by three pressure springs. The supporting lever continues to pivot until the control rack is in the "stop" or "no fuel" delivery position.

At this position, the supporting lever stops moving and the three pressure springs become tensioned, as the shut-off lever moves to the limit of its travel.

As the engine speed decreases, the tension on the three springs lessens, pushing the supporting lever (with the lower end of the fulcrum lever) farther forward, as the flyweights close. The upper end of the fulcrum lever remains nearly stationary to the fuel control rack in the "stop" position.

Speed Droop

Speed droop is the variation in engine speeds between full-load and no-load speed range, and is usually expressed as a percentage of rated speed.

For example, if the engine has been operating at full load, and the load is suddenly removed, the engine speed will increase to fast idle speed.

The amount of permissible speed increase is determined by governor design, but is usually no

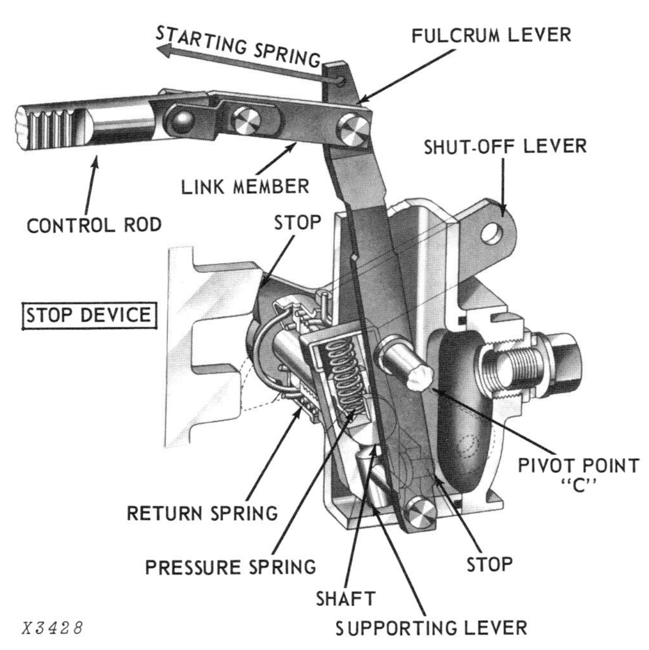

Fig. 43 — Governor Stop Device for Stopping Engine

more than ten percent. For example, if the maximum full-load engine speed is 2200 rpm, the no-load speed may be 2420 rpm.

ANEROID

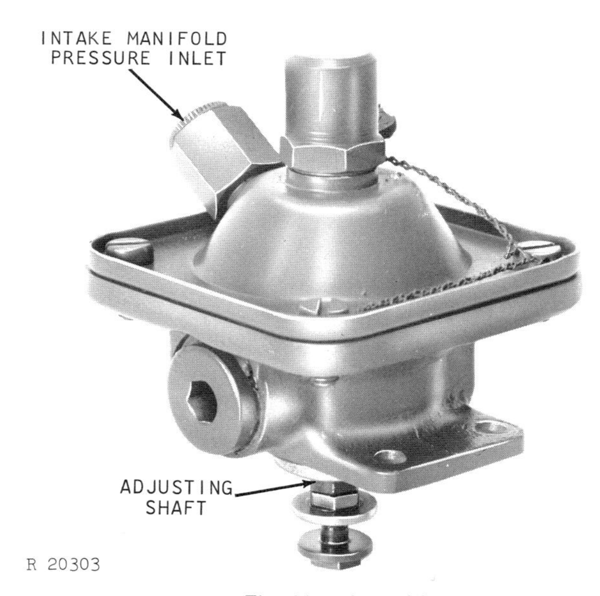

Fig. 44 — Aneroid

Description

The **aneroid** (Fig. 44) is a device mounted on top of the pump governor housing. It limits the fuel supply to the engine and thereby prevents excessive smoke.

The device is operated by the action of intake manifold pressure on a diaphragm (Fig. 45). The aneroid fuel control shaft engages the control rack either mechanically or hydraulically.

Adjustment is provided to obtain satisfactory acceleration with a minimum amount of smoke.

Operation (Mechanical Activator)

Without aneroid control, when the speed control lever is moved to increase the engine speed, even if only partway, the injection pump rack immediately moves to maximum fuel delivery position. Now, since there is no immediate increase in exhaust pressure to increase the speed of the turbocharger (to supply more air), there is not enough air in the engine cylinders to burn all the injected fuel. The result is a cloud of dense black smoke from the engine exhaust.

As we said, the job of the *aneroid* is to prevent this smoke.

The aneroid works as follows:

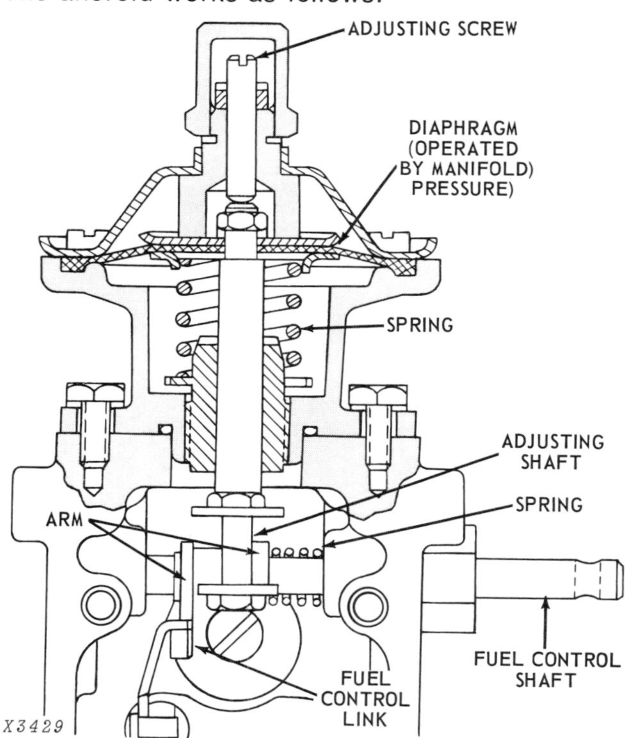

Fig. 45 — Cutaway View of Aneroid

When the engine stop knob is pulled to stop the engine, the pump fuel control rack is moved to the position where no fuel is injected into the engine. At the same time, the aneroid fuel control link (Fig. 45) is moved to where it no longer contacts the arm on the fuel control rack.

When the operator advances the engine speed control lever to start the engine, the starter spring (Fig. 46) moves the fuel control rack to the "excess fuel" position.

While the starter is cranking the engine, the injection pump is supplying excess fuel to the engine cylinders.

Once the engine starts, the centrifugal force of the governor flyweights overcomes the starter spring tension, even before slow idle speed is reached.

As the fuel control rack is moved from the excess fuel position, the spring on the aneroid fuel control lever shaft (Fig. 45) moves the control link back to its original position.

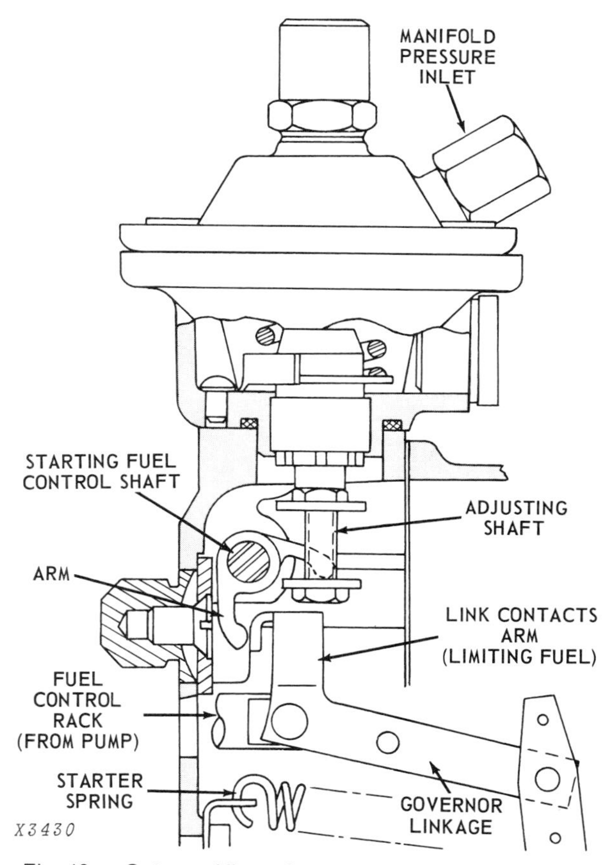

Fig. 46 — Cutaway View of Aneroid Operating Mechanism

The arm on the fuel control rack contacts the aneroid fuel control link before the rack has moved much over one-half of its full travel (Fig. 46). This prevents more fuel being injected into the engine than there is air to burn it.

The lever will hold the rack in this position until the engine turbocharger has created enough intake manifold pressure through the diaphragm to move the lever to release it. This action occurs every time the engine is accelerated.

As a result, this action limits the amount of fuel

injected into the engine during the acceleration period, thereby limiting the amount of black smoke emitted from the exhaust.

Operation (Hydraulic Activator)

This mechanism hydraulically disengages the aneroid from the control rack during engine start-up. It also engages the aneroid with the control rack during acceleration.

With the engine shut off, engine oil pressure to the hydraulic activator is zero. Spring pressure moves the piston and control shaft to the left (Fig. 47) disengaging aneroid arms from injection pump control rack. This prevents aneroid operation.

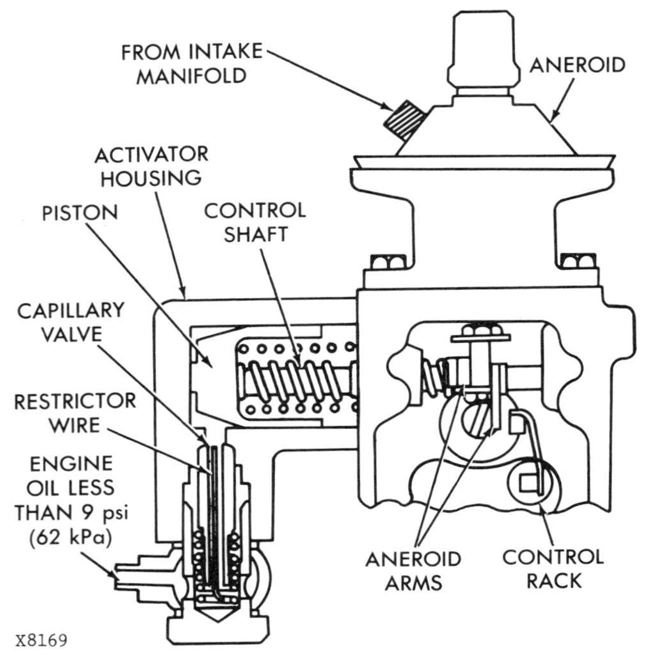

Fig. 47 — Hydraulic Activator Operation (Engine Off)

During initial start-up, the capillary valve momentarily restricts pressure oil movement into the aneroid and delays aneroid engagement. The length of delay depends on ambient temperature and engine oil viscosity.

When engine oil pressure reaches 9 psi (62 kPa), oil moves through the capillary valve into the aneroid (Fig. 48). The pressure causes the piston to move to the right. This movement forces the fuel control shaft and aneroid arms to contact the control rack, preventing more fuel being injected into the engine than there is air to burn it.

Again, the arms hold the rack in this position until the turbocharger creates adequate pressure to move the arms and release the rack. This occurs each time the engine is accelerated.

INTRODUCTION TO INJECTION PUMP SERVICE

Remember these points when servicing an injection pump:

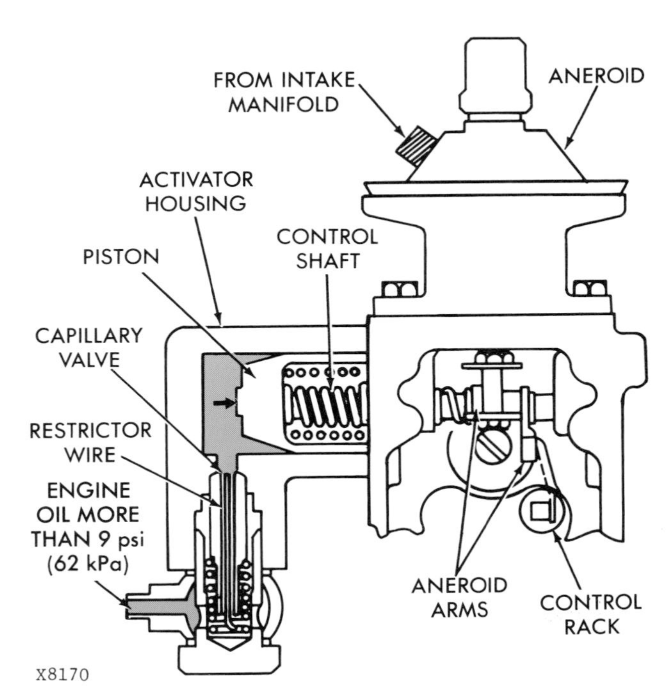

Fig. 48 — Hydraulic Activator Operation (Engine On)

1) Disassemble and inspect the injection pump only if there is a clear sign it is malfunctioning.

2) Be sure to trouble shoot the fuel system and pinpoint the failure before you begin to service a component. (See the "Trouble Shooting" section at the end of this chapter.)

3) The pump is a precision instrument and can be easily damaged by careless service.

4) Only a qualified mechanic with the proper tools should disassemble or service the injection pump.

5) The story which follows is not meant as a detailed instruction for servicing and testing the injection pump. The Technical Manual will provide this information.

SERVICING AND TESTING THE PUMP

To perform quality service on an injection pump, you need these items:

1. A clean, well-lighted work area isolated from the rest of the shop.

2. Tools that are clean and in good condition. (Many injection pumps require special tools for servicing.)

3. Accurate test equipment as prescribed by the pump's Technical Manual.

4. Technical publications that give detailed service procedure and specifications for the particular pump.

Service Procedures

Remember, the services described here are general and are only for a *typical* injection pump.

Get detailed service and specifications from the pump Technical Manual.

Prior to removal, clean the pump exterior and the area around it.

A word of caution—**never spray a warm pump with cold water during operation.** This could cause the pump to seize. Steam cleaning can also harm the pump at these times.

When removing injection lines from the pump, cap the lines as well as the pump ports to prevent dirt from entering.

Inspection Of Pump Parts

1. Sticking parts may have a coating of light varnish on them. If so, use a soft brush and lacquer thinner to remove it.

2. Some matching parts such as rotors and hydraulic heads must be replaced as a set. (The fit on some of these parts is so close that heat from your hand will expand them enough to temporarily make them oversize.)

3. Replace all seals and gaskets regardless of their condition. Use special tools and lubricants where recommended.

4. Flush all parts in clean fuel as they are reassembled.

Testing and Adjusting The Injection Pump

The only proper method of testing the injection pump is with an injection pump test stand (Fig. 49). Since each tester has its own set of instructions, be sure to follow them.

After the pump is connected to the tester, run it for 10 minutes. This will warm the pump and bleed air, through loosened injection lines.

LEAKAGE TEST

With all connections tight, run the pump at full-load rated speed for several minutes. Clean the pump exterior and dry it with compressed air. Then check the exterior for any leaks and repair, if necessary.

VACUUM TEST

The vacuum test used on some pumps helps determine if air is being drawn into the pump.

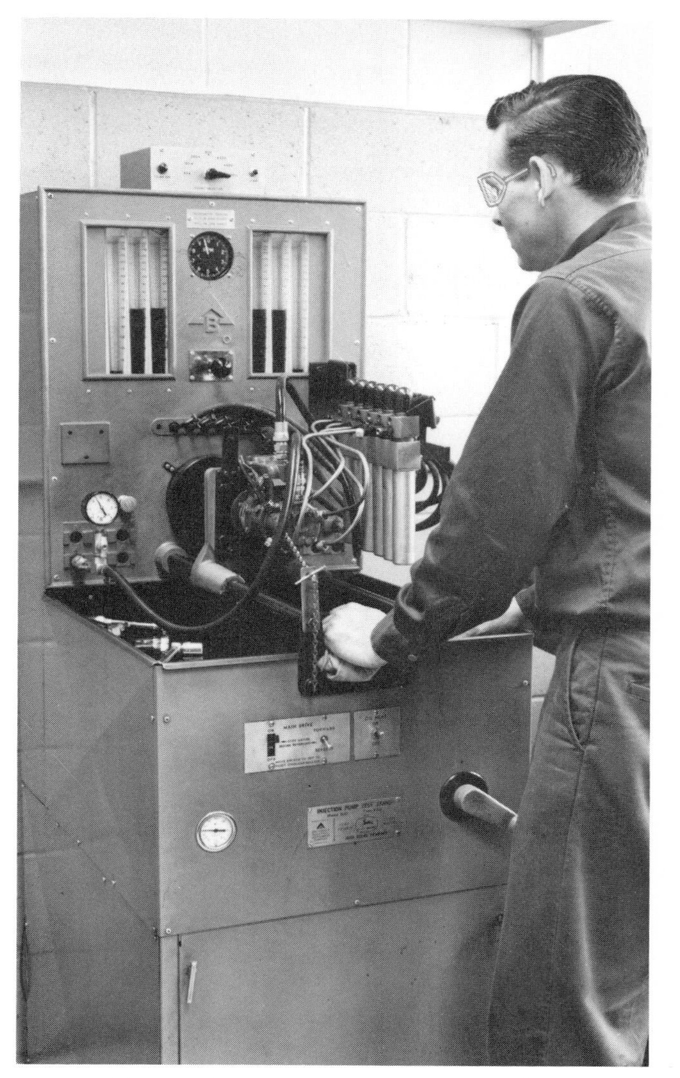

Fig. 49 — Using An Injection Pump Test Stand

With the fuel supply cut off, the test gauge should read a specified number of inches of mercury (Hg). If it doesn't, there is an air leak.

TRANSFER PUMP PRESSURE TEST

Transfer pump pressure is tested at various speeds. The correct pressures are given in the pump specifications.

FUEL DELIVERY TEST

Graduated glass tubes in the pump test stand measure the fuel at various speeds. The amount of fuel injected into the test tubes is recorded at various speeds and the average is compared to specifications.

IDLE ADJUSTMENT

Slow and fast idle adjustments can also be made to specifications on most test units. Once these

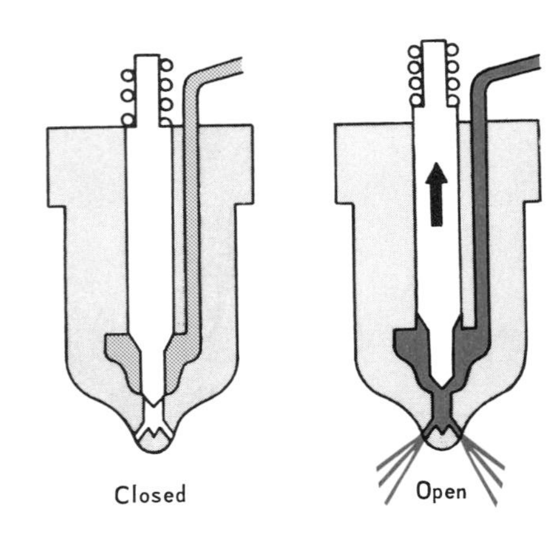

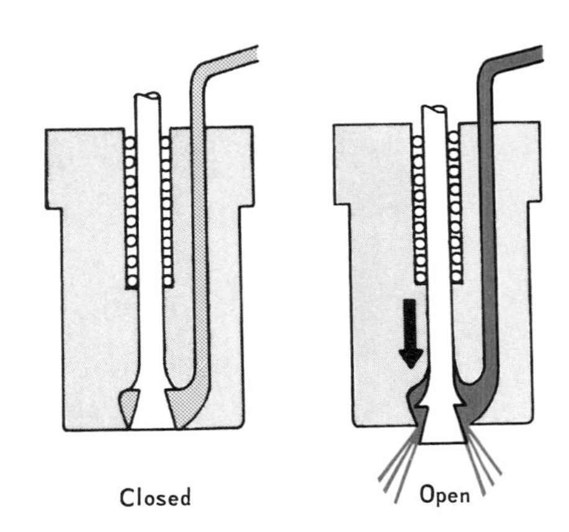

OUTWARD–OPENING NOZZLE

Fig. 50 — Nozzles Compared

adjustments are made, the pump is ready to install on the engine.

TORQUE CONTROL ADJUSTMENT

If the pump is equipped with a torque adjustment, make it while the pump is on the test stand.

Installing And Timing The Pump

When installing the pump, be sure it is timed to the engine. Various timing tools are used to line up the pump with the engine.

The advance mechanism (if used) should also be adjusted after the pump is installed. This will prevent misfiring or loss of power during engine operation.

INJECTION NOZZLES

The injection nozzle must do two jobs:

1) Atomize the fuel for better combustion

2) Spread the fuel spray to fully mix it with air

In addition, all nozzles in multi-cylinder engines must inject fuel equally for smooth power.

TYPES OF NOZZLES

Nozzles are simple devices. They use a spring to oppose fuel pressure until the right instant for injecting fuel, when the nozzle valve opens. The opening pressure usually can be adjusted.

Most nozzles today are **closed** types—that is, the fuel pressure acts on only one side of the nozzle valve.

Closed-type nozzles are classified as follows:

INWARD-OPENING NOZZLES

- *Differential Pressure*
- *Hole Type*
- *Pintle Type*
- *Pintaux Type*

OUTWARD-OPENING NOZZLES

- *Poppet Type*
- *Pintle Type*
- *Multi-Orifice Type*

In INWARD-OPENING NOZZLES, the fuel pressure acts on the lower end of the valve, moving it inward to release a fuel spray (Fig. 50). A spring at the upper end of the valve is normally adjusted to set the opening pressure.

In OUTWARD-OPENING NOZZLES, the fuel pressure acts on the valve only in the direction of flow (Fig. 50). Since combustion chamber pressures tend to close the valves, the opening pressure is relatively low. No fuel leak-off is required.

Let's discuss each nozzle under the two broad types—inward- and outward-opening.

INWARD-OPENING NOZZLES

Differential pressure nozzles are operated hydraulically. When fuel pressure to the nozzle rises to the set level, the nozzle valve opens against spring pressure and injects fuel. As pressure drops again, the valve closes and injection stops. The valve shown at left in Fig. 50 operates like this. This nozzle can be adjusted by spring tension for accurate operation.

Hole-type nozzles provide some extra features. The number and size of the spray holes can be varied for the exact spray pattern needed. These nozzles are best for engines with open combustion chambers.

Pintle-type nozzles use a tapered valve which seats in a single orifice in the valve body (see at right in Fig. 50). The pintle-type is fairly simple but produces a hollow spray and is not as versatile as the hole type.

Pintaux-type nozzles are a variation on the pintle type. An extra hole is added in the head to spray out fuel at an angle during low engine speeds. This gives an aid for starting the engine (as in cold weather). The extra hole is angled toward the "hottest" part of the combustion chamber for faster firing.

OUTWARD-OPENING NOZZLES

In all these valves, the opening pressures are fairly low, as explained above.

A **poppet-type nozzle** that opens outward sprays with a fine vapor, but has too little penetration for most diesels.

A **pintle-type nozzle** that opens outward is shown at right in Fig. 50. Fuel pressure acts on the tapered head to unseat the valve at a preset pressure. As the valve moves, the extra shoulder is exposed to fuel pressure causing the valve to open rapidly for a good spray without dribbling fuel. This nozzle is fairly simple and very lightweight.

Multi-orifice nozzles are a variation on the pintle type. The pintle fits closely on its seat and when it starts to move the fuel sprays out a ring of holes instead of past the pintle and out the head.

OPERATION OF A TYPICAL NOZZLE

Let's see how a typical nozzle works. In Fig. 51 we show a differential-pressure nozzle which opens inward.

Parts Of The Nozzle

The nozzle body contains the moving parts—the valve and spring.

Nozzle opening pressure is controlled by the adjusting screw on the valve spring.

How far the valve moves when injecting fuel is controlled by the lift adjusting screw.

How The Nozzle Works

Metered fuel at high pressure from the pump enters the fuel inlet (Fig. 51).

This fuel surrounds the nozzle valve and forces the valve from its seat at a preset pressure. A measured amount of fuel then sprays out the tip into the engine combustion chamber at high velocity.

As the fuel is injected, pressure drops and the spring is able to close the valve very rapidly.

A small amount of fuel leaks past the nozzle valve and lubricates the working parts.

The excess lubricating fuel is removed from the top of the nozzle at the fuel leak-off and returns to the tank.

SERVICE AND TESTING OF NOZZLES

Injection nozzles are precision parts operating rapidly in conditions of heat and pressure, so they can fail.

Usually, nozzle failure is caused by dirt clogging the orifices and passages. This can be easily remedied by thorough cleaning.

However, a nozzle does not have to stop injecting fuel completely to be considered a failure. On the next page is a list of causes of nozzle failures.

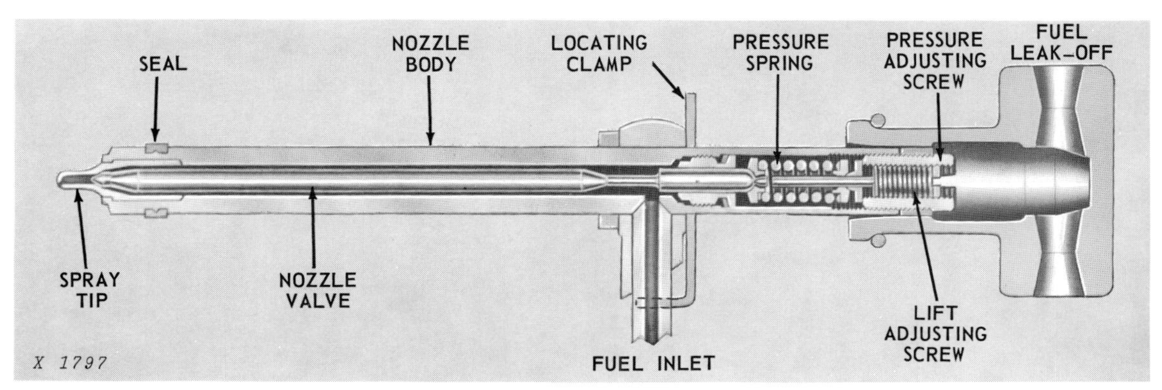

Fig. 51 — Typical Injection Nozzle Which Opens Inward and Operates By Pressure Differences

Nozzle Trouble Shooting Chart

Problem	*Possible Cause*
Nozzle opens at wrong pressure	Wrong spring pressure adjustment
	Broken spring
Nozzle will not open	Plugged orifices (spray tip only)
Poor spray pattern	Plugged or chipped orifices (spray tip only)
	Chipped or broken pintle end
	Deposits on pintle seat
	Chipped pintle seat
Poor misting fuel	Plugged or chipped orifices (spray tip only)
	Valve not free
	Cracked tip
Valve operates erratically	Valve spring misaligned
	Spring broken
	Deposits on pintle seat
	Bent valve
	Distorted body
Valve will not operate	Valve spring misaligned
	Spring broken
	Varnish on valve
	Deposits in seat area (pintle tip only)
	Bent valve
	Valve seat eroded or pitted
	Distorted body
Too much fuel leaks off	Valve guide wear
Too little fuel leaks off	Varnish on valve
	Not enough clearance between valve and guide

Removing The Nozzles

First clean the area around the nozzle. Next, remove and cap the injection and leak-off lines. Then remove the nozzles from the engine. (Some nozzles may require a special tool for removal).

Cleaning The Nozzles

Soak the entire nozzle assembly in clean solvent or fuel after discarding the outer seals. Clean the spray tip and nozzle body with a **brass** wire brush. **Never use emery cloth or a steel wire brush to clean nozzles as the precision tips will be damaged.**

Testing Of Nozzles

When testing the nozzles, always follow the Technical Manual.

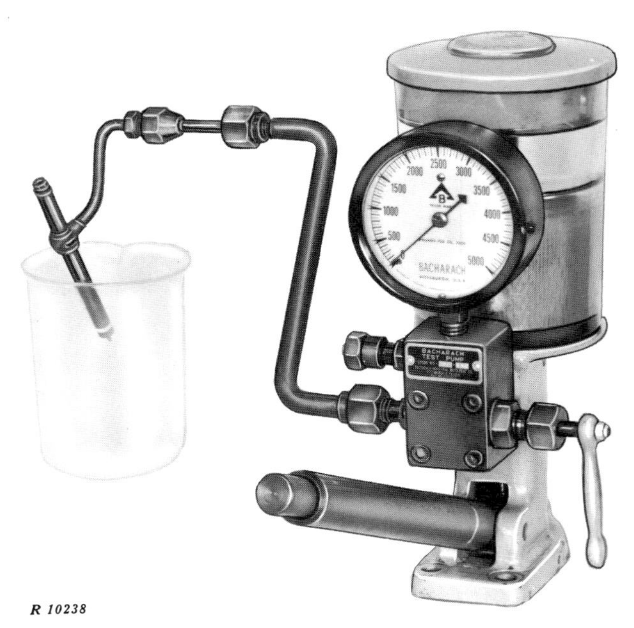

Fig. 52 — Nozzle Tester

A nozzle tester is used in making these tests (Fig. 52). It is a high-pressure hand pump which forces fuel through the nozzle. Usually the nozzle is tested for three things:

1) Chatter and spray pattern

2) Opening pressure

3) Valve leakage

A word of caution when using this type of tester: The fuel comes out of the nozzle at extremely high-pressure — which can penetrate clothing and skin and cause injury. Always keep the nozzle pointed away from you or, better yet, enclose it in a transparent beaker as shown in Fig. 52.

After the nozzle is attached to the tester, flush out the nozzle by rapidly pumping the tester (with the gauge closed).

Spray Pattern Check

Check the spray pattern first. The fuel should be finely atomized and distributed evenly. There should be no stream or large visible drops.

If the spray pattern is poor, look for a clogged or eroded orifice or a bent valve.

Opening Pressure Test

The pressure needed to open or "crack" the nozzle is checked by pumping the tester steadily until the tester gauge needle falls rapidly. Check this pressure reading against the specifications.

If the pressure is too low, a weak or broken spring might be the cause or the spring pressure may need adjustment.

If the pressure is too high, the tip may be plugged or the valve may be binding in the valve guide.

Summary: Testing Of Nozzles

If all tests prove that the nozzle is working okay, don't disassemble the nozzle any further.

If the tests show a failure, the nozzle must then be disassembled, inspected, cleaned or repaired.

Introduction To Nozzle Service

1) All nozzles require careful handling. See your nozzle or engine Technical Manual for the exact service procedures.

2) All nozzles require a special tool kit to perform any service. Never attempt to service them without it.

3) When working on several nozzles, DO NOT mix nozzle parts.

Cleaning Nozzle Parts

Most nozzle servicing is just cleaning off carbon and dirt deposits. A good soaking in clean solvent

Fig. 53 — Nozzle Orifices Eroded By Use Of A Steel Wire Brush For Cleaning

will remove much of the residue. Usually, a **brass** wire brush will clean off the remainder. **DO NOT use a steel wire brush.** It can scratch smooth surfaces or erode the spray orifices (Fig. 53).

Special tools from the nozzle kit are often needed to clean internal seats, orifices, and passages. For example, the tiny spray orifices in some nozzles tips are very critical in size and require special gauged cleaning wires.

Inspecting Nozzle Parts

Normal wear on valve parts can be ignored unless it is causing the valve to fail.

Inspect the parts primarily for abnormal wear, chipped edges, scratches, misalignment, and broken parts. The trouble shooting chart on page 5—32 gives you an idea of where to look for them.

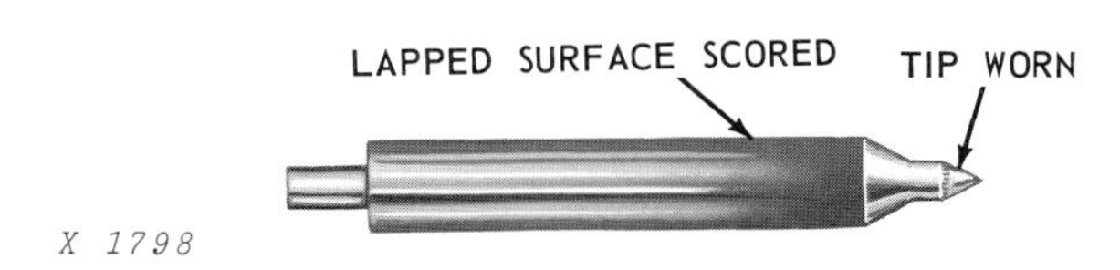

Fig. 54 — Nozzle Valve Damaged Beyond Repair

A magnifying glass will help in inspection.

Repair Of Nozzles

If cleaning does not correct the nozzle failure, replace the defective parts.

Examine parts closely because some nozzle parts can be salvaged. For example:

 Litho in U.S.A.

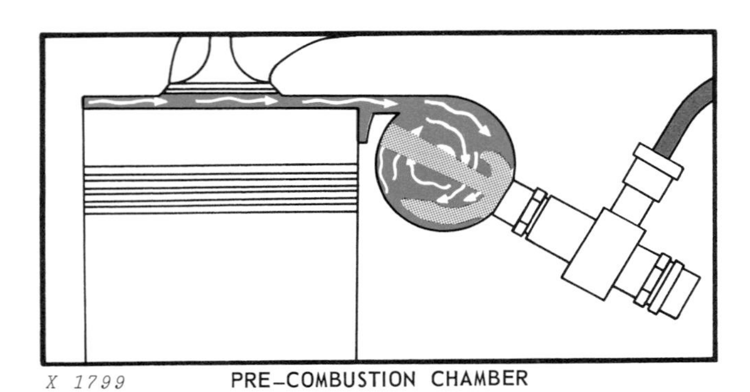

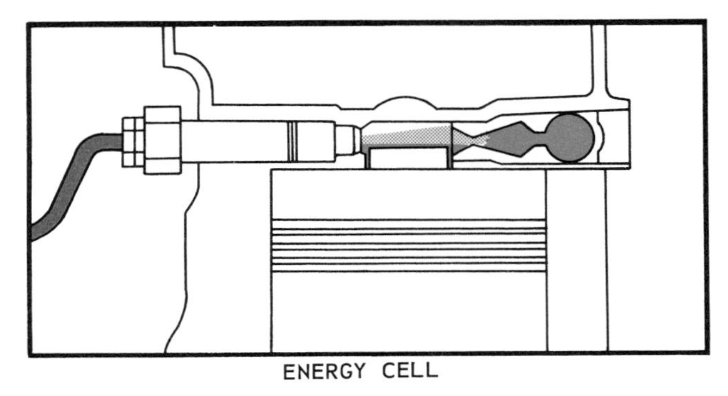

Fig. 55 — Combustion Chambers

1) If the valve in some types of nozzles is sticking and it is not bent, a little polish or lapping compound around the valve guide area may be all that is needed.

2) Other nozzle parts may also be corrected by lapping their surfaces to remove tiny scratches or burrs. Always be careful when lapping any parts to avoid excess wear on the parts.

Parts that are chipped, broken or badly bent should be replaced. If damage is extensive, replace the entire nozzle.

NOTE: Some nozzle parts are sold as matched sets. In this case, replace both parts if either one is damaged.

Reassembly Of Nozzles

Flush each part in clean fuel before reassembly. Then place it on the nozzle tester for retesting.

Any spring pressure adjustment is usually made before retesting.

If the tests still show a malfunction, then the nozzle will have to be serviced again.

COMBUSTION CHAMBERS

We have already mentioned that the injection nozzle and its atomization helps the fuel to mix with the air. However, the piston crown and cylinder head design also play an important part in mixing fuel and air.

In "open" combustion chambers (a chamber in which all fuel and air are confined to one area) the piston crown is concave like a saucer. The chamber is then like half a ball. The piston comes close to the cylinder head and makes it easier for the fuel to spray evenly into the chamber. The concave in the piston crown also sets up a turbulence of the compressed air. This speeds up the air and makes it easier to mix with the fuel.

OTHER TYPES OF COMBUSTION CHAMBERS

In some small, high-speed diesel engines, the combustion is too slow and without help all the fuel would not be burned.

These engines use specially designed chambers to compensate for this combustion loss. The chambers are of three basic types:

- **Turbulence Chambers**
- **Precombustion Chambers**
- **Energy Cell Chambers**

TURBULENCE CHAMBER—The fuel is injected into a small spherical chamber connected to the cylinder. The chamber is shaped to produce a high turbulent condition. As the piston starts the compression stroke, the air is forced into the chamber, setting up a rotary motion. Near the top of the piston stroke, the fuel is injected into the swirling air and good mixing results.

PRECOMBUSTION CHAMBER—It is similar to the turbulence chamber. However, only part of the air charge is forced in. When the fuel starts to burn with an insufficient amount of air, it ignites, forcing the burning fuel into the cylinder. At this velocity, the fuel readily mixes with the remaining air during combustion.

ENERGY CELL—An energy cell is illustrated in Fig. 55. Fuel is forced into the cell and combustion takes place there. The engine piston forces this air in during the compression stroke. At the end of the stroke the nozzle sprays fuel across the top of the piston toward the energy cell. About one-half of the fuel is burned in the main chamber at this time and forces the remaining fuel into the energy cell. Fuel and air mix and ignite in the cell, then blow out into the main chamber. The turbulence created by combustion in the energy cell mixes and ignites any remaining fuel and air in the main chamber. In this case, combustion occurs in *three* different stages, which completes the burning of the fuel.

The chamber and cells we have just described are special devices and are not used on most engines.

GENERAL PROBLEMS OF DIESEL COMBUSTION

The primary goal of the diesel fuel system is to deliver fuel to the engine cylinder for combustion.

However, good combustion depends on four conditions:

1. The fuel must be good-quality and the correct type.

2. The fuel must be finely atomized.

3. There must be high temperatures.

4. The fuel and air must be completely mixed.

Let's look at each of these briefly.

CORRECT TYPE OF FUEL

Diesel fuel is graded for various types of engine service. Always follow the operator's manual for the engine when selecting the fuel.

Other factors are sulfur content, water and sediment content, and pour point.

The cetane number is also important as a measure of the ignition quality of diesel fuel—at what pressure and temperature the fuel will ignite and burn.

ATOMIZING THE FUEL

The injection nozzle atomizes the fuel, which helps the fuel mix with the air. However, the nozzle also helps to make a **vapor** of the fuel.

Vaporizing the fuel is important. **Remember, only a gas or vapor will burn readily.** In this case, the heat of the compressed air vaporizes the fuel. But, because so little time is allowed for combustion, the fuel particles must be as small as possible so that they will vaporize faster.

This does not mean that the fuel becomes a gas and this gas mixes with the air and then ignites. Actually, each fuel particle absorbs heat and then throws off vapor. The vapor then mixes with the nearest air molecule and ignites. All the combined fuel vapor-air particles do not ignite at the same time. However, the speed of this process is so fast that it appears that they do.

HIGH TEMPERATURES NEEDED FOR COMBUSTION

We learned in Chapter 1 that the heat generated by compressed air does the same job that the spark does in a gasoline engine. The amount of heat is determined by the compression ratio of the engine, the size of the cylinders and the efficiency of the cooling system. Diesel engines with small cylinders have a *higher* compression ratio than a gasoline engine; and the higher the compression, the more heat for combustion.

In summary: Higher diesel compression = more heat = combustion without a spark.

STARTING A COLD DIESEL ENGINE

Diesel fuel does not efficiently ignite in a cold engine. There are two main types of cold start aids that can be used to increase cylinder turbulence chamber heat. They are:

1. Cold weather starting fluid added to the intake manifold.

2. Thermal heating elements that warm parts of the engine.

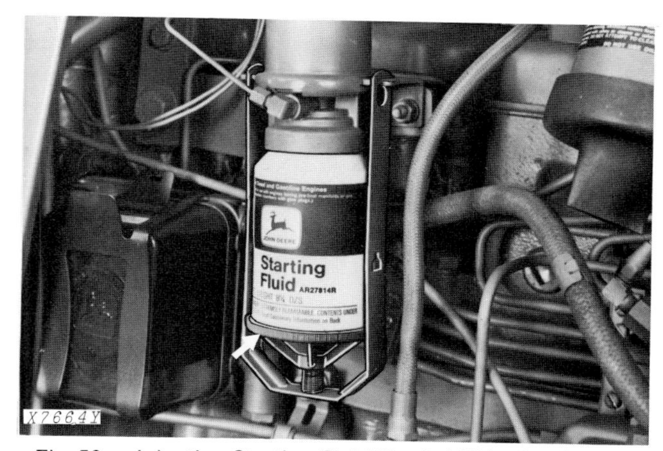

Fig. 56 — Injecting Starting Fluid For Cold-Weather Starting Of Diesel Engine

Always follow the engine operator's manual when any of these starting aids are used.

Starting fluid is not affected by low temperatures because it has a low combustion point. When it is injected into the intake manifold (Fig. 56) and compressed in the cylinder head it ignites without the use of thermal heating elements. Starting fluid can be manually sprayed into the air cleaner element from an aerosol can.

Litho in U.S.A.

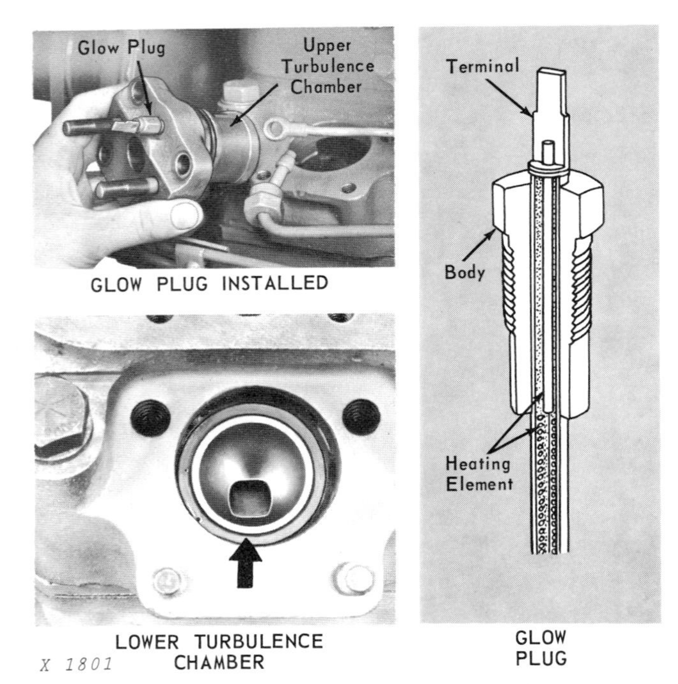

Fig. 57 — "Glow" Plug For Cold Weather Starting Of Diesel Engine

CAUTION: Do not use thermal starting aids in the cylinder head or intake manifold when starting fluid is used to start a cold diesel engine. Glow plugs in the turbulence chamber, intake manifold heating coils and other thermal aids can cause starting fluid to ignite prematurely. This can cause personal injury as well as damage to the engine.

If you have a starting fluid can mounted in the fuel system (Fig. 56), spray starting fluid into the engine *only* while cranking with the starter. Too much starting fluid can cause an uncontrolled explosion in the cylinder chambers and damage the engine. Starting fluid must never come into contact with a thermal starting aid.

Thermal starting aids include:

- **Glow plugs**
- **Lubricant coolant heating coils**
- **Intake air heater coils**

They can be used individually or in a combination. Thermal starting aids are coils of wire that carry electric current much like an electric blanket. Heat is created by the resistance of the coil to the electric current. This heat is transfered to the air intake, coolant or lubricant reservoirs.

The most common thermal starting aid is the "glow" plug (Fig. 57). It differs from other thermal aids because it is mounted in the turbulence chamber of the piston (Fig. 57). The injection nozzle also is in the chamber.

The glow plug is turned on to preheat the cylinder chamber. It is turned off before the starter motor is engaged. After the initial ignition, heat is generated by diesel fuel combustion and cylinder movement.

If starting fluid is sprayed onto this glow plug when it is on or hot, the starting fluid could explode. Starting fluid contacting heating elements in the air intake also is dangerous. Do not use starting fluid and air intake starting aids simultaneously.

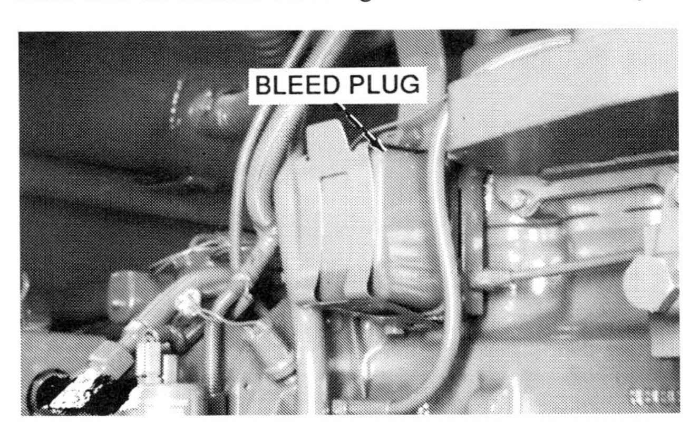

Fig. 58 — Bleed Plug on Fuel Filter

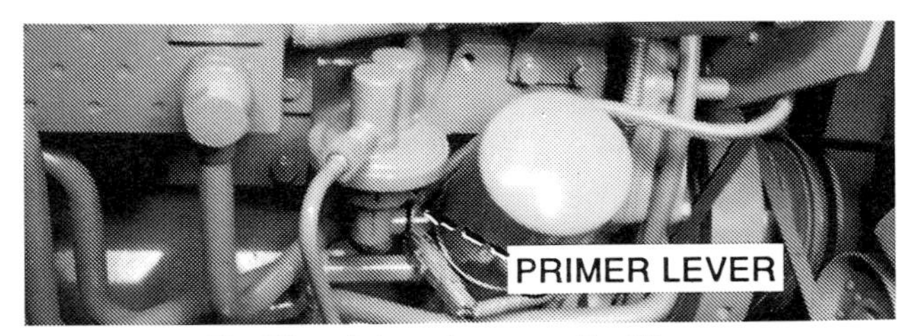

Fig. 59 — Primer Lever on Fuel Pump

BLEEDING AIR FROM DIESEL FUEL SYSTEMS

When the diesel fuel system has been opened or has run dry, a lot of air gets into the system. If this air is left in the lines, it may form an air lock which will prevent fuel from reaching or going through the injection pump. Result: the engine may not start or it may misfire and lose power.

Be sure to bleed the system to remove this trapped air before operating the engine.

Here is a general guide to bleeding the system:

1. Fill the fuel tank with the proper diesel fuel.

2. Open the fuel shut-off valve at the tank.

3. Loosen the bleed plug on the fuel filter (Fig. 58). Pump the primer lever (Fig. 59) on the fuel pump until a solid stream of fuel (free of air bubbles) flows from the opening. (If the primer lever will not pump fuel and no resistance is felt at upper end of stroke, turn the engine with the starter to change

the position of fuel pump cam.) Tighten the plug. On engines with dual filters, repeat the bleeding operation on the other filter. When bleeding is completed, be sure to leave the primer lever at lowest point of its stroke.

If the air lock is still present after bleeding air from the filter bleed plug, you will also have to bleed the injection lines.

4. Hold the injection line nuts with a wrench and loosen the injection lines on at least two injection nozzles. Crank the engine until fuel without foam flows around the connectors. Tighten the line connections carefully. Avoid bending the line connection. Tighten the lines only until snug and free of leaks.

 **CAUTION: Loosen injection line connectors only one turn to avoid excessive spray.**

TROUBLE SHOOTING THE DIESEL FUEL SYSTEM

Since we have already tested and serviced the individual components, we will only give general trouble shooting for the whole system here.

This chart only has causes of typical failures for the diesel fuel system. For trouble shooting on the complete engine, refer to the charts on Chapter 11, "Diagnosis and Testing of Engines".

ENGINE WILL NOT START, STARTS HARD, OR MISFIRES

1. No fuel in tank; leaky or clogged supply line or filter; water in fuel.

2. Fuel transfer pump not functioning.

3. Air lock in injection pump.

4. Governor linkage to pump loose or broken.

5. Drive shaft or key sheared; pump plungers or distributor seized.

6. Delivery valves not seating properly due to dirt or broken springs.

7. Pump out of time.

8. Nozzles not functioning because of stuck valves or plugged orifices.

ENGINE WILL NOT IDLE SMOOTHLY

1. Governor idling adjustment not set properly or throttle linkage worn.

2. Pump rack or control sticky or stuck; stuck plunger.

3. Nozzle opening pressure incorrect; stuck nozzle valves.

4. Pump out of calibration, loose control sleeves; leaky delivery valves, pump timing incorrect.

5. Dirty fuel filter.

ENGINE SMOKES AND KNOCKS

1. Pump out of time.

2. Dirty or fouled nozzles; valves stuck open; opening pressure too low.

3. Fuel stop setting incorrect so that fuel delivery is excessive.

4. Other possible causes: air cleaner dirty, broken engine valves, scored engine pistons, or turbocharger speed reduced by leaky manifold gaskets.

ENGINE LACKS POWER

1. Pump timing retarded.

2. Pump plungers or distributor worn.

3. Faulty nozzles.

4. Governor out of adjustment.

5. Plugged air or fuel filters.

ELECTRONICALLY CONTROLLED FUEL INJECTION SYSTEM

The following paragraphs describe the basic operation of the electronically controlled fuel injection system.

When the key switch is turned to the "ON" position, the **engine controller** receives power and the fuel shutoff solenoid is powered (opening the valve).

When the key switch is turned to the "START" position, the controller powers the actuator solenoid moving the rack to starting fuel positions based on fuel temperature and engine speed. This starting mode is triggered either by a controller input which senses the key switch "START" position or by an engine speed greater than 60 rpm. Starting fuel quantity is not affected by throttle position.

Once the engine has started, fuel delivery is controlled by the engine controller based on various inputs (primarily throttle and engine speed).

The engine controller controls rack position by adjusting the current level to the actuator solenoid until the rack position signal from the injection pump matches the commanded signal. When no fuel is desired, the controller turns off current to the

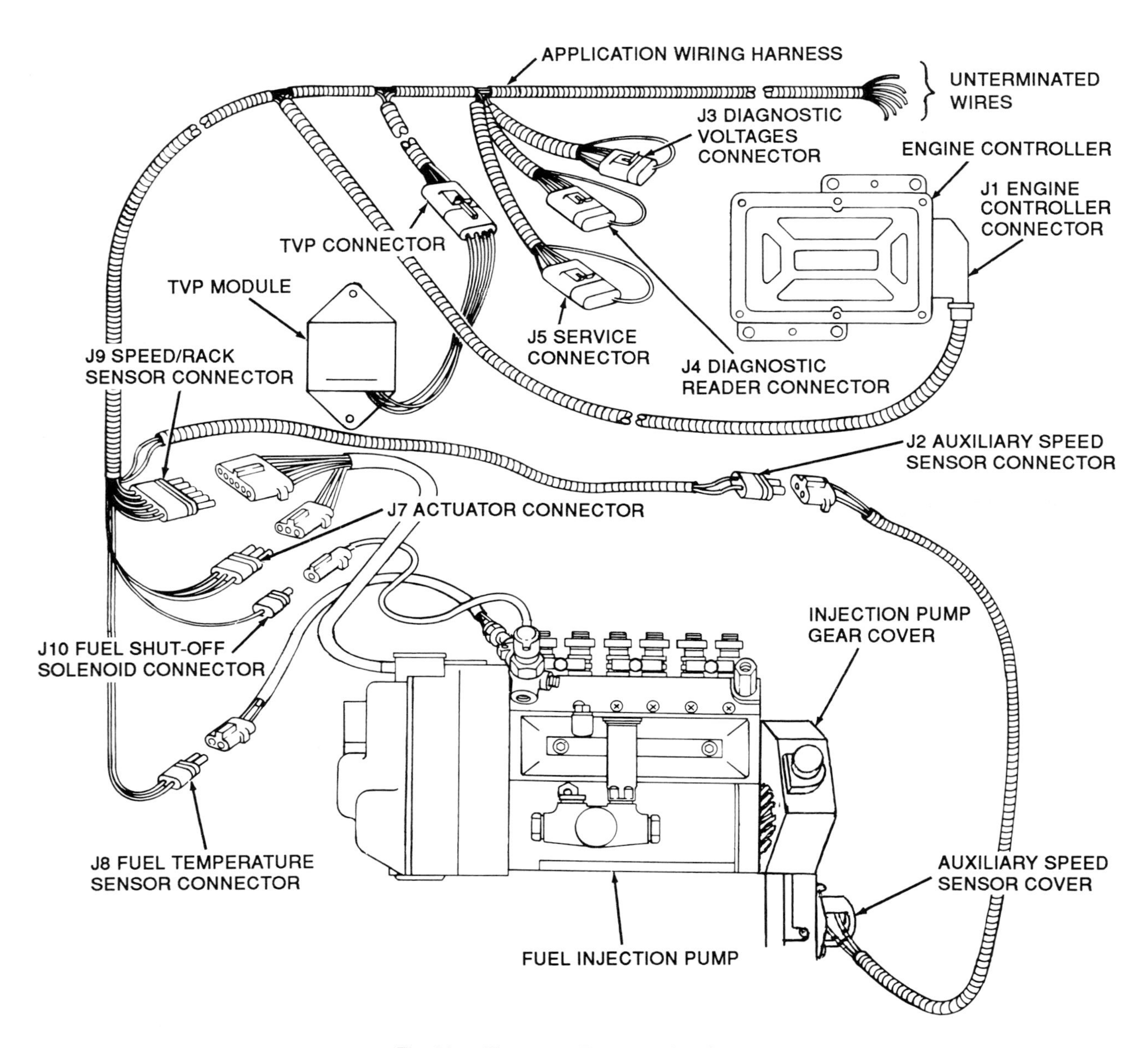

Fig. 60 — Electronic Fuel Injection System

actuator solenoid. If a problem occurs where the controller cannot control rack position, the fuel shut-off solenoid is turned off in addition to the actuator solenoid being turned off.

The engine controller uses engine speed and initial fuel temperature to control rack position during starting. This permits use of excess fuel and retarding for cold temperatures but less fuel and no retard for hot starts. Thus, cold starting is improved and black smoke can be greatly reduced on hot starts.

SUMMARY OF MAJOR COMPONENTS

Most electronic fuel injection systems consist of the following major components:

- **Engine Controller Unit**
- **Injection Pump/Actuator Assembly**
- **Auxiliary Speed Sensor**
- **Transient Voltage Protection Module**

ENGINE CONTROLLER UNIT (ECU)

The engine controller is a self-contained module with mounting brackets which contains electronic circuitry and a computer program which performs governor and diagnostic functions. The controller is remotely located from the engine in a protected environment and connects to the engine application through the wiring harness.

The engine controller (Fig. 61) controls the fuel delivery as a function of engine speed and throttle command. The controller also controls the fuel limiting for torque curves and the governing for speed control. Aneroids are eliminated and for most applications the controller can control fuel delivery to limit smoke without using additional sensors. For more stringent requirements, manifold air density can be monitored by the controller to limit smoke.

Fig. 61 — Engine Controller Unit

INJECTION PUMP/ACTUATOR ASSEMBLY

The injection pump/actuator assembly consists of the following:

- **Injection Pump**
- **Actuator Solenoid**
- **Rack Position Sensor**
- **Primary Speed Sensor**
- **Fuel Shut-off Solenoid**
- **Fuel Temperature Sensor**

Injection Pump

The electronically controlled in-line injection system uses the same basic hydraulic pumping mechanism used in mechanically governed in-line pumps. The mechanical governor mechanism is replaced with the actuator assembly which includes the actuator solenoid to move the control rack, rack position sensor, primary speed sensor, and a toothed speed wheel. The throttle lever mechanism used on mechanical pumps is removed and its function is implemented by a throttle position sensor input to the engine controller. Fuel is transferred from vehicle or engine fuel tanks with a fuel supply pump. The injection pump fuel inlet connection is located at the rear of the pump on the fuel inlet assembly, which includes the fuel shut-off solenoid and the fuel temperature sensor (Fig. 62).

Fig. 62 — Injection Pump

Actuator Solenoid

The control rack is spring loaded to the fuel shutoff position "zero rack." As increasing current is supplied to the actuator solenoid (Fig. 63) from the controller, the rack is driven toward full rack position. The engine controller has the capability to control the current to the solenoid in order to position the solenoid and control rack anywhere between zero rack and full rack.

Rack Position Sensor

The rack position sensor (Fig. 63) within the actuator housing supplies rack position information to the controller so that a specific rack position can be controlled. The sensor includes an electronic module mounted in the actuator housing which provides a voltage to the controller indicating the position of the rack and is used to control rack position for all operating conditions. If this critical sensor were to fail, the controller would be forced to shut down the engine due to loss of control.

The rack position sensor can only be serviced by an authorized shop because of the recalibration which is required if the actuator housing is removed from the pump.

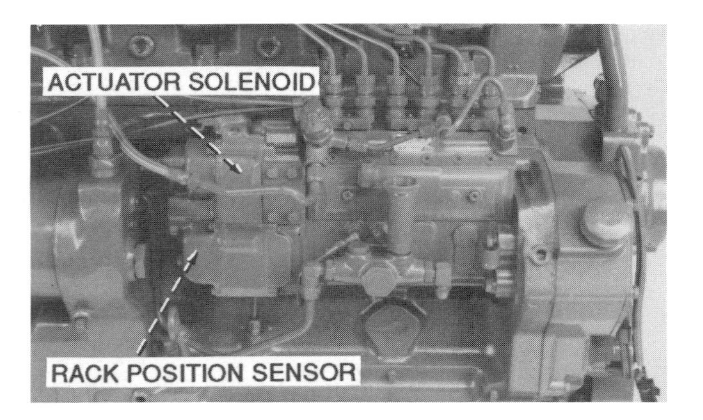

Fig. 63 — Actuator Solenoid and Rack Position Sensor

Primary Speed Sensor

The primary speed sensor (Fig. 64) is also located within the actuator housing. It is a magnetic pickup which generates voltage pulses to the controller as the teeth on the speed wheel pass by the tip of the sensor. If this sensor were to fail completely, the controller would use the signal coming from the auxiliary speed sensor to get engine speed information.

Because the primary speed sensor is located inside the actuator housing, it can only be serviced by an authorized repair shop.

Fig. 64 — Primary Speed Sensor

Fuel Shut-off Solenoid

The fuel shut-off solenoid (Fig. 65) will shut off fuel to the injection pump when the key switch is turned off or when the engine controller detects a rack position error.

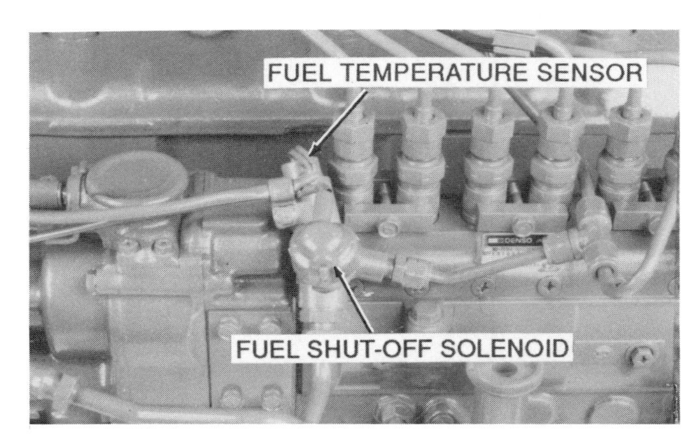

Fig. 65 — Fuel Shut-off Solenoid and Fuel Temperature Sensor

Fuel Temperature Sensor

The fuel temperature sensor (Fig. 65) is located at the fuel inlet to the pump. It is used to determine the optimum fuel delivery for starting, and depending on the application, is used to maintain constant power over a predetermined temperature range. If this sensor were to fail, a low temperature would be assumed by the engine controller. In warm weather this might result in a slight drop in maximum torque and smokier starts.

AUXILIARY SPEED SENSOR

This engine speed sensor (magnetic pickup) (Fig. 66) serves as a back-up speed sensor in the event of complete failure of the primary speed sensor.

Fig. 66 — Auxiliary Speed Sensor

TRANSIENT VOLTAGE PROTECTION MODULE (TVP)

The main function of the transient voltage protection (TVP) module is to limit high energy voltage transients (from the charging system) to a maximum of 40 volts to protect the electronic circuitry in the engine controller.

The TVP module (Fig. 67) provides a relay, a circuit breaker, and a transient voltage suppressor in an environmentally tested package. The module is normally used with the engine controller in applications which would not otherwise provide this type of function (such as OEM engines).

The relay contacts provide battery power to the engine controller through the 20 amp circuit breaker. The relay should be energized when the system is in the Run or Start modes.

Fig. 67 — Transient Voltage Protection Module

TROUBLE SHOOTING ELECTRONIC FUEL INJECTION

Checking Wiring and Connectors

When diagnosing electrical system problems, take special note of the condition of wiring and connectors since a high percentage of problems originate here. Check first for loose, dirty or disconnected connectors. Inspect the wiring routing looking for possible shorts caused by contact with external parts (for example, rubbing against sharp sheet metal edges). Inspect the connector vicinity looking for wires that have pulled out of connector terminals, damaged connectors, poorly positioned terminals, and corroded or damaged terminals.

Running Engine at Different Speeds

Run the throttle control(s) at slow-to-medium rate between the slow idle and the fast idle stops with the engine running. If engine speed "catches" or "drops," this may indicate a problem with throttle adjustment, throttle sensor operation, or the wiring between the engine controller and the throttle sensor(s). Using this method of changing engine speed, you may be able to identify problems which only occur at certain speeds or throttle settings. For machines with 3-stage throttle, switch between speed selections to see if problem is speed dependent.

Self-diagnosis and Back-up Features

NOTE: If the operator is able to keep the engine running during a fault condition, no damage to the engine should result. The problem should be fixed at the earliest convenience.

The engine controller self-diagnoses as many system faults as is practical. This includes determining if any of the sensor input voltages are too high or too low, if the engine speed signals are valid, and if the control rack is responding properly.

The controller also monitors its own operation for problems. In most cases, the controller will output a diagnostic code to indicate the specific problem which has been detected. The controller can also flash a fault lamp.

The controller will automatically switch to another mode of operation as a back-up whenever possible or will shut down the engine if control cannot be guaranteed. In some cases, little or no degradation in performance is noticed. For example, the engine will continue to operate normally using the auxiliary speed sensor in the event of a primary speed sensor failure.

Looking for Intermittent Problems

An intermittent code may mean that something is going outside of its normal operating range, but then returning to normal. This may be caused by a marginal adjustment (on a throttle sensor, for example) or by a poor connection. Check the condition of the wiring and connectors as described in the first paragraph. If codes 11, 12, 13 or 14 are intermittent and the wiring appears to be OK, then the throttle sensor(s) may need to be adjusted.

An intermittent code may also mean that something is beginning to fail. Some failures are related to vibration.

Some ECU systems have the following diagnostic error codes (numbers) displayed on the tach when the ECU system detects a failure (refer to Engine Technical Manual for detailed information on codes):

28 - A/D Converter Error
29 - Sensor Excitation Voltage Out-of-Range
33 - Actuator Solenoid Output-Shorted High
34 - Rack Position Error
35 - Rack Position Voltage Out-of-Range High
36 - Rack Position Voltage Out-of-Range Low
37 - Fuel Temperature Input Out-of-Range High
38 - Fuel Temperature Input Out-of-Range Low
39 - Primary Speed Input Error
41 - Start Signal Missing
42 - Engine Overspeed
44 - Auxiliary Speed Input Error
47 - Derated Torque Curve Selected
51 - Negative Voltage - Analog Throttle Input
52 - Negative Voltage - Manifold Air Pressure Input
53 - Negative Voltage - Manifold Air Temperature Input
54 - Negative Voltage - Rack Position Voltage Input
55 - Negative Voltage - Fuel Temperature Input
56 - Negative Voltage - Fuel Limit Select Input
57 - Negative Voltage - Speed Regulation Select Input
58 - Negative Voltage - 3-Stage Throttle Input
59 - Negative Voltage - +5V Sensor Excitation
71 - Diagnostic Codes Output Stuck High
72 - Diagnostic Codes Output Stuck Low

TEST YOURSELF

QUESTIONS

1. True or False? "Diesel combustion takes place when fuel and hot air are ignited by a spark."

2. Why is the distributor injection system well suited for a multi-cylinder engine?

3. How is the distributor injection pump lubricated?

4. (Fill in the blanks.) Injection nozzles are simple mechanisms; they use ______________ pressure to oppose ________________pressure.

5. What kind of a brush should be used to clean deposits from nozzles?

6. How does fuel flow through two series filters as compared to two parallel filters?

7. Which type of filters clean fuel the best?

8. Name the two most common types of electric fuel gauges.

(Answers in back of text.)

 Litho in U.S.A.

INTAKE AND EXHAUST SYSTEMS / CHAPTER 6

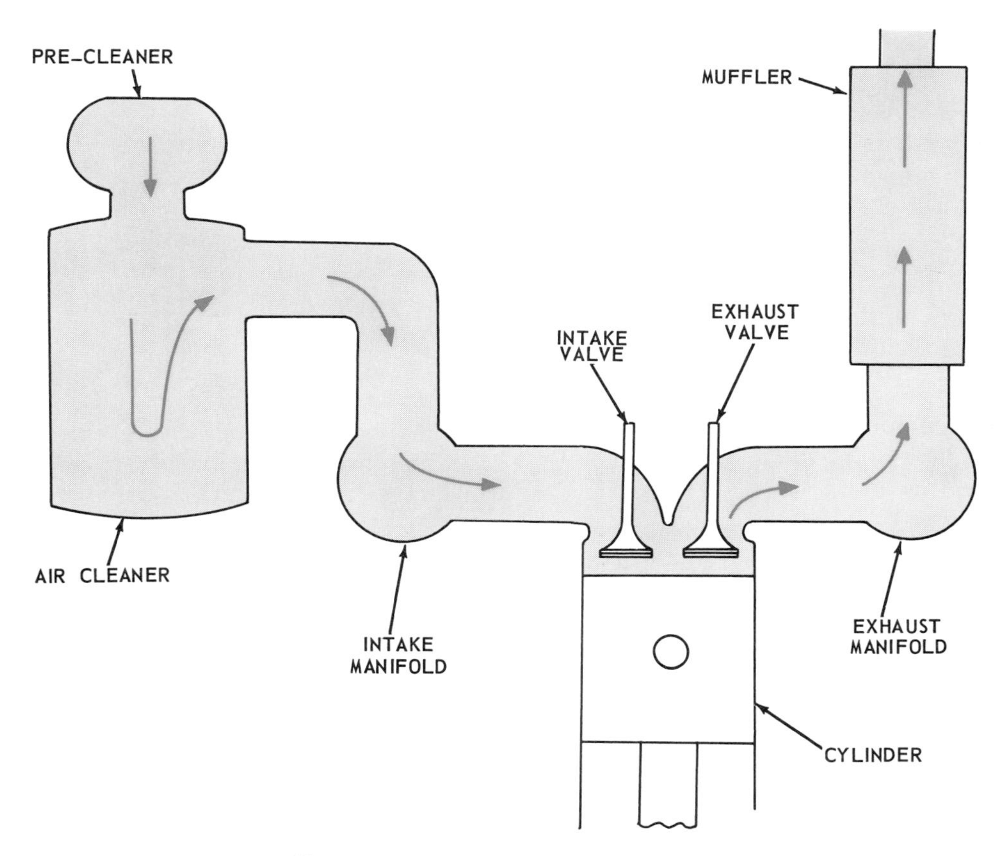

Fig. 1 — Intake And Exhaust System

The intake system carries the fuel-air mixture into the engine, while the exhaust system carries exhaust gases away.

AIR INTAKE SYSTEMS

The air intake system supplies the engine with the proper quantity of clean air, at the right temperature, with the correct amount of fuel which will be mixed for best combustion.

There are two basic types of air intake systems:

1) Cylinder Feed

2) Crankcase Feed

CYLINDER FEED—Fuel is mixed with air as it passes through the carburetor. This fuel-air mixture is passed into the cylinder through the intake valve.

CRANKCASE FEED—Some two-cycle engines (see Chapter 1) draw the fuel-air mixture into the crankcase before it is moved, under pressure, into the cylinder.

The intake system consists of:

- **Air cleaners, pre-cleaners**
- **Supercharger or turbocharger (if used)**
- **Intake manifold**
- **Carburetor air inlet**
- **Intake valves**

AIR CLEANERS filter dust and dirt from the air passing through them enroute to the carburetor. Pre-cleaners prevent larger particles from reaching the air cleaner and plugging it.

SUPERCHARGERS increase horsepower by packing more air or fuel-air mixture into the engine cylinders than the engine could take in by itself.

INTAKE MANIFOLDS transport the fuel-air mixture (air only on diesel engines) to the engine cylinders.

Litho in U.S.A.

CARBURETORS mix fuel with incoming air in the proper proportion for combustion.

INTAKE VALVES admit air to diesel engines and the fuel-air mixture to spark-ignition engines. See Chapter 2 for engine information.

AIR CLEANERS

Clean air is essential to satisfactory performance and long engine life. The air cleaner must be able to remove fine materials such as dust and blown sand as well as chaff, or lint from the air.

A cleaner must also have a reservoir large enough to hold material taken out of the air so operation over a reasonable period of time is possible before cleaning and servicing is necessary.

A buildup of dust and dirt in the air cleaner passages will eventually choke off the air supply, causing incomplete combustion and heavy carbon deposits on valves and pistons.

Multiple air cleaner installations are sometimes used where engines are operated under extremely dusty air conditions or where two small air cleaners must be used in place of a single large cleaner.

The most common types of air cleaners are:

1) *Pre-cleaners*

2) *Dry type cleaners*

3) *Dry element type cleaners*

4) *Viscous-impingement cleaners*

5) *Spiral rotor type cleaners*

6) *Oil bath type cleaners*

(1) PRE-CLEANERS

Pre-cleaners are usually installed at the end of a pipe extended upward into the air from the air cleaner inlet. This places them in an area relatively free of dust.

Pre-cleaners are simple devices which remove larger particles of dirt or other foreign matter from the air before it enters the main air cleaner. This relieves much of the load on the air cleaner and allows longer intervals between servicing.

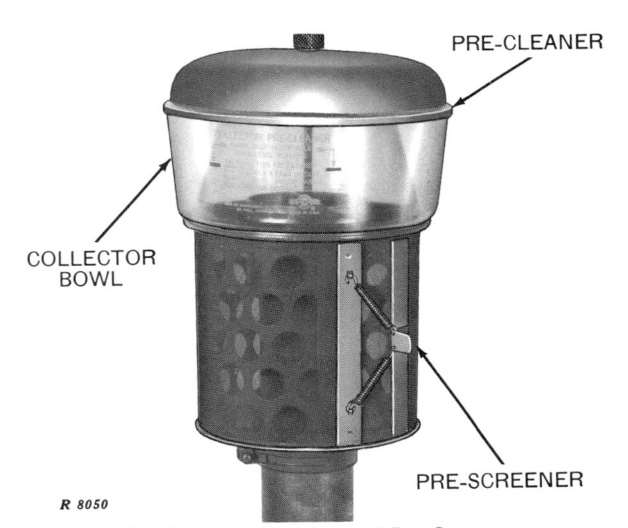

Fig. 2 — Pre-Cleaner and Pre-Screener

Most pre-cleaners have a *pre-screener* which prevents lint, chaff, and leaves from entering the air intake.

(2) DRY TYPE AIR CLEANERS

Dry type cleaners are attached directly to the carburetor or manifold. They are only used on engines in which the demand for air is small.

Fig. 3 — Simple Dry Air Cleaner

Dry type air cleaners or filters (Fig. 3) clean the air by passing it through layers of cloth or felt. It is most effective in removing larger dirt particles from the air and is used mostly on small engines.

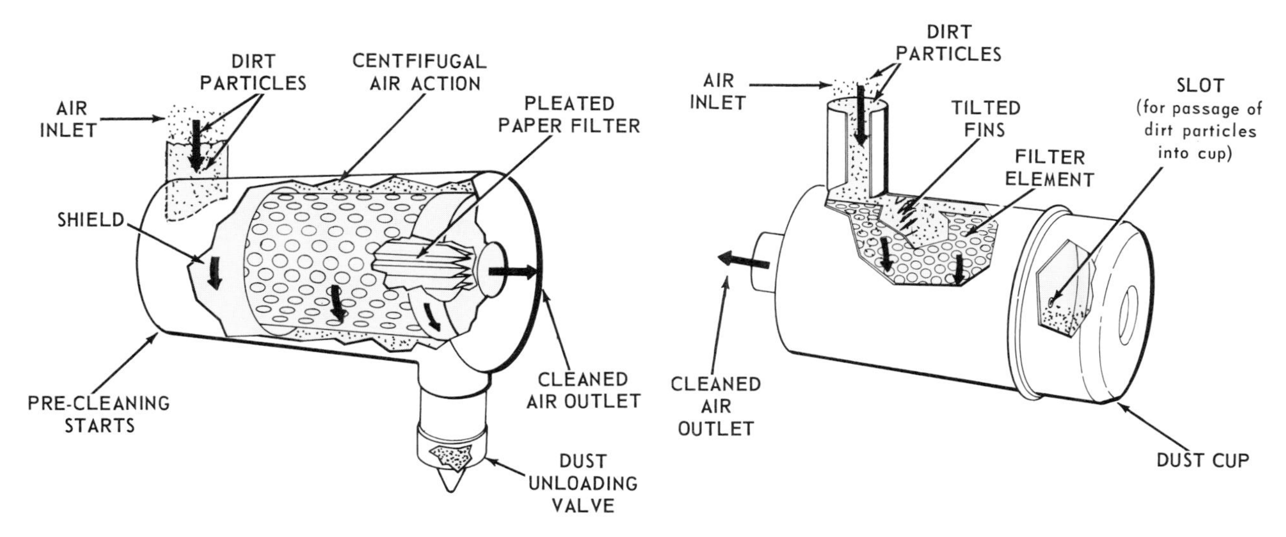

Fig. 4 — Dry Element Type Air Cleaner

(3) DRY ELEMENT TYPE AIR CLEANERS

Two major types of dry-type cleaners are used at the present time (Fig. 4).

Dry air cleaners are built for two-stage cleaning:

1) *Precleaning*

2) *Filtering*

This first stage (precleaning) directs the air into the cleaner at high speed so that it sets up centrifugal rotation (cyclone action) around the filter element.

The cleaner shown in Fig. 4A directs the air into the *pre-cleaner* so it strikes one side of the metal shield. This starts the centrifugal action which continues until it reaches the far end of the cleaner housing. At this point, the dirt is collected into a dust cap, or dust unloader, at the bottom of the housing.

The cleaner shown in Fig. 4B conducts the air past *tilted fins* which start the centrifugal (cyclone) action. When the air reaches the end of the cleaner housing, the dirt passes through a slot in the top of the cleaner and enters the dust cup.

In both types, *this precleaning action removes from 80 to 90 percent of the dirt particles* and greatly reduces the load on the filter.

The partially cleaned air then passes through the holes in the metal jacket surrounding the pleated-paper filter. *Filtering* is done as the air passes through the paper filter. It filters out almost all of the remaining small particles. This is the second stage of cleaning.

Some heavy-duty cleaners use a spiral rotor device for precleaning the air. Others have a small safety element built into the unit in case the main element fouls.

If the air cleaner has a *dust cup* (Fig. 4B) it should be emptied daily. If an *automatic dust unloader* is used in place of a cup (Fig. 4A), it is usually recommended that it be checked at least once daily to make certain it doesn't become clogged. The dust unloader is a rubber duck-bill device that is held closed by engine suction while the engine is running. When it stops, the weight of the accumulated dirt helps open the flaps so the dirt can drop out.

(4) VISCOUS-IMPINGEMENT AIR CLEANERS

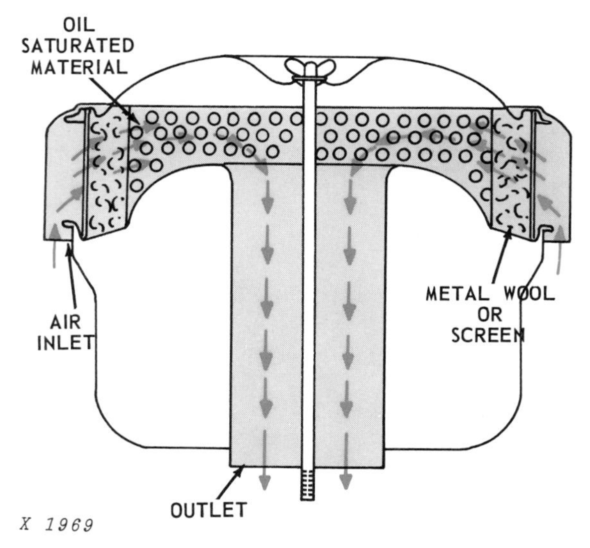

Fig. 5 — Viscous-Impingement Air Cleaner

In **viscous-impingement air cleaners** (Fig. 5), the air flows through a maze of metal wool, wire or screens saturated with oil.

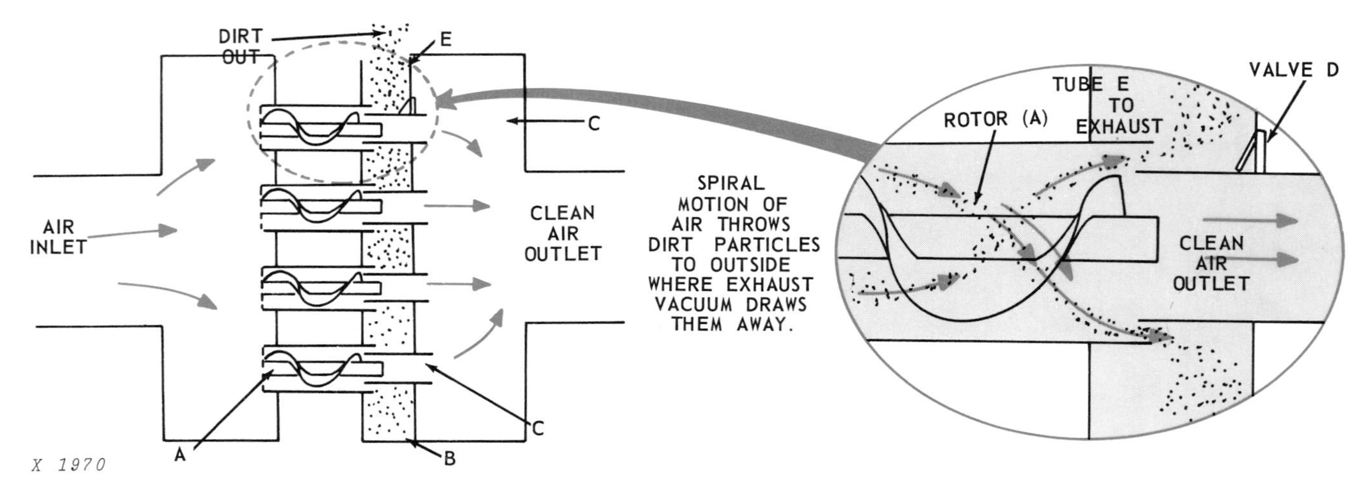

Fig. 6 — Spiral Rotor Air Cleaner

This type is used for automotive applications where dust conditions are relatively light. The lower portion of the unit acts as a hollow chamber to quiet the air intake noise.

Viscous-impingement type air cleaners are now normally used only on small engines. They utilize an oil-impregnated element for filtering.

(5) SPIRAL ROTOR AIR CLEANERS

Fig. 6 illustrates a recent innovation in air intake systems—the **spiral rotor air cleaner.**

It consists of a number of spiral rotors (A) installed in tubes in the air inlet. These tubes protrude into a compartment (B).

Smaller tubes (C) extend into the larger tubes. The smaller tubes lead to the main air inlet.

A valve (D) allows air to flow out of the main air cleaner inlet compartment but not into it.

The tube (E) connects compartment (B) to the muffler assembly. The muffler provides a constant vacuum, or suction.

Air passing through the tubes is spun by the rotors, causing the dirt particles to be thrown against the walls of the larger tubes and drawn off by the exhaust vacuum at (E).

Clean air passes on through the smaller tubes to compartment (C) and the main air outlet.

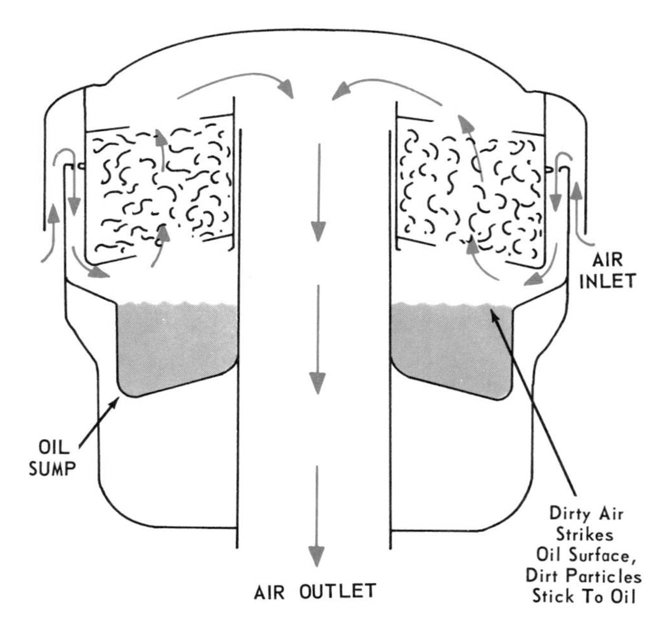

Fig. 7 — Light-Duty Oil Bath Air Cleaner And Silencer

Incoming air reverses when it strikes the surface of the oil, causing most of the dirt to become trapped by the oil and settle out in the sump.

The air then passes upward through the screen element where more dust, and suspended oil, is removed and drains back into the sump.

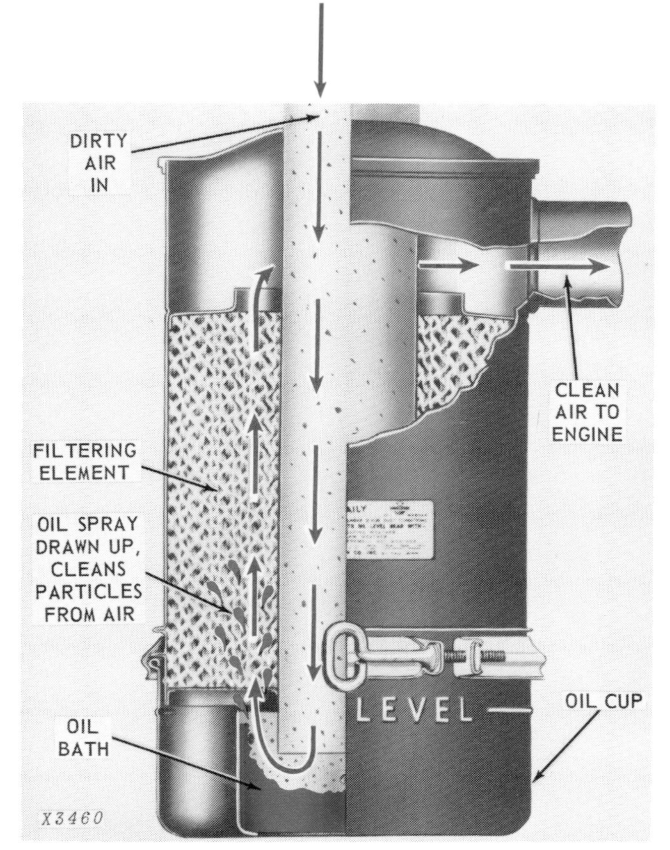

Fig. 8 — Medium-Duty Oil Bath Air Cleaner

Medium-duty oil bath air cleaners (Fig. 8) draw air down a center tube where it strikes the surface of oil in a partly filled cup. The impact causes a mixture of air and oil spray to be carried up into the element of baffles and wire mesh. The separating element breaks up the dust-laden air and fine dust particles are trapped by the oil film. The particles are then washed down as the oil later drains back into the cup. Clean air continues through the element and on to the engine.

The washing down of the oil keeps the filter element fairly clean.

The major services are (1) to keep the oil cup filled to the proper level with the correct weight of oil, and (2) to replace the oil when it gets dirty or thickens, reducing its ability to clean particles from the air.

However, at least once each year, the air cleaner should be disassembled and the filter element cleaned in solvent.

NOTE: Today's detergent oils may hold the dirt in suspension. For this reason the need for oil cup service cannot be determined by the layer of dirt which has settled into the bottom of the cup. Instead, see if the oil has thickened due to dirt.

Some medium-duty air cleaners are fitted with an oil trap. This trap catches the oil which otherwise would be thrown out of the cleaner when the engine backfires.

Heavy-duty oil bath air cleaners (Fig. 9) operate the same as medium-duty ones. They are used on larger engines, especially diesels, because these engines use more air.

AIR SILENCERS

Air silencers eliminate the noise caused by the pulsations of incoming air created by the engine blower. A screen between the air cleaner and the blower housing keeps out foreign matter.

An open space at the air outlet end of the air cleaner housing also acts as an intake air silencer.

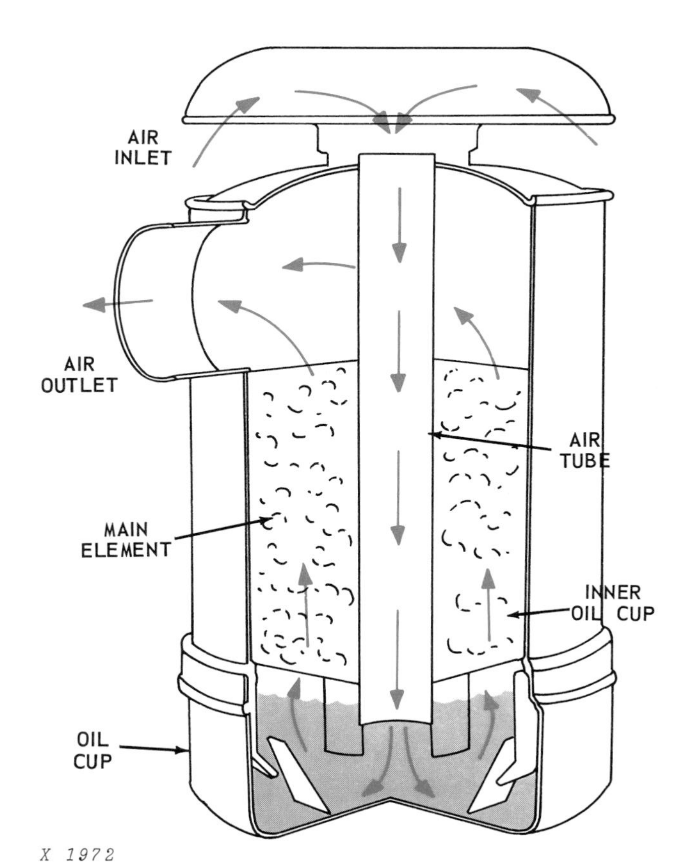

Fig. 9 — Air Flow Through Heavy-Duty Oil Bath Air Cleaner

Litho in U.S.A.

AIR CLEANER MAINTENANCE

Air cleaner efficiency depends upon proper maintenance and service.

Oil bath cleaners must be properly maintained or the oil cup will become filled with sludge, preventing the screens and elements from cleaning the air properly.

This will restrict air flow to the engine (it will cause the same effect as choking the engine) and may allow unclean air to enter.

Restricted air flow through the air cleaner will eventually cause incomplete combustion, increased carbon formation and crankcase dilution.

Leaks in the connecting pipes, loose hose connections, or damaged gaskets which permit dust-laden air to enter the engine, can defeat the efficiency of the air cleaner.

Dirty air going directly into an engine cylinder is abrasive and will cause premature wear of moving parts.

Use the following maintenance rules for good air cleaner operation:

1. Keep air cleaner-to-engine connections tight.

2. Keep air cleaner properly assembled so all joints are oil and air tight.

3. Make careful periodic inspections of the entire air intake system. Repair damaged parts at once. Enough dust-laden air can pass through an almost invisible crack over a period of time to severely damage the engine.

4. Inspect cleaner frequently under dusty conditions. Dust collections on the element gradually restricts the air flow and decreases engine efficiency.

5. Service oil bath cleaners often enough to prevent oil from becoming thick with sludge.

6. Use correct quality of oil. Keep oil at the proper level in the cup. **Do not overfill.**

NOTE: Oil from an overfilled cup can be drawn into an engine. This oil acts as fuel in a diesel engine over which there is no control and can cause the engine to run away (overspeed), resulting in severe damage.

AIR CLEANER SERVICE

No hard and fast rule can be given for servicing an air cleaner since this depends upon the type of cleaner, air condition, and type of application.

Normal service intervals are designated by the manufacturer, but frequent inspection can tell whether or not this is adequate for the conditions under which the engine is operated.

A cleaner operating in severe dust conditions will require more frequent service than one operating in clean air.

Always refer to the operator's manual for service information.

Pre-Cleaner and Pre-Screener Service

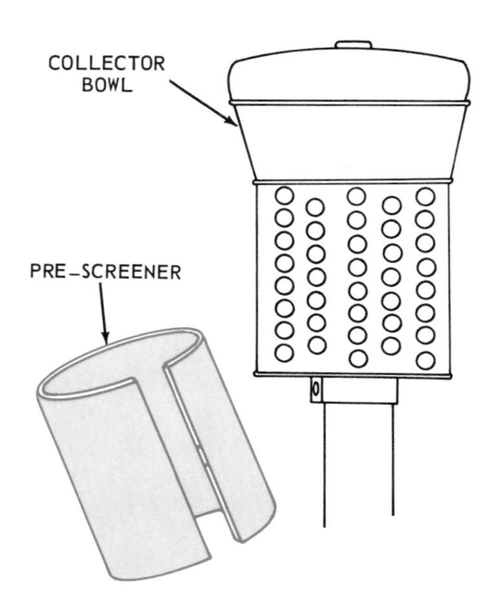

Fig. 10 — Pre-Cleaner And Pre-Screener

Remove the pre-screener (Fig. 10) and blow or brush off any accumulation of lint, chaff, or other foreign matter.

Some pre-cleaners use a removable collector bowl. Remove and thoroughly clean it.

If too much dirt is allowed to collect, the pre-cleaner becomes clogged and a greater load is placed on the main cleaner.

 Litho in U.S.A.

Dry Type Air Cleaner Service

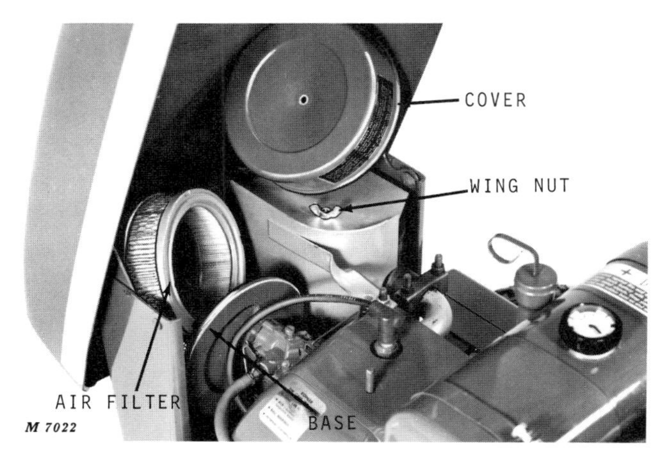

Fig. 11 — Simple Dry Type Air Cleaner

Remove the filter element and shake vigorously to remove most of the dust. Use compressed air or a vacuum cleaner to complete the job.

Wash the element with water and detergent if it is excessively dirty. Dry with compressed air.

Replace the element if it is too dirty to clean, has been damaged, has been washed not more than six times or has been in service for one year.

IMPORTANT: Never wash dry element in fuel oil, gasoline or solvents. Do not oil element.

Viscous-Impingement Type Air Cleaner Service

Remove the filter element and clean by soaking in solvent. Swish the element up and down in the solvent to thoroughly clean it. Dry with compressed air.

After cleaning and drying, immerse the element in engine oil and hang up to allow excess oil to drip out.

Wipe out the inlet tube with a lint-free cloth.

Install the element, making sure that all connections are air-tight.

Light-Duty Oil Bath Air Cleaner Service

Remove the cleaner assembly from the air inlet pipe. The upper section of the cleaner contains the metal wool or screen filtering element. The lower section is made up of the oil cup and inlet tube.

Clean the filter element by soaking it in solvent or diesel fuel. Flush the loosened dirt out with solvent.

Clean the oil cup with solvent and wipe out the inlet tube with a clean, lint-free cloth. **Never** use cotton waste to wipe the center tube.

Fill oil cup to level line with clean engine oil. Assemble and install on engine.

Medium-Duty Oil Bath Air Cleaner Service

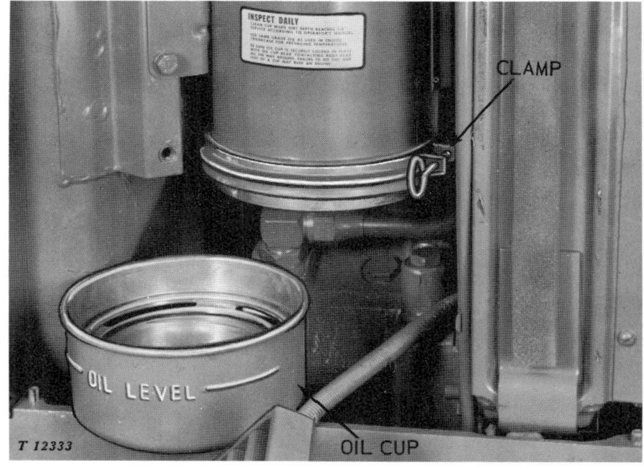

Fig. 12 — Medium-Duty Oil Bath Air Cleaner

Remove the oil cup, pour out the oil, remove the sediment, and thoroughly clean the cup.

Inspect the under surface of the fixed element for a collection of lint, trash, or other foreign matter. If any of these are present, the cleaner should be removed and cleaned.

To clean, soak the element in solvent to loosen accumulated dirt. Flush thoroughly by running solvent through the element from the air inlet end. Allow excess solvent to drip out. Blow out with compressed air.

IMPORTANT: Never attempt to clean the element with a steam cleaner. The force of the steam cannot be maintained throughout the element and will only force the dirt to the center of the element.

Wipe out the center tube with a clean lint-free cloth.

Inspect the inside of the air cleaner-to-manifold pipe for accumulation of oil and dirt. Remove the pipe and clean it, if necessary.

Install the air intake pipe and air cleaner on the engine. Attach oil trap to cleaner.

Fill oil cup to level line with the same weight and quality of oil as used in the engine. Install cup on cleaner. *Be sure that it is tightly secured in position with a leak-proof joint.*

Also be sure all joints are air-tight.

Heavy-Duty Oil Bath Air Cleaner Service

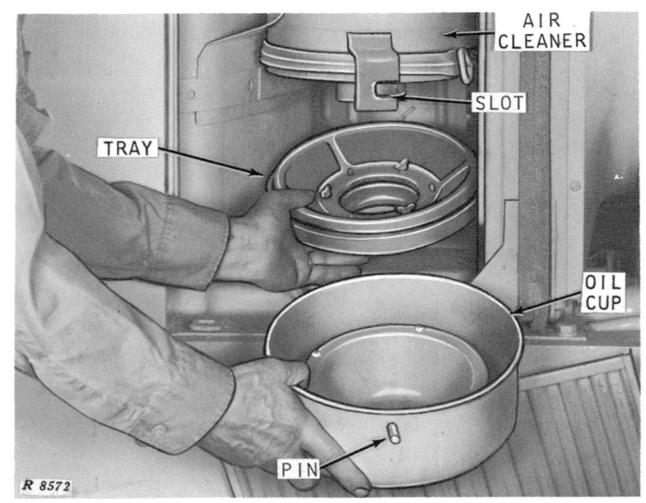

Fig. 13 — Heavy-Duty Air Cleaner Oil Cup And Tray

Remove and clean oil cup and tray. Refer to "Medium-Duty Oil Bath Air Cleaners Service."

If filter element is dirty enough to restrict air flow, remove the air cleaner and soak or run solvent through in the opposite direction that air normally passes.

Reinstall air cleaner. Fill oil cup to proper level with oil of the correct weight and quality and install cup and tray on the air cleaner.

Check all tubes and joints to be sure they are tight.

Since air cleaners vary in construction, always follow the instructions in the operator's manual for the particular cleaner being serviced.

In some heavy-duty air cleaners a collector screen, attached to the inlet tube, will be visible. Remove by loosening the wing nuts and rotating the tray to unlock it from the tube. In other models, (Fig. 13), the screen rests on the lip of the inner cup and is not secured by wing nuts.

This screen catches chaff and other foreign matter. Clean the screen each time the air cleaner is serviced.

Some screens may be separated for dirt and lint removal.

Wash screens in solvent and blow out with compressed air.

When a clean screen is held up to the light, an even pattern of light should be visible. If not, repeat the cleaning operation or replace the screen.

NOTE: The fixed elements of heavy-duty air cleaners are self-cleaning. However, it may be necessary to remove and clean these elements periodically. See the operator's manual for complete information.

OIL TRAP

An oil trap prevents oil from blowing out when an ignition engine backfires.

Remove and clean the oil trap each time the air cleaner is serviced.

Dry Element Type Air Cleaner Service

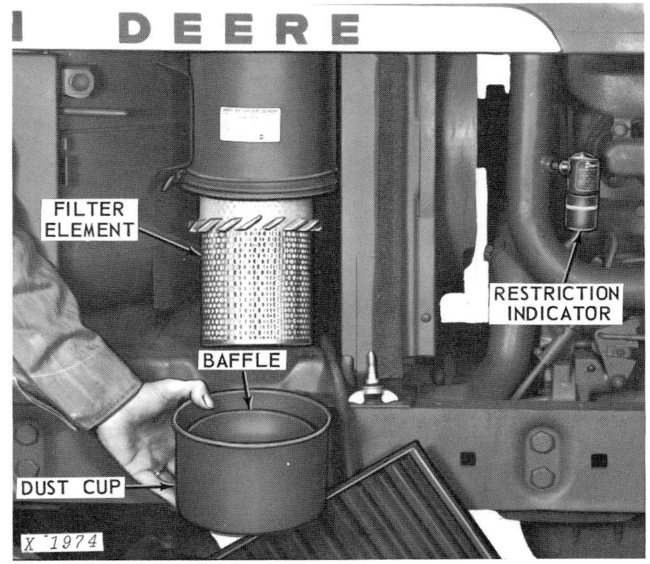

Fig. 14 — Dry Element Type Air Cleaner

When the engine smokes too much or loses power, there may be a restriction in the dry air cleaner.

Clean air filter element at the following times:

1) Units with restriction indicator: Clean element whenever indicator signal tells of a restriction.

2) Units without indicator: Clean element at recommended intervals—more often during dusty or unusual operation.

Dusty Element: Remove filter element and tap element on heel of hand to remove dust. NOT ON A HARD SURFACE. Rotate element while tapping it.

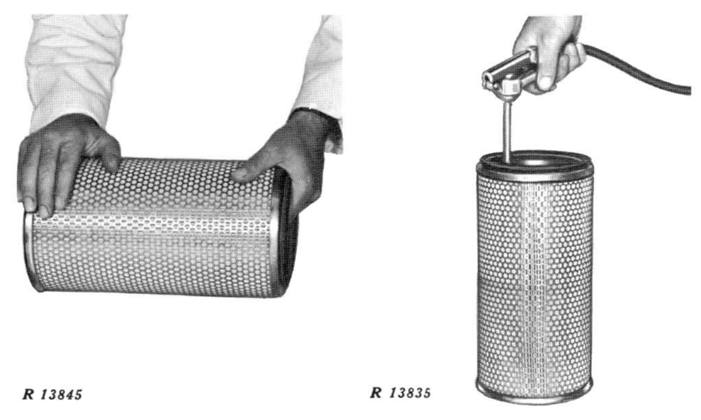

Patting The Element — Blowing The Element

Fig. 15 — Cleaning The Dry Element

If tapping does not remove the dust, use a dry element compressed air cleaning gun (pressure not to exceed 30 psi) (200 kPa) to clean the element (Fig. 15). Direct clean dry air up and down pleats, blowing from inside to outside.

IMPORTANT: BE CAREFUL NOT TO RUPTURE THE ELEMENT. Do not try to clean off outside of air cleaner with an air hose.

To clean with water: First blow out dirt with compressed air as described above. Then attach garden hose to cleaning gun and flush remainder of dirt from inside to outside of element. Allow element to dry.

Washing The Element — Rinsing The Element

Fig. 16 — Washing And Rinsing The Dry Element

Oily Or Sooty Elements: Blow dust from element with compressed air or flush with clean water. Soak and gently agitate element in a solution of lukewarm water and commercial filter element cleaner or an equivalent nonsudsing detergent. (Fig. 16).

Rinse element thoroughly with clean water from hose (Fig. 16) having maximum pressure of 40 psi (275 kPa) or less. Shake excess water from element and allow it to air dry (requires 24 hours or longer). Protect from freezing. Keep a spare element to use while the washed one is drying.

IMPORTANT: Never wash a dry element in fuel oil, gasoline, or solvent. Never use compressed air to dry the element.

After cleaning the element, inspect it for damage by placing a light inside it. Discard any element that shows the slightest damage.

Inspect filter element gasket for damage. Replace element if gasket is damaged or missing.

IMPORTANT: Replace the filter element:

(1) If damaged.

(2) After a recommended service period (such as one year).

(3) When attempts to clean it fail.

Thoroughly clean inside of cleaner body with a clean, damp cloth. Install element in cleaner body, gasket and fin end first. Be sure gasket is in place.

Draw cover tight on cleaner. *Reset restriction indicator* (if used).

Servicing Safety Element (If Used)

Some large engines have a safety element in case the primary element ruptures and fails. Normally, these elements are not cleaned, only replaced once each season. However, check the condition of the safety element during service; if it is very dirty, the primary element has probably failed and must be replaced.

DUST UNLOADING VALVE SERVICE

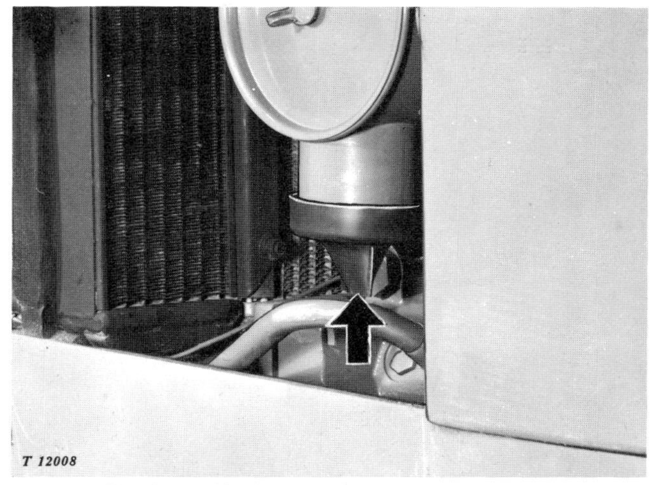

Fig. 17 — Air Cleaner Dust Unloading Valve

Inspect rubber dust unloading valve (if used) for cracks, clogging, or deterioration (Fig. 17). Clean or replace if necessary.

INTAKE SYSTEMS

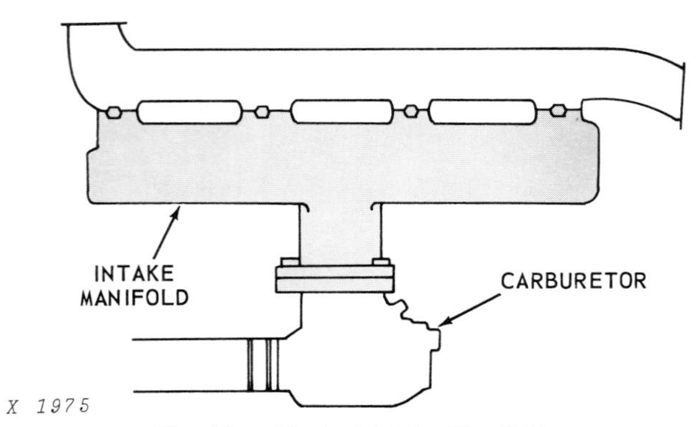

Fig. 18 — Typical Intake Manifold

The fuel-air mixture is carried from the carburetor to the engine intake valves by the manifold.

Diesel engine intake manifolds are similar to spark-ignition engine manifolds except that they do not require heat and are usually located on the side of the engine opposite the exhaust manifold.

Some turbocharged diesel engines have an intercooler which cools heated air from the turbocharger before it enters the intake manifold.

LP-gas fuel vaporizes at normal temperatures so a mixing valve is used instead of a carburetor.

Even though heat transfer from the exhaust manifold is not required to vaporize fuel, the same intake system (following the carburetor) may be used for LP-gas as for gasoline systems.

Fig. 18 illustrates a typical engine manifold. Heat is supplied either by having the exhaust manifold touch the intake manifold at various spots or by having exhaust passages built into the intake manifold.

INTEGRAL INTAKE MANIFOLDS

The intake manifold is an *integral* part of the cylinder head on some engines.

The advantages of this are:

1) The intake passages are surrounded by the engine coolant and the incoming air is warmed before entering the engine cylinders.

2) There are fewer joints, requiring fewer gaskets, and so less danger of leakage.

However, *external* intake manifolds have the advantage of being easier to remove when repairs are necessary.

CRANKCASE INTAKE

Two-cycle engines (see Chapter 1) ordinarily provide intake through the crankcase (Fig. 19). The carburetor is located directly on the crankcase.

On the most simple engines, the fuel-air mixture is drawn into the engine each time the piston moves upward on the compression stroke.

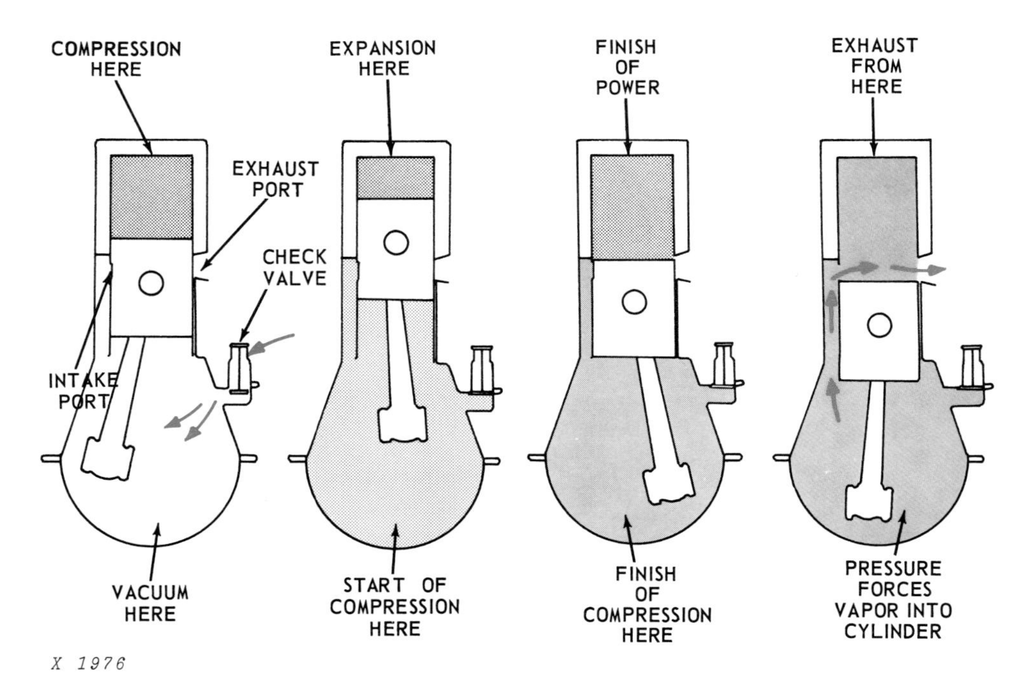

Fig. 19 — Crankcase Intake System

Litho in U.S.A.

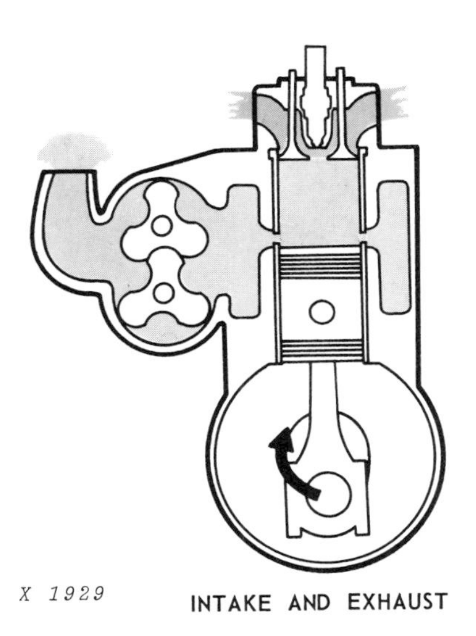

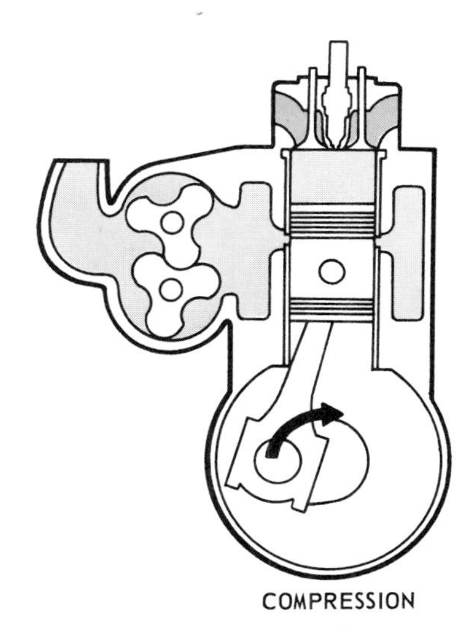

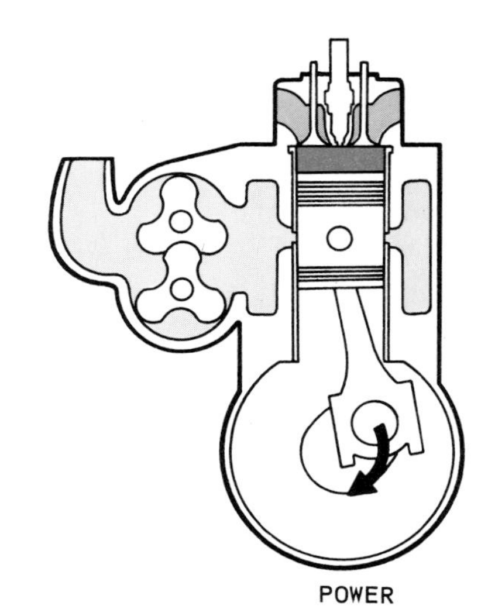

Fig. 20 — Two-Cycle Engine With Blower

A check valve (Fig. 19) prevents the vapor from being forced back out the carburetor on the power stroke.

As the piston moves downward on the power stroke, pressure is built-up in the crankcase until the *intake port* and *exhaust port* are uncovered.

The pressurized vapor flows into the cylinder, forcing the burned gases out the exhaust port.

The action of the crankcase is two stage: **vacuum** to pull in fuel-air; **compression** to push this mixture into the cylinder.

TWO-CYCLE ENGINE

Another type of two-cycle engine is one which uses a Roots type blower or supercharger (Fig. 20).

The Roots is a positive displacement blower which compresses air by trapping it at the inlet port between the rotor lobes and the housing, as the impellers turn. When the trapped air reaches the outlet port, it is forced out of the blower into the engine. The same amount of air is trapped between the lobes with each turn, regardless of impeller speed.

The fuel-air mixture is forced directly into the cylinder each time the piston moves downward far enough to uncover the intake port. The exhaust valve releases exhaust gases during the exhaust portion of the power/exhaust stroke (Fig. 20). The crankcase is never pressurized on this type of engine.

MANIFOLD HEAT CONTROL VALVE

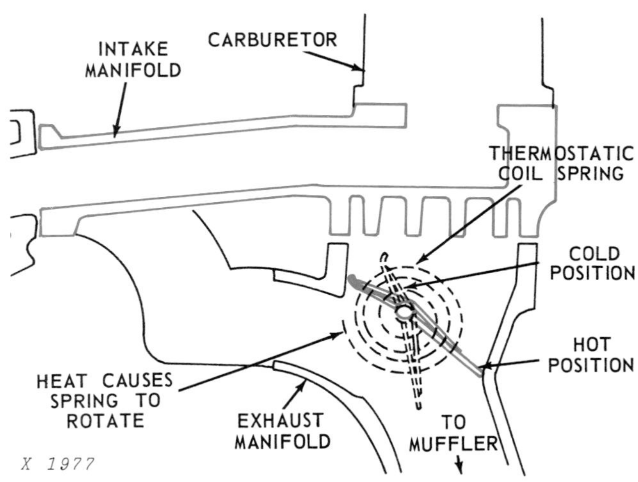

Fig. 21 — Heat Control Valve And Spring

Several types of thermostatic heat control valves are used to direct exhaust gases against some area of the intake manifold when the engine is cold, and deflect gases away from the intake manifold when the engine has warmed up. See Fig. 21.

The thermostatic bimetallic coil spring holds the valve open until exhaust heat causes it to expand. Force from the expanding spring closes the valve and shuts off heat to the intake manifold.

Acids from the burned fuel, together with heat from the gases tend to seize the valve. Use a special heat-control lubricating oil to keep the valve free at all times.

DIESEL ENGINE MANIFOLDS

Diesel engines do not require heat from the exhaust manifold to assist in keeping intake fuel vaporized, since fuel is injected directly into the cylinder as required.

For this reason, diesel intake and exhaust manifolds are usually separated.

SERVICE OF INTAKE MANIFOLDS

Check the intake manifold for restrictions and air leaks. Inspect intake passages of integral manifolds by removing the cylinder head. Replace warped manifolds.

EXHAUST SYSTEMS

The exhaust system collects exhaust gases from the engine and carries them away. This system:

1) Removes heat

2) Muffles engine sounds

3) Carries away burned and unburned gases

The exhaust system consists of:

- **Exhaust valve**
- **Exhaust manifold**
- **Turbocharger (if used)**
- **Muffler**

EXHAUST VALVES seal the burning gases within the cylinder until most of the energy has been expended, then open so the cylinder can clear before the next fuel-air charge is admitted. See Chapter 2.

EXHAUST MANIFOLDS receive burned gases from each cylinder and carry them away from the engine. Some heat from the exhaust manifold is used on gasoline engines to maintain the intake manifold at the proper temperature.

TURBOCHARGERS use exhaust gases to drive the intake supercharger.

MUFFLERS carry away exhaust gases and heat, and muffle engine noise.

EXHAUST MANIFOLD

Fig. 22 — Exhaust Manifold (Gasoline Shown)

Exhaust manifolds (Fig. 22) collect the spent gases from the engine cylinders and conduct them to the muffler or to some other location away from the engine.

The exhaust ports in the engine and the passages in the exhaust manifold are large to allow free flow and expansion of the escaping gases. This is important as it permits better scavenging of the engine cylinders.

If any burned gases are left in the cylinders following the exhaust stroke, the amount of fuel-air mixture which can be taken in on the next intake stroke is limited. This reduces engine power and increases fuel consumption.

Exhaust Manifold Service

Keep inner surfaces of the passages free of carbon buildup. Remove by scraping, with carbon solvent, or a combination of both.

MUFFLERS

There are two common types of mufflers:

- **Straight-through**
- **Reverse-flow**

STRAIGHT-THROUGH mufflers (Fig. 23) consist of a perforated inner pipe enclosed by an outer pipe roughly three times larger in diameter. The space between the pipes is sometimes filled with a sound-absorbing and heat-resistant material.

REVERSE-FLOW mufflers (Fig. 23) are hollow chambers using short pieces of pipe and baffles to force the exhaust gases to travel a back-and-forth path before being discharged.

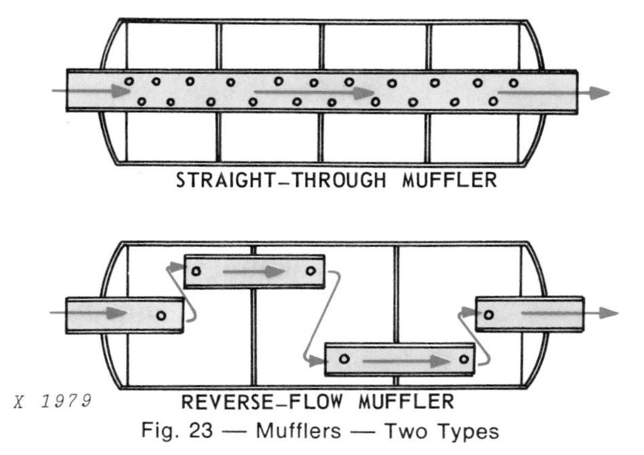

Fig. 23 — Mufflers — Two Types

The muffler acts as an expansion chamber, reducing the noise of the spent gases.

It also serves as a spark arrester, to eliminate fire hazards when operating near combustible material.

Muffler Service

Exhaust systems are designed to provide the least amount of restriction. Excessive restrictions cause back-pressure, resulting in incomplete cylinder scavenging. This, in turn, causes loss of power and increased fuel consumption.

It has been estimated that each 2 psi (14 kPa) of back pressure causes a loss of about 4 engine horsepower (3 kilowatts).

CAUTION: One of the products of combustion is carbon monoxide. This is a deadly, odorless, poisonous gas. Provide good ventilation anytime the engine is operating.

The entire system must also be free of leaks. When they do occur, repair immediately. This is particularly true when the engine is in a machine equipped with a cab or other enclosure.

TURBOCHARGERS

What is a Turbocharger?

A **turbocharger** is an exhaust-driven turbine which drives a centrifugal compressor wheel.

The compressor is usually located between the air cleaner and the engine intake manifold, while the turbine is located between the exhaust manifold and the muffler.

The prime job of the turbocharger is, by compressing the air, to force more air into the engine cylinders. This allows the engine to efficiently burn more fuel, thereby producing more horsepower.

Operation Of Turbocharger

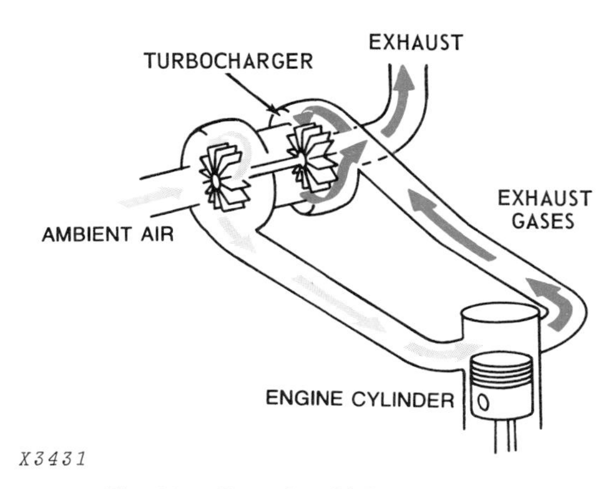

Fig. 24 — Operation Of Basic Turbocharger

All of the engine exhaust gases pass through the *turbine* housing (Fig. 24). The expansion of these gases, acting on the turbine wheel, causes it to turn. After passing through the turbine, the exhaust gases are routed to the atmosphere. In many cases, the turbine muffles the exhaust sound, so no muffler is needed.

The turbine also functions as a spark arrester. For example, it is recognized by the U.S. Department of Agriculture as providing a spark arrester function adequate for forestry operations.

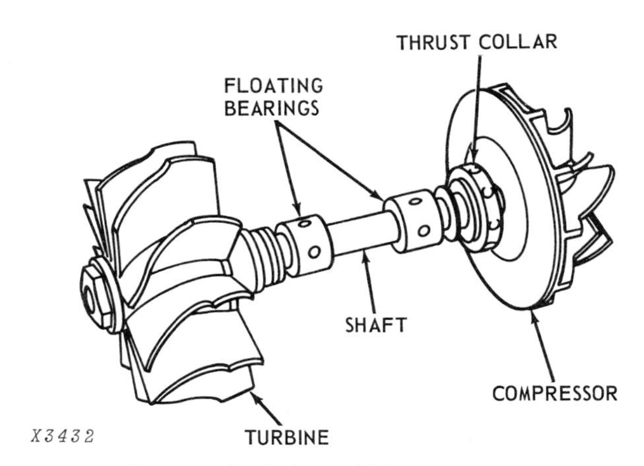

Fig. 25 — Basic Parts Of Turbocharger

The *compressor* is directly connected to the turbine by a shaft (Fig. 25). The only power loss from the turbine to the compressor is the slight friction of the journal bearings.

Air is drawn in through a filtered air intake system, compressed by the wheel, and discharged into the engine intake manifold.

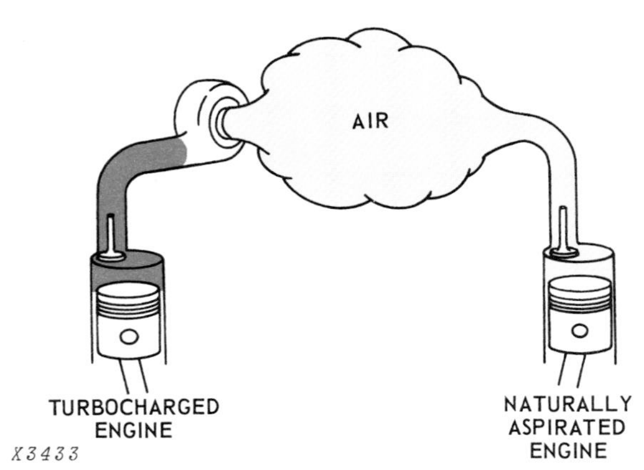

Fig. 26 — Air Intake Comparison

The extra air provided by the turbocharger allows more fuel to be burned, which increases horsepower output. Lack of air is one factor limiting the engine horsepower of naturally-aspirated engines (Fig. 26).

As engine speed increases, the length of time the intake valves are open decreases, giving the air less time to fill the cylinders. On an engine running at 2500 rpm, the intake valves are open less than 0.017 second. The air drawn into a naturally-aspirated engine cylinder is at less than atmospheric pressure. A turbocharger packs the air into the cylinder at greater than atmospheric pressure.

The flow of exhaust gas from each cylinder occurs intermittently as the exhaust valve opens. This results in fluctuating gas pressures (pulse energy) at the turbine inlet. With a conventional turbine housing, only a small amount of the pulse energy is used.

To better utilize these impulses, one design has an internal division in the turbine housing and the exhaust manifold which directs these exhaust gases to the turbine wheel. There is a separate passage for each half of the engine cylinder exhaust (Fig. 27).

On a six-cylinder engine, there is a separate passage for the front three cylinders and another passage for the rear three cylinders.

By using a fully divided exhaust system combined with a dual scroll turbine housing, the result is a highly effective nozzle velocity. This produces higher turbine speeds and manifold pressures than can be obtained with an undivided exhaust system.

The turbocharger offers a distinct advantage to an engine operating at high altitudes. The turbocharger automatically compensates for the normal loss of air density and power as the altitude increases.

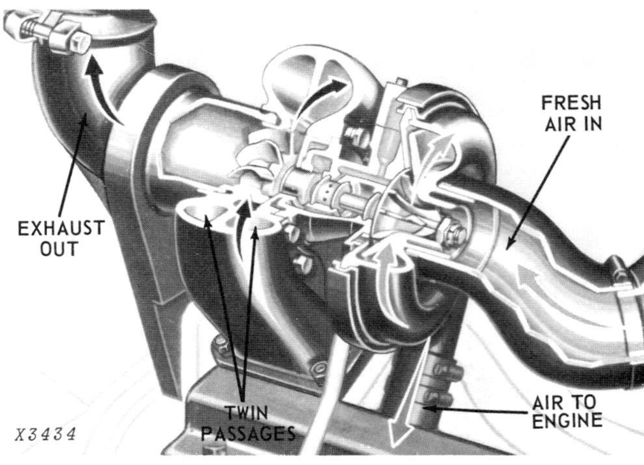

Fig. 27 — Twin Passage Turbine

With a naturally aspirated engine, horsepower drops off 3 percent per 1000 ft. (300 m) because of the 3 percent decrease in air density per 1000 ft. (300 m). If fuel delivery is not reduced, smoke level and fuel dilution will increase with altitude.

With a turbocharged engine, an increase in altitude also increases the pressure drop across the turbine. Inlet turbine pressure remains the same, but the outlet pressure decreases as the altitude increases. Turbine speed also increases as the pressure differential increases. The compressor wheel turns faster, providing approximately the same inlet manifold pressure as at sea level, even though the incoming air is less dense.

However, there are limitations to the actual amount of altitude compensation a turbocharged engine has. This is primarily determined by the amount of turbocharger boost and the turbocharger-to-engine match.

All turbochargers operate at a very high speed. This can range from 40,000 to 130,000 rpm or more.

The appearance, construction, and operation of the **altitude compensator** (Fig. 28) is the same as that of a turbocharger. However, the *purpose is different.*

The purpose of a *turbocharger* is to increase the power output of an engine by supplying compressed air to the engine intake manifold so increased fuel can be utilized for combustion.

The purpose of the *altitude compensator* is to maintain consistent power output and efficiency of an engine operating at all altitudes. This is done by supplying compressed air to the engine intake manifold at a pressure about equal to that at sea level.

ALTITUDE COMPENSATORS

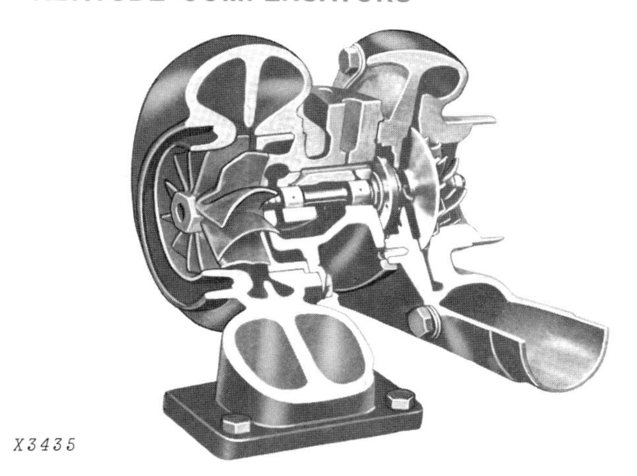

Fig. 28 — Altitude Compensator

There is no increase of fuel for combustion and consequently no increase in basic horsepower of the engine. However, the extra air provided by the altitude compensator normally increases combustion efficiency, which generally will improve fuel economy and reduce smoke level.

WHY USE TURBOCHARGERS?

There are five basic reasons for using an engine turbocharger:

1. **To increase horsepower output of a given displacement engine.** Where the engine compartment of a machine is of a given size, a turbocharged engine can be used to provide increased horsepower without having to enlarge the engine compartment for a larger displacement engine (Fig. 29).

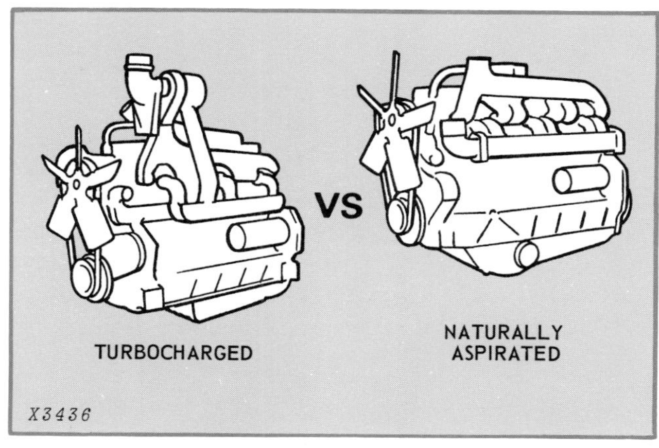

Fig. 29 — Engine Sizes Compared

2. **To reduce weight.** Turbocharged engines have more horsepower per pound than non-turbocharged engines.

3. **To keep down costs.** Initial cost of turbocharged engines, on a dollar per horsepower basis, is less than for a naturally aspirated (N.A.) engine, and the differential increases with the rate of turbocharging. It all adds up to more horsepower per dollar.

4. **To maintain power at higher altitudes.** The altitude compensator also falls in this category, giving vital machine productivity at high altitudes.

5. **To reduce smoke.** Turbocharging can be an effective way to reduce exhaust density by providing excess air. However, using a turbocharger does not insure this, as many other components also affect exhaust density and these must be properly designed and matched to provide an acceptable smoke level.

SUPERCHARGERS

The power delivered by an internal combustion engine is determined by the amount of fuel and air which can be packed into each cylinder. The more fuel-air mixture, the more power the engine develops.

A **supercharger** is a type of *air pump.* Located either ahead of or behind the carburetor, it gives an engine a higher overall compression than it would normally have.

It does this by taking air from the atmosphere, compressing it and packing it into the engine cylinders.

Pumping air into an engine also aids in getting exhaust gases out. This too results in more efficient combustion.

A NOTE ON SUPERCHARGER "KITS"

Installing a supercharger on most engines *will* increase horsepower by about 30%.

But adding a supercharger will also cause:

- **A 51% increase in peak cylinder pressure.**
- **Much higher cylinder temperatures.**
- **Increased air intake volume.**
- **Increased wear on the engine and drive components.**
- **Increased oil consumption.**

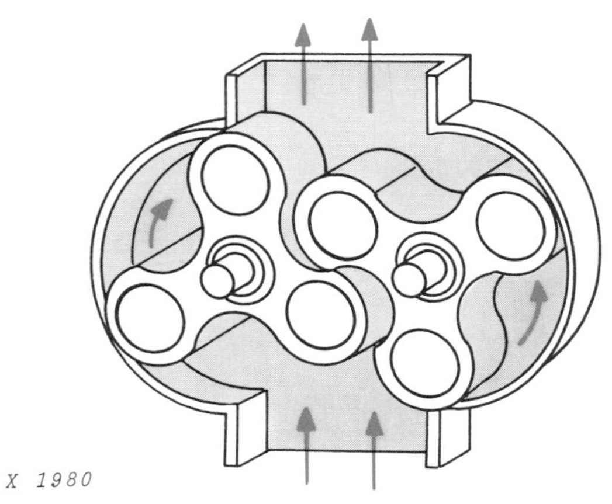

Fig. 30 — Roots-Type Supercharger

Before mounting a supercharger on an engine, make certain it also has been equipped with the following:

1) Heavy-duty engine and drive components

2) Larger cooling system

3) Larger lubrication system

4) Bigger air cleaner

TWO TYPES OF SUPERCHARGERS

Roots-type superchargers or blowers (Fig. 30) are positive-displacement compressors. They resemble oil pumps in design.

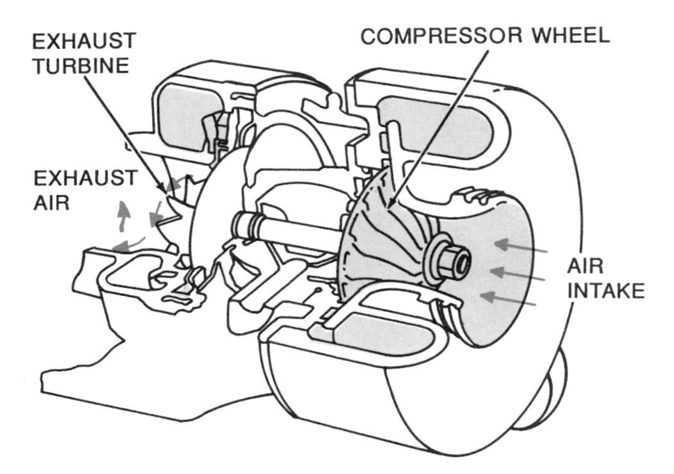

Fig. 31 — Centrifugal-Type Supercharger (Turbocharger Shown)

In operation, air enters the housing due to rotation of the rotors and passes between the lobes of the rotors and housing.

The air is then forced out through the outlet opening of the unit.

This type of blower is either chain, belt or gear driven.

Centrifugal-type superchargers (Fig. 31) have an impeller rotating at high speed inside a housing. Normal speed is about 30 times engine speed.

The impeller moves air by "flinging" it off the tips of the impeller blades.

The type may be driven by the engine or engine exhaust, or by a separate motor. The advantage of a separately driven supercharger is that it does not use engine power.

Centrifugal-type superchargers are less expensive, require less maintenance, and do not make as much noise as Roots type units.

The **turbocharger,** an exhaust-driven centrifugal type supercharger, is most commonly used.

SUPERCHARGERS FOR TWO-CYCLE ENGINES

An air pump of some type must be used on two-cycle engines to provide scavenging. (On four-cycle engines, this is done by the exhaust and intake strokes.)

There are three methods of providing this scavenging through a direct engine coupling. They are:

1) Crankcase scavenging

2) Power-piston scavenging

3) Pump or blower scavenging

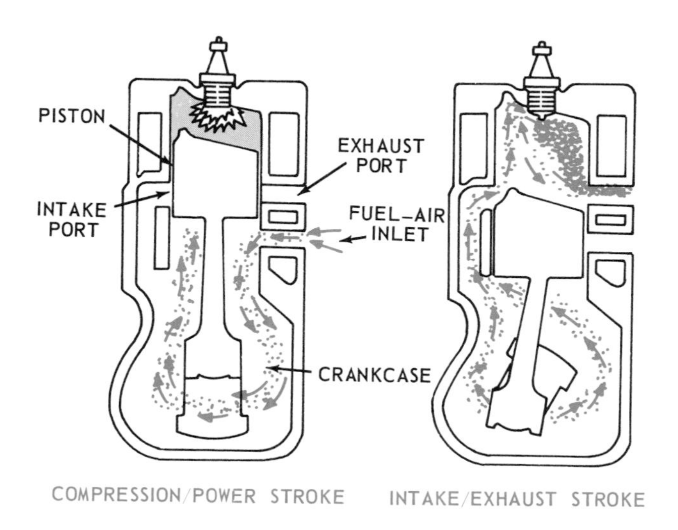

X 1982

Fig. 32 — Crankcase Scavenging In Two-Cycle Engine

Crankcase scavenging is used when the fuel-air vapor enters the engine through the crankcase (Fig. 32).

Each downward movement of the piston compresses the vapor within the crankcase until the intake port or valve opens.

The compressed vapor then escapes into the cylinder at a pressure nearly equal to atmospheric pressure.

Power-piston scavenging uses a separate piston and cylinder, driven by the engine crankshaft, to push the vapor into the cylinder as the intake port or valve opens (Fig. 33).

With **blower scavenging** (Fig. 34), a positive-displacement rotary blower, driven by the engine, compresses the fuel-air vapor into an air chamber surrounding the intake ports.

This type of blower has an advantage over an engine-driven centrifugal blower because it delivers practically the same amount of air per revolution regardless of speed or working pressure.

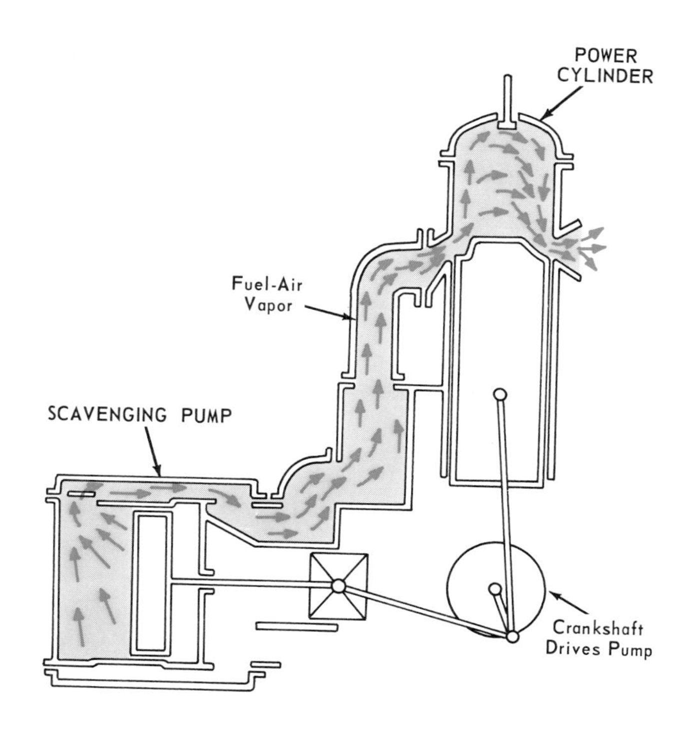

X 1983

Fig. 33 — Power-Piston Scavenging

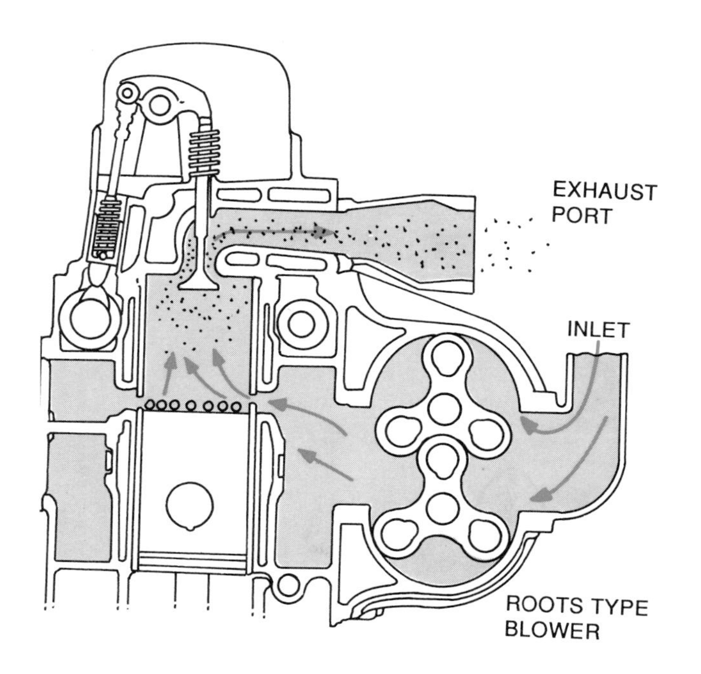

Fig. 34 — Positive-Displacement Blower

Litho in U.S.A.

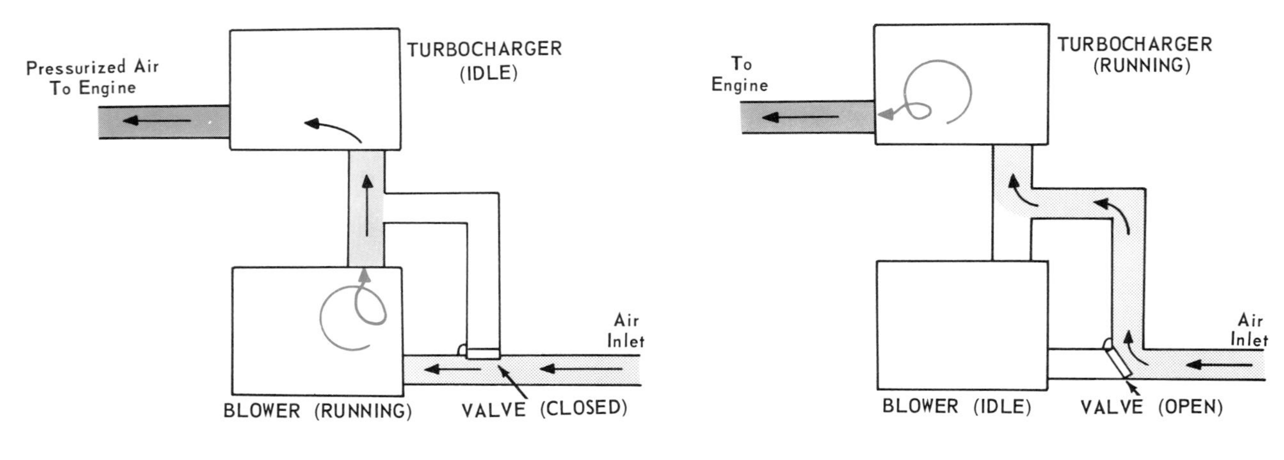

Fig. 35 — Scavenging Pump And Turbocharger

Scavenging Pump And Turbocharger

A turbocharger may be used in addition to the regular scavenging system. In this case the air drawn into the scavenging pump or blower is compressed to scavenging pressure in the normal manner and then passed to the turbocharger where it is raised to supercharged pressure (See Fig. 35).

At light loads when there is little energy available to drive the turbocharger, the mechanically-driven blower alone puts the scavenging air into the cylinders.

At increased loads, the turbocharger speeds up and takes in so much air that its inlet pressure drops to atmospheric level, causing the blower check valve to open (Fig. 35).

At this engine speed, the blower becomes unloaded (saving engine power) and the turbocharger enters the load range where it alone can provide scavenging *and* supercharging.

Under the most favorable conditions the engine starting air contains enough energy to start the turbocharger and also supply enough combustion air to burn the fuel.

However, usual practice today is to equip the turbocharger with some method for supplying additional scavenging air while the engine is being started and while it is running at slow speed.

Two of the methods are:

1) *Mechanical*—At starting and slow speeds both the blower and turbocharger are driven from the crankshaft. When speed picks up, the drive coupling disconnects and the turbocharger operates on exhaust gas only.

2) *Jet Air Starting*—Used only for starting when high pressure air is available. Air is blown through jets into the turbocharger turbine or compressor. Air passing through the compressor passes on into the engine to assist in scavenging.

Aftercoolers

When the turbocharger compresses the engine intake air, the air becomes heated (due to compression) and expands. When the heated air expands it becomes less dense. The result is that part of the purpose of the turbocharger is defeated; that is, due to heat expansion, less air is forced into the engine.

To overcome this condition, some turbocharged engines are equipped with an **aftercooler** (Fig. 36). This is installed between the turbocharger and the engine intake manifold.

The aftercooler reduces the temperature of the compressed air by 80 to 90°F (44 to 50°C). This

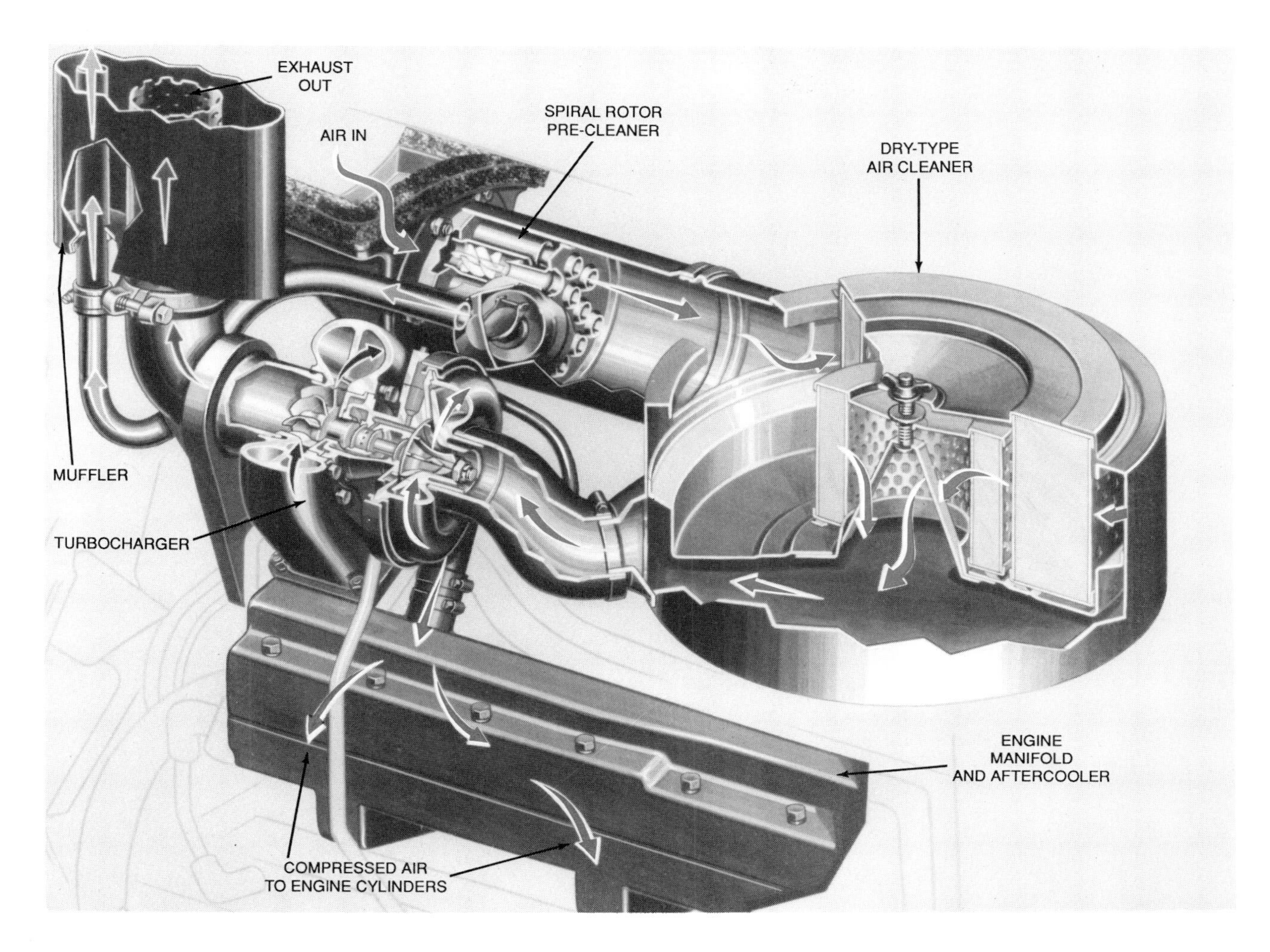

Fig. 36 — Intake-Exhaust System On Turbocharged, Aftercooled Engine

makes the air denser, allowing more to be packed into the combustion chambers.

The result is:

1. More power

Sufficient air is provided to burn the fuel, resulting in higher horsepower.

2. Greater economy

The fuel is burned more completely, giving more power from a given amount of fuel.

3. Quieter combustion

By controlling warm air for fuel-air mixing, there is a smoother pressure rise in the engine cylinder.

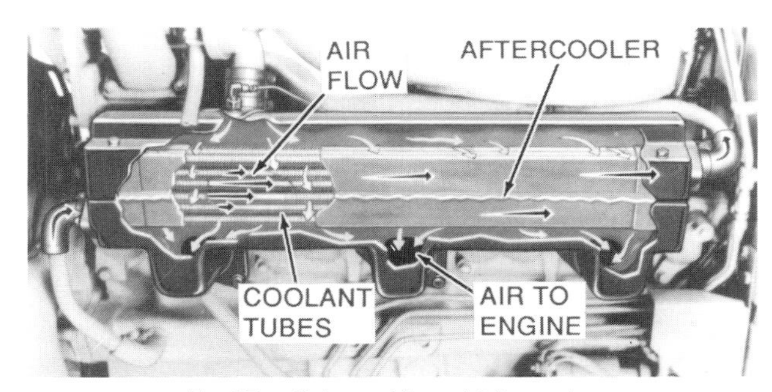

Fig. 37 — Cutaway View of Aftercooler

The aftercooler is nothing more than a heat exchanger. The intake air flows over a series of tubes through which engine coolant is circulated (Fig. 37). This provides the necessary cooling.

Special Instructions For Turbocharged Engines

Follow these special operating instructions when operating a turbocharged engine.

1. After starting the engine do not accelerate or apply load until there is positive indication of oil pressure.

2. After starting during cold weather, allow the engine to run five minutes at half throttle to insure oil pressure at the turbocharger before putting the engine under load.

3. Before stopping the engine, allow it to run at near slow idle speed for a few minutes to allow internal engine temperature to normalize. Failure to do this can damage the turbocharger (as well as the engine) due to distortion and "coking" of the oil in the passages.

4. Should the engine stall when operating at normal operating temperature, restart it immediately to prevent "temperature soaking" of the turbocharger. The turbocharger gets hot in operation. If stopped or stalled, oil in the center section may "coke", causing oil passages to clog.

Important: *When transporting an idle turbocharged engine with the exhaust outlet exposed, cover the exhaust outlet to prevent entrance of foreign matter and possible rotation of the turbocharger. Rotation of the turbocharger could damage the rotor bearings due to lack of lubrication.*

On other occasions, when the engine is not operating, cover the exhaust outlet to prevent entrance of water or other foreign matter.

Lubrication Of Turbochargers

The center housing may contain lubrication passages through which engine oil, under pressure, is directed to the journal bearings and thrust washers. It may also contain fittings for oil inlet and outlet connections.

If the unit has floating sleeve type bearings, they provide oil clearance between the bearing and housing as well as oil clearance between the bearing and the shaft. When the turbocharger is operating, this allows the bearing to turn as the shaft rotates.

Since all parts of the rotating assembly are protected by a film of oil, no metal contact occurs. Consequently no wear occurs. If a constant supply of clean oil is supplied to the unit, bearing life will be indefinite.

All clearances in the turbocharger are controlled by closely maintained machine tolerances of detailed parts. This makes it imperative that only clean engine oil be supplied to the unit.

Periodic Inspection Of Turbochargers

1. Inspect the mounting and connections of the turbocharger to be certain they are secure and there is no leakage of oil or air.

2. Check the engine crankcase to be sure there is no restriction to oil flow.

3. Operate the engine at approximate rated output and listen for unusual turbocharger noise. If a shrill whine (other than normal) is heard, stop the engine immediately. The whine means that the bearings are about to fail. Remove the turbocharger for inspection.

NOTE: Do not confuse the whine heard during "run down", as the engine stops, with a bearing failure during operation.

Other unusual turbocharger noises could mean improper clearance between the turbine wheel and housing. If such noises are heard, remove the turbocharger for inspection. See the engine Technical Manual.

4. Check the turbocharger for unusual vibration while engine is operating at rated output. If necessary, remove the turbocharger for inspection.

5. Check engine under load conditions. Excessive exhaust smoke indicates incorrect fuel-air mixture. This could be due to engine overload or turbocharger malfunction.

6. Inspect and service air cleaner according to instructions in the operator's manual.

Inspection Of Damaged Parts

If the impeller or turbine wheel are damaged as shown on this page, replace them. Also be sure to eliminate the *cause* of failure.

FOREIGN MATERIAL IN INTAKE SYSTEM (Fig. 38)

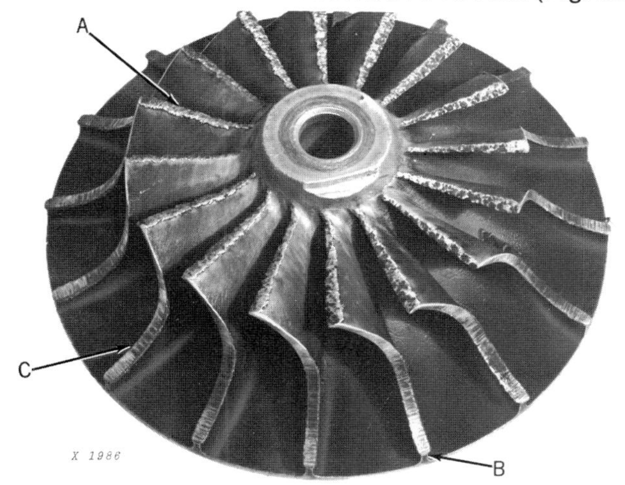

Fig. 38 — Impeller Damaged By Foreign Material In Intake System (Replace It)

Appearance

a. Leading edges of compressor wheel badly nicked.

b. Blade tips usually intact.

c. Compressor wheel may or may not have contacted compressor housing.

d. (Not shown) Imbalance of rotating assembly may cause bearing extrusion and heavy seal wear.

e. (Not shown) Inlet section of compressor housing will be rough and pitted.

Probable Causes

a. Leak in connections of joints.

b. Loose material left in duct.

c. Welding slag not removed from duct.

d. Ice formation in duct from water pulled into air ducting.

e. Loose wire particles from air cleaner.

SAND BLASTING OF IMPELLER (Fig. 39)

Appearance

a. Nicked leading edges.

b. Blade tips thinned from peening.

c. Blade contour deeply eroded.

Fig. 39 — Impeller Eroded By Sand Blasting Problem

NOTE: Sand blasting problems may or may not introduce imbalance and cause bearing failure.

Probable Causes

a. Loose connections at joints of air cleaner or ducting.

b. Dry-type paper element split.

c. "Channeling" in oil bath air cleaner.

FOREIGN MATERIAL IN EXHAUST SYSTEM (Fig. 40)

Fig. 40 — Turbine Wheel Damaged By Foreign Material In Exhaust System

Appearance

a. Blade tips of turbine wheel chewed and battered.

b. In extreme cases, imbalance will destroy seals and bearings.

c. Shaft journals in good condition.

NOTE: Thrust bearing may or may not be damaged. Pounding and heat may cause shaft break.

Probable Causes

a. Piston ring breakage.

b. Engine valve breakage.

c. Loose material left in manifold, or broken injector tips.

See the engine Technical Manual for disassembly instructions.

Trouble Shooting of Turbochargers

As has been explained, most turbocharger failures start in some other area of the engine. The turbocharger is a relatively simple device and, if properly serviced, will operate with little or no attention.

TURBOCHARGER TROUBLE SHOOTING CHART

Trouble	*Possible Cause*	*Remedy*
Noisy operation or vibration.	Bearings not being lubricated.	Supply required oil pressure. Clean or replace oil line. If trouble persists, overhaul the turbocharger.
	Leak in engine intake or exhaust manifold.	Tighten loose connections or replace manifold gaskets.
Engine will not deliver rated power.	Clogged manifold system.	Clear all ducting.
	Foreign matter lodged in compressor, impeller, or turbine.	Disassemble and clean.
	Excessive buildup in compressor.	Thoroughly clean compressor assembly. Clean air cleaner and check for leaks.

TURBOCHARGER TROUBLE SHOOTING CHART (Continued)

Trouble	*Possible Cause*	*Remedy*
Engine will not deliver rated power (cont.)	Leak in engine intake or exhaust manifold.	Tighten loose connections or replace manifold gaskets.
	Rotating assembly bearing seizure.	Overhaul turbocharger.
Oil seal leakage.	Failure of seal.	Overhaul turbocharger.
	Restriction in air cleaner or air intake creating suction.	Remove the restriction.

TESTING THE AIR INTAKE SYSTEM

When the air flow into the engine is restricted, there is more vacuum or suction in the cylinders. This can cause oil to be drawn in around the valve stems and pistons and so increase oil consumption.

The test given here will tell if there is a restriction in the air intake system.

INTAKE VACUUM TEST (DIESEL ENGINES)

Test as follows:

1. Warm up the engine.

2. *On engines with restriction indicators,* remove the indicator, install a pipe tee fitting and reinstall the indicator. Connect the gauge to the tee fitting.

3. *On engines without restriction indicators,* connect the gauge to the intake manifold.

4. Set engine speed at fast idle and note the reading on the gauge. **Too high a reading means that there is a restriction in the air intake system.** Check the engine Technical Manual for correct specifications.

5. On engines with restriction indicators, check the operation of the indicator. Use a board or metal plate to very slowly cover the air intake opening. Note the action of the indicator in relation to the reading on the gauge. If the indicator does not operate properly, replace it.

Fig. 41 — Connecting The Vacuum Gauge

MANIFOLD DEPRESSION TEST (SPARK-IGNITION ENGINES)

Use a vacuum gauge calibrated in inches of mercury to perform this test.

1. Connect the vacuum gauge to the intake manifold (Fig. 41).

2. Warm up the engine and operate it at idle speed.

3. Note the reading on the vacuum gauge. Check the engine Technical Manual for exact specifications.

4. Interpret the gauge reading as follows:

• **If the reading is steady and low,** loss of power in all cylinders is indicated. Possible causes are late ignition, bad valve timing, or loss of compression at valves or piston rings. A leaky carburetor gasket will also cause a low reading.

• **If the needle fluctuates steadily,** a partial or complete loss of power in one or more cylinders is indicated. This can be due to an ignition defect, or loss of compression due to stuck piston rings or a leaky cylinder head gasket.

• **Intermittent needle fluctuation** indicates occasional loss of power due to an ignition defect or a sticking valve.

• **Slow needle fluctuation** is usually caused by improper carburetor idle mixture adjustment.

• **A gradual drop in the gauge reading** at idle engine speed indicates back pressure in the exhaust system due to a restriction.

Be very careful in analyzing abnormal readings since the gauge readings can indicate more than one thing.

TEST YOURSELF

QUESTIONS

1. What are the two basic types of air intake systems?

2. What are the two types of superchargers? Which type delivers the same amount of air per revolution at all engine speeds?

3. What are the three parts of the exhaust system?

4. What three things does the exhaust system do?

(Answers in back of text.)

Litho in U.S.A.

LUBRICATION SYSTEMS / CHAPTER 7

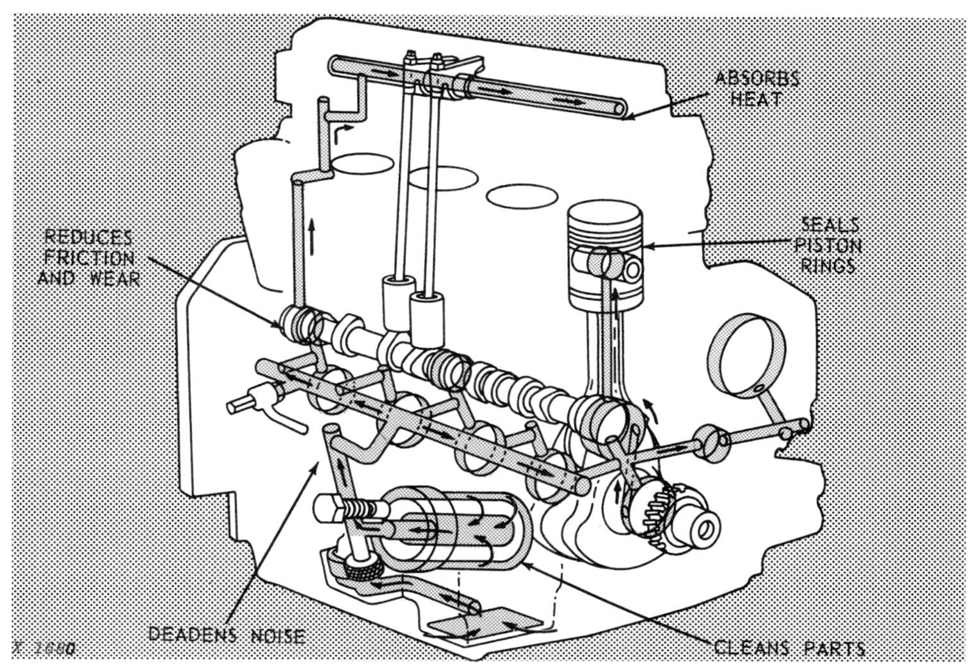

Fig. 1 — Lubrication System for Typical Engine

INTRODUCTION

The lubrication system does these jobs for the engine:

1) **Reduces friction between moving parts**

2) **Absorbs and dissipates heat**

3) **Seals the piston rings and cylinder walls**

4) **Cleans and flushes moving parts**

5) **Helps deaden the noise of the engine**

With lubricating oil, the system is able to perform all these jobs at once; without it, the engine would soon wear out or burn up.

TYPES OF LUBRICATION SYSTEMS

Engine lubrication systems may be classified as:

- **Circulating Splash**
- **Internal Force Feed and Splash**
- **Full Internal Force Feed**

The type of system used depends largely upon the size and design of the engine.

Let's discuss each system in detail.

CIRCULATING SPLASH SYSTEM

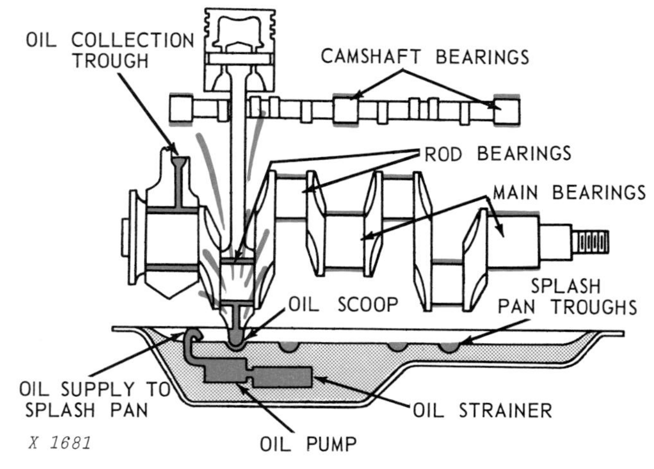

Fig. 2 — Circulating Splash System

In the **circulating splash system,** an oil pump supplies oil to a splash pan located under the crankshaft (Fig. 2). As the connecting rods revolve, scoops on the ends of the rods dip into troughs in the splash pan, creating the oil splash.

The splashing oil lubricates the moving parts nearby. Other parts are lubricated by oil splash which builds up in collecting troughs and is gravity fed through channels or lines.

The upper parts of the cylinders, pistons, and pins are lubricated more by oil mist than by the oil splash itself. This mist is created as the connecting rods spin.

The circulating splash system must have:

1) Proper oil level in the pan

2) Suitable oil for good splashing

There must be enough oil in the troughs of the splash pan for the connecting rods to splash. The oil pump must be working properly to provide this oil.

Because the oil must splash and flow freely, heavy oil will not work. Use only oil of the viscosity recommended by the engine manufacturer.

INTERNAL FORCE FEED AND SPLASH SYSTEM

In the **internal force feed and splash system,** the pump forces oil directly to a main oil gallery in the engine block rather than to a splash pan.

From the main oil gallery, the oil is forced through passages to the main bearings, connecting rod bearings, camshaft bearings, rocker arm shaft, filter, and pressure sending unit.

The oil escaping from the bearings creates a mist which also lubricates the upper cylinder walls, pistons, and pins.

Pressure of the lubricating oil can usually be adjusted in these systems.

FULL INTERNAL FORCE FEED SYSTEM

The **full internal force feed system** goes one step farther than the system above. Oil is forced not only to the crankshaft bearings, rocker arm shaft, filter and pressure sending unit, but also to the piston pin bearings.

The piston pin bearings are lubricated through drilled passages in the connecting rods (Fig. 3). The cylinder walls and pistons are lubricated by oil escaping from the piston pin bearings or the connecting rod bearings.

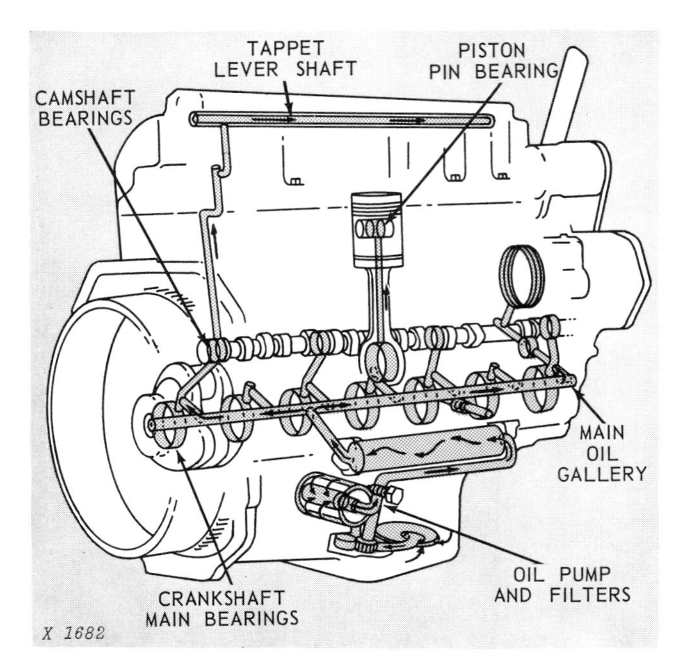

Fig. 3 — Full Internal Force Feed System

The force feed system has one prime need: **good oil pressure.** This system is used (with a full-flow filter) in most agricultural and industrial machines.

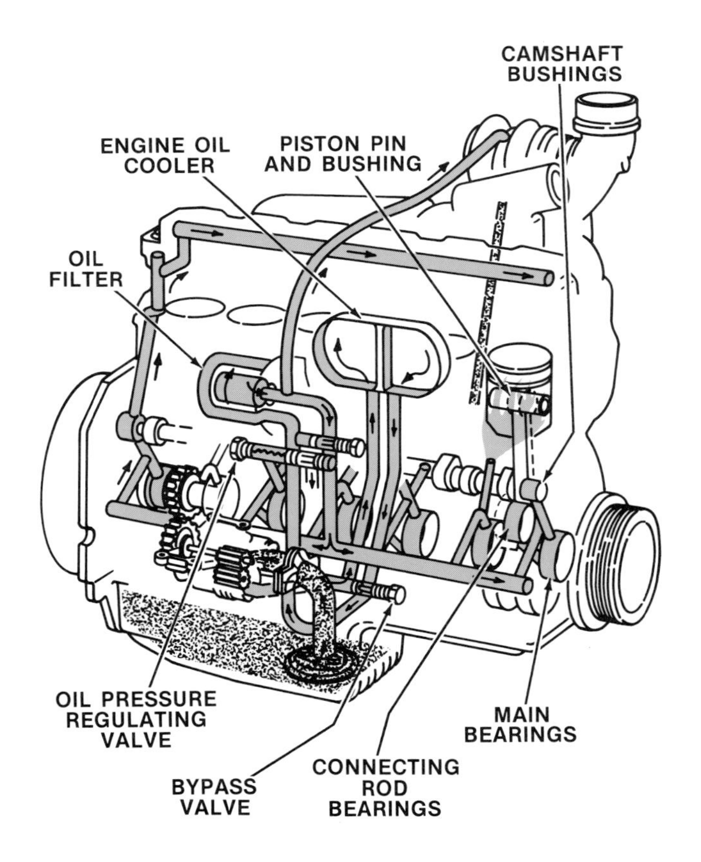

Fig. 3A — Crankshaft Driven Oil Pump

EXTERNAL GEAR PUMP

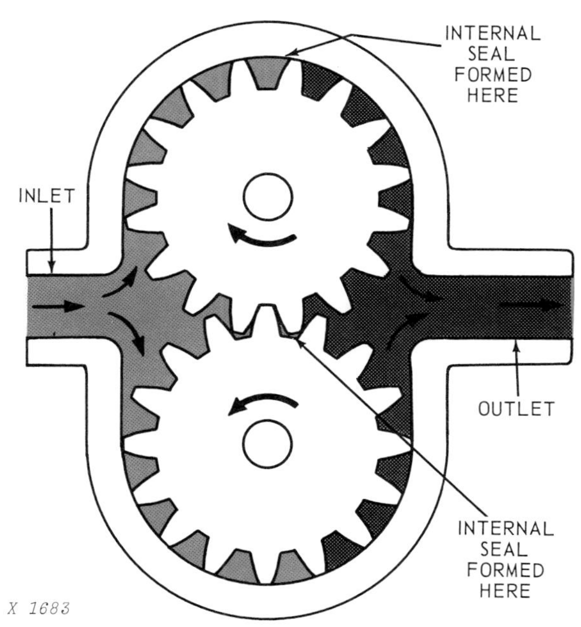

Fig. 4 — External Gear Pump

As the force feed system requires oil to be pumped long distances and through many passages, sufficient oil pressure is required. The required pressure for most engines is 25-40 psi (170-275 kPa), but it may go as high as 65 psi (450 kPa).

OIL PUMPS

Pumps for engine lubrication are usually one of two types:

- **External Gear Pumps**
- **Rotor Pumps**

The external gear type is the most common.

Normally, the oil pump is mechanically driven by the engine. Usually, the external gear pump is driven from the camshaft, while the rotor type is driven from the crankshaft. In some large engines, an electric motor is used to drive an auxiliary pump.

The **external gear pump** has two gears in mesh, closely fitted inside a housing (Fig. 4). The drive shaft drives one gear, which in turn drives the other gear. The machined surfaces of the outer housing are used to seal the gears.

As the gears rotate and come out of mesh, they trap inlet oil between the gear teeth and the housing. The trapped oil is carried around to the outlet chamber. As the gears mesh again, they form a seal which prevents oil from backing up to the inlet. The oil is forced out at the outlet and sent through the system.

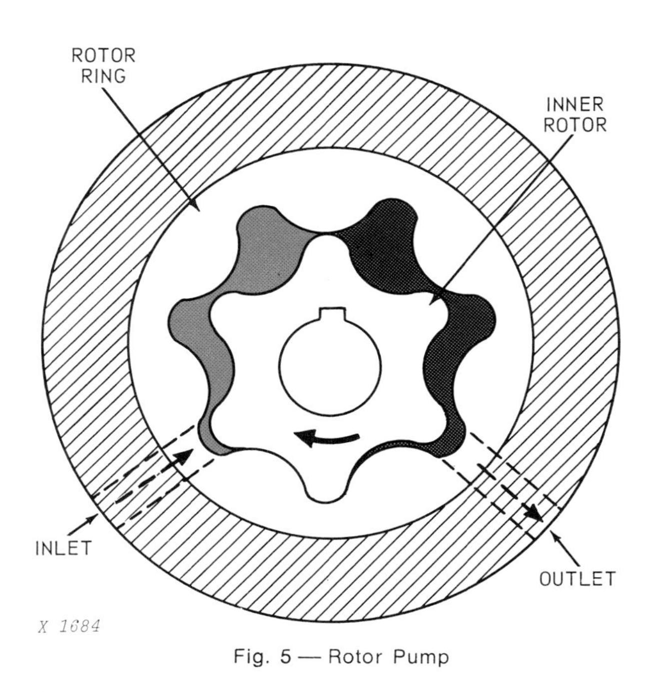

Fig. 5 — Rotor Pump

The **rotor pump,** which is a variation on the *internal* gear pump, is also relatively simple in design. An inner rotor turns inside a fixed rotor ring (Fig. 5).

In operation, the inner rotor is driven inside the rotor ring. The inner rotor has one less lobe than the ring, so that only one lobe is engaged with the outer ring at any one time. This allows the other lobes to slide over the outer lobes, making a seal to prevent back-up of oil. As the lobes slide up and over the lobes on the outer ring, oil is drawn in. As the lobes fall into the ring's cavities, oil is squeezed out.

SERVICING OIL PUMPS

When servicing an oil pump, be careful not to damage the mounting surfaces of either the housing or cover. Many pumps do not use a gasket, and these machined surfaces form the only seal.

Check the machined surfaces of the pump housing and cover for warping. They must be perfectly flat if they are to seal.

In an external gear pump, most wear will occur between the teeth of the gears.

In a rotor pump, most wear will occur between the lobes on the inner rotor and the lobes on the rotor ring.

Use a micrometer to measure the widths of these parts. Check the results against the specifications for the pump.

Also measure the pump drive shaft and compare it with the specifications.

 Litho in U.S.A.

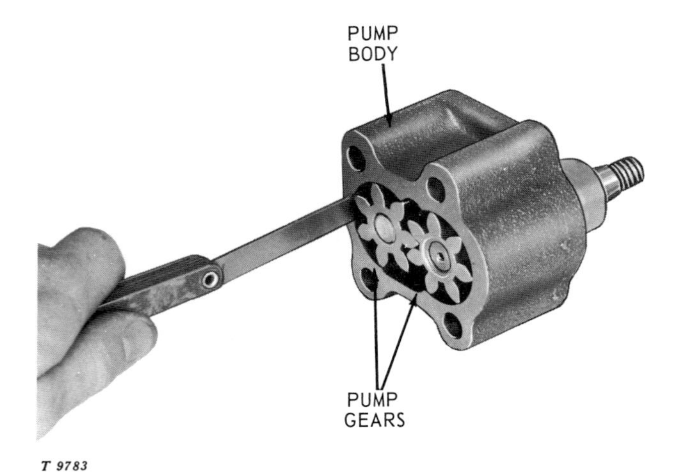

Fig. 6 — Measuring Clearance Between Pump Gears and Housing

With the gears mounted in the housing, use a feeler gauge to measure the clearance between the gears and the housing (Fig. 6). If the clearance is more than specifications, replace the gears.

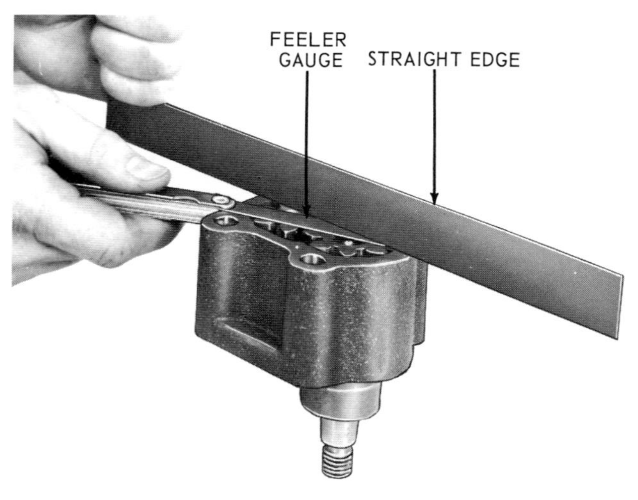

Fig. 7 — Measuring Clearance Height of Pump Gears

Place a straight edge across the top of the pump housing (to represent the cover) and measure the clearance between the gears and the straight edge (Fig. 7). If the clearance is excessive, replace the gears.

Almost all pumps use a screen over the iniet to strain out foreign material. Where possible, remove the screen from the inlet pipe, and clean both with a solvent. Use compressed air to dry the parts.

Inspect all bushings in the housings and replace those with excessive wear.

OIL FILTERS AND FILTRATION SYSTEMS

Oil contamination reduces engine life more than any other factor. To help combat this, oil filters are designed into all modern engine lubrication systems.

There are two basic types of filters—**surface** and **depth** and they are used in two basic types of filtering systems—**bypass** and **full flow.** Some large engines use a combination bypass and full flow filtering system.

Let's look first at the two types of filters, then at the two types of filtering systems.

TYPES OF FILTERS

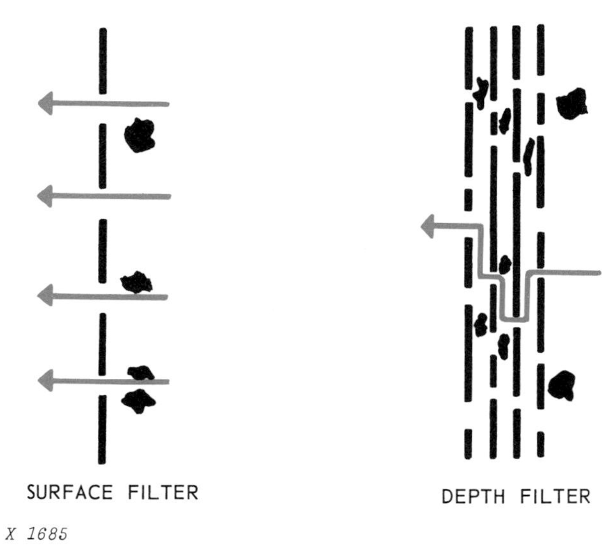

Fig. 8 — Surface and Depth Filters Compared

Filters are classified as either **surface**-type filters or **depth**-type filters depending on the way they remove dirt from the oil (Fig. 8).

SURFACE FILTERS have a single surface that catches and removes dirt particles larger than the holes in the filter. Dirt is strained or sheared from the oil and stopped outside the filter as oil passes through the holes in a straight path. Many of the large particles will fall to the bottom of the reservoir or filter container, but eventually enough particles will wedge in the holes of the filter to prevent further filtration. Then the filter must be cleaned or replaced.

A surface filter may be made of fine wire mesh (Fig. 9), stacked metal or paper disks, metal rib-

 Litho in U.S.A.

bon wound edgewise to form a cylinder (Fig. 10), cellulose material molded to the shape of the filter, or accordian-pleated paper (Fig. 11).

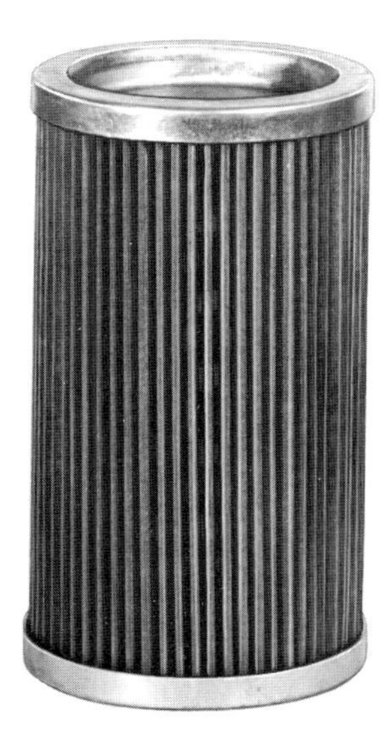

Fig. 9 — Wire Mesh Filter

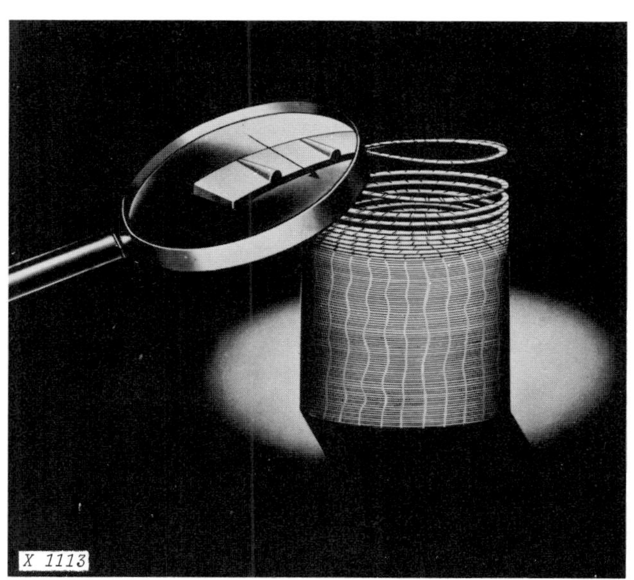

Fig. 10 — Metal Edge Filter

DEPTH FILTERS, in contrast to the surface type, use a large volume of filter material to make the oil move in many different directions before it finally gets into the lubrication system. The filter made of cotton waste shown in Fig. 12 is an example of a depth filter.

Depth filters can be classified as either absorbent or adsorbent, depending on the way they remove dirt.

Fig. 11 — Pleated Paper Filter

Absorbent filters operate mechanically like a sponge soaking up water. Oil passes through a large mass of porous material such as cotton waste, wood pulp, wool yarn, paper or quartz, leaving dirt trapped in the filter. This type of filter will remove particles suspended in the oil and some water and water soluble impurities.

Adsorbent filters operate the same way as absorbent filters but also are chemically treated to attract and remove contaminant. This filter may be made of charcoal, chemically-treated paper or fuller's earth. It will remove contaminating particles, water soluble impurities and, because of its chemical treatment, will also remove contamination caused by oil oxidation and deterioration. Adsorbent filters may also remove desirable additives from the oil and for this reason are not often used in lubrication systems.

Fig. 12 — Depth Filter — Cotton Waste Type

DEGREES OF FILTRATION

In addition to the type of filter, the degree of filtration is also important to a lubrication system.

It is the degree of filtration that tells just how small a particle the filter will remove. The most common measurement used to determine degree of filtration is a micron — one micron is approximately 0.00004-inch or 400 millionth of an inch (0.001 mm). To get an idea of how big a micron really is, 25,000 particles of this size would have to be laid side by side to total just one inch (25.4 mm).

The smallest particle that can normally be seen with an unaided eye is about 40 microns (0.04 mm), so much of the dirt that is filtered out of lubrication system is invisible.

Some filters, such as those made of wire mesh, may allow particles as big as 150 microns (0.152 mm) to pass. Although they do not provide as fine a cleaning action as some other types of filters, wire mesh offers less resistance to oil flow and is often used on pump inlet lines to prevent the possibility of starvation.

A filter will pass small solids and stop larger ones but the actual amount of filtering done is difficult to determine. Because the material that is stopped by the filter is not taken away continuously, the size of the openings in the filter will usually decrease with use.

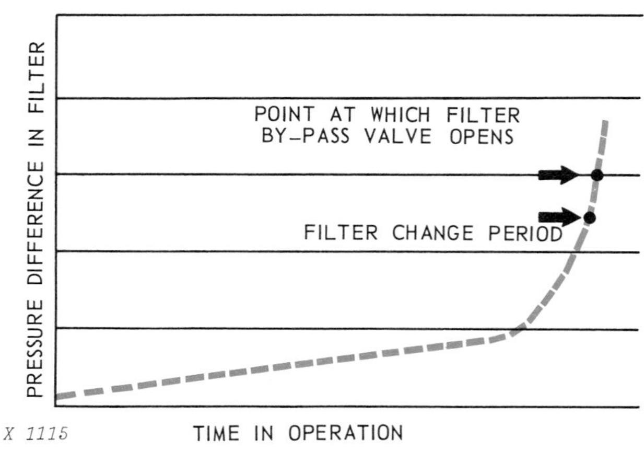

Fig. 13 — Life of a Filter Element

Two or more particles smaller than the holes in the filter may approach at the same time and become wedged in the hole. The result is that the hole will now remove dirt particles far smaller than it would when new. As more dirt wedges in the filter, the holes become smaller and eventually are plugged solid.

Fig. 13 shows the gradual reduction in the size of the filter pores until, near the end of filter life, the pressure difference between the inside and outside of the filter rises sharply. At this point, the filter stops operating and should be replaced.

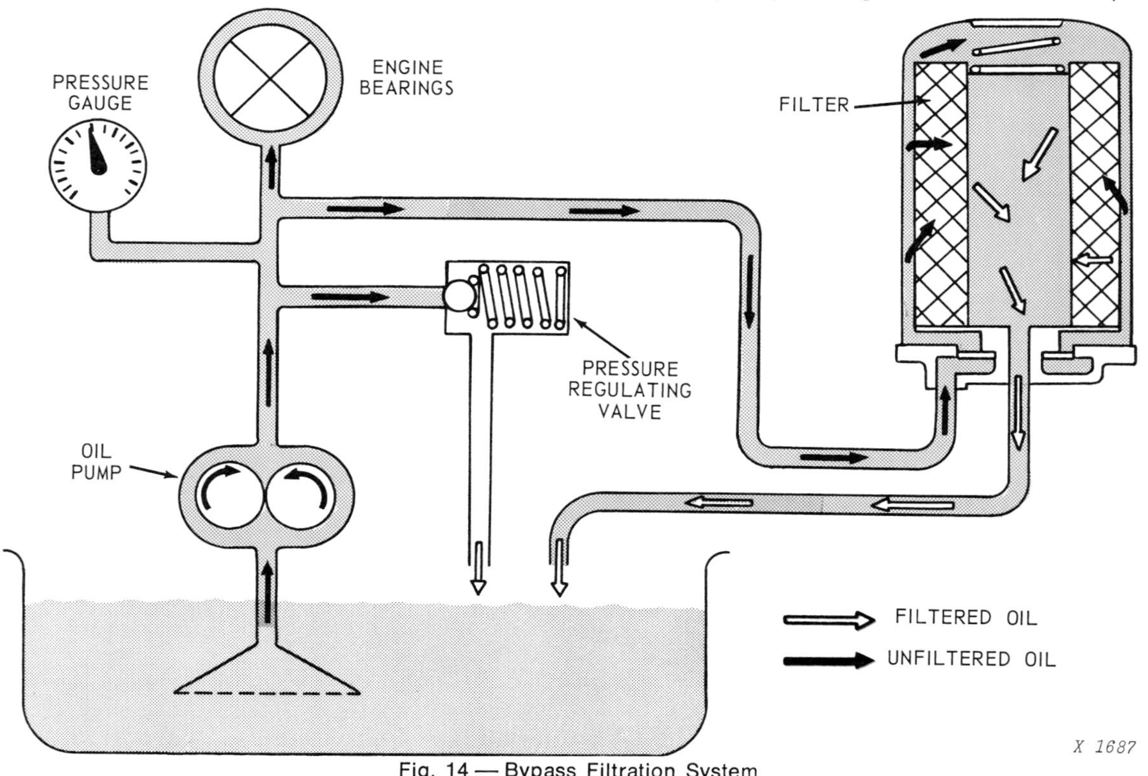

Fig. 14 — Bypass Filtration System

 Litho in U.S.A.

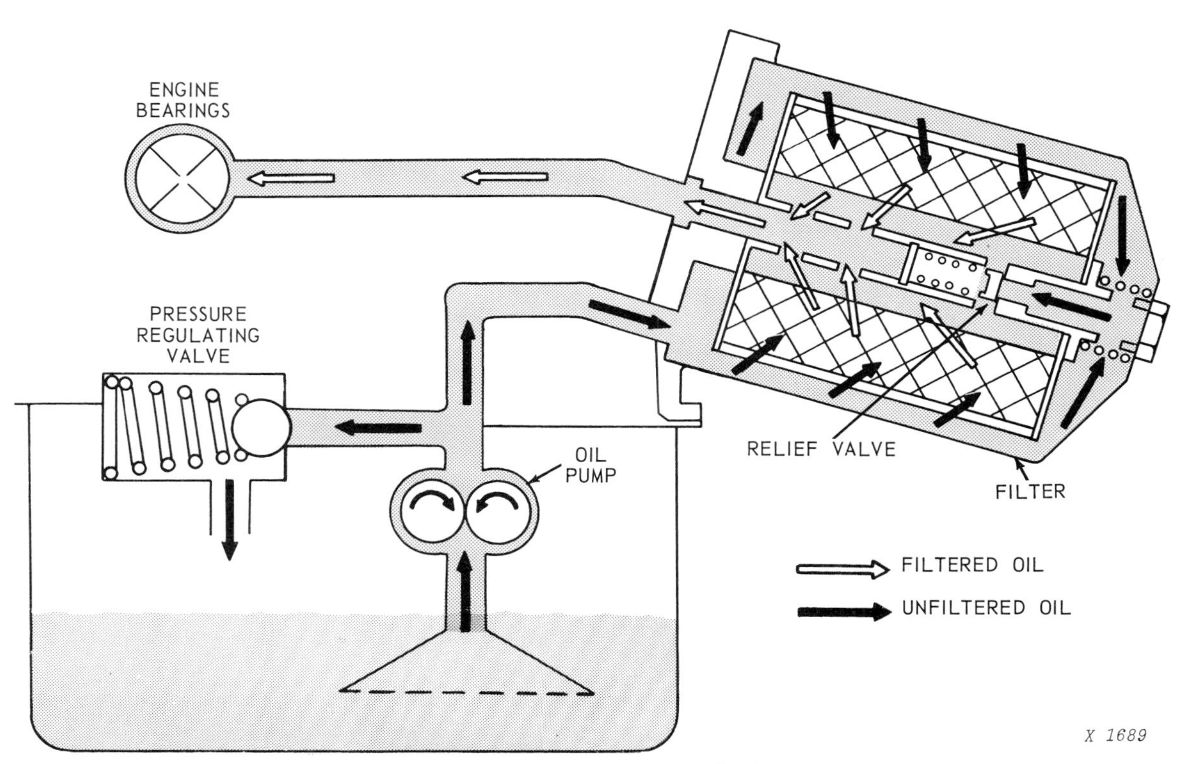

Fig. 15 — Full Flow Filtration System

Now let's look at the two types of filtration systems—**bypass** and **full flow.**

BYPASS FILTRATION SYSTEM

In the **bypass** filtration system (Fig. 14) there are two separate oil flows, one to the bearings and one to the filter.

In this sytem, five to ten percent of the oil delivered by the pump is routed or *bypassed* to the filter instead of the bearings. After filtering, the oil is returned to the crankcase. This system is sometimes called a "partial flow" system as only part of the supply oil is filtered at one time.

The volume of oil bypassed through the filter is initially controlled by a restriction in the filter outlet. However, as the filter passages become clogged, the volume of oil through the filter is reduced. This, of course, reduces the volume of filtered oil being returned to the crankcase. Because of the two separate oil flows, the oil pressure at the bearings is constant regardless of the condition of the filter.

This means that the filter and the oil must be changed regularly to prevent loss of filtering.

FULL-FLOW SYSTEM

The **full-flow** filtration system (Fig. 15) is a variation of the bypass system. In this case, there is only *one* oil flow to the filter. Within the filter case, part of the oil goes through the filter to the bearings while the remainder is shunted directly to the bearings.

Initially, the volume of filtered oil is greater than the shunted oil. But as the filter becomes clogged, less and less filtered oil reaches the bearings, until finally no oil is filtered. However, oil still reaches the bearings even though it is not filtered.

FULL FLOW FILTRATION SYSTEM

In the **full flow** filtration system, (Fig. 15) there is only one oil flow—from the pump, to the filter, and then to the bearings. A pressure gauge and pressure regulating valve are used just as in the bypass system.

Notice the relief valve, within the filter. When the filter is new, there is very little pressure drop through the filter. However, if the filter gets clogged, the resulting pressure will open the relief valve and allow unfiltered oil to bypass to the bearings. This relief valve may be located in the base of some filter housings.

FLOW THROUGH DUAL FILTERS AND OIL COOLER

Full-flow systems with dual filters and oil cooler have an oil flow similar to that shown in Fig. 16.

 Litho in U.S.A.

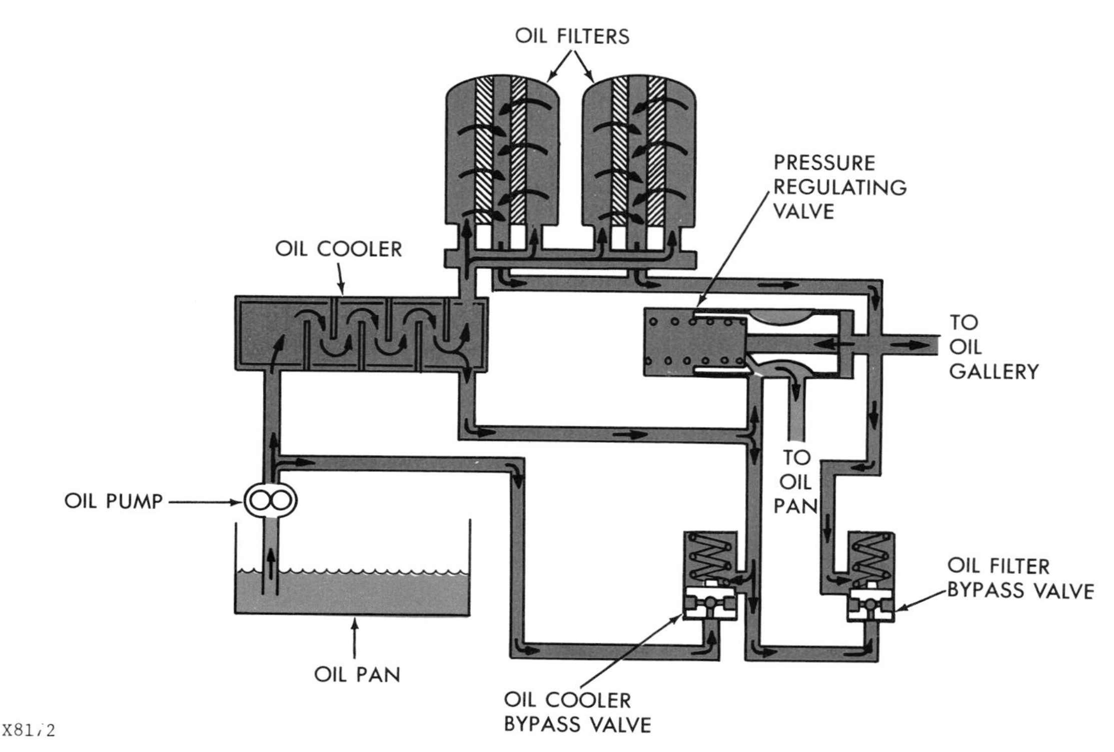

Fig. 16 — Oil Flow Through System With Dual Filters and Oil Cooler

Oil is pumped to the oil cooler and the oil cooler bypass valve. If the cooler is unrestricted, the oil flows through the cooler to the oil filters. Cool oil to the filters is also directed to the head of the oil filter bypass valve, the spring end of the oil cooler bypass valve, and the spring end of the pressure regulating valve.

Cooled oil that passes through the filter is routed to the spring end of the oil filter bypass valve, the pressure regulating valve, and the engine oil gallery.

The cooler bypass valve opens when the cooler is restricted to the point of creating enough pressure on the head of the valve to open it. When open, oil flows through the valve and bypasses the cooling element by flowing through an opening at one end of the cooler to the filters.

If the filters are restricted, pressure oil at the head of the filter bypass valve causes it to open. This allows unfiltered oil to flow to the oil gallery for lubrication.

When oil pressure in the oil gallery exceeds a predetermined pressure, the pressure regulating valve spool opens allowing cooled unfiltered oil to go to the oil pan. The pressure regulating valve has orifices that provide dampening for the spool.

SERVICING OF OIL FILTERS

Most engines today use a replaceable oil filter element.

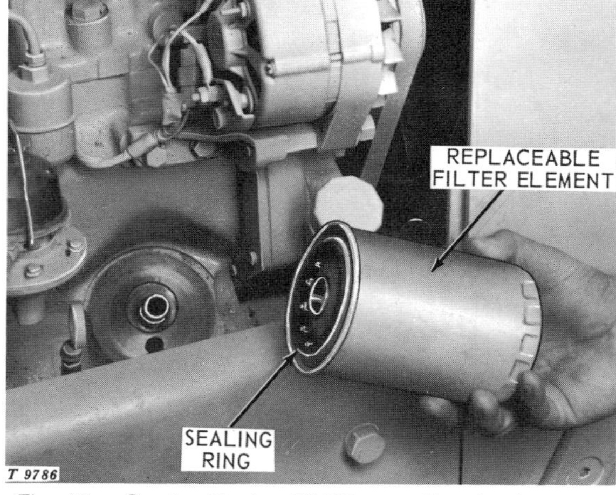

Fig. 17 — Service Engine Oil Filters at Regular Intervals

The filter may be enclosed in a disposable housing with a permanent base.

Service for these filters is simple — replace it at the regular intervals specified by the engine manufacturer (Fig. 17).

If the filter element is contained in a removable housing, use a solvent to clean the housing and the mounting base. Dry thoroughly with compressed air. Likewise, clean and dry the relief valve if it is removable.

Edge-type metal filters should be cleaned with solvent and a soft brush. Hard varnish-like deposits may be removed by soaking the filter overnight in a strong solution of lye water. Flush the filter thoroughly after soaking to remove the loosened particles.

When servicing any filter, use new gaskets and seal rings. *Tighten the housing firmly,* **but not too tightly.** Run the engine until oil pressure registers and check for leaks. Leaks are caused by the housing being too loose or the gasket installed improperly.

Often the crankcase oil is changed at the same time as the filter is serviced. This assures both a clean oil supply and filter for the lubricating system.

IMPORTANT: Many full flow filters have a special bypass valve. Replace them only with a genuine duplicate of the original filter.

LUBRICATING VALVES

Valves have two uses in the lubrication system:

- **To regulate oil pressure**
- **To bypass oil at filters and oil coolers**

The operation of the valve is the same in either case.

Basically, the valve has a poppet held in place over an opening by a spring.

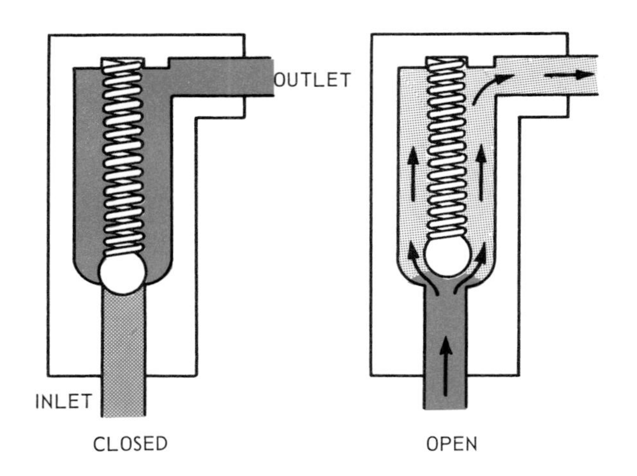

X 1690 Fig. 18 — Operation of Lubricating Valve

As shown in Fig. 18, the valve is **closed** when the spring tension is greater than the oil pressure at the inlet. The spring tension holds the poppet securely in position.

The valve **opens** when pressure at the oil inlet exceeds that of the spring. This pushes the poppet off the inlet hole and oil flows through the valve.

Now let's see how this valve is used.

OIL PRESSURE REGULATING VALVES

The oil pressure regulating valve maintains the correct pressure in the lubrication system regardless of the engine speed or the temperature of the oil. Most regulating valves are adjustable.

Most oil pumps can push out more oil than needed for engine lubrication. At idling speeds, or on older engines, this is no problem. But what happens to the excess oil when the engine is new and operating properly?

This is where the **pressure regulating valve** does its job. When the oil pressure exceeds the valve setting, the valve opens and returns the excess oil to the crankcase.

The regulating valve is usually connected to the main oil gallery through passages in the block. However, it may be a part of the oil pump.

CHECKING AND ADJUSTING OIL PRESSURE

Before checking the engine oil pressure, always check the condition of the oil filter. A dirty filter will limit the flow of filtered oil.

Many engines have tapped holes in the block as oil pressure test points. On other engines, the oil pressure sending unit hole is used for testing.

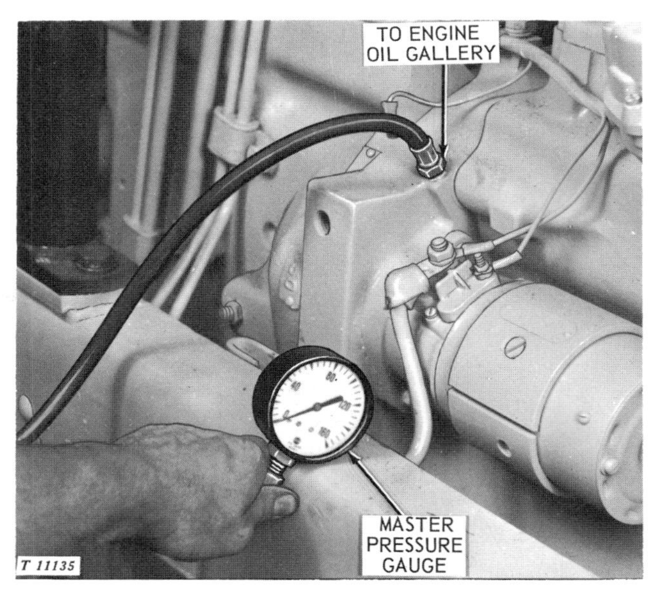

Fig. 19 — Checking The Engine Oil Pressure

Install a master gauge as shown in Fig. 19. Start and run the engine at fast-idle speed. When the engine is warmed up to operating temperature, record the pressure reading on the gauge. Compare this reading with the specifications for the engine.

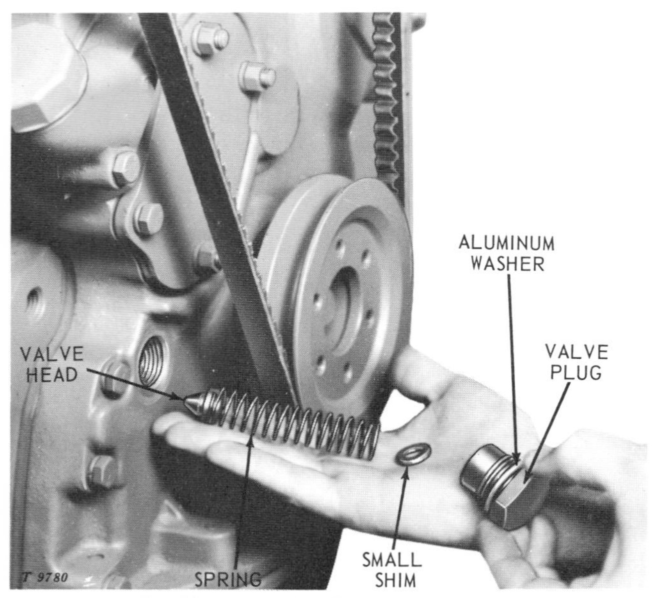

Fig. 20 — Adjusting Engine Oil Pressure (Shim-Adjusted Valves)

For adjusting pressure, shims and washers are used in some regulating valves (Fig. 20).

To raise pressure, shims are added behind the spring to increase the tension of the spring. This in turn increases the setting at which the valve will open.

To lower pressure, shims are removed to decrease the tension of the spring and allow the valve to open at a lower setting.

Fig. 21 — Adjusting Engine Oil Pressure (Screw-Adjusted Valves)

Some regulating valves are adjusted by an adjusting screw (Fig. 21). Normally, turn the screw in to increase the spring tension and the pressure setting. Turn the screw out to decrease the pressure setting.

CAUSES OF TOO-LOW ENGINE OIL PRESSURES

1. *Oil level in crankcase too low*
2. *Oil in crankcase too thin*
3. *Worn engine bearings*
4. *Worn oil pump*
5. *Filter or pump leaks*
6. *Regulating valve spring failed*
7. *Regulating valve needs adjustment*

For remedies, see Chapter 11, "Diagnosis and Testing".

CAUSES OF TOO-HIGH ENGINE OIL PRESSURES

1. *Oil in crankcase too heavy*
2. *Defective pressure gauge or sender*
3. *Stuck regulating valve*
4. *Regulating valve needs adjustment*

For remedies, see Chapter 11, "Diagnosis and Testing".

FILTER BYPASS VALVES

Every filter in a full flow lubrication system must have a **bypass** valve. This valve bypasses oil around the filter when the filter becomes clogged.

In the full flow system, all of the oil delivered by the pump passes either through the filter to the bearings or through the pressure regulating valve to the crankcase. Let's see what would happen if a bypass valve were not provided.

As the filter becomes completely clogged, pressure would build up on the inlet side of the filter (see Fig. 15). This would cause the pressure regulating valve to open completely, allowing all of the oil to return directly to the crankcase. Result—a burned-up engine.

The bypass valve, then, is a safety device to ensure that the oil, filtered or dirty, will get to the bearings. The valve is usually set to open before the filter becomes completely clogged.

Fig. 22 shows the filter bypass valve in two locations — inside the filter and in the filter mounting pad.

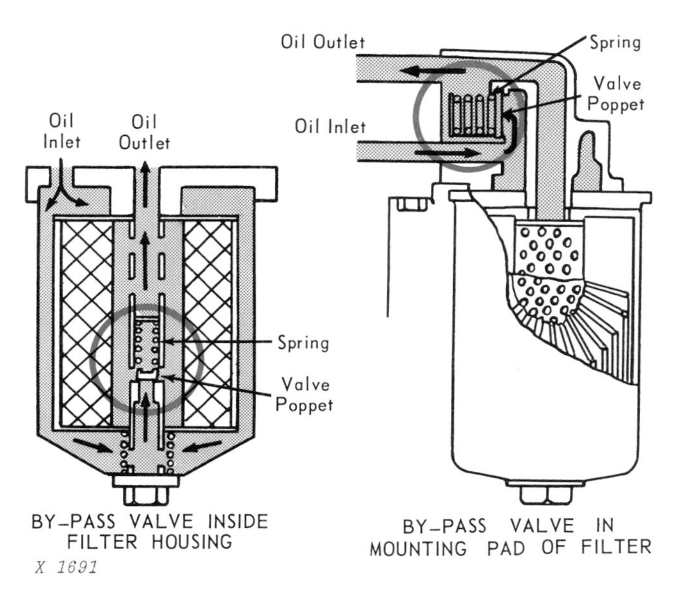

Fig. 22 — Filter Bypass Valves — Two Locations

Many of the "spin-on" filter elements have the bypass built into the element.

When replacing a filter, **be sure to use only the recommended filter.** Another type may not have the built-in bypass valve.

Bypass valves are often used with oil coolers for the same reason that they are used with filters. If the oil cooler becomes clogged, oil flows through the valve and back into the lubrication system.

SERVICING OF LUBRICATING VALVES

Most important in servicing of lubricating valves is to clean them. Where possible, completely disassemble the valve and wash the parts in solvent. Clean the bore in which the valve slides. Use compressed air to dry the parts.

Use a spring tester to check the strength of the spring. A broken or weakened spring will cause the valve to open at a much lower setting.

Inspect the valve poppet for wear and nicks that might cause it to hang up in the bore. *It is very important that the valve slides freely in the bore.*

Some regulating valves use a bushing in the bore as the seat for the poppet. Inspect the seat for wear and replace it if necessary.

Always check and adjust the engine oil pressure after servicing a lubricating valve.

OIL COOLERS

Lubrication systems of many engines use an oil cooler to help dissipate the heat created by the engine. Coolers may be oil-to-air or oil-to-water types. Oil-to-water coolers (Fig. 24, left) can obtain oil temperatures only as low as temperatures in the radiator. In many cases this temperature is acceptable and water-to-oil coolers use engine coolant to dissipate heat from the oil.

Oil-to-air coolers must be used if lower than radiator oil temperatures are required. These are almost always of the fin and tube design and appear much like an automobile radiator (Fig. 24, right).

The oil cooler may be mounted **internally** in the crankcase or **externally** on the outside of the engine block.

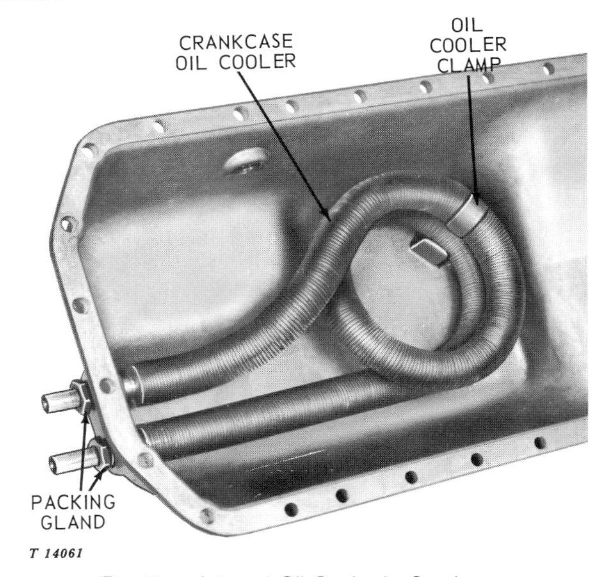

Fig. 23 — Internal Oil Cooler in Crankcase

When the cooler is mounted in the crankcase (Fig. 23), coolant is pumped by the water pump through the cooler to the radiator of the cooling system. The heat of the oil in the crankcase is conducted through the fins of the cooler and is absorbed by the coolant. The heat is then dissipated in the radiator.

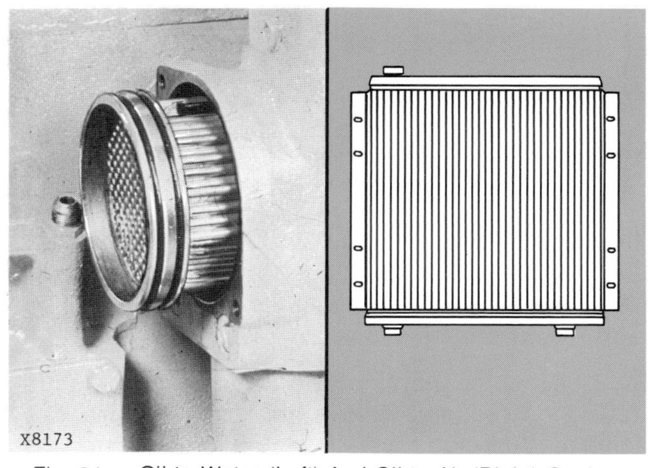

Fig. 24 — Oil-to-Water (Left) And Oil-to-Air (Right) Coolers

When the cooler is mounted externally (Fig. 24), both coolant and lubricating oil are pumped through it.

 Litho in U.S.A.

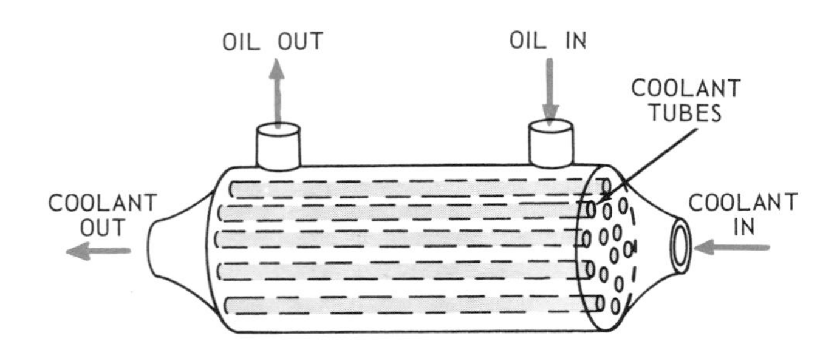

Fig. 25 — Operation of Engine Oil Cooler

As shown in Fig. 25, coolant flows through the tubes in the cooler and oil circulates around the tubes. Heat from the oil is conducted through the tubes to the coolant which carries it to the radiator for dissipation.

A similar cooler uses a small radiator core, instead of tubes within the housing. The oil is pumped through the core and the coolant is circulated around it.

A bypass value is used with some oil coolers to assure circulation of the oil if the cooler should become clogged.

MAINTENANCE OF OIL COOLERS

Normal maintenance of the cooling system (Chapter 8) will keep the oil cooler passages clean.

When cleaning the lubrication system, remove the cooler and clean the oil passages with a solvent.

When assembling or installing an external cooler, use new gaskets and be sure the cap screws are tight.

OIL PRESSURE INDICATING SYSTEMS

A pressure indicating system is essential for monitoring the lubrication system. These systems may be classified as:

- **Mechanical**
- **Electrical**

MECHANICAL INDICATING SYSTEMS

The **mechanical** indicating system uses a Bourdon tube gauge which is attached by tubing to the pressure source. This is usually the main oil passage in the engine block.

The basic components of the Bourdon tube gauge are a tube made of spring bronze or steel and a pinion and sector mechanism (Fig. 26). One end of the tube is permanently attached to the pressure inlet of the gauge. The other end is attached to the pinion and the sector mechanism.

Oil pressure inside the tube tends to straighten the tube out. This movement acts upon the pinion and sector mechanism, causing the pointer to rotate on the face of the gauge. The pressure can be read directly on the gauge.

Maintenance

1) If a mechanical gauge will not register pressure, loosen the oil line at the engine block (but do not remove it).

2) If oil leaks from the connection while the engine is running, there is pressure in the lubricating system. This means the trouble is in the oil line of the gauge itself.

3) Check the oil line for leaks. The pressure must get to the gauge before it can register.

4) If you suspect that the gauge is defective, replace it with one that is known to be accurate. If the new gauge works correctly, this proves that the original is defective.

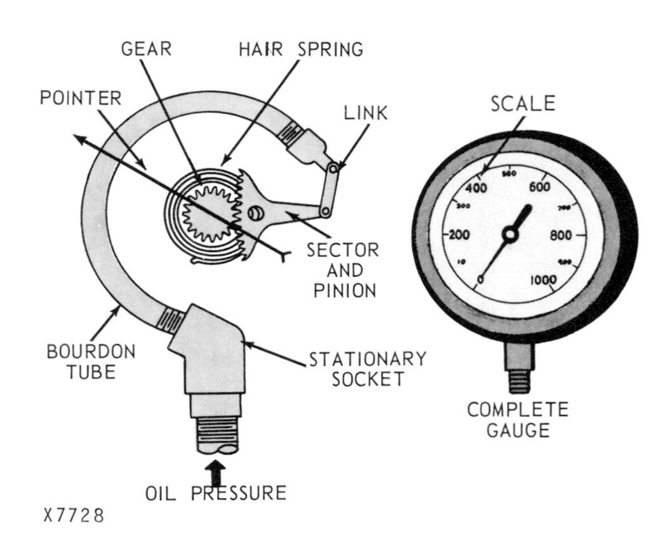

Fig. 26 — Bourdon Tube Oil Gauge

ELECTRICAL INDICATING SYSTEMS

An **electrical** oil pressure indicating system has two parts:

1) Sending unit—at pressure source

2) Indicating gauge—on control panel

The **sending unit,** actuated by the pressure of the lubricating oil electrically actuates the indicating unit. The two units are connected together by a single wire, and both must be grounded.

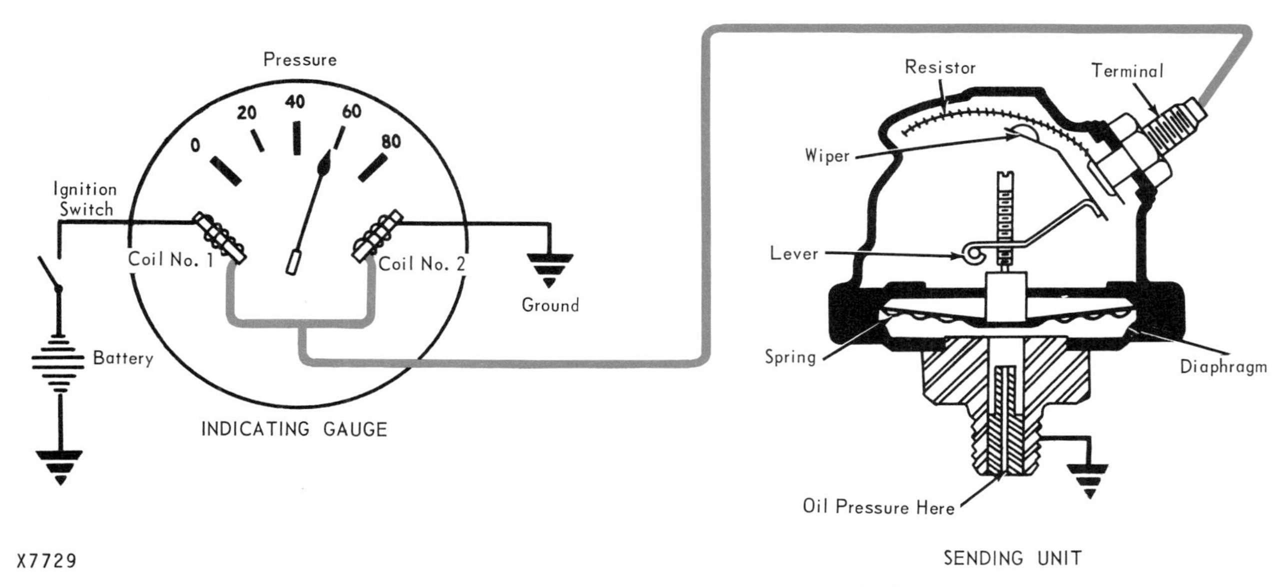

Fig. 27 — Electromagnetic Coil System For Indicating Oil Pressure

There are three kinds of electrical indicating systems:

- **Electromagnetic coil type**
- **Heating coil type**
- **Pressure switch type**

Let's look at the operation of each one.

Electromagnetic Coil System

The **sending unit** of the electromagnetic system consists of a resistor, a sliding contact called the wiper, an actuating lever, and a spring-loaded diaphragm enclosed in a housing (Fig. 27).

The pressure of the lubricating oil works against the diaphragm causing the lever to move. The movement of the lever changes the position of the wiper which varies the resistance of the resistor.

The **gauge** contains two coils and an armature with a pointer. The magnetic fields of the two coils creates another field which controls the rotation of the armature. This, in turn, controls the rotation of the pointer on the face of the gauge.

Current flows from the battery to coil No. 1 in the gauge as shown, then through a parallel circuit to coil No. 2 and to the resistor in the sending unit.

As pressure against the diaphragm in the sending unit changes, the resistance of the resistor also changes. The change of the current in coil No. 2 causes its magnetic field to be stronger or weaker than that of coil No. 1.

The armature, of course, is attracted to the stronger of the two fields.

Heating Coil System

The **sending unit** of the heating coil system has a heating coil wound around a bimetal strip. (See upper diagram in Fig. 28). The pressure of the lubricating oil deflects the diaphragm, causing the contact to close. The current flowing through the coil then creates heat.

The heat deflects the bimetal strip until the contact is opened. The bimetal then cools and returns to its original position, again closing the circuit. This cycle of closing and opening, heating and cooling is repeated continuously.

The **gauge** contains a similar heating coil wound around a bimetal strip. This coil is connected in series with the coil in the sending unit. As the coil of the sending unit heats up, so does the coil of the gauge. The bimetal strips in each unit then deflect at the same time.

The pointer of the gauge is linked to the bimetal strip in the gauge. The deflection of the bimetal strip causes the pointer to rotate on the face of the gauge.

As oil pressure increases, the diaphragm of the sending unit is deflected more (see lower diagram in Fig. 28). A greater amount of current is then required to heat the bimetal strip enough to open the contact. The increased current in the coil of the gauge, likewise, causes a greater deflection of the bimetal strip in the gauge. This, in turn, causes the pointer to register the increased oil pressure.

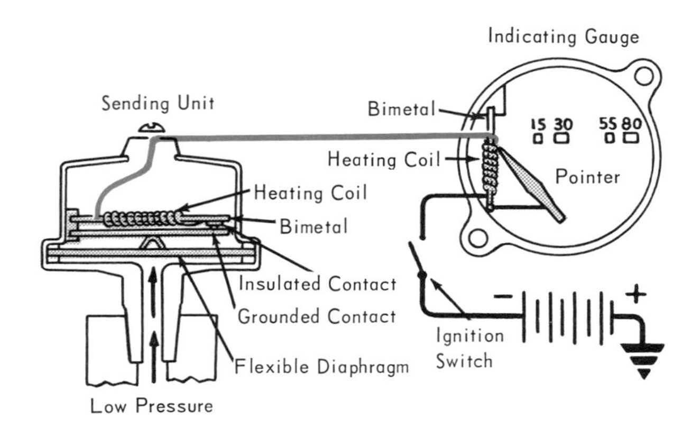

OPERATION WITH LOW OIL PRESSURE

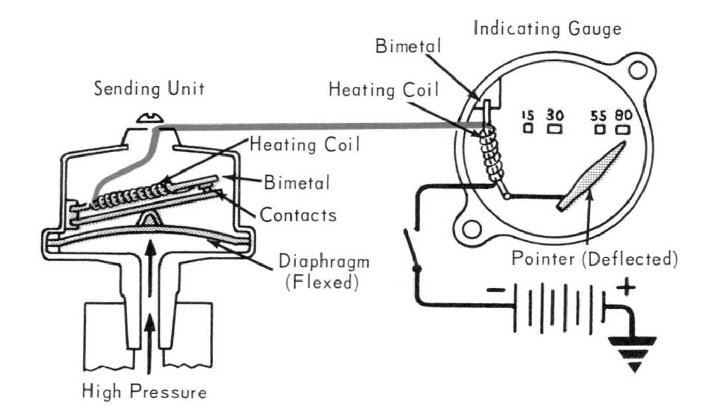

OPERATION WITH HIGH OIL PRESSURE

X 1695

Fig. 28 — Heating Coil System for Indicating Oil Pressure

Pressure Switch System

The pressure switch system is quite different from the electromagnetic coil and heating coil types.

A sending unit is used, but a light bulb is used as the indicator rather than a gauge (Fig. 29). This system can only warn of low oil pressure, it cannot tell the actual pressure of the system.

Basically, the sending unit consists of a set of contacts and a diaphragm. When the ignition switch is on and before the engine is started, the contacts are closed and the indicating bulb is lit.

When the engine is started and as the oil pressure builds, the diaphragm is pushed up. This separates the contacts and breaks the circuit, causing the bulb to go out.

If the oil pressure drops, the resulting deflection of the diaphragm will close the contacts and again complete the circuit. The lighted bulb warns that

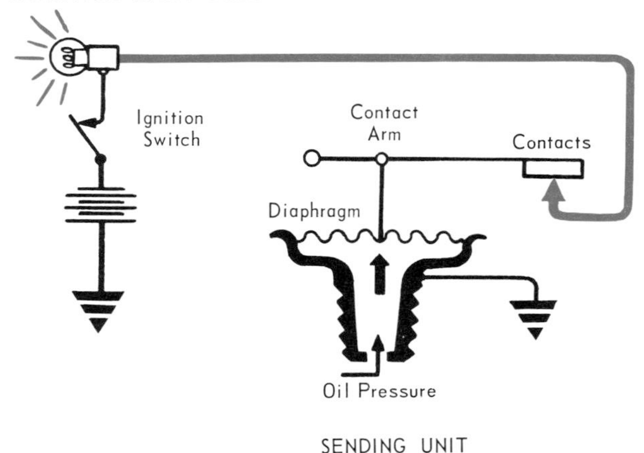

SENDING UNIT
(PRESSURE SWITCH)

X 1696

Fig. 29 — Pressure Switch System for Indicating Oil Pressure

the pressure of the lubrication system has dropped below the low pressure setting.

Maintenance

Look for these things when servicing an electrical oil pressure indicating system:

1. Poor grounds

2. Broken or poor connections

3. Defective units

Specific tests vary with the different types of units. See the engine technical Manual for details.

CONTAMINATION OF OIL

Oil contamination will reduce engine life more than any other factor. Some sources of contamination are obvious while others are not.

Let's review some of these sources of contamination and what can be done about them.

1) The most obvious source of contamination is the **storing and handling of the oil** *itself (see Fig. 30). If at all possible, store lubricants in a clean, enclosed storage area. Keep all covers and spouts on containers when not in use. These practices not only keep dirt out of the oil, but also reduce condensation of water caused by atmospheric changes.*

2) Another obvious source is **dust that is breathed into the engine** *with combustion air. It is very important that the air cleaner be cleaned or replaced regularly. At the same time, clean or replace the breather on the oil filler. See Chapter 6 for details.*

3) A major source of contamination is a **cold engine.** *When the engine is cold, its fuel burning*

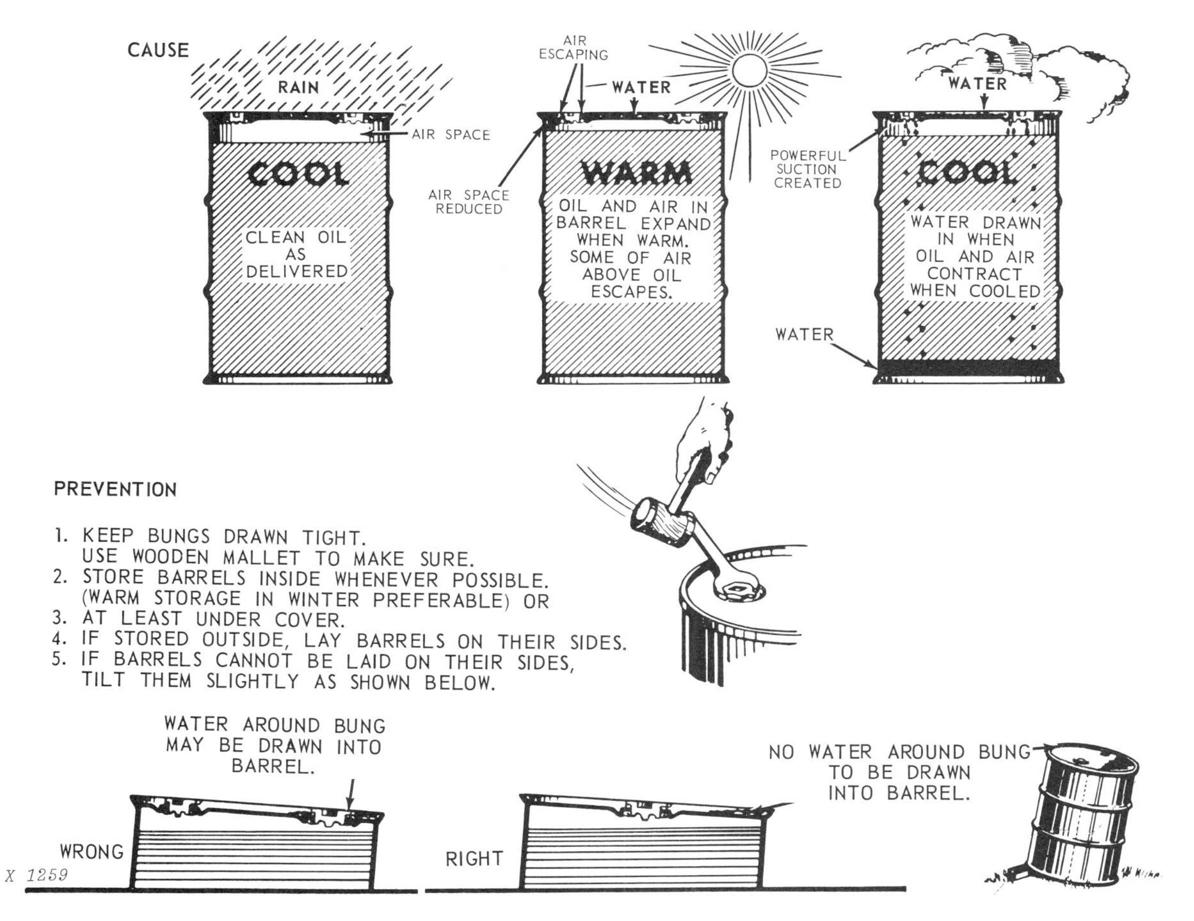

Fig. 30 — Storage Practices Which Prevent Contamination of Oil

efficiency is greatly reduced. Partially burned fuel blows by the piston rings and into the crankcase. Oxidation of this fuel in the oil forms a very harmful varnish which collects on engine parts (Fig. 31). An overchoked or misfiring engine will also create this contamination from unburned fuel.

Fig. 31 — Varnish Build-Up on Pistons from Contamination by Fuel

Water created by a cold engine contaminates the oil too (Fig. 32). Water vapor, a normal product of combustion, tends to condense on cold cylinder walls. This condensation is also blown by the rings into the crankcase. The engine must warm up before this condensation problem is eliminated.

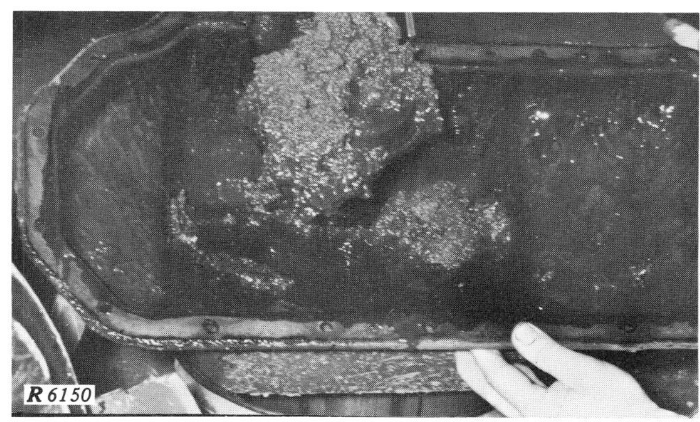

Fig. 32 — Ice in Crankcase Following Cold Engine Operation

Water not only causes rusting of steel and iron surfaces, but it can combine with oxidized oil and carbon to form sludge. This sludge can very effectively plug oil screens and passages.

Contamination from a cold engine can be prevented by:

(a) Properly warming up the engine before applying a load.

(b) Making sure the engine is brought up to operating temperature each time it is used.

(c) Using the proper thermostat to warm up the engine as quickly as possible.

Fig. 33 — Antifreeze in Crankcase Will Create Sludge

4) **Antifreeze** *can be another sludge-forming source of contamination (Fig. 33). To guard against antifreeze contamination:*

(a) Torque head bolts to specifications during an overhaul.

(b) Use a cooling system sealer when filling the cooling system.

(c) Guard against detonation and improper use of starting fluids in diesel engine (both can result in head gasket damage).

Other problems which causes oil contamination are:

5) **Oxidation** *is not an obvious source of contamination, but it is a very real one. Oxidation occurs when the hydrocarbons in the oil combine with oxygen in the air to produce organic acids. Besides being highly corrosive, these acids create harmful sludges and varnish deposits.*

6) **Carbon** *particles are another contaminant created by the normal operation of the engine. The particles are created when oil around the upper cylinder walls is burned during combustion. Excessive deposits can cause the piston rings to stick in their grooves.*

7) **Engine wear** *also creates a contaminant. Tiny metal particles are constantly being worn off bearings and other parts. These particles tend to oxidize and deteriorate the oil.*

All sources of contamination cannot be eliminated. **What, then, can be done to protect the engine?**

PROTECTING THE ENGINE AGAINST OIL CONTAMINATION

Start by using a good-quality oil which has additives. Additives are put in the oil for a specific reason, based on the service expected for the oil.

Here are some of the important **additives:**

ANTI-CORROSION ADDITIVES protect metal surfaces from corrosive attack. These work with oxidation inhibitors.

OXIDATION INHIBITOR ADDITIVES keep oil from oxidizing even at high temperatures. They prevent the oil from absorbing oxygen, thereby preventing varnish and sludge formations.

ANTI-RUST ADDITIVES prevent rusting of metal parts during storage periods, downtime, or even overnight. They form a protective coating which repels water droplets and protects the metal. These additives also help to neutralize harmful acids.

DETERGENT ADDITIVES help keep metal surfaces clean and prevent deposits. Particles of carbon and oxidized oil are held suspended in the oil. The suspended contaminants are then removed from the system when the oil is drained. Black oil is evidence that the oil is helping to keep the engine clean by carrying the particles in the oil rather than letting them accumulate as sludge.

However, remember that **additives eventually wear out.**

To prevent this, drain the oil *before* the additives are completely depleted.

Also service all the filters at regular intervals.

Finally, keep the fuel, cooling, and ignition systems in good condition so that the fuel is efficiently burned.

OIL CONSUMPTION

Some oil consumption is natural during normal operation of internal combustion engines.

As we have already learned, lubricating oil provides a seal between the piston rings and cylinders. It is only natural that during the combustion process some of this sealing oil is burned up.

Excessive oil consumption can be caused by several conditions.

To properly diagnose the problem, follow this procedure:

1) First be sure the correct weight and grade of oil for the type of service and climate is being used. Oil that is too thin may "flood" the piston rings, while the oil that is too thick may "starve" the rings.

2) Be sure the engine has run long enough ***under load*** *to insure that the rings have had a chance to seat. Some variation in oil consumption can be expected during break-in, but it should be stabilized before 250 hours of operation. If not, assume that a problem exists and correct it at once. (See Chapter 2 for "Engine Break-In".)*

3) Check the engine oil pressure and, if necessary, adjust the pressure regulating valve. High engine oil pressure can cause oil consumption by flooding the rings and valves with oil. Also check the crankcase breather. A plugged breather can increase the crankcase oil pressure.

4) Check for external oil leaks. What appears to be a small oil leak can add up to a considerable loss of oil. Drops of oil lost externally can add up to quarts of oil between oil change periods. Check the front and rear oil seals, all gaskets, and the filter attaching points.

5) Check for engine blow-by. Watch the fumes expelled through the crankcase vent tube. Fumes should be barely visible with the engine at fast idle with no load. If possible, compare the blow-by with that of identical engines. This can be a guide to determine if the blow-by is excessive.

Excessive blow-by indicates that piston rings and cylinder liners have worn to the point where the rings cannot seal off the combustion chambers. An overhaul with new rings is then required. New or reconditioned pistons and liners may also be necessary.

6) If blow-by is not excessive, check to see if the oil is being lost through the valve guides. A generous supply of oil is usually maintained in the rocker arm are to lubricate the rocker arms, valve stems, and valve guides.

As shown in Fig. 34, gravity, inertia, vacuum, and an atomizer effect all combine to force oil down

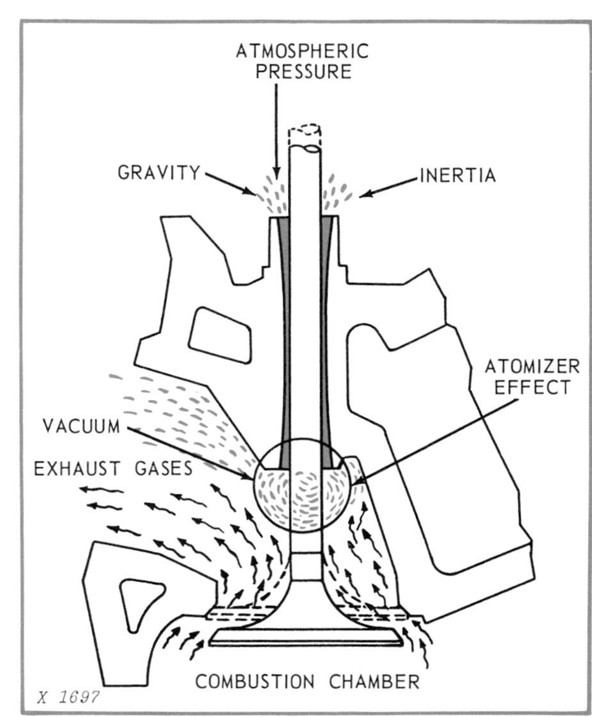

Fig. 34 — Problems Which Cause Oil Loss Through Valve Guides

between the valve stem and guide. Valve guide seals are usually used to prevent excessive oil from entering the combustion chambers. If these seals become defective, they create a definite oil consumption problem.

To check for this condition, start the engine and warm it up. Let it idle slowly for ten minutes. Then remove the cylinder head, and inspect the valve ports and the undersides of the valve heads. If these are wet with oil, the oil has been drawn through the valve guides.

At the same time, check the piston heads. If they are wet, the oil has been drawn past the piston rings. If excessive blow-by has indicated worn rings, this procedure can be used to confirm it.

7) Worn connecting rod bearings can also contribute to oil consumption. When worn, the bearings throw an excess of oil onto the cylinder walls. The action of the piston forces part of this excess oil into the combustion chambers where it is burned.

8) Excessive engine speeds are another common cause of oil consumption. Observe the fast idle limits set by the manufacturer.

CRANKCASE VENTILATION

During normal operation, unburned fuel vapor and water vapor are created in the engine. If allowed to condense, these vapors become contaminating liquids that drain into the crankcase. The purpose of the ventilation systems is to circulate fresh air through the engine to carry away these harmful vapors.

As with all ventilating systems, there must be:

1) **An air inlet**

2) **An air and vapor outlet**

3) **A means of circulating air between the two**

The oil filler cap is probably the most common inlet. A small air cleaner built into the cap filters the air as it is drawn into the system. Some systems use the main air cleaner as the inlet.

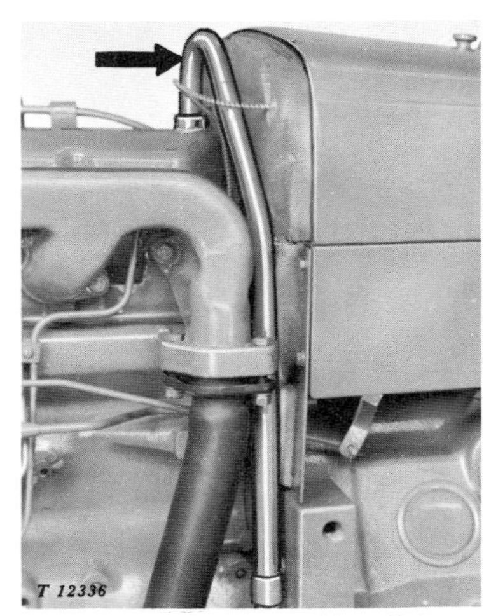

Fig. 35 — Crankcase Vent Tube

The use of a vent tube is one means of circulating the air through the system (Fig. 35). As the machine moves, the air moving past the tube opening creates a lower pressure than that at the breather. Air in the engine naturally flows to this low pressure area, and the tube becomes an outlet for the system. Some vent tubes use a wire mesh screen to filter oil particles out of the vapor.

INTAKE MANIFOLD VENTILATION SYSTEM

The intake manifold can serve the double function of circulating the air and being the vent outlet. In this type of ventilation system, a tube connects the crankcase to the intake manifold. The intake vacuum draws the air through the engine, into the manifold, into the cylinders, and out the exhaust system. A ventilation valve is used to regulate the flow of air into the manifold.

At full throttle speed, there is maximum vapor in the crankcase and minimum vacuum in the intake manifold. The valve is help open by the spring, permitting full air flow through the system.

At idle speed, there is minimum vapor in the crankcase and maximum vacuum in the manifold. As the pressure of the air is greater than that of the vacuum, the spring is compressed and the valve is closed. Air flow is restricted by the bleed hole in the valve.

Some ventilating valves do not have a bleed hole. This design eliminates the possibility of a backfire traveling through the ventilating system to the crankcase.

VENTILATION SYSTEM WITH CIRCULATING PUMP

Another type of system uses a vane or impeller pump to circulate the air. The air is taken from the main air cleaner and pumped through the engine. The outlet can be either a vent tube or the intake manifold.

SERVICING THE VENTILATION SYSTEM

Begin with the inlet when servicing the ventilation system. If the oil filler cap is the inlet, periodically clean it with a solvent. If the main air cleaner is the inlet, service it regularly as recommended by the engine manufacturer.

If the system uses a ventilating valve, service it regularly. Some manufacturers recommend cleaning the valve with solvent, while others recommend replacing the valve rather than cleaning it.

If a vent tube is used in the system (Fig. 35), periodically remove and clean it with a solvent. If a filter is used with the vent tube, be sure to clean it too.

CHANGING ENGINE OIL

The statement that "oil doesn't wear out" is false. Oil loses many of its good lubricating qualities as it gets dirty and its additives wear out.

Acid formations, sludge, varnish, and engine deposits contaminate oil and make it unfit for continued use.

On the other hand, just because crankcase oil is black, doesn't mean it's time for an oil change.

Additives in the oil are supposed to clean and hold deposits in suspension for removal when the oil is drained.

Change the oil sometime **before** *the additives wear out which protect the engine against sludge.* For the average person, this time is practically impossible to determine. That's why the best policy is to follow the manufacturer's recommendations on oil and filter changes.

New or rebuilt engines require oil and filter changes after a specified break-in period. Performing this service on time is very important since foreign materials accumulate in the oil at a faster rate during initial operation than later when the engine is broken in.

When changing the oil and filter on any engine, always warm up the engine first. This way the contaminants and foreign materials are mixed with the oil and are drained out with it.

Replacing some oil filters requires installing new gaskets or sealing rings. Be sure the sealing surfaces on the engine and filter are clean.

After installing the filter and filling the engine with oil, run the engine and check for possible filter leaks.

Keep a record of all oil and filter changes to be sure of regular engine service.

DIESEL ENGINE OIL

Use oil viscosity based on the expected air temperature range during the period between oil changes.

Some oil is specially formulated to provide superior protection against high temperature thickening and wear as well as exceptional cold weather starting performance; these properties may result in longer engine life.

IMPORTANT: For industrial products, do not extend drain interval beyond recommendations in the operator's manual.

For agricultural and consumer products, the oil and filter change interval may be extended by 50 hours when using certain manufacturer's oil and oil filter.

Other oils may be used if they meet one or more of the following:

- API Service Classification CE or CD
- Military Specification MIL-L-2104D or MIL-L-2104C

SAE 5W20, SAE 5W30, and arctic oil viscosity grades meeting API Service Classification CC may be used, but oil and filter must be changed at one-half the normal interval.

Oils meeting Military Specification MIL-L-46167A may be used as arctic oils.

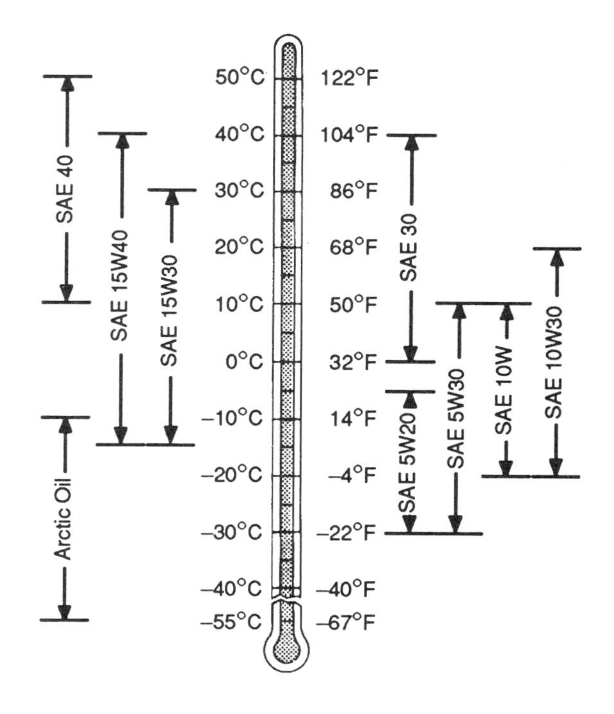

Fig. 36 — Temperature Range For Diesel Engine Oil

TEST YOURSELF

QUESTIONS

1. Name any three of the five jobs which the lubrication system does for the engine.

2. What does a bypass valve do for a full flow lubrication system?

3. What does a good lubricating oil contain which helps to prevent rust, corrosion, oxidation, and foaming?

4. True or false? "Oil doesn't wear out".

(Answers in back of text.)

COOLING SYSTEMS / CHAPTER 8

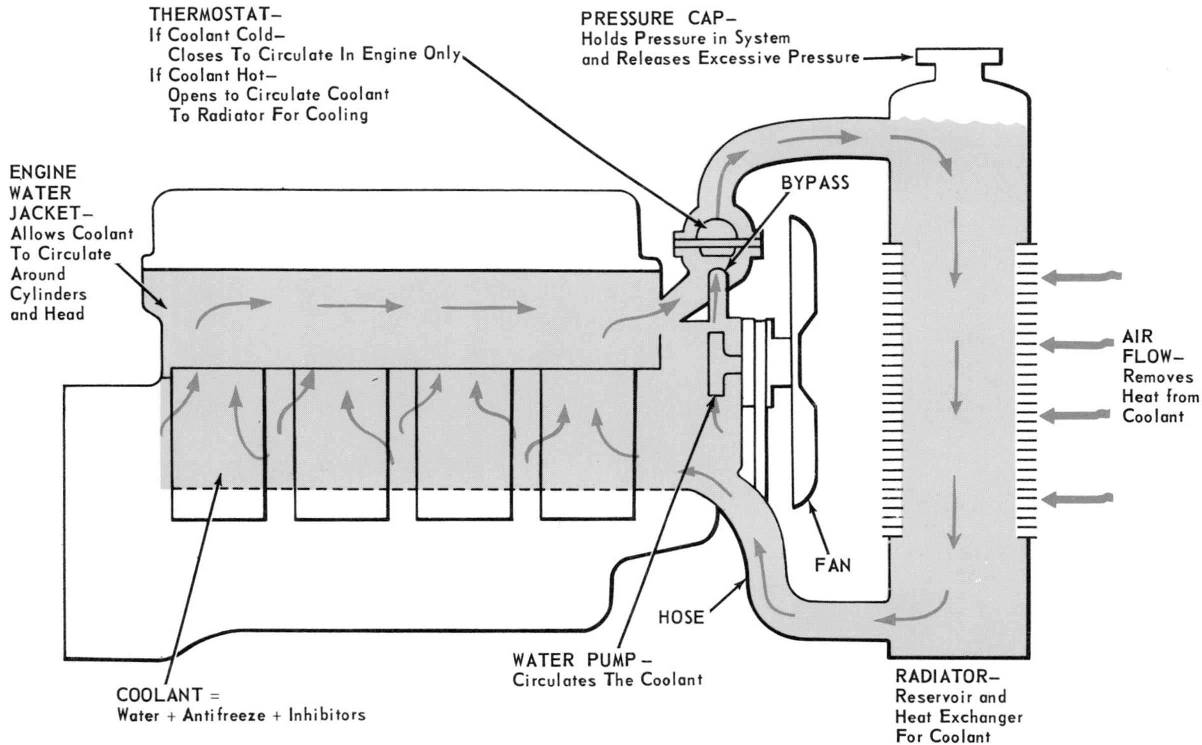

Fig. 1 — Liquid Cooling System

INTRODUCTION

The engine cooling system does two things:

1) Prevents overheating

2) Regulates temperatures

OVERHEATING could burn up the engine parts. Some heat is necessary for combustion, but the engine generates too much heat. So the cooling system carries off the excess heat.

REGULATING TEMPERATURES keeps the engine at the best heat level for each operation. During starting, the engine must be warmed up faster. During peak operations, the engine must be cooled.

Running the engine too hot can cause:

- *Preignition*
- *Detonation*
- *Knock*
- *Burned pistons and valves*
- *Lubrication failure*

Running the engine too cold can cause:

- *Unnecessary wear*
- *Poor fuel economy*
- *Accumulation of water and sludge in the crank-case*

TYPES OF COOLING SYSTEMS

Two types of cooling systems are used in modern engines:

- **Air Cooling System—uses air passing around the engine to dissipate heat.**
- **Liquid Cooling System—uses a liquid to cool the engine and air to cool the liquid.**

AIR COOLING is used only on small engines as it is difficult to route air to all the heat points of larger engines. Metal baffles, ducts, and blowers are used to aid in distributing air.

LIQUID COOLING normally uses water as a coolant. In cold weather, antifreeze solutions are added to the water to prevent freezing. The water circulates in jackets around the cylinders and other friction parts. As heat radiates, it is absorbed into the water, which then flows to the radiator. Air flow through the radiator cools the water and so dissipates heat into the air. The water then recirculates into the engine to pick up more heat (Fig. 1).

In this chapter we will cover only the *liquid* cooling systems used on most modern engines.

LIQUID COOLING SYSTEMS

A liquid system (Fig. 1) may consist of the following:

- **Radiator and Pressure Cap**
- **Fan and Fan Belt**
- **Water Pump**
- **Engine Water Jacket**
- **Thermostat**
- **Engine Oil Cooler**
- **Connecting Hoses**
- **Liquid or Coolant**

The RADIATOR is one of the major components of any liquid cooling system. It is here that heat in the coolant is released to the atmosphere. It also provides a reservoir for enough liquid to operate the cooling system efficiently.

The FAN forces cooling air through the radiator core to more quickly dissipate the heat being carried by the coolant in the radiator.

The WATER PUMP circulates the coolant through the system. The pump draws hot coolant from the engine block and forces it through the radiator for cooling.

Some engines have distribution tubes and some have transfer holes which direct extra coolant flow to "hot" areas, such as exhaust valve seats.

CONNECTING HOSES are the flexible connections between the engine and other parts of the cooling system.

The THERMOSTAT is a heat-operated valve. It controls the flow of coolant to the *radiator* to maintain the correct operating temperatures.

The FAN BELT transmits power from the engine crankshaft to drive the *fan* and *water pump.*

COOLANT is the liquid that circulates through the cooling system carrying heat from the *engine water jacket* into the *radiator* for transfer to the outside air. The coolant then flows back through the engine to absorb more heat.

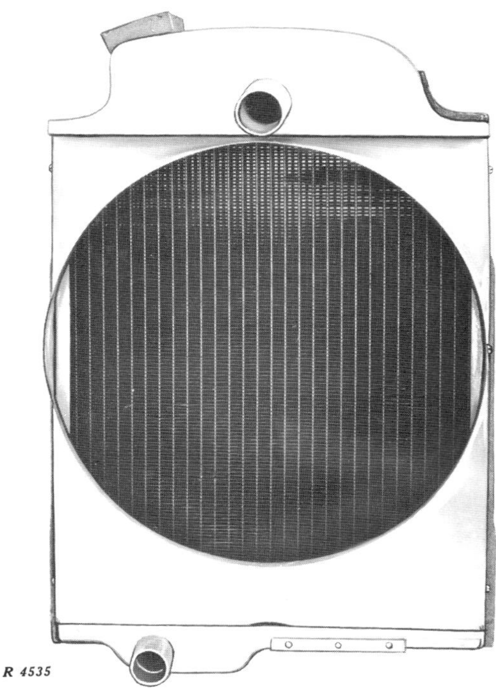

Fig. 2 — Typical Radiator

RADIATORS

Two types of radiators are used in liquid cooling systems:

- **Cellular-Type Core Radiators—used where air speeds are high and resistance to air flow must be low, as in aircraft or racing cars.**
- **Tubular or Tube-and-Fin Type Core Radiators—used in most other machines.**

In both types of radiators, coolant from the engine enters the radiator by way of the top tank, then passes down through a series of small tubes surrounded by fins and air passages.

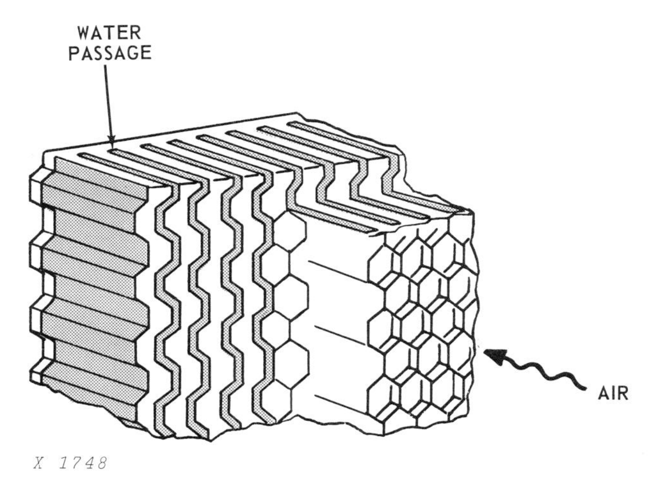

Fig. 3 — Cellular-Type Core Radiator

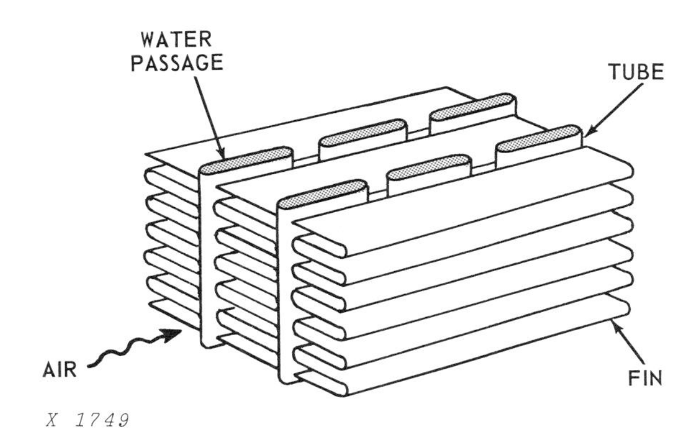

Fig. 4 — Tube and Fin-Type Core Radiator

Figs. 3 and 4 illustrate how air flowing around the fins and tubes removes the heat.

Cooled liquid reaching the bottom tank is picked up by the water pump to repeat the cycle.

TESTING THE RADIATOR

Because most modern cooling systems are pressurized, all of the components must be tight and in good condition before the system can operate properly.

Overheating and **loss of coolant** will result unless pressure in the system is maintained.

Test the entire cooling system before servicing it.

Install a *pressure tester* on the radiator according to the manufacturer's instructions (Fig. 5).

With the tester installed, carefully inspect the *radiator, water pump, hoses, drain cocks,* and *cylinder block* for leakage.

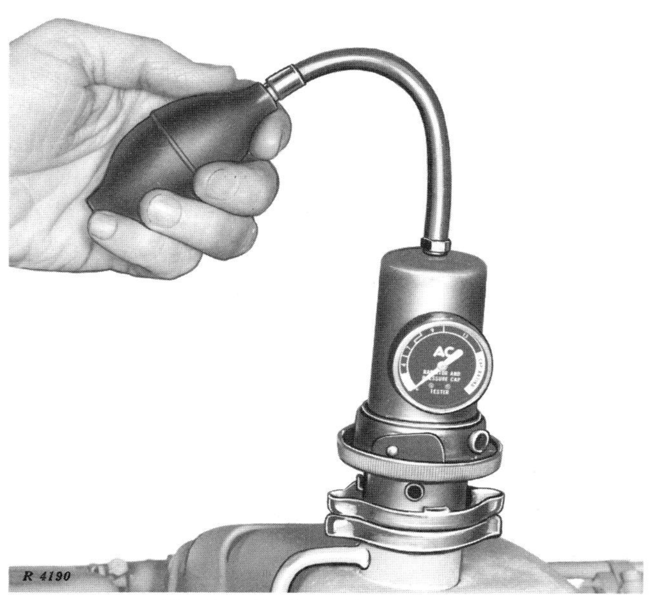

Fig. 5 — Radiator Pressure Test

Mark all leaks plainly to help locate them when repairs are made.

SERVICING THE RADIATOR

NOTE: Repairs should be done only by experienced radiator repairmen.

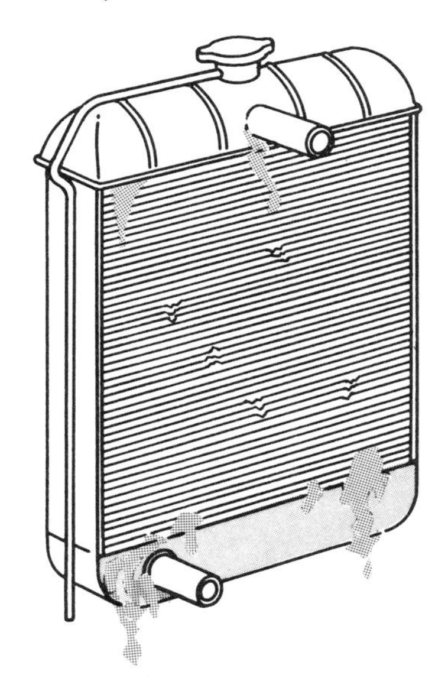

Fig. 6 — Damaged Radiator

Inspect the radiator for bent fins (Fig. 6). Inspect the tubes for cracks, kinks, dents, and fractured seams.

If radiator leaks and source of leak cannot be determined visually, test it as follows:

1. Install radiator cap. Plug overflow tube and outlet pipe and attach air hose to inlet connection.

 Litho in U.S.A.

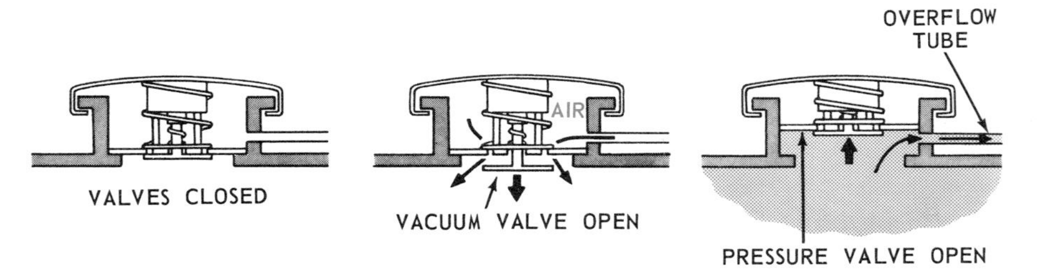

Fig. 7 — Pressure Control Radiator Cap

2. Fill radiator with compressed air [not more than 7 to 10 psi (50 to 70 kPa), depending upon size of radiator] and submerge in a tank of water.

3. Look for bubbles which tell the location of leaks.

Straighten any bent fins. Repair leaks by soldering, after thoroughly cleaning surfaces. Inspect rubber cushion washers. Replace as necessary.

RADIATOR CAP

The pressure system permits operating the engine at a higher temperature without boiling the coolant or losing it by evaporation.

An increase in pressure of one psi will raise the boiling temperature of pure water about 3°F (1.67°C).

A special *radiator cap* is required on most modern cooling systems. The cap has two functions:

- **Allows atmospheric pressure to enter the cooling system.**
- **Prevents coolant escape at normal pressures.**

A *pressure valve* (Fig. 7) in the cap permits the escape of coolant or steam when the pressure reaches a certain point.

The *vacuum valve* (Fig. 7) in the cap opens to prevent a vacuum in the cooling system.

Hotter operation is desirable for efficient combustion and for evaporating contaminants from the crankcase.

CAUTION: Always remove the radiator cap slowly and carefully to avoid a possible fast discharge of hot coolant.

TESTING THE RADIATOR CAP

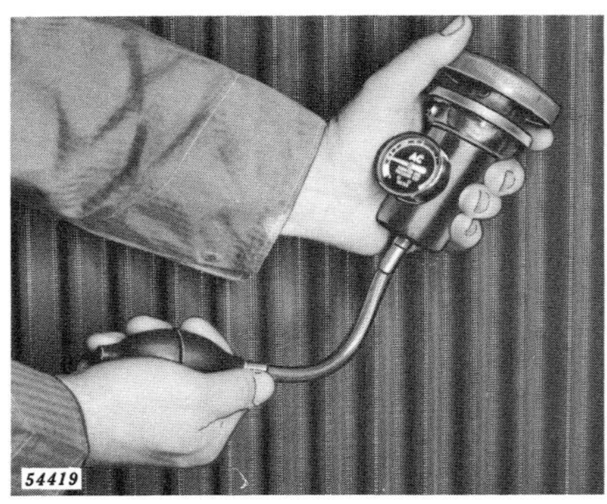

Fig. 8 — Testing The Radiator Cap

A radiator and pressure cap tester (Fig. 8) can be purchased from local tool jobbers. Use instructions with the tester to see if the radiator cap can build and hold pressure.

If the pressure cap is defective, replace it. A radiator cap cannot be repaired.

To prevent damage to the cooling system from either excessive pressure or vacuum, check both valves periodically for proper opening and closing pressures. Service or replace the cap as required.

FAN AND FAN BELTS

The cooling system fan is usually located between the radiator and engine, mounted on the front of the engine block. The fan is generally driven by a V-belt from the engine crankshaft.

The fan can be either a **suction**-type or a **blower**-type (Fig. 9), depending upon the design of the system.

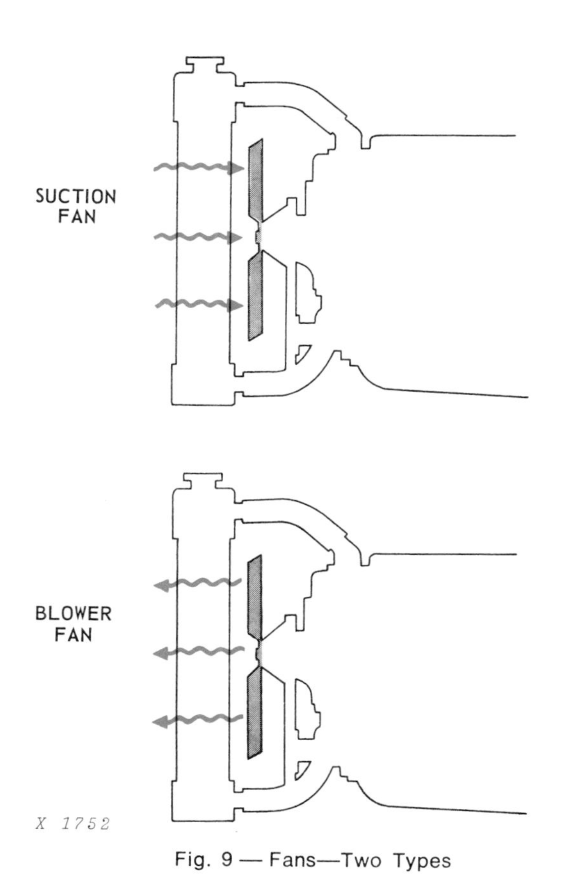

Fig. 9 — Fans—Two Types

• **Suction-type fans** *pull air through the radiator and push it over the engine. This design permits the use of a smaller fan and radiator than is required for blower-type fans. It is used when machine motion aids air movement through the radiator.*

• **Blower-type fans** *pull air across the engine, then push it through the radiator. They are used in slow-moving machines, and on machines where harmful materials might be drawn into the radiator with a suction-type fan.*

In both cases when the engine runs, the fan moves air across the radiator core to cool the liquid in the radiator. The ideal location of the fan is approximately 2-½ inches (60 mm) from the radiator core.

SERVICING THE FAN

The only service on the fan is to be certain the fan blades are straight and are far enough from the radiator so they do not strike the core.

Bent blades reduce the efficiency of the cooling system and throw the fan out of balance.

FAN BELT

The fan belt should be neither too tight nor too loose.

Too tight a belt puts an extra load on the fan bearings and shortens the life of the bearings as well as the belt.

Too loose a belt allows slippage and lowers the fan speed, causes excessive belt wear and leads to overheating of the cooling system.

The condition of the belt and its tension should be checked periodically. Adjust fan belt tension as specified by the manufacturer.

WATER PUMP

The water pump is normally a *centrifugal* type and might be called the "heart" of the cooling system (Fig. 10).

When the pump fails to circulate the coolant, heat is not removed from the engine and overheating damage may occur.

Some pumps turn to 4,000 revolutions a minute and pump as much as 125 gallons (470 L) of coolant per minute.

Most pumps today have self-lubricated, sealed ball bearings.

TESTING AND SERVICING WATER PUMPS

A typical water pump is made up of a *housing, impeller,* and *shaft* (Fig. 11).

When inspecting a pump, check for leaks in the housing, broken or bent vanes on the impeller, and damaged seals and bearings. Replace all damaged parts.

Always install new seals and gaskets when reassembling the pump.

ENGINE WATER JACKET

The engine cylinder block and head both usually contain passages for coolant to flow around the cylinders and valves. Together, they make up the *water jacket.*

The water jacket holds only a small amount of the total coolant. The advantages of this are:

• *Rapid engine warm-up—while the thermostat is closed*

• *Efficient cooling—when the thermostat opens*

 Litho in U.S.A.

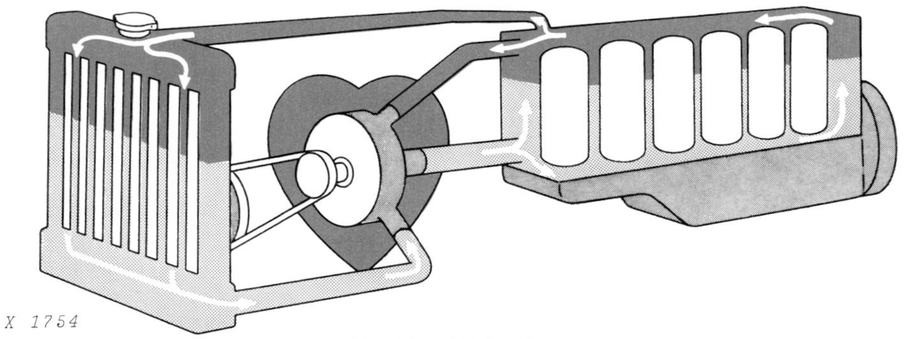

Fig. 10 — Water Pump

NOTE: As coolant is heated in the engine block, dissolved minerals separate out in solid form and coat the metal surfaces. The thicker this coating, the less heat transferred from the engine to the coolant. Prevent rust by adding anti-rust compounds each time coolant is changed.

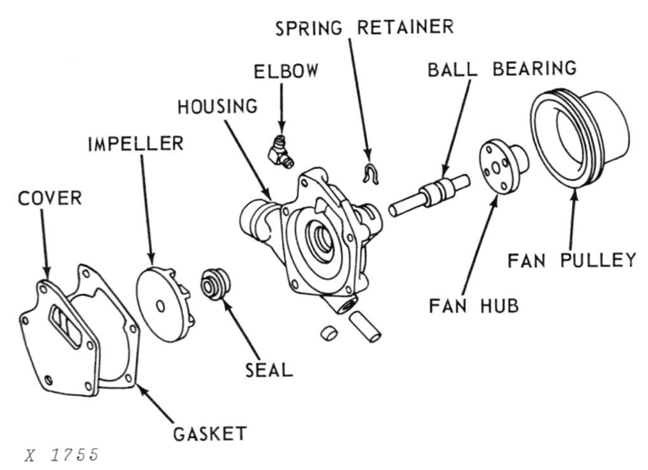

Fig. 11 — Water Pump Disassembled

COOLANT FILTER

Some engines now use a filter in the cooling system (Fig. 12).

The coolant filter does two jobs. The outer paper element filters out rust, scale, or dirt particles in the coolant. The inner element releases chemicals into the coolant to soften the water, maintain a proper acid/alkaline condition, prevent corrosion, and suppress cavitation erosion.

The chemicals released into the coolant by the inner element forms a protective film on the cylinder liner surface. The film acts as a barrier against collapsing vapor bubbles and reduces the quantity of bubbles formed.

The coolant filter in Fig. 12 is a spin-on type and should be replaced periodically according to operator's manual recommendations.

COOLANT ADDITIVES — DIESEL ENGINES

Some diesel engines are modified or originally designed to work with coolant additives instead of the

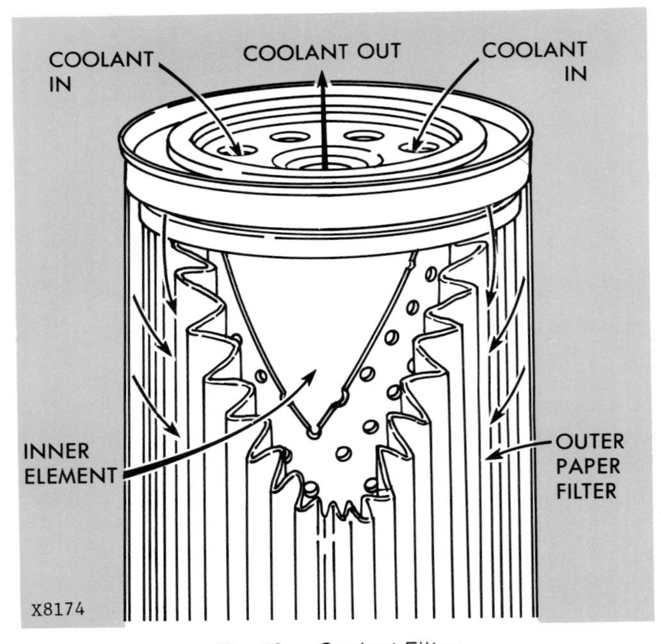

Fig. 12 — Coolant Filter

coolant filter. This type of conditioner is added directly to the radiator of the machine

It is commercially distributed in a powder or pre-mix liquid form. Coolant additives generally give the same protection as the coolant filter. They also contain other anti-rust and pitting inhibitors that cannot be included in the filter alone.

The coolant additives are poured directly into the cooling system as the cooling system is being filled.

 Litho in U.S.A.

They are designed to go to work immediately to help prevent system corrosion, cavitation, and contamination.

The liquid coolant additive may be more reliable than the filter method. Spin-on filters may be worked beyond their capacity. Always see the machine manufacturer's specifications before adding coolant additives to your machine.

CAUTION: Do not use a liquid coolant conditioner with a spin-on coolant filter unless the manufacturer specifications require simultaneous use. Over-concentration may occur and damage the machine cooling system.

THERMOSTATS

The thermostat provides automatic control of engine temperature at the correct level. This is necessary in order to get the best performance from an engine.

Some larger engines use dual thermostats for temperature control. The function and operation is the same as for a single thermostat system but allows for more capacity.

Only a small part of the engine's cooling capacity is required under the light loads, even during warm weather.

During warm-up the thermostat remains closed (Fig. 13). The water pump circulates coolant through the engine water jacket only, by way of the bypass.

The engine quickly warms up to its operating temperature before the thermostat opens.

When the thermostat opens, hot coolant flows from the engine to the radiator and back.

High-temperature thermostats, which open at 180°F (82°C), improve engine operation, reduce crankcase sludging and corrosive wear of engine parts.

An engine operating in this temperature range is hot enough to:

- *Improve combustion*
- *Burn impurities out of oil in crankcase*
- *Thin oil to provide good lubrication*

Do not use low-boiling-point alcohol or methanol antifreeze with high-temperature thermostats.

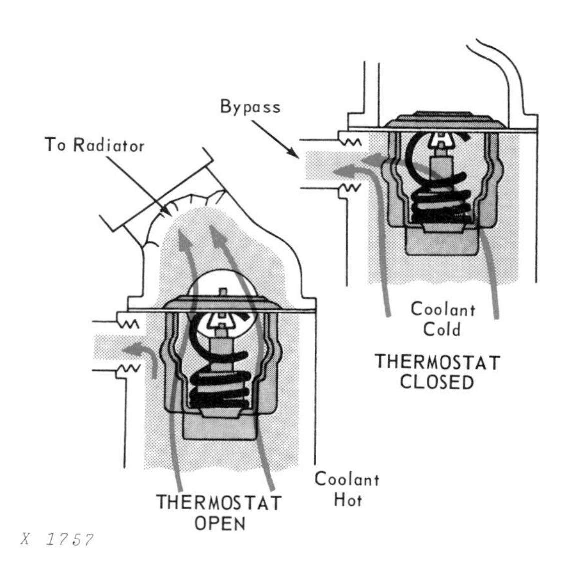

Fig. 13 — Cooling System Thermostat

When the thermostat is not operating properly, the engine may run too hot or too cold. Overheating may damage the thermostat so the valve will not function properly. Rust can also interfere with thermostat operation.

Always keep the thermostat in good working condition.

Never operate the engine without a thermostat.

Always use the thermostat design specified for the make and model of engine being used.

THERMOSTAT ACTUATOR

The actuator controls the inner workings of the thermostat. It can be made of special wax and granular copper that is sealed in a flexible rubber boot. It also can be made of bimetallic strips or a bellows tube filled with a liquid such as ether.

In all cases, heat expands the actuator which in turn forces the thermostat valve open. Cool temperatures contract the actuator and close the valve. The water pump in the cooling system does not create enough pressure to force the valve open.

The temperature that is required to open the thermostat is usually stamped on the spring strap holder. Replacement thermostats must be the same opening temperature as original equipment specifications. If they are not compatible, the engine could run cold or overheat. Both conditions put excessive wear on the engine.

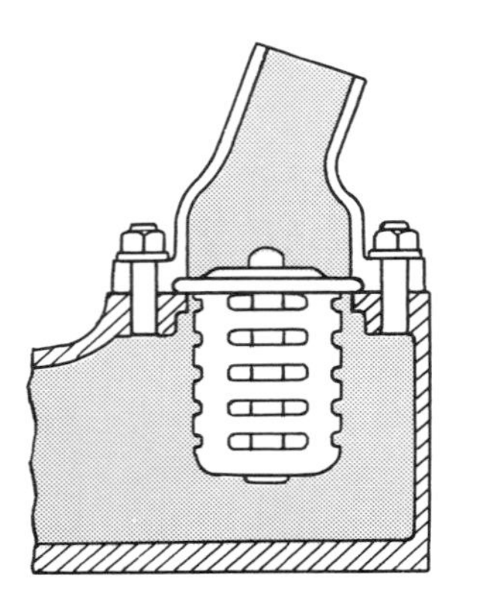

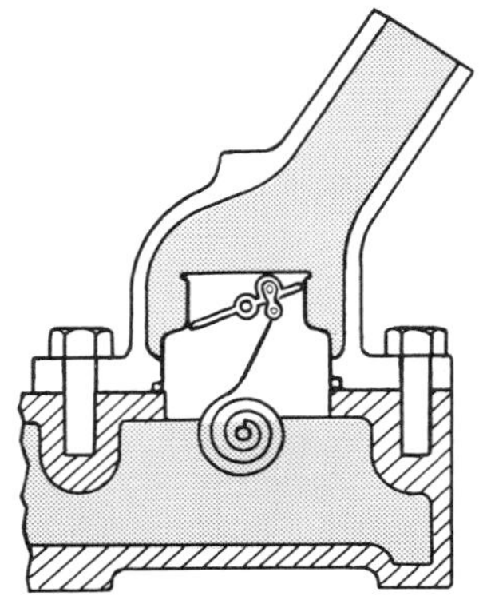

Fig. 14 — Types of Thermostats

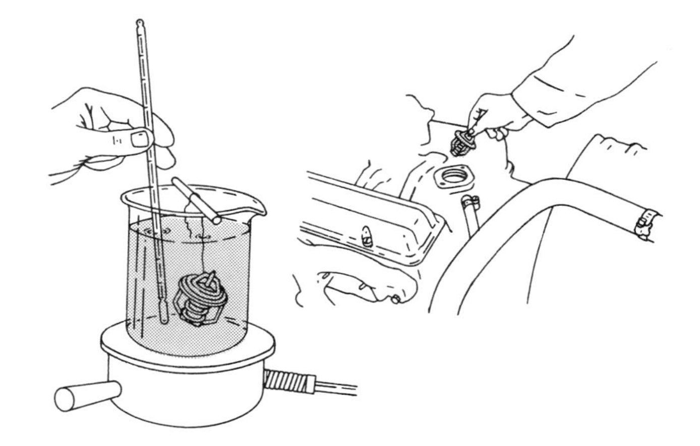

Fig. 15 — Testing A Thermostat

INSPECTION AND TESTING OF THERMOSTATS

If a thermostat is broken or corroded, discard it.

If it is not of the proper temperature and type, as indicated on an application chart, replace it. The number stamped on it is the approximate temperature that it will reach before starting to open.

• *Never use a high-temperature thermostat with an alcohol-base antifreeze.*

• *Never use a bellows thermostat in high-pressure cooling systems [9 pounds (60 kPa) or higher].*

Test the thermostat as follows:

1. Suspend the thermostat and a thermometer in a container of water (Fig. 15). Do not let them rest against the sides or bottom.

2. Heat and stir the water.

3. The thermostat should begin to open at the temperature stamped on it, plus or minus 10°F (5.5°C). It should be fully open — approximately 1/4-inch (6 mm) at 22°F (-12°C) above the specified temperature.

4. Remove thermostat and observe its closing action.

5. If the thermostat is defective, discard it.

INSTALLING THERMOSTATS

When installing a thermostat in the engine water jacket, position the thermostat with the expansion element toward the engine.

Some thermostats are marked with arrows that point to the radiator or to the engine block, or are marked "top" or "T". "Front" is indicated on some models. **The frame must not block the water flow.**

To prevent leakage, clean the gasket surfaces on the thermostat. Use a *new* gasket; normally it need not be cemented.

When the outlet casting and gasket have been properly located, tighten the nuts evenly and securely.

ENGINE OIL COOLER

Coolant flows through coolant tubes to the cylinder block. As oil flows through the cooler, heat is transferred to the coolant tubes and coolant.

Some engines are not equipped with engine oil cooler.

COOLING SYSTEM HOSES

Flexible hoses are used in connecting cooling system components because they stand up under vibration better than rigid pipes do. However, hoses have weak points, too.

Radiator hoses can be damaged by air, heat and water in two ways:

• **Hardening or cracking**—*destroys flexibility, causes leakage, and allows small particles of rubber to jam the radiator.*

• **Softening and swelling**—*produces lining failure and hose rupture.*

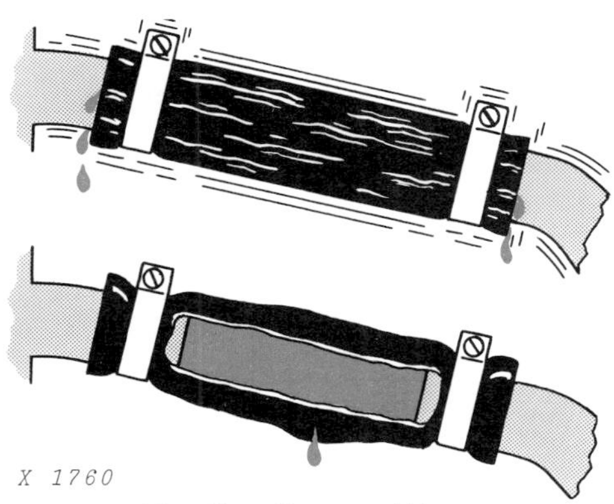

Fig. 16 — Damaged Hoses

Replace hoses often enough to be sure they are always pliable and able to pass coolant without leakage.

Examine hoses at least twice a year for possible replacement or tightening.

Check any reinforcing springs inside the hoses for corrosion.

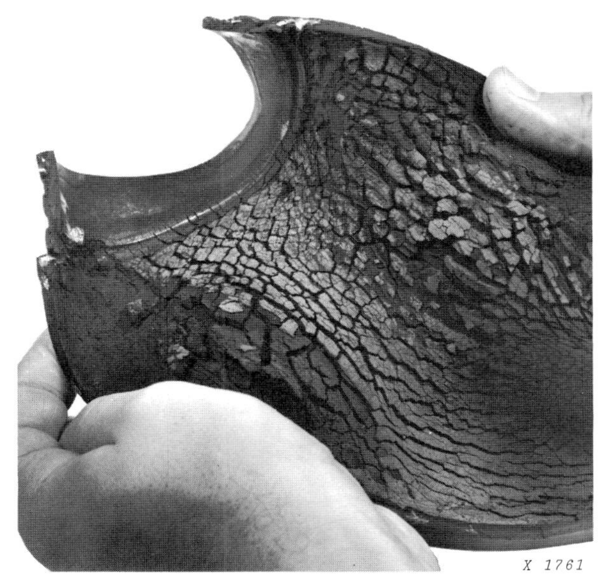

Fig. 17 — Interior Of Damaged Hose

Hoses can deteriorate on the inside and still appear all right on the outside (Fig. 17).

When hoses are removed, check them for wear. Hoses may harden and crack, allowing the system to leak, or they may soften and then collapse. A softened hose may also collapse during high-speed operation or restrict circulation enough to cause overheating.

The quality of the hose will affect its service life more than the cooling liquid. **Use only the best hoses available.**

INSTALLING HOSES

Fig. 18 — Connecting Hoses

Clean the pipe connections and apply a thin layer of non-hardening sealing compound when installing hoses (Fig. 18).

Locate the hose clamps properly over the connections as shown to provide a secure fastening. A pressurized cooling system will blow off an improperly installed hose.

OTHER RUBBER COMPONENTS

Heat and water can also damage other rubber sealing parts, such as the rubber seals in water pumps and O-rings which seal the lower end of wet cylinder sleeves.

Damage or improper installation of these parts will result in serious coolant leakage.

Litho in U.S.A.

COOLING LIQUID

Water is a good coolant because:

- *It is plentiful and readily available*
- *It absorbs heat well*
- *It circulates freely at all temperatures between its freezing and boiling points — 32°F to 212°F (0 to 100°C).*

But water has the following disadvantages:

- *It freezes readily when cold*
- *It boils and evaporates when hot*
- *It may corrode metal parts*
- *It may cause deposits in water jackets*

Modern pressurized systems overcome the low boiling temperature of water, and additives can offset many of its harmful properties.

Antifreeze is used to reduce the high freezing point of water.

PREPARING THE SYSTEM FOR USE

WARM WEATHER OPERATION

1. Fill with soft water whenever available.
2. Add a reliable brand of radiator sealer if required.
3. Add a can of rust inhibitor.
4. Keep system filled to a level midway between the radiator core and bottom of filler neck.

COLD WEATHER OPERATION

1. Fill with a mixture of antifreeze (with rust inhibitor) and water for protection against freezing to the lowest temperature anticipated.
2. Add a reliable brand of radiator sealer if required.
3. Run engine several minutes after thermostat opens to circulate coolant.
4. Recheck system for leaks.

ANTIFREEZE SELECTION

Use only one of these types of antifreeze:

- **Alcohol base**
- **Ethylene glycol base**

Modern engines usually operate at a temperature above the boiling point of alcohol so the permanent type (ethylene glycol) is recommended by most engine manufacturers.

The term "Permanent antifreeze" means only that the solution will not boil away at normal engine operating temperatures and does not mean that it is good for use for more than one season.

SERVICING THE COOLING SYSTEM

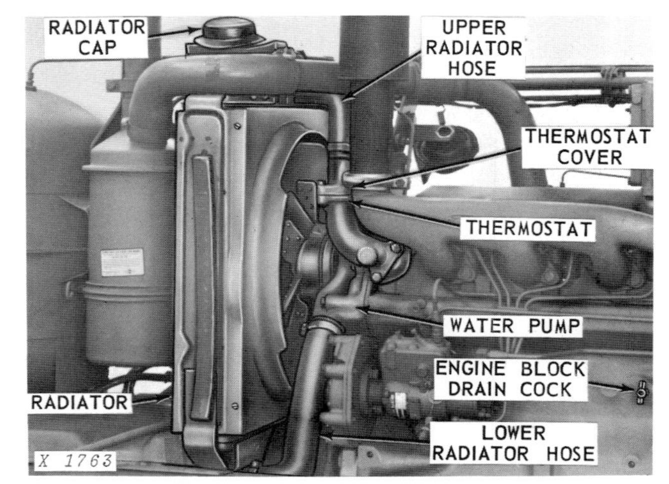

Fig. 19 — Engine Cooling System

Correct cooling system servicing is vital for a smooth-running engine.

Overheating is the big danger. It can be caused by:

- *Clogging of cooling system*
- *Lack of coolant*
- *Defective water pump or thermostat*

Check the coolant level and temperature frequently. Service the entire cooling system at least *twice a year.*

CLEANING THE COOLING SYSTEM

Efficient operation of the cooling system requires an occasional cleaning, particularly at seasonal changes when antifreeze solution is added or removed.

Several methods of cleaning the system are available, the proper one depending upon the amount of corrosion present in the system.

RUST-PROOFING THE SYSTEM

Rust is formed by corrosion of iron parts:

IRON + WATER + OXYGEN = RUST

Make it a rule to check the cooling liquid spring and fall.

FOR BEST COOLING SYSTEM PERFORMANCE, DRAIN THE SYSTEM AT LEAST ONCE A YEAR.

Use a *cooling system cleaner* if the liquid drains out rusty. Otherwise, use a *radiator flush* or *plain water.* Corrosion inhibitors do not clean out rust already formed.

In the spring, if the antifreeze is drained, renew the rustproofing by adding a reputable anti-rust compound to a filling of fresh water, or by installing a new filling of antifreeze.

Regardless of the summertime practice, corrosion protection can best be provided in winter by inhibitors built into the antifreeze coolant.

ADDING RUST INHIBITORS OR FRESH ANTIFREEZE TO USED SOLUTIONS WILL NOT RESTORE FULL STRENGTH CORROSION PROTECTION. SOME MIXTURES MAY EVEN DO HARM.

Some manufacturers recommend the year-round use of a highly inhibited antifreeze coolant. However, remember that the service life of the inhibitor may be drastically reduced as the machine gets older and the coolant may have to be replaced more often.

Some of the specific factors involved are:

- *Accumulation of contaminants, including rust and corrosion products.*
- *Dilution with water required to replace leakage losses, and possible absorption of air and exhaust gases.*

For these reasons, be sure to:

1. *Check the cooling system and coolant periodically.*
2. *Maintain at least a 25% solution concentration year-round for adequate corrosion protection.*
3. *Install new antifreeze coolant at least once each year.*

RUST CLOGGING

Rust clogging is a fairly common cause of cooling system trouble. It can be avoided entirely by periodic rustproofing and cleaning when necessary.

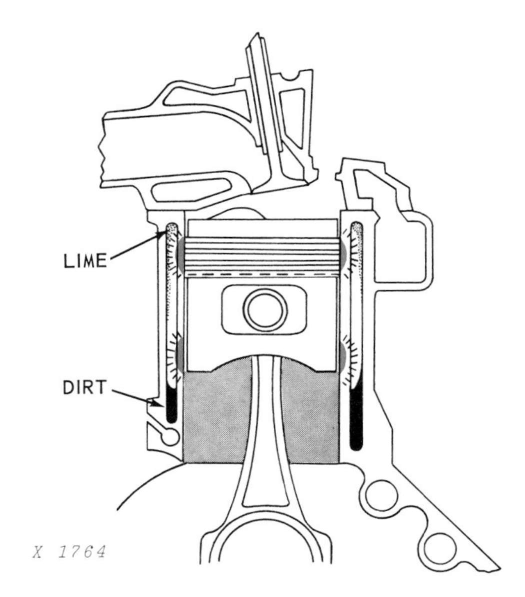

Fig. 20 — Mineral Deposits

The most common clogging materials are:

- **Rust**
- **Scale**
- **Grease**
- **Lime**

Rust accounts for 90% of the clogging. It forms on the walls of the engine water jacket and other metal parts.

Grease and oil enter the system through:

- *Cylinder head joint*
- *Water pump*
- *Leaking oil cooler*

Coolant circulation loosens rust particles which settle in the water jacket and build up in layers inside the radiator water tubes.

As the rust layer becomes thicker it keeps cutting down heat transfer from the radiator until the engine overheats and boiling starts in the water jacket.

This boiling stirs up more rust in the block and forces it into the radiator—eventually clogging it.

MINERAL DEPOSITS

Mineral deposits form rapidly at hot spots in the engine (Fig. 20).

Overheating, knock, and eventually engine damage can result from the build-up of rust and mineral scale on the water side of the combustion chamber (Fig. 20).

To avoid excessive formation of rust or scale deposits:

1. Keep the cooling system free of leaks.

2. Avoid adding too much hard water.

3. Maintain full-strength corrosion protection at all times.

COOLING SYSTEM CLEANERS

Well-maintained cooling systems seldom, if ever, require corrective cleaning.

However, if periodic rustproofing or other mechanical preventive maintenance is neglected, deposits will build up and cleaners must be used to restore cooling capacity.

Use a cleaner to remove:

- *Hard rust deposits*
- *Scale*
- *Grease*

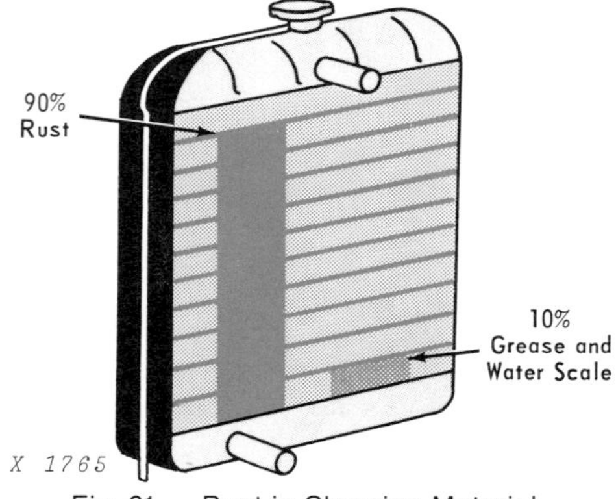

Fig. 21 — Rust in Clogging Material

Iron rust and water scale build up together in the cooling system. Clogging material is usually made up of grease, rust and water scale, but is *90% iron rust.*

Remove both hard rust-scale and grease with a double-action cleaner. It is harmless to cooling system metals and hose connections when used according to directions.

If rust and grease are not completely neutralized and flushed out, they can destroy the corrosion inhibitors in later fills of antifreeze or anti-rust solutions.

NOTE: Some radiators are made of aluminum or other metals which may be affected by certain compounds used for immersion cleaning. Follow manufacturers' instructions for cleaning these radiators.

FLUSHING THE COOLING SYSTEM

Incomplete flushing, such as hosing out the radiator, closes the thermostat and prevents thorough flushing of the water jacket. For complete flushing, take the following steps.

1. Fill the system completely with fresh water.

2. Run the engine long enough to open the thermostat (or remove the thermostat).

3. Open all drain points to drain the system completely.

Clean out the *overflow pipe* and remove insects and dirt from *radiator air passages, radiator grille* and *screens.*

Also check the thermostat, radiator pressure cap, and the cap seat for dirt or corrosion.

COOLING SYSTEM LEAKAGE

Leakage is the most common problem in a cooling system. During the winter it can involve the loss of valuable antifreeze.

But leakage actually does increase during the winter due to *metal shrinkage* and *cooling system pressure.*

Air pressure leakage testers can be helpful in locating external leaks but they cannot be depended upon to locate small combustion leaks.

Only practical experience enables a serviceman to tell whether the location and size of a leak can be corrected with a sealing solution.

All leakage exposed to combustion pressures must be repaired mechanically.

Follow instructions when using sealing solutions. Some sealing solutions react chemically with antifreeze and rust inhibitors, and seriously affect coolant performance.

Radiator Leakage

Most radiator leakage is due to mechanical failure of soldered joints. This is caused by:

- *Engine vibration*

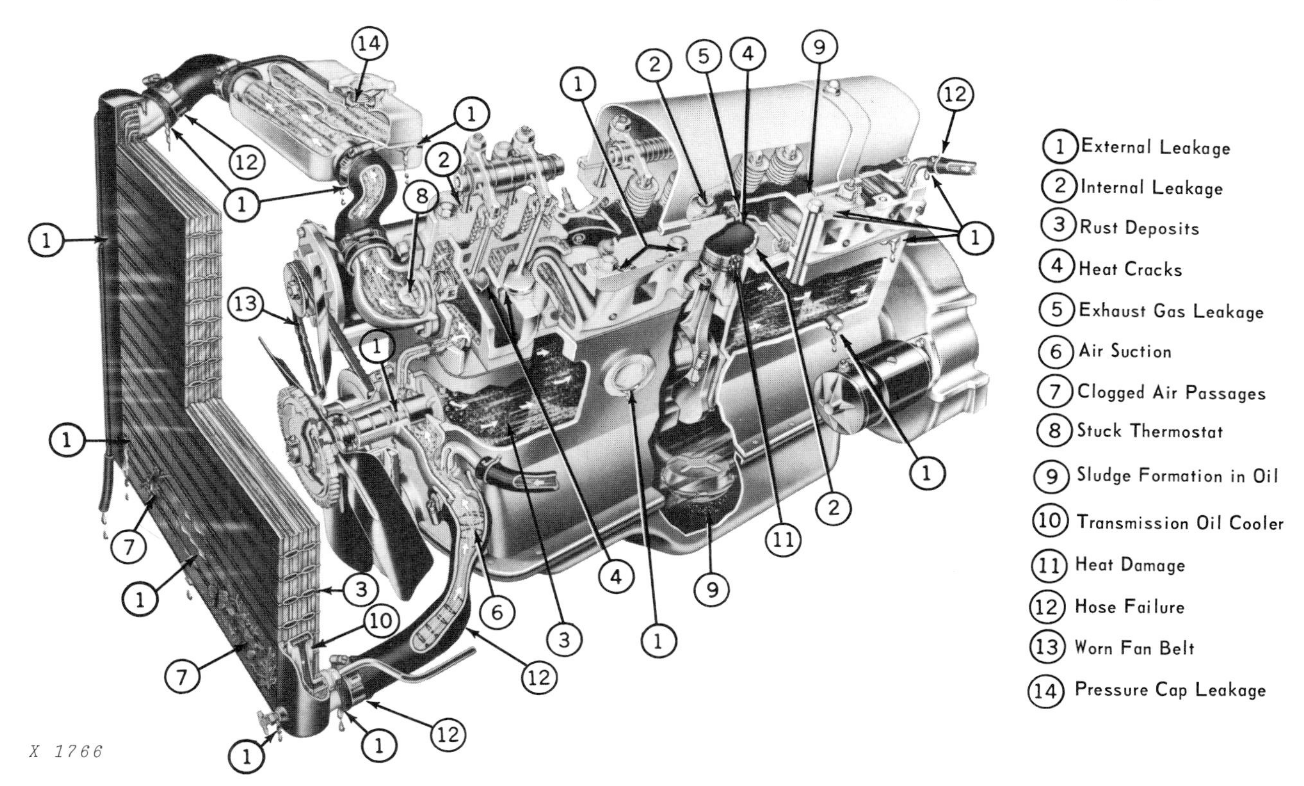

Fig. 22 — Maintenance Problems Of the Cooling System

- *Frame vibration*
- *Pressure in the cooling system*

Examine radiators carefully for leaks before and after cleaning. Cleaning may uncover leakage points already existing but plugged with rust.

White, rusty or colored leakage stains indicate previous radiator leakage. These spots may not be damp if water or alcohol is used since such coolants evaporate rapidly, but ethylene glycol antifreeze shows up because it does not evaporate.

Always stop any radiator leakage before installing antifreeze coolant.

Leakage Outside Engine Water Jacket

Inspect the engine cylinder block before and after it gets hot while the engine is running.

Leakage of the engine block is aggravated by:

- *Pump pressure*
- *Pressurized cooling systems*
- *Temperature changes of the metal*

Remember: Small leaks may appear only as rust, corrosion or stains, due to evaporation.

Watch for these trouble spots:

1. **Core-Hole Plugs.** Remove old plug. Clean plug seat, coat with sealing compound. Drive new plug into place with proper tool.

2. **Gaskets.** Tighten joint or install new gasket. Use sealing compound when required.

3. **Stud Bolts And Cap Screws.** Apply sealing compound to threads.

Water Jacket Leakage Into Engine

Coolant leaks into the engine through:

- *A loose cylinder head or sleeve joint*
- *A cracked or porous casting*
- *The push rod compartment*

Water or *antifreeze* mixed with *engine oil* will form **sludge** which causes:

1) Lubrication failure

2) Sticking piston rings and pins

3) Sticking valves and valve lifters

4) Extensive engine damage

 Litho in U.S.A.

Fig. 23 — Water Jacket Leakage Into Engine

The amount of damage depends upon *the amount and duration of leakage, service procedures,* and *seasonal operating conditions.*

High-temperature thermostats let the engine heat remove moisture (formed from blow-by) from the crankcase.

Head gasket leakage is more likely to occur in winter than summer because of more metal contraction and expansion in winter.

Give special attention to the cylinder head joint gasket no matter what type or material.

An improperly installed gasket can cause:

- *Coolant and oil leakage*
- *Overheating*

It may be necessary to pressurize the cooling system and tear down part of the upper part of the engine to find a coolant leak in the push-rod compartment.

Replacing Head Gaskets

To replace a cylinder head gasket:

1. Use only a new gasket designed for this particular engine.
2. Make sure the head and block surfaces are clean, level and smooth.
3. See Chapter 2 for details.

IMPORTANT: Leading engine manufacturers strongly recommend the use of torque wrenches to tighten cylinder heads.

Always follow the engine manufacturers' instructions on cylinder head bolt tension and order of tightening.

OVERHEATING DAMAGE

In modern high-performance engines, the intense heat of combustion can cause engine components such as valves, pistons and rings to operate near their critical temperature limits—even when the cooling system is operating normally.

Overheating severely affects the lubrication of the engine. High metal temperatures can destroy the lubricating film, accelerate oil breakdown, and cause formation of varnish.

These in turn may cause:

1) Excessive wear

2) Scoring

3) Valve burning

4) Seizure of moving parts

Continued operation above the normal temperature range may result in:

- *Lubrication failure*
- *Heat distortion*
- *Engine knocking*

Cylinder heads and engine blocks are often warped and cracked (Fig. 24) by terrific strains set up in the metal by overheating, especially when followed by rapid cooling.

Hot spots in an overheated engine can cause engine knocking. If allowed to continue, knocking results in:

1) "Blown" head gaskets

2) Damaged pistons

3) Ring and bearing failure

See Chapter 2 for details.

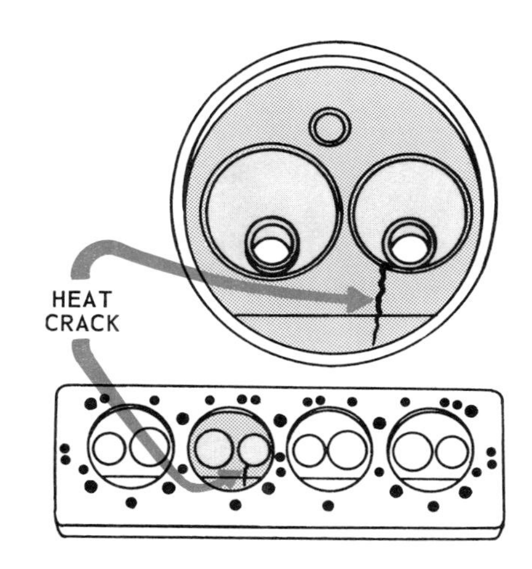

Fig. 24 — Cracks Caused By Overheating

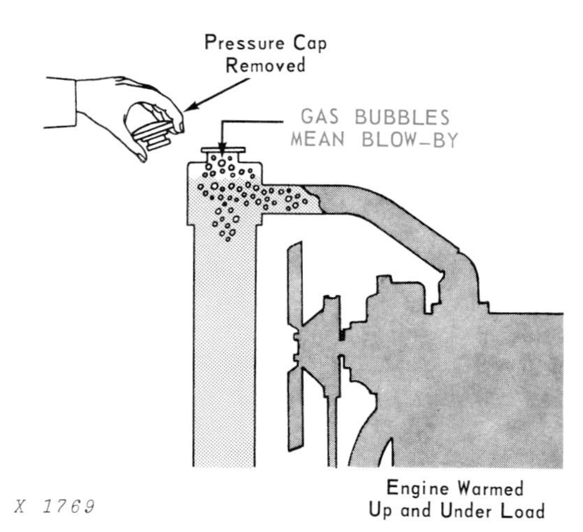

Fig. 25 — Test For Exhaust Gas Leakage (Blow-By)

EXHAUST GAS LEAKAGE

A cracked head or a loose cylinder head joint allows hot exhaust gas to be blown into the cooling system under combustion pressures; even though the joint may be tight enough to keep liquid from leaking into the cylinder.

The cylinder head gasket itself may be burned and corroded by escaping exhaust gases.

Exhaust gases dissolved in coolant destroy the inhibitors and form acids which cause corrosion, rust and clogging.

Excess pressure may also force coolant out the overflow pipe.

Testing For Exhaust Gas Leakage (Blow-By)

1. Warm up the engine and keep it under load.

2. Remove the radiator cap and look for excessive bubbles in the coolant (Fig. 25).

3. Either bubbling or an oil film on the coolant is a sign of blow-by in the engine cylinders.

NOTE: Make this test quickly, before boiling starts, since steam bubbles give misleading results.

CORROSION IN THE COOLING SYSTEM

Water solutions will corrode the cooling system metals when they are not protected by inhibitors added to water for summer operation.

Corrosion coating on metal surfaces reduces heat transfer at the walls of the engine water jacket and in the tubes of the radiator even before clogging occurs.

Galvanic action is a form of corrosion which can occur when two different metals (in contact) are suspended in a liquid which will carry a current.

It is similar to what happens in a battery when electric current flows from one area to another, removing metal from one electrode and depositing it on another in the process.

Galvanic corrosion or direct attack by the coolant will affect the following metals in the cooling system: stainless steel, brass, solder, aluminum, and copper.

AERATION IN COOLING SYSTEM

Aeration, or mixing of air with water, speeds up the formation of rust and increases corrosion of cooling system metals.

Aeration may also cause:

- *Foaming*
- *Overheating*
- *Overflow loss of coolant*

Air may be drawn into the coolant because of:

1) A leak in the system

2) Turbulence in the top tank

3) Too-low coolant level

Litho in U.S.A.

Check the cooling system for exhaust gas leakage and air suction when you find these conditions:

- *Rusty Coolant*
- *Severe Rust Clogging*
- *Corrosion*
- *Overflow losses*

Testing For Air In The Cooling System

If air leaks in the cooling system are suspected, the following checks can be made.

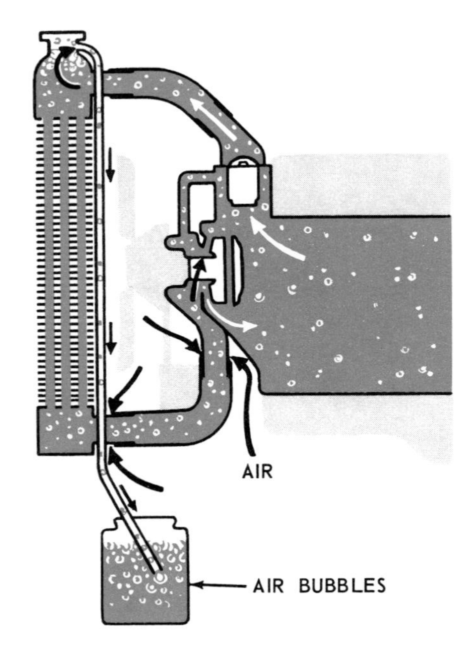

Fig. 26 — Testing For Air In Cooling System

1. Adjust coolant to correct level.
2. Replace pressure cap with a plain, air-tight cap.
3. Attach rubber tube to lower end of overflow pipe. Be sure radiator cap and tube are air tight.
4. With transmission in neutral gear, run engine at high speed until temperature gauge stops rising and remains stationary.
5. Without changing engine speed or temperature, put end of rubber tube in bottle of water.
6. Watch for a continuous stream of bubbles in the waste bottle, showing that air is being drawn into the cooling system (Fig. 26).

TEST YOURSELF

QUESTIONS

1. What are the eight parts of the liquid cooling system?

2. (True or false?) "Permanent antifreeze should be changed every season."

3. What is the basic difference between a BELLOWS-Type Thermostat and a BIMETALLIC-Type?

4. What part of the cooling system could be referred to as the "heart" of the system?

5. (True or false?) "Thermostats are not required when operating the engine in hot weather."

(Answer in back of text.)

GOVERNING SYSTEMS / CHAPTER 9

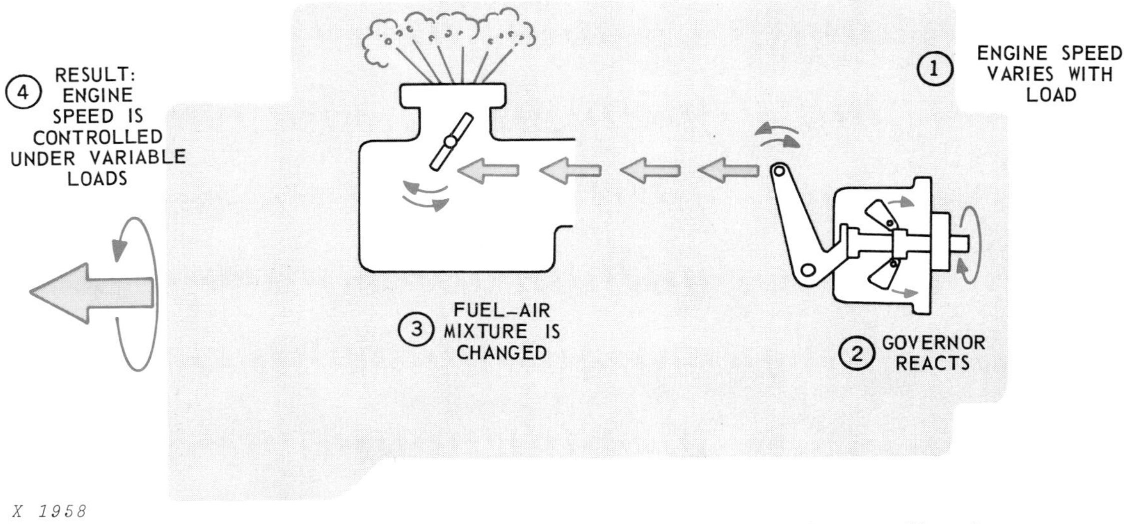

Fig. 1 — Governing System (Gasoline Engine With Centrifugal Governor Shown)

INTRODUCTION

The *governor* is a device that automatically controls the speed of an engine under varying loads.

Governors can do any of three jobs:

- **Maintain a selected speed**
- **Limit the slow and fast speeds**
- **Shut down the engine when it overspeeds**

Note in Fig. 2 how the governor keeps the engine at a constant speed when the tractor is going both uphill and downhill.

Most modern governors are the centrifugal spring-loaded type and adjust fuel intake according to demand on the engine (Fig. 1). Others use a vacuum to control a throttle valve.

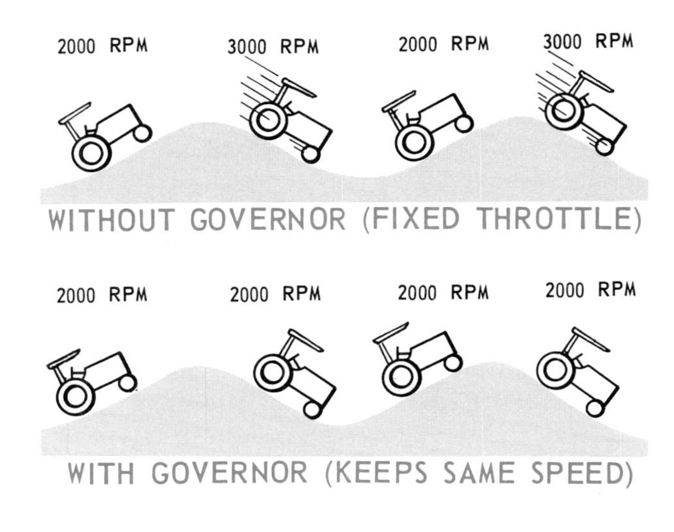

Fig. 2 — What The Governor Does For The Engine

FEATURES OF GOVERNORS

Let's discuss some of the main features of governing:

Stability—to maintain a desired engine speed without fluctuations. Indicated by the number of corrective movements the governor makes and the time required to correct engine speed for a given load change.

Lack of stability will result in *hunting,* or oscillations due to over-correction. Too much stability will result in a *dead-beat* governor, or one that does not correct sufficiently.

Another term for this ability of the governor to keep a steady speed is *isochronous* governing.

 Litho in U.S.A.

Sensitivity—the percent of speed change required to produce a corrective movement of the fuel-control mechanism.

Promptness of Response—the time in seconds required for movement of the fuel control from no-load to full-load position.

Speed Drift—a very gradual deviation of the main governed speed above or below the desired governed speed.

Work Capacity—the work which the governor can perform to overcome the resistance in the fuel controls, expressed in inch-pounds or foot-pounds.

Load-Limit Changer—an adjustable device for limiting maximum fuel flow to the engine, thus limiting power output to any desired maximum value.

Speed Changer—a device for adjusting the speed-governing system to change engine speed, or power-output relationship, with other engine generator units when in parallel operation. Normally this can be adjusted while the engine is in operation.

Speed-Regulation Changer—a device by which the steady speed regulation (speed droop) can be adjusted while the engine is in operation.

Speed Droop—All mechanical governors and some hydraulic governors have permanent speed droop and are non-isochronous because the engine's steady speed is slightly different at different speeds.

Speed droop is the variation in governor rotating speed from no load to full load (sometimes referred to as *steady-state speed regulation).*

Speed droop refers to a change in the steady speed when the load on an engine is reduced from rated output to zero output, without adjustment of the governor. It is usually expressed as a percent of rated speed.

For example: An engine equipped with a governor which has been set for a no load speed of 2,000 rpm will not run this fast under load. The difference between the governor setting (no load speed) and full load speed is called *speed droop.* This difference may be as much as 10 percent less than the no load speed.

In this example, no load speed will be 2,000 rpm but all engine speeds under partial loads will be somewhat less than this and full load speed will drop down to 1,800 rpm.

Speed droop is useful in preventing overshooting or hunting by the governor when the engine loads change.

Now let's discuss the governing of engines in two parts—first spark-ignition, then diesel.

GOVERNING OF SPARK-IGNITION ENGINES

Governing of spark-ignition engines is more critical than for diesel engines. This is partly due to the design of the air and fuel control system, but chiefly to:

- *The small variation in fuel-air ratio which is permissible.*
- *The metering lag of nearly one cycle for complete response to governor action.*

The governor quickly feels a reduction in speed, and increases fuel to the engine about one cycle before air is supplied. This results in erratic operation.

If the engine over-speeds, a quick reduction in fuel causes the engine to misfire. Therefore, the governor response must be slow enough to ignore quick speed variations caused by occasional misfiring, yet be sensitive to the sustained speed changes resulting from variations of load.

Most gasoline engines use a *throttle* governor.

The throttle governor permits the engine to fire steadily regardless of the load but varies the charge of fuel-air and so the intensity of the combustion. This speeds up or slows down the engine to allow for different loads.

The governor actuates a throttle valve in the intake passage between the carburetor and the engine manifold (see Fig. 1). The position of the throttle valve at any given time controls the mixture to the engine (see Chapter 3 for details).

The two most common types of throttle governor systems are:

- **Automatic or vacuum governors**
- **Centrifugal governors**

Let's discuss each type.

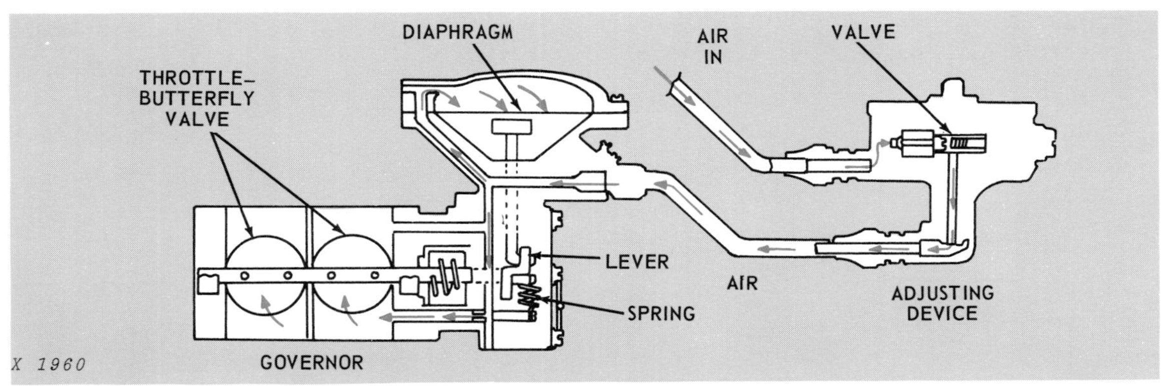

Fig. 3 — Vacuum Governor

AUTOMATIC OR VACUUM GOVERNORS

Many farm and industrial machines are equipped with a governor for maintaining a set maximum operating speed.

An automatic or vacuum governor is used for this purpose (Fig. 3). Located between the carburetor and the intake manifold, it has no mechanical connection to any other parts of the engine.

This type of governor consists of a housing and a throttle-butterfly valve connected to a cam-and-lever mechanism that is spring-controlled.

The unit operates by closing the throttle-butterfly valve, causing a decrease in the fuel flow, as engine speed and suction increase.

When engine speed and suction decrease, the spring opens the throttle-butterfly valve and the fuel flow increases.

A simple adjustment of spring tension maintains the desired speed range.

Efficient operation of this governor depends largely upon precise machining, freedom of movement of the parts, and proper installation and adjustment of the engine.

CENTRIFUGAL GOVERNORS

The centrifugal governor has two weights, called flyweights, suspended on a weight carrier, which is mounted on the governor shaft (Fig. 4). The governor is driven directly from the engine.

As the governor shaft rotates, the weights are thrown outward by centrifugal force, pushing a thrust sleeve against the governor fork which in turn actuates the governor lever and, through governor-to-carburetor linkage, controls the amount of fuel and air supplied to the engine (Fig. 4).

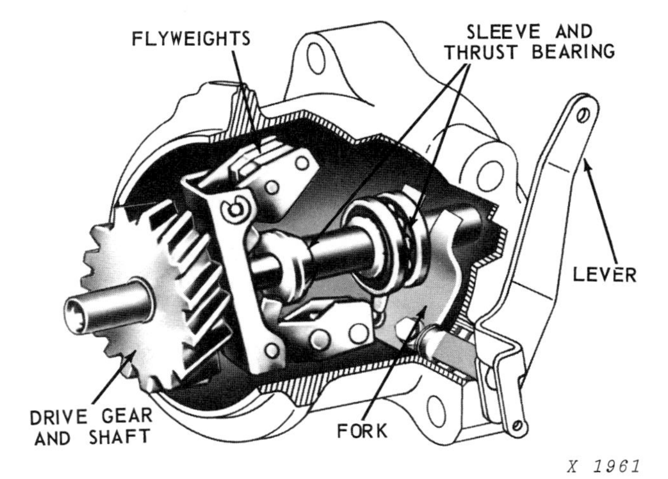

Fig. 4 — Cutaway Of Centrifugal Governor

This control of fuel-air mixture keeps the engine at the desired speed under varying loads.

The governor is linked to the hand throttle and will maintain any engine speed, unless the engine is over-loaded.

The centrifugal force of the weights decreases as the speed falls and the speeder spring pushes them inward until a new balance point is reached (Fig. 5). For any given speed, the flyweights will always take a definite position a certain distance from the axis of rotation.

"Hunting" Of Centrifugal Governors

A problem with centrifugal governors is *hunting.*

Frequently when an engine is first started or after it is warmed up and is working under load, its speed will become uneven or irregular. The engine speeds up quickly, the governor suddenly responds, the speed drops quickly. The governor again suddenly responds, the speed drops quickly, the governor responds again, and the action is repeated. This is known as *hunting.*

 Litho in U.S.A.

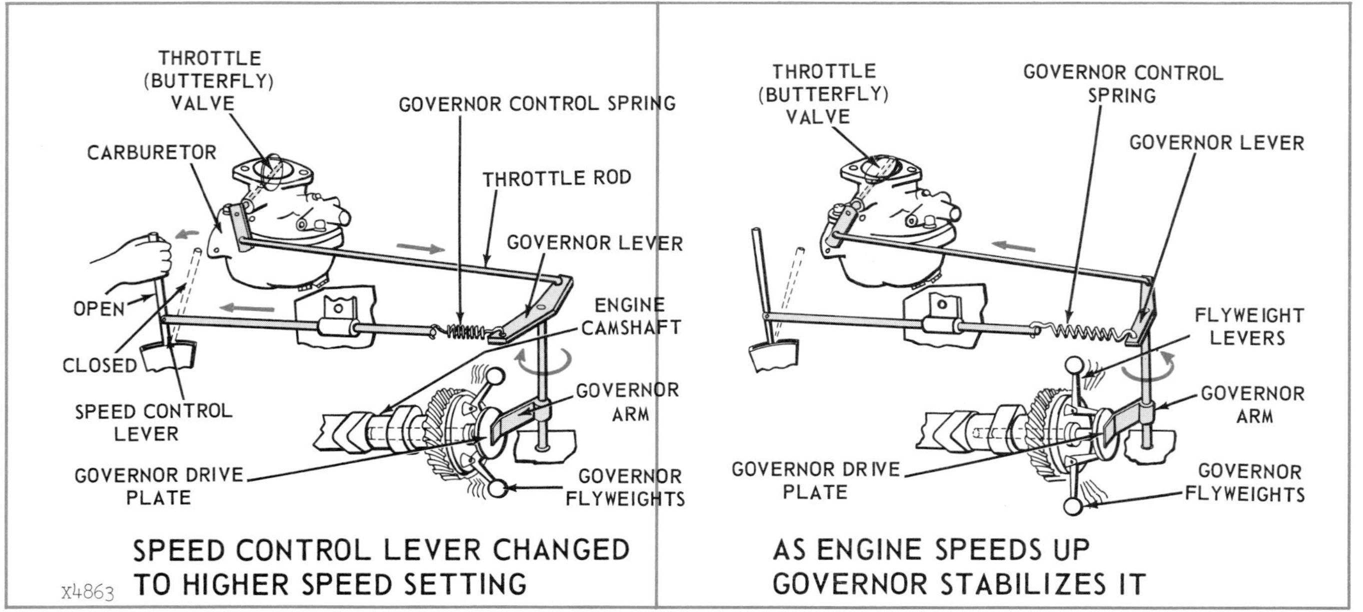

Fig. 5 — Operation Of Centrifugal Governor

Hunting is usually caused by the wrong *carburetor adjustment* and can be corrected by making the mixture either slightly leaner or slightly richer. The governor itself may also cause hunting by being too stiff or by striking or binding at some point so that it fails to act freely.

GOVERNOR LINKAGE AND ADJUSTMENT

Remember these two points when connecting or adjusting the governor to the engine:

1) **Amount of work**—*Governors are capable of doing a specified amount of work, say 8 foot-pounds, when moving through full control travel.*

2) **Speed droop**—*Governors may have an adjustable speed droop up to 10 percent for full control travel.*

If the linkage is set to give full control of fuel during one-quarter of governor lever travel, the governor can do only one-quarter of its maximum work. So the maximum speed droop obtainable will be *one-quarter* of *rated* maximum.

For smoother operation of the engine, change the lengths of the governor and fuel control rods so that 70 to 80 percent of full governor lever travel is used between no load and full load. This means that the governor can perform 70 to 80 percent of its work rating.

MAINTENANCE OF GOVERNOR

The two chief requirements for trouble-free operation of a governor are:

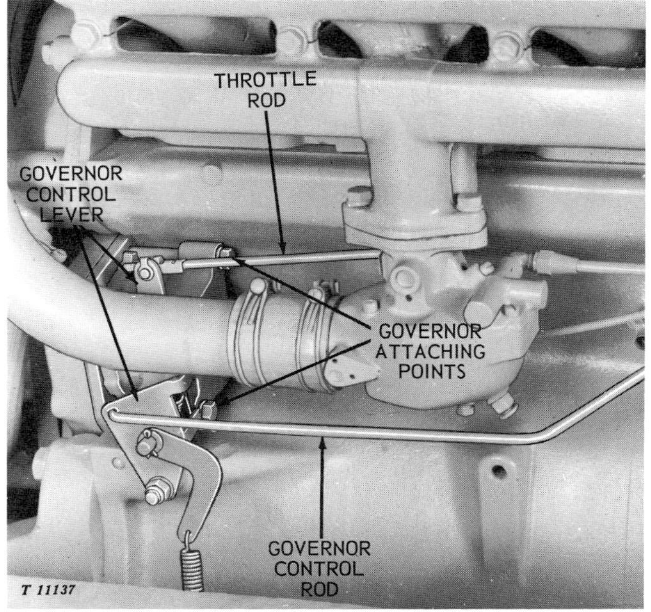

Fig. 6 — Governor And Linkage (Centrifugal Type Shown)

- **Vibration-free drive**
- **Cleanliness of moving parts**

The governor is a sensitive device which demands a non-fluctuating speed drive with no vibration. This is the reason why many governors are driven through a spring or rubber coupling to smooth out any roughness in the engine drive.

The smoothest governor drive is obtained through gearing from the crankshaft. The free end of the camshaft will give the roughest drive, the greatest variation in speed, and the most vibration.

 Litho in U.S.A.

GOVERNING OF DIESEL ENGINES

The principles of governing a diesel engine are the same as for gasoline engines, but there are several different ways in which governors perform their jobs:

- **Mechanical**—a speed measuring device, which directly actuates the linkage to the throttle.
- **Hydraulic**—the speeder rod actuates a small valve controlling fluid under pressure; this fluid operates a piston or "servo-motor" which actuates the fuel control.
- **Variable speed**—maintains any selected engine speed from idling to maximum speed.
- **Over-speed**—shuts down the engine in case it runs too fast. It is only a safety device.
- **Torque converter regulator**—controls the maximum speed of the output shaft of a torque converter attached to an engine, independently of engine speed.
- **Pressure regulator**—used on any engine driving a pump, to maintain a constant inlet or outlet pressure on the pump.
- **Load control**—adjusts the amount of load applied to the engine to suit the speed at which it is set to run. This actually improves fuel economy and reduces engine strain when the engine is not under load.
- **Electric load sensing**—detects variations in the electrical load and quickly adjusts fuel quantity to match. This governor is most generally used in electrical power plants.
- **Electric speed**—controls speed by measuring the frequency in cycles-per-second of the electrical current produced by a generator driven by the engine.

MECHANICAL GOVERNORS (DIESEL)

Mechanical governors provide full fuel for starting when the speed control lever is in the idle or run position. Immediately after starting, the governor moves the injector rack to the position required for idling.

Mechanical governors are simple centrifugal types, with few moving parts. They work best when engine speed is not critical.

Disadvantages are:

1) *Dead Band*—the change in speed the engine must make before the governor makes a corrective movement of the throttle. Caused mainly by friction.

2) *Speed Droop*—the change in governor speed required to cause the throttle rod to move from full-open to full-closed or full-closed to full-open.

Types Of Mechanical Governors

The two types of mechanical governors are:

- *Speed Limiting*—used when a maximum and minimum speed plus manually-controlled intermediate speed is required.
- *Speed Varying*—used for a near-constant engine speed under varying loads, with speeds adjustable by the operator.

HYDRAULIC GOVERNORS (DIESEL)

Hydraulic governors have more parts and are generally more expensive than mechanical governors. They can also be made isochronous because they are more sensitive and have greater power to move the fuel control mechanism than mechanical governors.

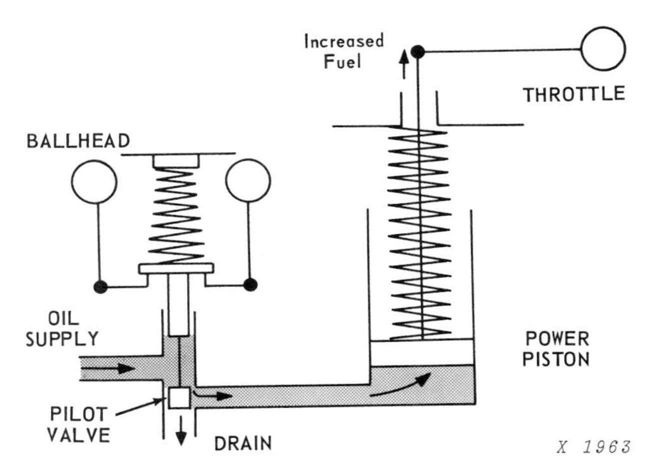

Fig. 7 — Hydraulic Governor (Diesel)

The power which moves the engine throttle does not come from the speed measuring device of hydraulic governors. Instead, it comes from a hydraulic power piston or "servo-motor". This is a piston which is acted upon by oil under pressure to operate the throttle (Fig. 7).

The speed measuring device is attached to a small pilot valve as shown. This valve slides up and down in a bushing containing ports which control the oil flow to the piston.

The force needed to slide the port valve is very small. Therefore, a small ballhead is able to control a large amount of power in the piston.

The engine should be equipped with a separate over-speed governor to prevent a runaway should the governor fail.

Hunting Of Hydraulic Governors

A basic fault of hydraulic governors is that they have a tendency to "hunt." The cause is the unavoidable time delay between the time the governor acts and the engine responds. The engine cannot instantly come back to the speed called for by the governor.

For example: If the engine speed is below the governor control speed, the pilot valve moves the power piston or servo-motor to increase the fuel flow.

By the time the speed has increased to the control setting so that the port valve is centered and the power piston or servo-motor dropped, the fuel has already been increased too much, but the engine continues to speed up.

This overspeed opens the port valve the other way to decrease the fuel, but by the time engine speed falls to the right level, the fuel control has again traveled too far. The engine over-speeds, and the whole cycle is repeated continuously.

To overcome "hunting", *speed droop* must be provided (Fig. 8).

VARIABLE SPEED GOVERNORS

A variable speed governor can control the engine speed at different settings between minimum and maximum. It resembles a constant-speed governor in its ability to hold the engine at steady speed. The difference is that the operator can adjust the governor to maintain any desired steady speed.

Speed adjustment is obtained in most variable speed governors by changing the spring force that resists the centrifugal force of the flyweights (see Fig. 4).

ISOCHRONOUS GOVERNORS

Governors which hold steady speeds are classed as constant speed governors. The only truly constant speed governors are *isochronous governors.*

An isochronous governor is able to maintain a constant speed without changing it. It does this by employing *speed droop* to give stability while fuel intake is being corrected.

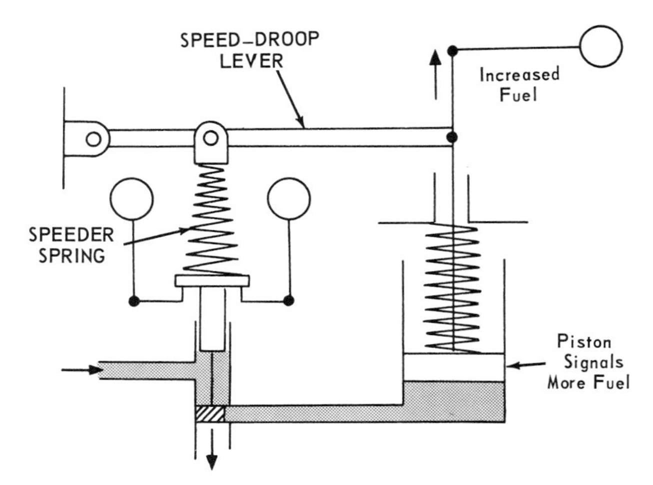

Fig. 8 — Hydraulic Governor With Adjustable Speed Droop

OVER-SPEED GOVERNORS

Over-speed governors are safety devices which protect engines from damage due to over-speeding.

When an engine is equipped with a regular speed governor, the over-speed governor will function only if the regular governor fails to operate. The over-speed governor merely slows the engine down while allowing it to keep on running at a safe speed.

Over-speed governors use a preloaded spring to over-balance the centrifugal force of the flyweights.

When the maximum speed is passed, centrifugal force overcomes the spring and activates the fuel controls which slow down the engine.

TORQUE CONVERTER GOVERNORS

A *torque converter governor* is used on a torque converter in conjunction with the engine governor.

This governor takes its speed reading from the output shaft of the torque converter.

Engine speed is governed by the engine converter as long as the torque converter speed is below its maximum level.

But if the torque converter overspeeds, its governor reduces the engine fuel and keeps converter speed from rising further.

PRESSURE REGULATING GOVERNORS

Pressure regulating governors are used on pipeline pumps to insure a constant pumping pressure.

An ordinary variable speed governor can be converted into a pressure regulating governor by adding a device which varies the speed setting of the governor in accordance with the pumping pressure.

The pumping pressure acts on a piston or diaphragm, balanced by a spring connected to a speed adjusted on the governor. If the pressure drops below the set level, the resulting movement of the diaphragm moves the speed adjuster to increase the speed.

When the speed is increased, the engine runs faster and pumping pressure is increased. When pumping pressure rises, engine speed is reduced; causing the pressure to return to normal.

LOAD LIMITING GOVERNORS

Load limiting governors prevent the loss of engine speed due to heavy loading by limiting the load to the capacity of the engine. They cannot always be used because it is not always practical to limit the load.

A prime application of load limiting is on diesel-electric locomotives. Here the load limit device is attached to the governor, which, when the governor terminal shaft reaches the maximum safe fuel position, operates an electric control, reducing the output of the electric generator.

The electric control is a rheostat, which is used to reduce generator excitation. This reduces the load on the engine enough to permit it to run at full speed and develop full power.

Load limiting governors are required when a series of various capacity engines are working together. The governor limits the fuel consumed by each engine to utilize its full power capability without the danger of overloading any of the others.

LOAD CONTROL GOVERNORS

Load control governors maintain the engine speed required and adjust the load limiting device in such a manner that the amount of load applied to the engine is the amount predetermined as suitable for each speed.

When partial power is required, the engine not only runs at a reduced speed, but also uses less fuel per stroke. This greatly reduces engine strain.

ELECTRIC LOAD-SENSING GOVERNORS

Electric load-sensing governors find their greatest use in electric power generating plants where an absolutely *stable frequency* must be maintained regardless of wide, rapid changes in load.

Load-sensing governors have an advantage over speed governors because they can divide the load among several generating units. Using this method, governors in all engines can be set for the same speed, and the load-sensing elements can be adjusted so the engines will always share the load, per their capacities, through the entire load range.

In operation, electric load-sensing governors detect changes in the load and adjust the fuel supply to maintain the same engine speed under the new load. They are able to respond faster to load changes than speed governors because *the load change itself, directly operates the governor mechanism.*

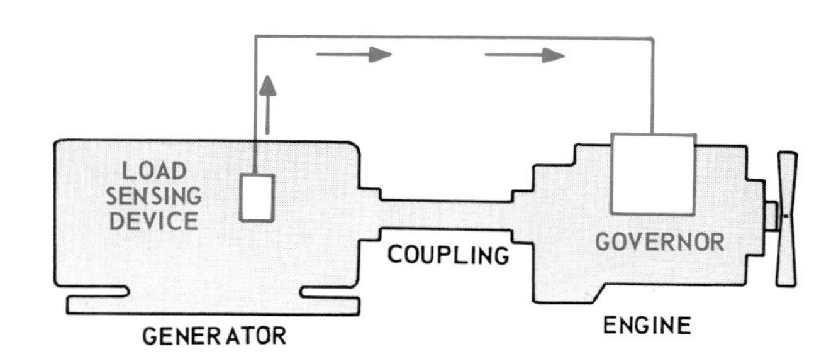

X 1965

Fig. 9 — Governor With Load Sensing

An electric load-sensing device consists of an electric network connected to an engine-driven generator (Fig. 9). The network measures the generator load and controls a solenoid within the governor which operates the fuel control mechanism.

Load-sensing governors always include *speed-sensing* and *load-sensing* elements since the speed would gradually float away from normal (due to the fact that "normal" is constantly changing), if the engine fuel were not controlled.

ELECTRIC SPEED GOVERNORS

Electric speed governors respond to engine speed by measuring the frequency of the electric current produced by the generator. When the measuring network notes that the frequency has departed from normal, it delivers an electric signal to the solenoid which operates the fuel control mechanism. The frequency measuring network replaces the flyweights and speeder spring used on mechanical governors.

MALFUNCTIONS OF GOVERNORS

Governors are relatively trouble-free, but need adjusting, servicing, etc.

Listed below are a few of the symptoms and causes of governor problems.

I. Erratic Engine Operation, Hunting, or Misfiring

Possible Cause	*Possible Remedy*
1. Governor idle spring missing or adjusted incorrectly.	1. Replace or adjust spring.
2. Governor control spring worn or broken.	2. Remove and replace.
3. Governor not operating—parts worn, sticking, binding, or incorrectly assembled.	3. Disassemble and inspect parts.
4. Wrong governor spring.	4. Remove and replace with proper spring.
5. Governor high idle adjustment incorrect.	5. Adjust governor to diesel pump specifications.
6. Governor adjusting screw not properly adjusted.	6. Adjust governor screw.

II. Engine Idles Erratically

1. Governor idle spring missing or adjusted incorrectly.	1. Replace or adjust spring.
2. Governor not operating—parts worn, sticking, binding, or incorrectly assembled.	2. Disassemble and inspect parts.
3. Wrong governor spring.	3. Remove and replace with proper spring.

III. Engine Not Receiving Fuel

1. Governor not operating—parts worn, sticking, binding, or incorrectly assembled.	1. Disassemble and inspect parts.

IV. Engine Does Not Develop Full Power Or Speed

1. Governor not operating—parts worn, sticking, binding, or incorrectly assembled.	1. Disassemble and inspect parts.
2. Wrong governor spring.	2. Remove and replace with proper spring.
3. Governor adjusting screw not properly adjusted.	3. Adjust governor screw.

SUMMARY: GOVERNORS

Here are some key facts that will help you understand governors:

1. A governor is a speed-sensitive device that controls the speed of an engine under varying loads.

2. Governors can do any of three jobs:

a. Maintain a selected speed

b. Limit the slow and fast speeds

c. Shut down the engine before it overspeeds

3. Isochronous governors maintain a constant engine speed independent of load requirements.

4. Governor speed droop is a change in the steady-state of speed.

5. Speed drift is a gradual deviation from desired governor speed.

6. A speed changer is a device for adjusting the speed governing system to change engine speed.

7. A governor's two functions are: to measure speed and to operate the throttle.

8. The two most common types of throttle governors are automatic or vacuum, and centrifugal.

9. The different types of diesel engine governors are: mechanical, hydraulic, variable speed, over-speed, torque converter, pressure regulator, load control, electric load-sensing, and electric speed.

10. Hydraulic governors are more expensive and have more parts than mechanical governors but are more sensitive to speed changes.

11. Variable speed governors maintain any selected engine speed from idle to maximum speed.

12. Over-speed governors are safety devices for controlling maximum engine speed.

13. A torque converter governor controls the maximum speed of the output shaft of a torque converter attached to an engine, independently of the engine speed.

14. A load control converter adjusts the amount of load applied to the engine to match the speed at which it is set to run.

15. A pressure regulator governor is used on any engine driving a pump for maintaining a constant pump pressure.

TEST YOURSELF

QUESTIONS

1. What are the three jobs which governors can do?

2. What is speed droop?

3. What is an isochronous governor?

(Answer in back of text.)

 Litho in U.S.A.

ENGINE TEST EQUIPMENT AND SERVICE TOOLS / CHAPTER 10

INTRODUCTION

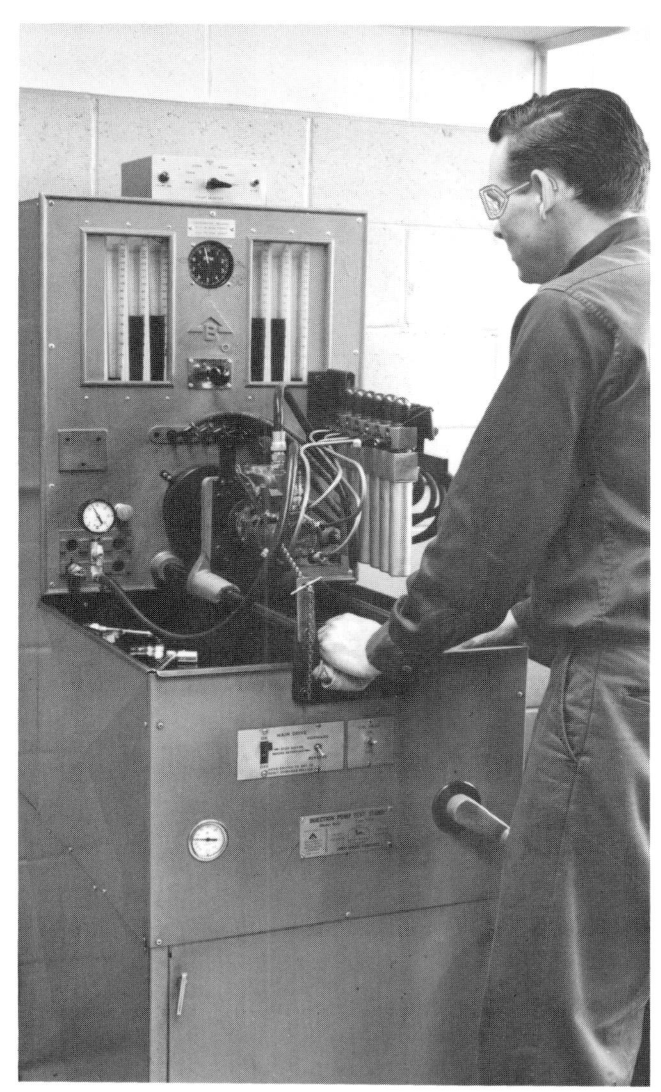

Fig. 1 — Good Test Equipment Does The Job Better And Faster

Many tools are available today for servicing engines, but they all have one purpose—to help the serviceman do his job better and faster.

Two general types of engine tools are used:

- **Testing Tools — to diagnose the engine**
- **Servicing Tools — to fix the engine**

Testing tools help to find out what must be done, while servicing tools help to do the actual service.

Here are the tools we will cover:

TESTING TOOLS	SERVICING TOOLS
Cylinder Compression Tester	*Valve Refacer*
Cylinder Bore Gauge	*Valve Seat Grinder*
Ring Groove Wear Gauge	*Valve Spring Compressor*
Piston Pin Bore Gauge	*Valve Guide Reamer*
Plastigage	*Valve Guide Knurling Tool*
Valve Spring Tester	*Cylinder Ridge Reamer*
Injection Nozzle Tester	*Cylinder Liner Puller*
Injection Pump Tester	*Cylinder Reboring Bar*
Dynamometer	*Cylinder Deglazer*
Water Manometer	*Piston Ring Groove Cutter*
Vacuum Gauge	*Piston Ring Groove Cleaning Machine*
Exhaust Smoke Analyzer	*Piston Knurling Tool*
Ignition Coil Meter	*Piston Ring Expander*
Distributor Tester	*Piston Ring Compressor*
Engine Analyzer	*Piston and Rod Aligner*
Spark Plug Tester	*Carburetor Service Tools*
Spark Plug Gap Gauge	*Injection Nozzle Tools*
Power Timing Light	*Nozzle Cleaning Kit*
Battery Hydrometer	
Oil Pressure Gauge	
Thermostat Tester	
Radiator and Cap Tester	

Fig. 2 — Servicing Tools Help To Fix The Engine After It Is Tested

 Litho in U.S.A.

TESTING TOOLS

Before servicing the engine, test it to locate the problem. Here are the prime tools you will need.

CYLINDER COMPRESSION TESTER

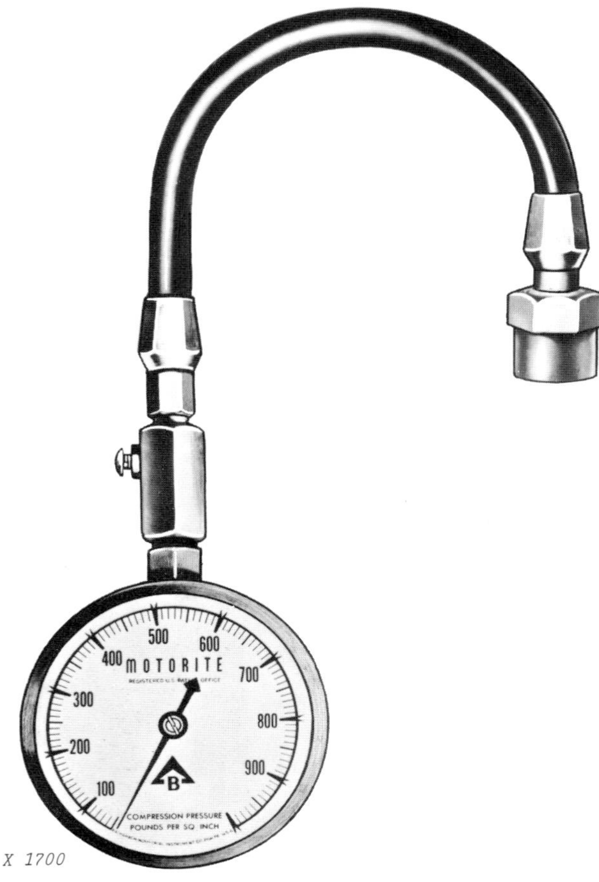

Fig. 3 — Cylinder Compression Tester

Use a cylinder compression tester to check the compression of each cylinder. Poor compression indicates bad piston rings or valves.

CYLINDER BORE GAUGE

The cylinder bore gauge (Fig. 4) checks the roundness and taper of a cylinder or cylinder liner bore. You can also use it to find the exact amount of oversize needed for honing or reboring cylinders.

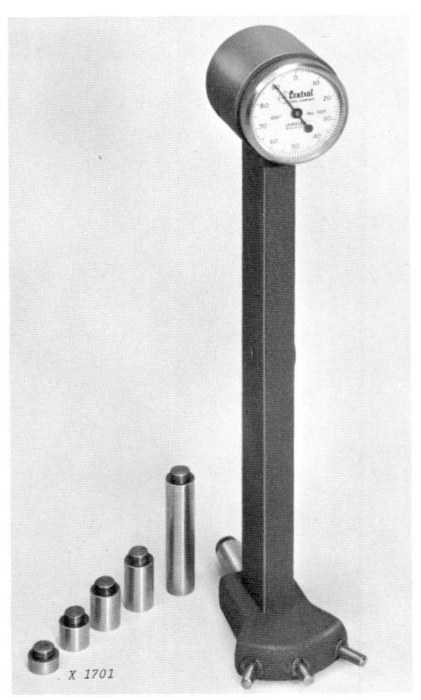

Fig. 4 — Cylinder Bore Gauge

RING GROOVE WEAR GAUGE

Fig. 5 — Ring Groove Wear Gauge

Use a wear gauge to make a fast check for wear in the piston ring grooves. Gauges are available for both regular and keystone grooves.

NOTE: A new piston ring and a feeler gauge can also be used to measure ring groove wear.

PISTON PIN BORE GAUGE

X 1703

Fig. 6 — Piston Pin Bore Gauge

A piston pin hole gauge is used to measure piston pin bores for precision pin fits. The gauge shown in Fig. 6 can also be used to measure bores in the journal end of piston rod.

PLASTIGAGE

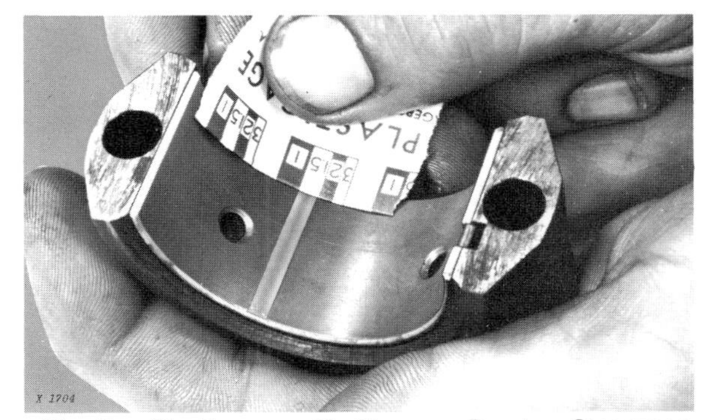

X 1704

Fig. 7 — Plastigage — To Measure Bearing Clearance

Using "Plastigage" is a fast and accurate way to check engine bearing clearance. Determine clearance by crushing the "Plastigage" between the bearing and the crankshaft, and then comparing it with a graduation scale provided with the package.

VALVE SPRING TESTER

Use the valve spring tester to check the strength of valve springs in service (Fig. 8).

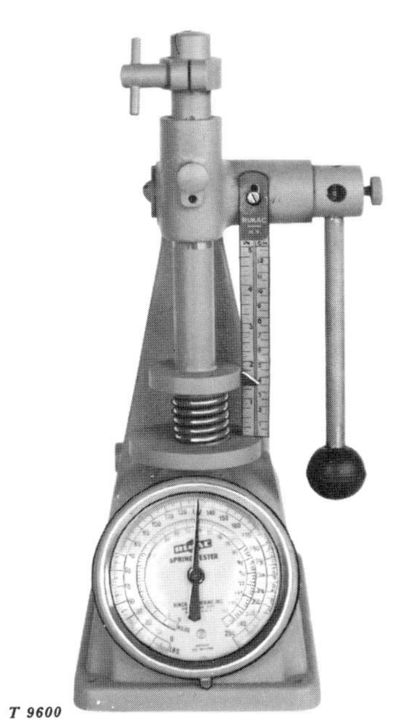

T 9600

Fig. 8 — Valve Spring Tester

INJECTION NOZZLE TESTER (Diesel)

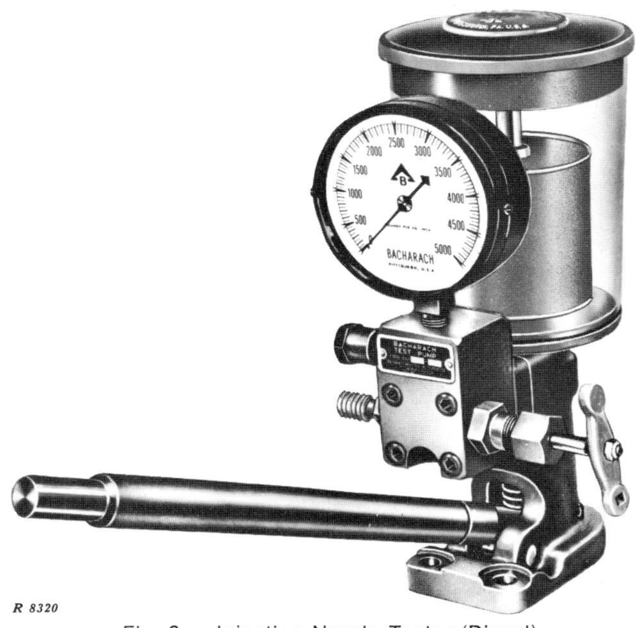

R 8320

Fig. 9 — Injection Nozzle Tester (Diesel)

An injection nozzle tester will check the condition of:

1) *Needle valve and seat within the nozzle.*
2) *Spray pattern of the nozzle tip.*
3) *Cracking pressure of the nozzle.*
4) *Leak-off through the nozzle.*
5) *Nozzle valve lift.*

INJECTION PUMP TESTER (Diesel)

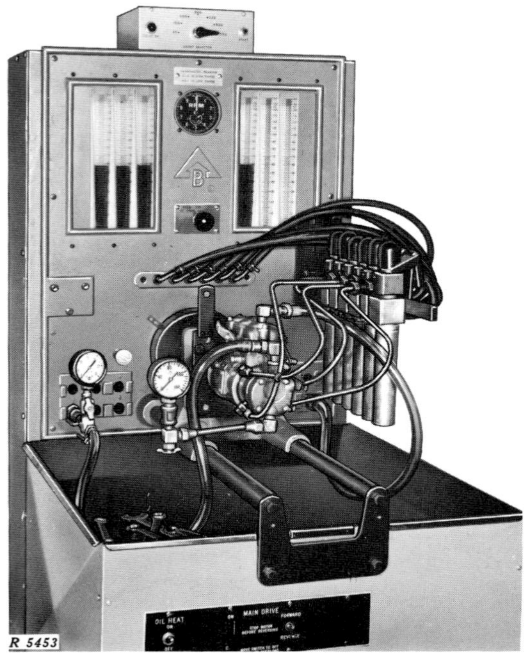

Fig. 10 — Injection Pump Tester (Diesel)

An injection pump tester is essential to properly test and calibrate a diesel fuel injection pump. The tester shown in Fig. 10 is used to test leakage, vacuum, pressure, and delivery. It is also used to make idle and torque control adjustments.

DYNAMOMETER

Fig. 11 — Dynamometer

Measure engine horsepower and fuel consumption with a dynamometer (Fig. 11). The results will tell you if the engine efficiency can be restored by a tune-up or whether reconditioning is required.

WATER MANOMETER

Fig. 12 — Water Manometer

Use a water manometer to check the air intake system of an engine. High depression readings indicate a restriction. A vacuum gauge calibrated in inches of water can also be used to make this check.

TACHOMETER

Fig. 13 — Portable Electronic Tachometer

Tachometers (Fig. 13) show the revolutions multiplied by time (as rpm). Another example of this is the common speedometer, which indicates wheel revolutions per hour, computed in miles per hour (mph). Tachometers on most machines, however, give engine speed in revolutions per minute (rpm).

EXHAUST SMOKE ANALYZER

X 1708

Fig. 14 — Exhaust Smoke Analyzer

An exhaust analyzer is used to check carburetor performance, combustion efficiency, and air-fuel ratio. The model shown in Fig. 14 includes a vacuum gauge.

IGNITION COIL AND CONDENSER METER

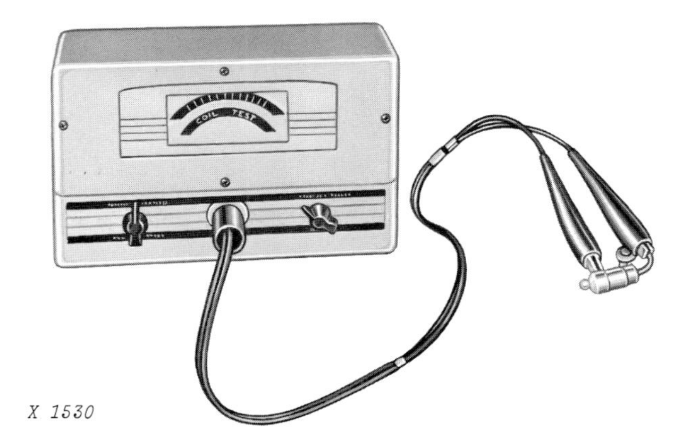

X 1530

Fig. 15 — Ignition Coil and Condenser Meter

This combination tester checks the condition of the coil and condenser in the ignition system.

DISTRIBUTOR TESTER

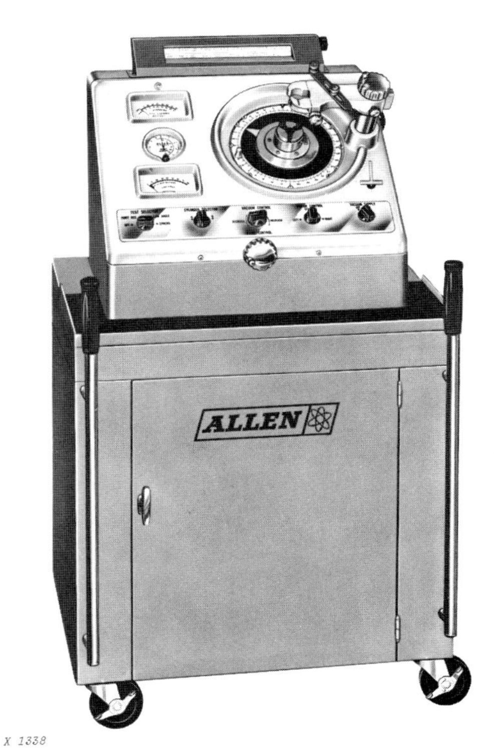

X 1338

Fig. 16 — Distributor Tester

The all-purpose distributor tester shown in Fig. 16 will test:

- **Centrifugal advance**
- **Vacuum advance**
- **Cam dwell angle**
- **Variations in cam lobes**
- **Breaker point contact resistance**
- **Breaker point synchronization**
- **Breaker point operation**
- **Breaker arm spring tension**
- **Vacuum chamber condition**
- **Distributor shorts**
- **Excessive wear**
- **Distributor high-speed performance**

These testers are also known as "synchrographs." Bench models of the tester are also available.

Litho in U.S.A.

ENGINE ANALYZER (Spark-Ignition)

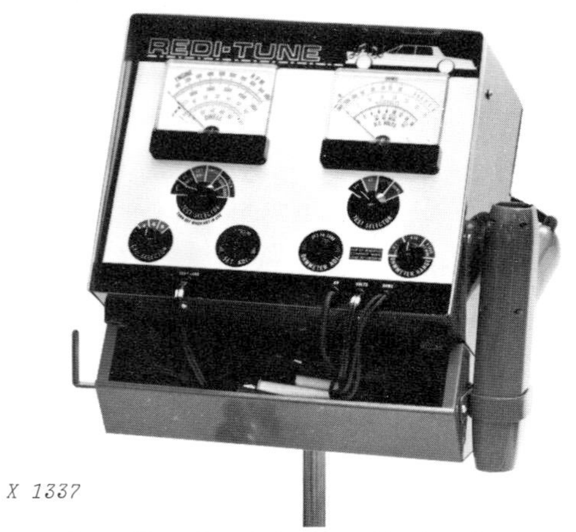

X 1337

Fig. 17 — Spark-Ignition Engine Analyzer

The spark-ignition engine analyzer can be used to test the entire ignition system:

- **Point resistance**
- **Cam angle**
- **Basic timing**
- **Generator voltage**
- **Starting motor voltage**
- **Mechanical condition of distributor**

SPARK PLUG TESTER

Use this tester to check the condition of spark plugs. Most of these testers are also equipped to clean the plugs.

SPARK PLUG GAP GAUGE

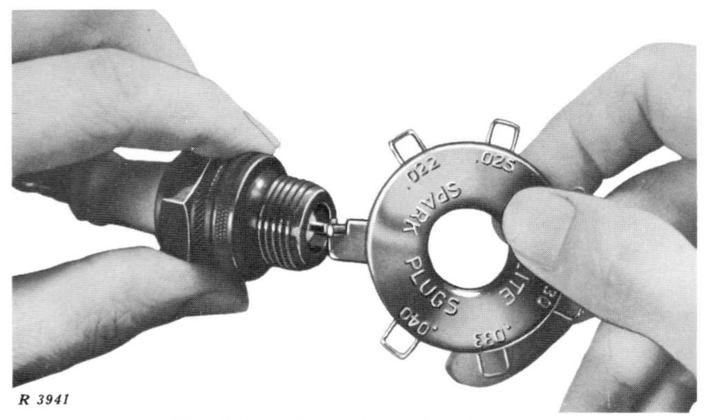

R 3941

Fig. 18 — Spark Plug Gap Gauge

Use a spark plug gap gauge to check and set the gap between spark plug electrodes. A plug can't do a good job unless it is set to exact engine specifications.

POWER TIMING LIGHT

X 1339

Fig. 19 — Power Timing Light

Use the power timing light to check the ignition timing while the engine is running.

BATTERY HYDROMETER

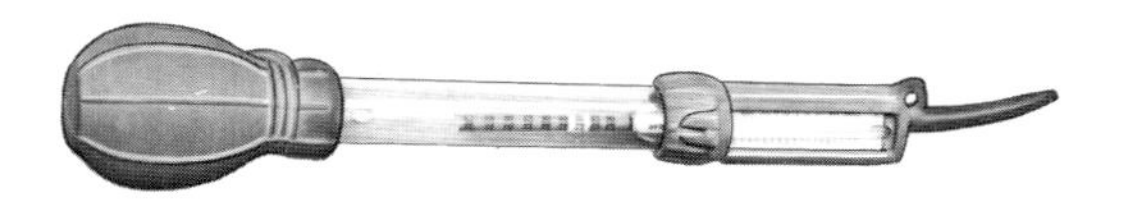

R 458

Fig. 20 — Battery Hydrometer

Use the hydrometer to check the specific gravity of the battery. All good hydrometers have a thermometer built in.

NOTE: Other electrical tools are listed in the FOS manual on "Electrical Systems."

OIL PRESSURE GAUGE

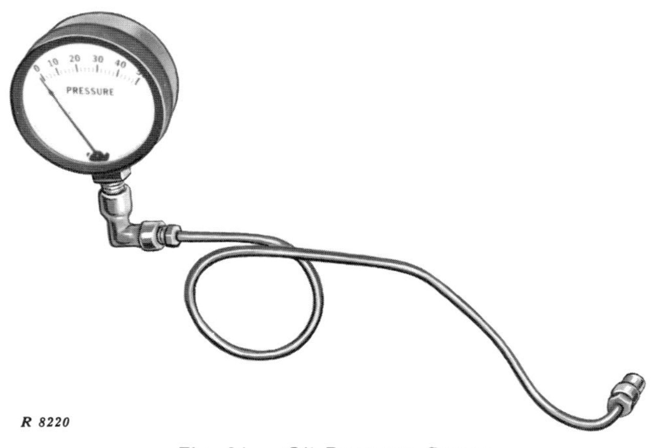

R 8220

Fig. 21 — Oil Pressure Gauge

As most engine lubricating systems now use an oil pressure indicating light, an oil pressure gauge is required to check the system pressure. A typical gauge with fittings is shown in Fig. 21.

 Litho in U.S.A.

THERMOSTAT TESTER

Fig. 22 — Thermostat Tester

A thermostat tester is used to check the temperature at which a thermostat valve starts to open. This is an important check because the proper engine operating temperature is maintained by the thermostat.

RADIATOR AND CAP TESTER

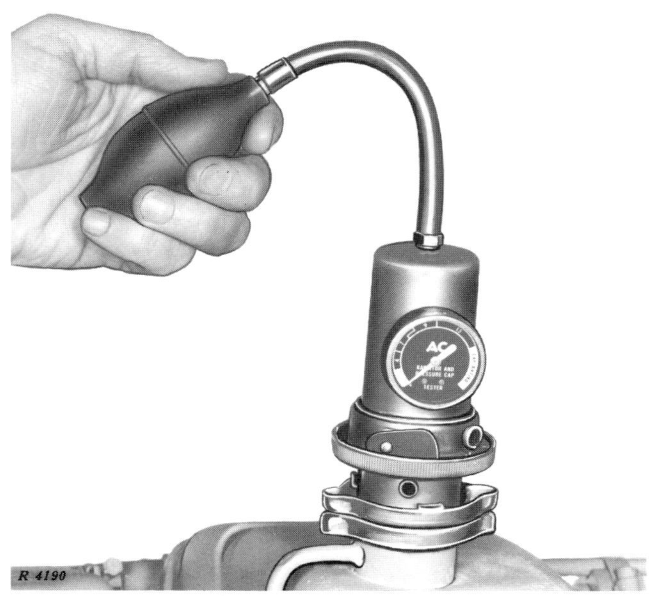

Fig. 23 — Radiator and Cap Tester

This tester is used to check the radiator and radiator cap for leaks and for correct opening pressure. An efficient pressure cooling system must be free from leaks. In Fig. 23, the tester is being used to check the radiator cap.

SERVICING TOOLS

After the engine is tested and diagnosed, service it using the proper tools. Here is a list of the major ones.

VALVE REFACER

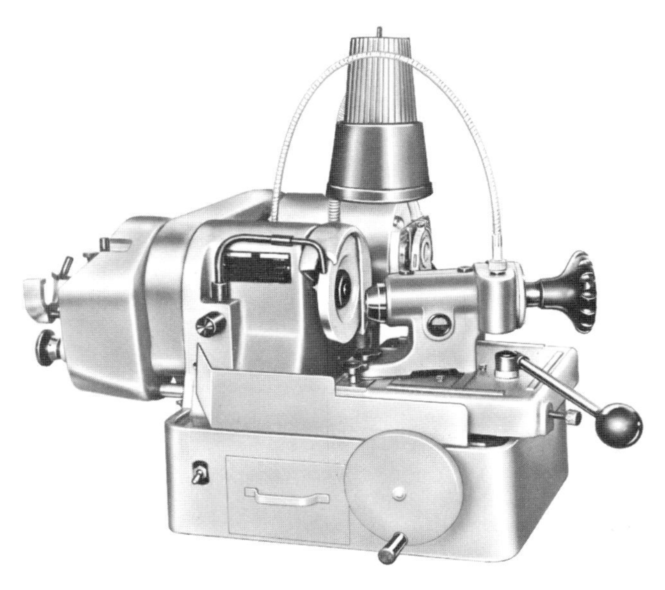

Fig. 24 — Valve Refacer

A valve refacer is used to grind an exact angle on the face of a valve. The accuracy of this machine is critical as the angle of the valve face must match the angle of the valve seat.

VALVE SEAT GRINDER

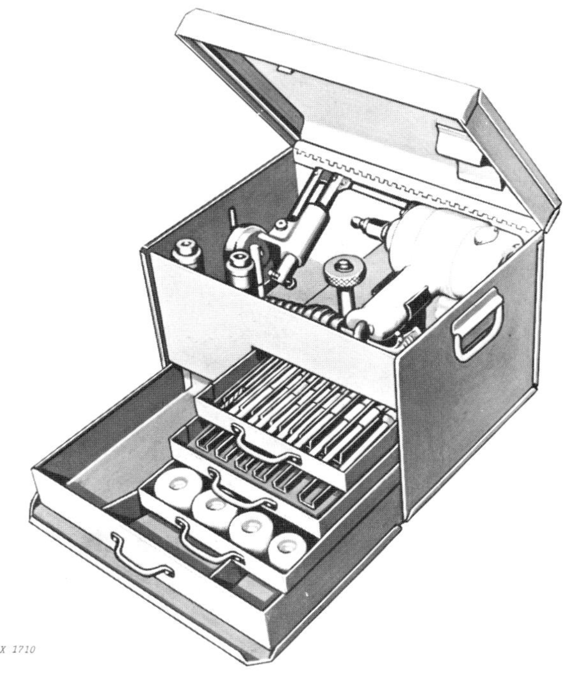

Fig. 25 — Valve Seat Grinder

The valve seat grinder (Fig. 25) is used to reseat the valves. Both rough and finish grinding stones are required.

VALVE SPRING COMPRESSOR

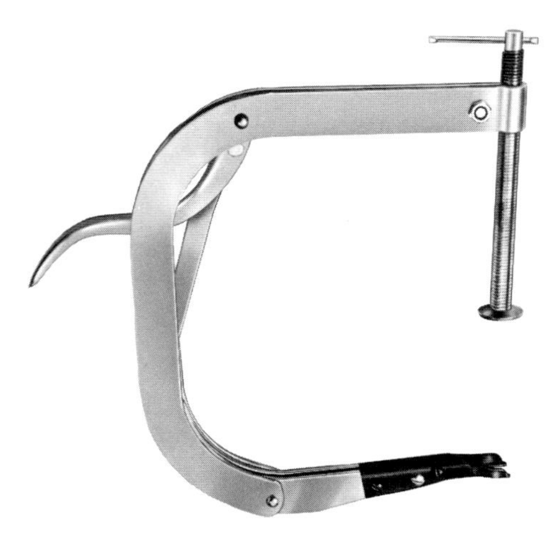

Fig. 26 — Valve Spring Compressor

This tool is used to compress the spring when removing or installing a valve. Various types of spring compressors are available.

VALVE GUIDE REAMER

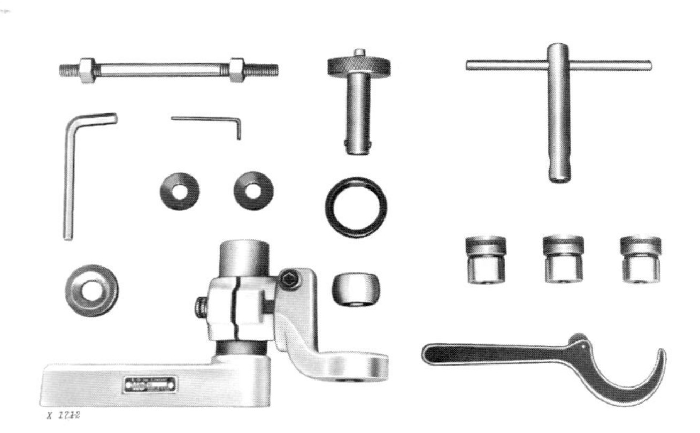

Fig. 27 — Valve Guide Reamer

Use a valve guide reamer to ream out old guides so that new guides can be used. This is also an ideal tool for cleaning carbon from guides.

VALVE GUIDE KNURLING TOOL

A worn valve guide can be re-sized with a valve guide knurling tool. This tool upsets the metal from the sidewalls toward the center of the bore. The bore is then reamed to accept a standard valve.

CYLINDER RIDGE REAMER

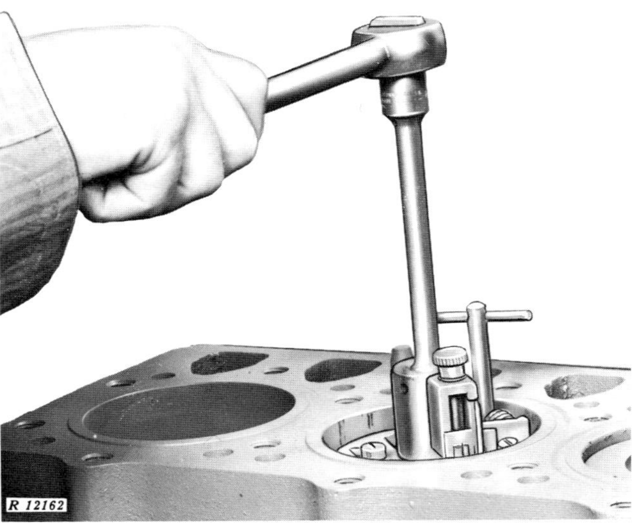

Fig. 28 — Cylinder Ridge Reamer

Use a cylinder ridge reamer to remove the ridges found at the end of the ring travel at the top of a cylinder or liner. Pistons can then be easily and safely removed.

CYLINDER LINER PULLER-INSTALLER

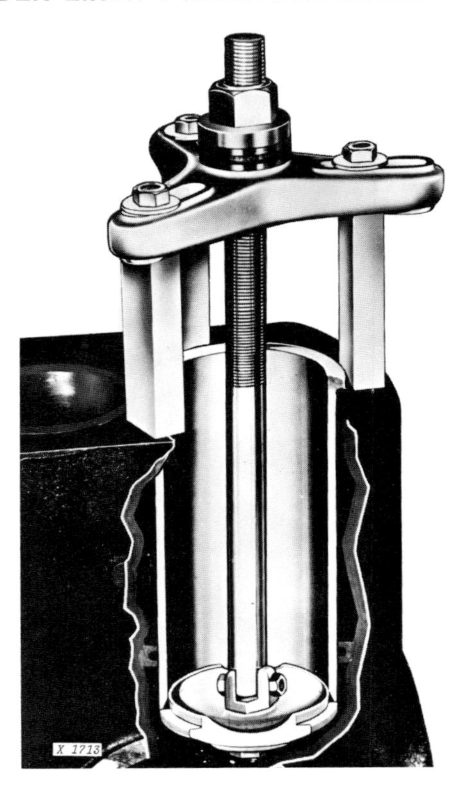

Removal and installation of cylinder liners or sleeves is easier and quicker with a puller-installer (Fig. 29).

Both hydraulic and manual models are available. A great variety of engines can be serviced by using different adapter plates.

CYLINDER REBORING BAR

Fig. 30 — Cylinder Reboring Bar

A reboring bar is used to machine the cylinders of large engines which have integral cylinders.

This boring operation is necessary if oversize pistons are to be used.

A typical reboring machine is shown in Fig. 30.

CYLINDER DEGLAZER

X 1715

Fig. 31 — Cylinder Deglazer

Use a cylinder deglazer to deglaze and finish the cylinder or cylinder liner bore. Stone, pad, and brush type deglazers are available for use with an electric drill. The brush type shown in Fig. 31 does an excellent job.

PISTON RING GROOVE CUTTING TOOL

Fig. 32 — Ring Groove Cutting Tool

Worn piston ring grooves on some pistons can be machined with a ring groove cutting tool. New standard width rings can then be used with flat steel spacers.

PISTON RING GROOVE CLEANING MACHINE

Glass bead cleaning has proven successful in cleaning piston ring grooves. The machine uses air blasting with glass beads and avoids scratching, as with hand scraping. Chemical soaking with a recommended piston cleaner is another good way to clean pistons. See Fig. 91, page 2-40.

PISTON KNURLING TOOL (Automotive Engines)

Fig. 33 — Piston Knurling Tool (Automotive Engines)

A piston knurling tool is used to resize automotive pistons. This tool increases the diameter of a piston by displacing the metal of the skirt thrust faces.

PISTON RING EXPANDER

Fig. 34 — Piston Ring Expander

Use a piston ring expander to remove and install piston rings without damaging them. The expander should limit the ring opening so that rings are not stretched or broken.

PISTON RING COMPRESSOR

Fig. 35 — Piston Ring Compressor

Use a piston ring compressor to compress the rings when installing pistons into the cylinders. This prevents breaking the rings.

PISTON AND ROD ALIGNING TOOL

Fig. 36 — Piston and Rod Aligning Tool

A piston and rod aligning tool is used to check the piston and connecting rod alignment. Bending bars and clamps are provided to correct any bend, twist, or offset in the rod.

CARBURETOR SERVICE TOOLS (Spark-Ignition Engines)

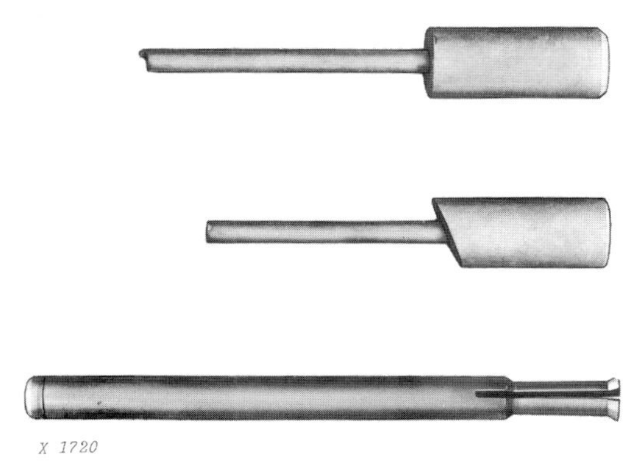

Fig. 37 — Carburetor Service Tools

Special tools are available to service most gasoline and LP-gas carburetors. Tools are usually designed for one particular carburetor, and some carburetors require more tools than others.

INJECTION NOZZLE REMOVAL TOOLS (Diesel)

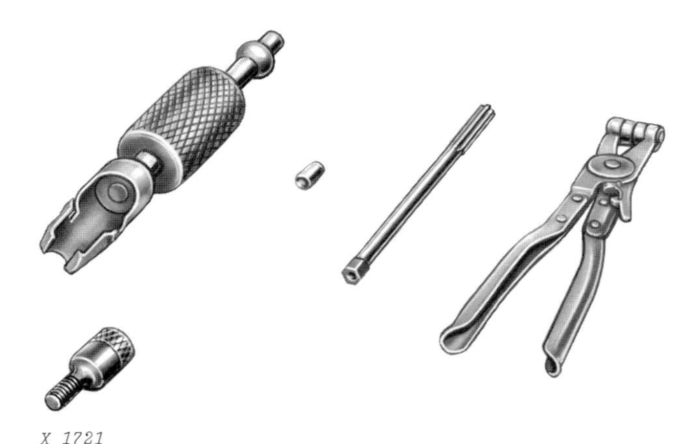

Fig. 38 — Injection Nozzle Removal Tools (Diesel)

Removal and installation kits are available for some injection nozzles. A typical kit includes hose clamp pliers, nozzle puller, bore cleaning tool, and a guide.

INJECTION NOZZLE CLEANING KIT (Diesel)

Fig. 39 — Injection Nozzle Cleaning Kit (Diesel)

Cleaning kits are usually designed to service just one particular make of nozzle. These kits contain such items as cleaning wires, brushes, drills, and lapping compounds.

Fig. 40 — Special Tools For Each Engine Are Also Needed To Completely Equip The Service Shop

VACUUM GAUGE

X 1707

Fig. 41 — Vacuum Gauge

A vacuum gauge, calibrated in inches of mercury, will help determine the condition of an engine. Various malfunctions can be identified by how they cause the gauge to react. The gauge shown in Fig. 41 has an auxiliary scale to check fuel pump pressure.

SUMMARY

Most of the tools we have listed can be used for any engine. However, there is another category of tools that we have not covered—the **Special Tools** designed for each engine.

Be sure that your shop has all the special tools needed to service the models of engines you normally work on. These tools are listed in the Technical Manual for each engine.

The tools we have shown in this chapter are basic to almost all engines. Some are simple tools, others complex testers. Yet each one, regardless of size or cost, is designed to make your job easier and faster.

TEST YOURSELF

QUESTIONS

1. What are the two major types of engine tools?

2. Which type of tool is normally used first?

3. What other category of tools is not covered in this chapter?

(Answers in back of text.)

 Litho in U.S.A.

DIAGNOSIS AND TESTING OF ENGINES / CHAPTER 11

INTRODUCTION

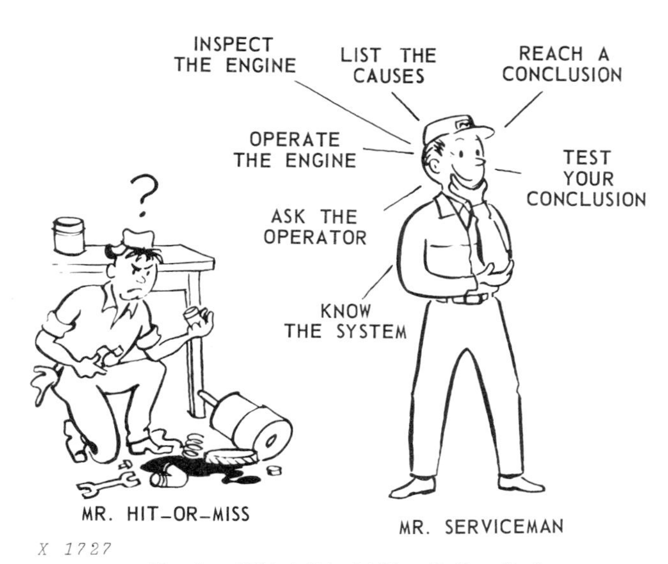

Fig. 1 — Which Would You Rather Be?

Mr. Hit-or-Miss or Mr. Serviceman—which would you rather be? Both have the title of Serviceman but don't be fooled by that.

Mr. Hit-or-Miss is a parts exchanger who dives into an engine and starts replacing parts helter-skelter until he finds the trouble—*maybe*—after wasting a lot of the customer's time and money.

Mr. Serviceman starts out by using his brain. He gets all the facts and examines them until he has pinpointed the trouble. Then he checks out his diagnosis by testing it and *only then* does he start replacing parts.

Mr. Hit-or-Miss is fast becoming a man of the past. What dealer can afford to keep him around at today's prices?

With the complex systems of today, diagnosis and testing by Mr. Serviceman is the only way.

SEVEN BASIC STEPS

A good program of diagnosis and testing has seven basic steps:

1. **Know the System**
2. **Ask the Operator**
3. **Inspect the Engine**
4. **Operate the Engine**
5. **List the Possible Causes**
6. **Reach a Conclusion**
7. **Test Your Conclusion**

Let's see what these steps mean.

1. KNOW THE SYSTEM

Fig. 2 — Know The System

In other words, "Do your homework." Study the engine Technical Manuals. Know how the engine works, how it can fail, and the three basic needs—*fuel-air mixture, compression,* and *ignition.*

Keep up with the service bulletins. Read them and then file in the proper place. The solution to your problem may be in this month's bulletin.

You can be prepared for any problem by knowing the engine.

2. ASK THE OPERATOR

Fig. 3 — Ask the Operator

A good reporter gets the full story from a witness —the operator.

He can tell you how the engine acted when it started to fail; what was unusual about it.

What work was the engine doing when the trouble occurred? Was the trouble erratic or constant?

What did the operator do after the trouble? Did he attempt to repair it himself?

Ask how the engine is used and when it was serviced. Many problems can be traced to poor maintenance or abuse of the engine.

Be tactful but get the full story from the operator.

3. INSPECT THE ENGINE

Go over the engine and inspect for all the things listed in Fig. 4.

Use your eyes, ears, and nose to spot any "tips" to the trouble.

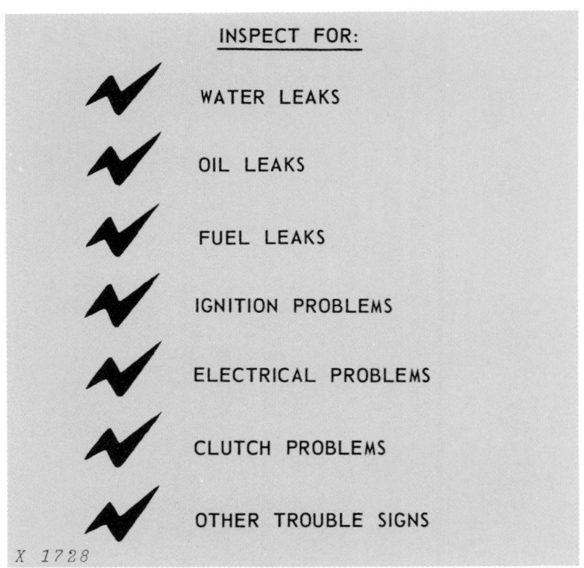

Fig. 4 — Checklist for Inspecting the Engine

- *Look for* **water leaks** *at the radiator, water pump, hoses, and around the cylinder head gasket.*

- *Check for* **oil leaks** *at the oil pan, drain plugs, and gaskets. Also inspect around crankshaft bores, oil seals, and tappet cover. Look inside the flywheel housing for signs of oil.*

- *Look for* **fuel leaks** *at the tank, lines, filters, and pumps. Also check for restrictions or evidence of water in the fuel.*

- *Inspect for* **ignition problems** *such as: Loose or bare wires, cracked distributor cap, shorted wires at plugs.*

- *Check for* **electrical problems** *like these: battery corroded, sulfated, or low on water; battery cables loose or corroded; low charge in battery.*

- *Inspect for* **clutch problems** *which might affect the engine. Free travel should be adequate.*

- *Look for* **other trouble signs** *which could lead to future problems if not corrected early.*

In general, look for anything unusual. *Keep a list of all the trouble signs.*

 Litho in U.S.A.

4. OPERATE THE ENGINE

Fig. 5 — Operate The Engine

If the engine can be run, start it and warm it up. Then run it through its paces. Don't completely trust the operator's story—check it yourself.

Test the engine on a dynamometer. This is the only way to get a full picture of the engine's condition. For details, see "Testing The Engine."

If a dynamometer is not available, look, smell, and listen for engine problems:

- *Are the gauges reading normal?*
- *Hear any funny sounds? Where? At what speeds?*
- *Smell anything? Any signs of unusual exhaust smoke? Are the breathers smoking?*
- *How do the engine controls work?*
- *How is the power under load?*
- *Does the engine idle okay?*

Use your common sense to find out how the engine is operating.

5. LIST THE POSSIBLE CAUSES

Now you are ready to make a list of the possible causes of the engine's troubles.

Fig. 6 — List The Possible Causes

What were the signs you discovered while inspecting and operating the engine?

- *Did the engine lack power?*
- *Any smoke from the crankcase vent?*
- *Did the engine run too hot or too cold?*
- *How was the oil pressure?*

Which of the signs tell you the most likely cause, which is second, etc.?

Are there any other possibilities? (One failure often leads to another.)

6. REACH A CONCLUSION

Look over your list of possible causes and decide which are most likely and which are easiest to verify.

Use the Trouble Shooting Charts at the end of this chapter as a guide.

Reach your decision on the leading causes and plan to check them first—after making the easy checks.

Fig. 7 — Reach A Conclusion

7. TEST YOUR CONCLUSION

Fig. 8 — Test Your Conclusion

Before you start repairing the system, test your conclusions to see if they are correct.

Many of the items on your list can be verified without further testing.

Maybe you can isolate the problem to one system of the engine—lubrication, cooling, etc.

But the location within the system may be harder to find. This is where testing tools can help.

The next part of this chapter will tell how to test the engine and locate failures.

But first let's repeat the seven rules for good trouble shooting:

1. **Know the System**
2. **Ask the Operator**
3. **Inspect the Engine**
4. **Operate the Engine**
5. **List the Possible Causes**
6. **Reach a Conclusion**
7. **Test Your Conclusion**

TESTING THE ENGINE

The use of engine test equipment is most effective once the failure has been isolated to one system or area of the engine.

Chapter 10 lists all the testing tools needed for normal engine trouble shooting. Other chapters of this manual show how to use most of this equipment.

DYNAMOMETER TEST

A complete dynamometer test is the best way to trouble shoot the engine.

Remember, however, the engine must be reasonably well-tuned to pass the dynamometer tests.

If the engine was tested on a dynamometer, the following tests should have been made and the results listed:

1. *Engine horsepower*
2. *Exhaust smoke analysis*
3. *Fuel consumption*
4. *Crankcase blow-by*
5. *Air cleaner restriction*
6. *Oil pressure*
7. *Battery condition*
8. *Alternator or generator output*
9. *Clutch operation*

 Litho in U.S.A.

Fig. 9 — Dynamometer Test

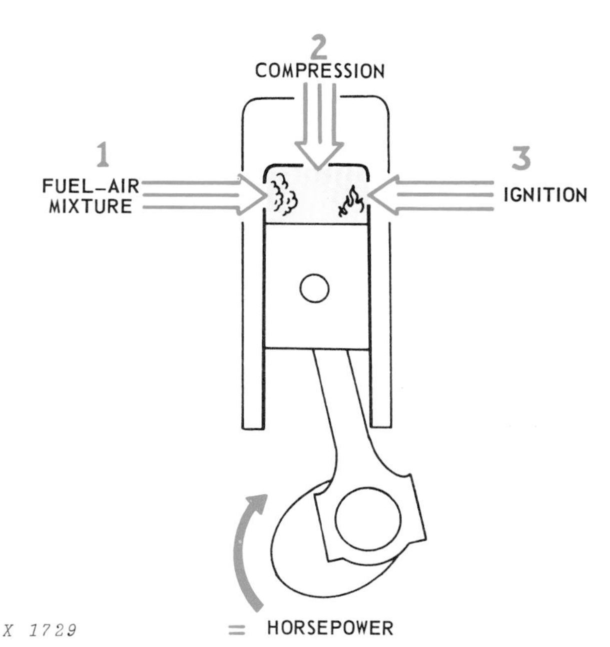

Fig. 10 — Three Basic Things Are Needed To Produce Engine Horsepower

Refer to Chapter 10 for the testing equipment used in the above tests. Chapter 12 gives a general guide to the tests.

Compare the test results with the correct readings given in the engine Technical Manual.

ENGINE HORSEPOWER

Horsepower is the fundamental measure of engine efficiency.

Three basic things are needed for the engine to produce horsepower:

- **Fuel-air mixture**
- **Compression**
- **Ignition**

A good supply of AIR is essential to engine combustion. The air cleaner must be kept clean and the air intake system kept free of restrictions.

There must be a good supply of FUEL. The tank, lines, filters, and pumps must be open and clean.

COMPRESSION must be adequate. Low compression can be from bad valves, leaking head gaskets, or blow-by at the pistons. In a diesel engine, compression is extra vital because the heat of compression rather than a spark is what ignites the fuel.

IGNITION must be adequate and properly timed. In spark-ignition engines, the battery and ignition circuit are vital in producing a strong and timely spark for ignition. In diesel engines, the injection pump must be timed properly.

Unless all three—*fuel-air, compression, and ignition*— are doing their job in the right sequence, engine horsepower will be low.

The basic dynamometer test will tell you whether the engine is low on horsepower. Further tests will isolate the cause to one of the three major areas.

MECHANICAL PROBLEMS

Beyond the problems of engine efficiency, there are many *mechanical* things which can cause an engine to fail.

Here are a few of the basic problems and their likely results:

Loose Bearings—can increase oil consumption, create noise, cause eventual failure of bearings and damage to crankshaft.

Worn Valve Guides—can increase oil consumption, eventually cause valves to fail.

Worn Pistons and Rings—also increase oil consumption, cause loss of compression and power.

Worn Camshaft Lobes—loss of power because valves open too late and close too early.

Litho in U.S.A.

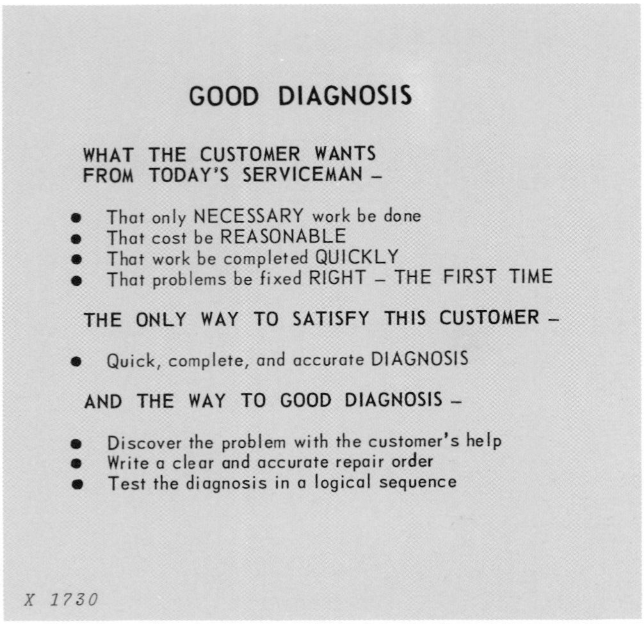

GOOD DIAGNOSIS

WHAT THE CUSTOMER WANTS
FROM TODAY'S SERVICEMAN –

- That only NECESSARY work be done
- That cost be REASONABLE
- That work be completed QUICKLY
- That problems be fixed RIGHT – THE FIRST TIME

THE ONLY WAY TO SATISFY THIS CUSTOMER –

- Quick, complete, and accurate DIAGNOSIS

AND THE WAY TO GOOD DIAGNOSIS –

- Discover the problem with the customer's help
- Write a clear and accurate repair order
- Test the diagnosis in a logical sequence

X 1730

Fig. 11 — The Serviceman's Guide To Good Diagnosis

SUMMARY: TESTING THE ENGINE

The tests we have given you are only basic guidelines.

Once you start testing actual engines, use the engine Technical Manual for detailed tests and test results.

And remember that the best testing tools have no value unless the man at the controls knows how to interpret the results.

TROUBLE SHOOTING CHARTS

Use the charts on the following pages to help in listing all the possible causes of trouble when you diagnose and test an engine.

Once you have located the cause, refer to the chart again for the possible remedy.

Some of the problems are discussed in detail following each chart.

Other chapters of this manual give more details on diagnosing a particular component.

Also use the Technical Manual for each engine for specific tests and specifications.

1. ENGINE HARD TO START OR WILL NOT START

No Fuel or Improper Fuel

Fill tank. If wrong fuel, drain and fill with proper fuel.

Water or Dirt in Fuel or Dirty Filters

Check out fuel supply. Replace or clean filters.

Air in Fuel System (Diesel)

Bleed air from the system.

Low Cranking Speed

Charge or replace battery, or service starter as necessary.

Poor Timing

Check distributor or injection pump timing.

Defective Coil or Condenser (Spark-Ignition)

Replace coil or condenser.

Pitted or Burned Distributor Points

Clean or replace points.

Cracked or Eroded Distributor Rotor (Spark-Ignition)

Replace rotor.

Distributor Wires Loose or Installed in Wrong Order (Spark-Ignition)

Push wires into sockets. Install in correct firing order.

Defective or Wrong Spark Plugs (Spark-Ignition)

Replace Plugs.

Poor Nozzle Operation (Diesel)

Clean, repair, adjust, or replace.

Liquid Fuel in Lines (LP-Gas)

Always turn on the vapor valve when starting the engine.

Too-Heavy Oil in Air Cleaner (LP-Gas)

Use correct weight of oil.

 Litho in U.S.A.

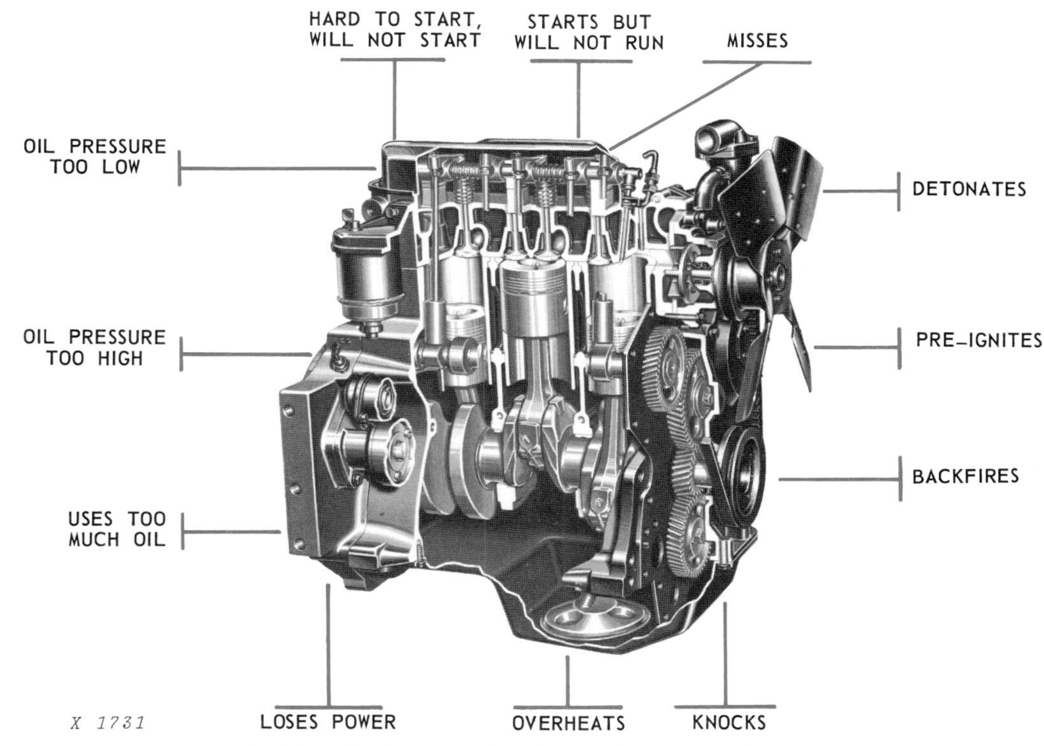

Fig. 12 — The Major Engine Trouble Shooting Problems

Fuel Problems

In *spark-ignition engines,* the **volatility** of the fuel has much to do with the starting ability of the engine.

NOTE: Volatility is what causes the fuel to vaporize for combustion.

Liquid gasoline or LP-gas must be in vapor form to mix with the air when it is taken into the engine.

If the flash point of the fuel is too low, not enough vapor is formed to mix with the air for easy ignition.

NOTE: Flash point is the temperature at which vapor from the fuel will ignite.

In *diesel engines,* since ignition is from the heat of compression, the **flash point** of the fuel affects the ability of the engine to start.

Generally speaking, the lower the grade of diesel fuel, the higher the flash point. This means that as the grade of fuel is lowered, the heat created in the engine must be hotter. This can contribute to hard starting.

Low Cranking Speed

Low cranking speed is a result of 1) low charge in the battery, 2) a defective battery or terminal connection, or 3) binding in the starter or engine.

In spark-ignition engines, when any of these conditions exist, so much of the battery voltage is used by the starter in cranking the engine that the remaining voltage is not sufficient to create an ignition spark.

If, while cranking the engine, the voltage of a 12-volt battery falls below 9 volts, the battery is either defective or needs recharging.

In diesel engines, when the battery is low or defective, the starter cannot turn the engine fast enough to create enough heat within the cylinders to provide ignition.

Also, worn starter bearings can allow the armature to drag on the field poles. This results in low cranking speed and more voltage drop at the battery.

Any binding within the engine will produce the same effect as starter drag.

It is also possible that the battery cables are too small or the terminals are corroded. In this case, the resistance of the cable is too high to carry the amount of current required to crank the engine at the proper speed.

Defective Coil or Condenser (Spark-Ignition)

When either the coil or condenser is defective, not enough spark will be created to start the engine.

 Litho in U.S.A.

Pitted or Burned Distributor Points (Spark-Ignition)

When the distributor points become pitted or burned, their ability to conduct current is impaired. This results in a poor spark from the ignition coil.

Poor Nozzle Operation (Diesel)

The nozzles must be in good operating condition. If they are partially clogged or if the cracking pressure is low, much of the fuel is injected in a solid stream. This will cause hard starting.

Too-Heavy Oil in Air Cleaner (LP-Gas)

Too-heavy oil in the air cleaner restricts the flow of air through the air cleaner, resulting in a choking action at the carburetor. This, in turn, creates a fuel-air mixture in the engine which is too rich to burn.

2. ENGINE STARTS BUT WILL NOT RUN

Fuel Problems—Dirt, Air Restrictions, Or Clogged Filters

Check fuel supply, bleed system (diesel), check for line restrictions, or clean or replace filters and screens.

Carburetor Idle Fuel Needs Adjustment

Adjust correctly.

Defective Coil or Condenser (Spark-Ignition)

Replace.

Defective Ignition Resistor or Key Switch (Spark-Ignition)

Replace resistor or key switch.

Fuel Problems

When the flow of fuel to the engine is restricted, not enough fuel can get to the engine to keep it running. See Group 1 for details.

Ignition Coil and Condenser (Spark-Ignition)

In some instances a coil or condenser will provide a spark to start the engine. Then, after the engine starts, due to a loose connection or internal short, it will not keep on running. Also a coil or condenser may function when cold and then fail when heated.

Defective Ignition Resistor Key Switch (Spark-Ignition)

When the key switch is turned to crank the engine, full battery voltage is impressed on the ignition coil for easier starting. When the key switch is released, the ignition current flows through a resistor, limiting the coil voltage to about 6 volts. So, if the resistor is burned out or has a loose connection, the engine will not continue to run when the key switch is released.

3. ENGINE MISSES

Water or Dirt in Fuel

Drain and refill with clean fuel.

Gasoline in Diesel Fuel

Drain and refill with proper fuel.

Dirty Spark Plugs or Faulty Cables (Spark-Ignition)

Clean or replace plugs. Replace cables.

Cracked Distributor Cap (Spark-Ignition)

Replace Cap.

Carburetor Not Adjusted Correctly (Spark-Ignition)

Adjust carburetor.

Air in Fuel System (Diesel)

Bleed the system.

Poor Nozzle Operation (Diesel)

Clean and check nozzle spray pattern.

Faulty Injection Pump (Diesel)

Check and calibrate the fuel injection pump.

Nozzles Not Seated Properly in the Cylinder Head (Diesel)

Reposition nozzles and tighten retaining screws to specified torque.

Gasoline in Diesel Fuel

Gasoline in diesel fuel can cause the engine to miss. The reason is this: The diesel fuel flowing through the nozzle provides lubrication for the nozzle valve. Gasoline does not have this lubricating quality and the valve sticks, causing misfiring. In many cases the nozzle valve will be scored and need replacement.

Air in Fuel System (Diesel)

Air in the diesel fuel system can cause the engine to miss. Fuel cannot be compressed while air can be. Consequently, if there is a bubble of air in the system, the air will compress and no fuel will be injected.

Nozzles Not Seated Properly (Diesel)

It is very important that the nozzles be correctly installed. They should seat squarely on the seating surface and the retaining clamp cap screws or nuts be tightened to the specified torque.

If the nozzle clamp is too loose, the nozzle will leak. This causes a build-up of carbon around the nozzle, making it very difficult to remove. In addition, the escape of burning gases will cause the nozzle to overheat, and the nozzle valve to stick, causing the cylinder to misfire.

4. ENGINE DETONATES (Gasoline)

Wrong Type of Fuel

Use fuel of correct octane.

Detonation of Gasoline Engine

Detonation in an engine is usually recognized by sharp "pinging" in the engine cylinders and erratic firing. The engine appears to be working against itself.

In gasoline engines, detonation is usually caused by the use of fuel with too-low octane rating.

NOTE: The **octane rating** *of a fuel is its ability to avoid detonation or "self ignition". The higher the octane rating, the more the mixture can be compressed without detonation when it burns.*

Detonation is "an explosion with sudden violence". When the compressed fuel mixture in the cylinder is ignited, instead of burning evenly across the combustion chamber, the entire charge literally explodes. This is partially due to the heat of compression. That is to say, when the mixture starts to burn and the pressure starts to rise, the entire charge ignites. This action is very harmful to the engine.

5. ENGINE PRE-IGNITES (Gasoline)

Distributor Timed Too Early

Retime distributor.

Distributor Advance Mechanism Stuck

Free up and test.

Carbon Particles in Cylinder

Remove carbon.

Faulty Spark Plugs or Spark Plug Heat Range Too High

Install proper plugs.

Pre-Ignition Of Gasoline Engines

Pre-ignition is the ignition of the fuel mixture before the proper time. Its effect on the engine is similar to that of detonation.

Pre-ignition can be caused by incandescent particles of carbon in the cylinder which will ignite the mixture as the heat rises due to compression.

If the distributor advance mechanism sticks in the advanced position, at certain engine speeds the spark will occur too early.

Pre-ignition can be caused by spark plugs which are eroded at the electrodes or by plugs with too high a heat range. In either case, the electrodes become red hot and the action is the same as that caused by incandescent carbon particles.

6. ENGINE BACKFIRES (Spark-Ignition)

Spark Plug Cables Installed Wrong

Install in correct firing order.

Carburetor Mixture Too Lean

Adjust carburetor for correct mixture.

Backfiring Of Gasoline And LP-Gas Engines

When the spark plug wires are incorrectly installed, the spark will not occur in the cylinder at the right time. This can cause backfiring.

Engine backfiring can also be caused by a too-lean carburetor mixture. A lean fuel mixture is a slow-burning mixture and may still be burning when the intake valve opens and admits a new charge.

7. ENGINE KNOCKS

Improper Distributor or Injection Pump Timing

Check and time correctly.

Worn Engine Bearings or Bushings

Replace.

Excessive Crankshaft Endplay

Adjust to specification.

Loose Bearing Caps

Tighten caps.

Foreign Matter in the Cylinder

Remove.

Knocking Of Engines

When an engine knocks, there is one of two causes:

1) Mechanical Interference Of Parts. The parts are striking each other due to loose parts, worn bearings, or a loose bolt or other object loose in the cylinder. The remedy is to repair the engine at once.

2) Improper Adjustment Of Timing, Fueling, Or Poor Fuels. If the ignition or injection timing is bad, the engine gets "out of rhythm" and so knocks. Using the wrong fuels can also cause knocking, especially with low-octane gasoline. The remedy is either using proper fuels or readjusting the timing or fuel settings.

8. ENGINE OVERHEATS

Defective Radiator Cap

Install new gasket or replace the cap.

Radiator Fins Bent or Plugged

Straighten fins or clean out dirt.

Defective Thermostat

Replace.

Loss of Coolant

Check for leaks and correct.

Loose Fan Belt

Adjust.

Cooling System Limed Up

Use good scale remover to clean, then flush.

Overloaded Engine

Adjust load.

Faulty Engine Timing

Time distributor or injection pump correctly.

Distributor Advance Mechanism Stuck (Spark-Ignition)

Free up and lubricate.

Engine Low on Oil

Add oil to the proper level or change oil.

Wrong Type of Fuel

Use correct fuel.

Overheating Of Engines

RADIATOR CAP

The radiator cap must be in good condition so that it does not leak. The pressure cap allows pressure to build up in the cooling system and this causes the boiling point of the coolant to rise.

Using the pressure cap thus prevents the coolant from boiling away in hot weather or during heavy loads.

RADIATOR

Foreign matter between the fins of the radiator core or bent fins can cause overheating.

THERMOSTAT

If the thermostat develops a leak or sticks closed, it stops the circulation of coolant through the radiator core. This will cause overheating.

ENGINE OVERLOADED

Overloading an engine often causes the peak firing pressure in the cylinders to be too high, creating more heat. The overload causes the engine speed to slow and the piston does not move in the cylinder as fast as it does under normal load. This can also contribute to overheating.

FAULTY ENGINE TIMING

When the ignition is timed too late, a normal engine load becomes an overload. This is because the engine does not develop as much power. Even if the load is light, the engine may still run hot, due to the late firing.

DISTRIBUTOR ADVANCE MECHANISM STUCK (Spark-Ignition)

If the distributor advance mechanism should stick in the retarded position, it has the same effect on the operation of the engine as late timing.

ENGINE LOW ON OIL

One of the functions of the oil in the engine is to help carry off the heat created in the cylinders. If the oil level is too low, if the oil pressure is too low, or even if the oil is too heavy, not enough oil will be circulated to carry off the heat of combustion. This can lead to overheating.

WRONG TYPE OF FUEL

Using the wrong type of fuel in an engine can cause overheating. As the fuel detonates or does not burn efficiently, loss of engine power also results.

9. LACK OF POWER

Air Cleaner Dirty or Otherwise Obstructed

Clean and check main element for restriction.

Restricted Air Flow in Intake System

Clean intake hoses and tubes.

Restriction in Fuel Lines or Filters

Disconnect line at carburetor or injection pump and check it. Remove any restriction. Also check filters.

Wrong Type of Fuel

Use correct type.

Frost at Fuel-Lock Strainer (LP-Gas)

Clean strainer.

Governor Binds (Spark-Ignition)

Free up and lubricate. Adjust if necessary.

Valve Failure

Recondition.

Incorrect Valve Tappet Clearance

Adjust.

Low Engine Speed

Adjust.

Crankcase Oil Too Heavy

Use correct weight of oil.

Low Compression

Check for leaky valves or worn or stuck piston rings. Replace or recondition.

Improper Hitching or Belting of Machine

Adjust or correct as required.

Incorrect Timing

Check and time correctly.

Faulty Carburetor, Float Level Too Low, Plugged Orifices, or Wrong Fuel Adjustment

Clean and repair.

Wrong Spark Plugs (Spark-Ignition)

Replace with proper plugs.

Distributor Points Burned (Spark-Ignition)

Replace and adjust.

Incorrect Camshaft Timing

Retime.

Low Operating Temperature

Check thermostat.

Faulty Injection Pump Delivery (Diesel)

Service pump.

Lack Of Power

AIR CLEANER

If the air cleaner is dirty or the air passages obstructed the engine will not receive enough air to develop its horsepower.

FUEL LINES

When the flow of fuel through the fuel lines is restricted the engine cannot get enough fuel to develop its power.

FROST AT FUEL-LOCK STRAINER (LP-GAS)

Frost at the fuel-lock strainer restricts the flow of gas, resulting in a power loss.

GOVERNOR BINDS (Spark-Ignition)

When there is binding in the governor, the action will be slow. Engine speed is pulled down before the governor can react to increase the fuel supply to carry the load.

WRONG TYPE OF FUEL

See Group 4 above.

LOW COMPRESSION

Leaky valves or blow-by at the pistons result in low compression. As a result, the burning of fuel in the cylinders cannot produce enough pressure to develop full power.

VALVE CLEARANCE

Valve adjustment can affect the compression pressure. If the adjustment is too tight the valve cannot entirely close, allowing leakage.

If the valve clearance is too great the valve opens late and closes early. This results in it not being open long enough for complete intake of air or air-fuel mixture of complete expulsion of exhaust gases.

TOO-HEAVY CRANKCASE OIL

Many times, when an engine is consuming too much oil, the operator will change to a heavier weight. This can affect the power output of the engine due to the increased drag on the moving parts.

IMPROPER HITCHING OR BELTING

Improper hitching or adjustment of equipment behind a tractor or other machine can result in an apparent loss of engine power. Actually the improper hitching really results in increasing the load on the machine.

Raising or lowering the hitch point can change this load.

For example, by correctly hitching a tool to the drawbar, a tractor may be able to operate in a higher transmission gear range.

Improper belting can also reduce the ability of the engine to pull its load.

If the driven pulley is too large, the engine can handle the load easily enough but the driven machine speed will be too slow.

If the driven pulley is too small, the machine speed will be too fast and much more engine power will be required to operate it.

10. ENGINE USES TOO MUCH OIL

Crankcase Oil Too Light

Use correct weight of oil.

Worn Pistons and Rings

Recondition.

Worn Valve Guides or Stem Oil Seals

Replace.

Loose Connecting Rod Bearings

Replace bearings.

External Oil Leaks

Eliminate leaks.

Internal Oil Leaks

Locate and correct.

Oil Pressure Too High

Adjust to specifications.

Engine Speed Too High

Adjust to specifications.

Crankcase Ventilator Pump Not Working

Repair or replace.

Restricted Air Intake or Breather

Remove the restriction.

Restricted Oil Return Passage From Valve Cover

Remove the restriction.

Engine Using Too Much Oil

PISTONS AND RINGS

Worn pistons and rings can cause an engine to use too much oil. Reconditioning is normally required.

However, many other conditions can cause excessive oil consumption, as given below.

VALVE GUIDES

Worn valve guides can be a cause of oil consumption. Every time the piston goes down in the cylinder it creates a partial vacuum. To equalize this vacuum, air tries to enter the cylinder. If the valve guides are worn, this leaves an opening for the air to enter. As it enters it carries oil from around the valve stem with it.

VALVE STEM OIL SEALS

Some engines are equipped with valve stem oil seals, usually on the intake valves. These seals prevent loss of oil at this point. If worn, these seals can make the engine use oil.

LOOSE CONNECTING ROD BEARINGS

When the bearings are worn, more oil is pumped through them. As a result, more oil than needed is thrown into the cylinder from the crankshaft. The extra oil escapes past the piston and rings.

Much the same action occurs when the engine oil pressure is too high.

LEAKAGE

External leaks can usually be seen and corrected. These include leaks at the crankshaft rear bearing, crankshaft front oil seal, valve cover gasket, and cylinder head gasket.

Internal leaks are not so easily detected. Generally, the engine must be disassembled while carefully checking points at which leakage could occur.

When an internal leak is causing oil consumption, it will usually be found that the leak is located so that the escaping oil strikes the crankshaft. This results in too much oil being thrown into the cylinders.

CRANKCASE VENTILATING PUMP

Most ventilating pumps draw in air, mix it with oil and force the mixture into the engine crankcase, pushing out vapors.

When the pump fails, the oil is drawn into the air intake and then on into the engine.

To find if the ventilating pump is at fault, remove the air intake pipe. If the pump is defective, the inside of the pipe will be oily.

Another method of checking the pump is to remove the pipe between the pump and the air intake pipe. Be sure to cap both openings.

Then fill the engine crankcase with a measured amount of oil and operate the engine under load for four or five hours. Drain the oil and measure it. If too much oil is lost, the engine is using oil and the vent pump is not the cause.

RESTRICTED AIR INTAKE SYSTEM

When the air flow into the engine is restricted, there is more vacuum or suction in the cylinders. This results in increased suction around the valve stems and pistons and rings and can cause excessive oil consumption.

RESTRICTED OIL RETURN PASSAGE FROM VALVE COVER

When this passage is restricted, the oil collects in the valve tappet area. Much of this oil enters the engine around the valve stems.

11. OIL PRESSURE TOO HIGH

Stuck Relief Valve

Free up valve.

Oil Pressure Too High

This condition rarely occurs unless it has been deliberately set too high.

However, if the relief valve sticks closed, the oil pressure will be too high.

Other possible causes could be a defective pressure gauge or sending unit.

12. OIL PRESSURE TOO LOW

Worn Bearings

Replace engine bearings.

Poor Relief Valve Seating

Replace valve, valve seat or both.

Too-Light Oil

Use correct weight of oil.

Worn Oil Pump

Repair.

Engine Low on Oil

Fill to proper level.

Loose Connections or Leaking Seals at Oil Filter, Pump, or Oil Cooler

Tighten connections or replace seals.

Oil Pressure Too Low

Worn bearings, poor relief valve action, loose connections, or leaking seals can all contribute to low oil pressure.

Any of these items allow the escape of so much oil that the pump cannot keep up the pressure.

WORN OIL PUMP

When the oil pump becomes worn, the oil tends to recirculate within the pump, preventing a pressure build-up.

TOO-LIGHT OIL

Oil is actually composed of many tiny globules. The heavier the oil, the larger the globules.

Therefore, a lighter oil with smaller globules can flow more freely through the bearings and other openings.

While lighter oil provides better lubrication, using *too* light an oil can contribute to high oil consumption.

However, do not attempt to eliminate excessive oil consumption by using a heavier oil than recommended.

TEST YOURSELF

QUESTIONS

1. List the seven basic steps for good trouble shooting.

2. During which of the seven steps should you begin replacing parts?

3. What test is the most comprehensive way to tell the condition of the engine?

4. What three basic things are needed for the engine to run?

(Answers in back of text.)

ENGINE TUNE-UP / CHAPTER 12

INTRODUCTION

Fig. 1 — Tune-Up Is Making Checks And Minor Adjustments

WHAT IS TUNE-UP?

Tune-up is the process of making checks and minor adjustments to improve the operation of the engine.

Tune-up is also **preventive maintenance.** Troubles can be caught early and prevented by checking out the engine before it actually fails.

WHEN SHOULD AN ENGINE BE TUNED?

Regularly! The intervals for tune-up may vary from 500 to 1000 hours or seasonally, depending upon the operating conditions. But **regularity** is the key to maintaining the engine so that major problems are prevented.

A badly worn engine cannot be tuned up. This is why the engine must first be tested to see if:

1) A tune-up will restore it, or

2) Major overhaul is needed.

Let's go through this visual inspection first and then we'll see how tune-up follows it.

VISUAL INSPECTION

By inspecting the engine before you tune it, you can learn a lot about its general condition.

Check out the following items:

1. OIL AND WATER LEAKAGE

Inspect the engine for any oil or water leaks. If the engine has been using too much oil, this often means an external oil leak. If the engine overheats, look for leaks in the cooling system.

2. ELECTRICAL SYSTEM

Inspect the **battery** for corrosion, cracked case, or leaks at the cell covers.

Remove the cell caps and examine the tops of the battery plates. If they are covered with a chalky deposit, this means one of three things:

1) Electrolyte level has been too low.

2) Battery charge has been too low, causing sulfation.

3) Battery was charged at too high a rate, boiling out water.

Fig. 2 — Visual Inspection Tells You Whether To Tune It Or Overhaul It

Litho in U.S.A.

Any of these conditions can reduce the life of the battery. If they have gone too far, the battery must be replaced.

Check the **battery cables and connections** for damage and looseness.

Be sure the cables are the right size. Many complaints of poor starting can be traced to battery cables that are too small.

To check for this, operate the starter with the engine cold. If the battery cable gets hot, the cable is probably too small.

Inspect the **wiring harnesses.** If they are too oil-soaked, frayed, or corroded, replace them.

On spark-ignition engines, check the **distributor** for a cracked cap, excessive grease, or other damage.

Check the operation of the alternator or generator **indicator light.** It should light when the starter switch is turned on.

Failure to light can be due to a burned-out bulb, an incomplete circuit, or the alternator or generator is not producing current. (Lack of current to the battery will show up a discharged battery.)

If the oil pressure indicator light does not go out when the engine is running, check for low or no oil pressure, or a short circuit.

Stop the engine at once and find the cause.

Lack of engine oil pressure can result in failure of expensive parts inside the engine due to lack of lubrication.

VISUAL INSPECTION CHECKLIST

- ☐ **OIL AND WATER LEAKAGE**
- ☐ **ELECTRICAL SYSTEM**
  - **Battery**
  - **Cables**
  - **Wiring**
  - **Indicator Lights**
- ☐ **COOLING SYSTEM**
  - **Water in Crankcase**
  - **External Leaks**
  - **Clogging**
- ☐ **AIR INTAKE SYSTEM**
  - **Air Leaks**
  - **Restrictions**
- ☐ **FUEL SYSTEM**
  - **Leaks**
  - **Restrictions**
  - **Clogged Filter**
- ☐ **STEAM CLEANING**

3. COOLING SYSTEM

Wait until the engine has been idle for several hours and the crankcase oil is cold; then loosen the crankcase drain plug and carefully turn it out to see if any water seeps out. If water is present, locate the cause of the cooling system leak.

Inspect the cooling system for leaks, deteriorated hoses, bent or clogged radiator fins, slipping fan belt, or any other condition which could result in improper cooling.

4. AIR INTAKE SYSTEM

Inspect the air intake system for possible leaks or restrictions. If the proper amount of clean air does not reach the engine, performance and durability will be affected.

5. FUEL SYSTEM

Check the fuel system for leaks and for bent or dented lines, which might cause a restriction.

Check the fuel transfer pump sediment bowl. On diesel engines, inspect the fuel filters for dirt, water, or other foreign matter.

6. STEAM CLEANING

After checking for leaks, steam clean the engine. This not only helps to recondition the engine, it makes tune-up easier and troubles easier to spot.

 Litho in U.S.A.

DYNAMOMETER TEST

If possible, test the engine on a dynamometer *before* it is tuned. This test gives you the horsepower output and fuel consumption of the engine as it is. This will help you to determine if a tune-up can restore the engine or whether an overhaul is needed.

Good performance by the engine depends on these basic things:

1) An adequate supply of clean air and fuel

2) Good compression

3) Proper valve and ignition timing for good combustion

Failure or low performance on any of these factors makes it impossible to fully restore the engine.

Therefore, if the dynamometer test shows that any of these three factors is bad, the engine will have to be reconditioned.

Make the dynamometer test as follows:

1. Connect the engine to the dynamometer using the manufacturer's instructions.

2. Operate the engine at about one-half load until the coolant and crankcase oil temperatures are up to normal. (This will take about 30 minutes—*but is very important to a good test.)*

3. Gradually increase the load on the engine until its speed is reduced to rated load speed as given in the engine Technical Manual.

4. Read the horsepower on the dynamometer.

5. Compare the horsepower with that given in the engine Technical Manual.

However, do not expect engines to always equal these specifications. But if the engine rates much lower than normal, this is a signal that service is needed.

While the engine is operating under load, note the outlet of the crankcase ventilating system. If too much vapor appears, also remove the crankcase oil filler cap.

Fig. 3 — A Typical Dynamometer Test

If an excessive amount of vapor or smoke appears here as well as at the vent, there is blow-by in the engine cylinders and they must be reconditioned before the engine will perform at its best.

NOTE: Instruments are available to measure the flow of air and gases through the crankcase ventilating system.

The normal rate of engine vapor flow is specified in the engine Technical Manual.

Any increase in flow over the specified amount indicates crankcase blow-by.

If the blow-by is excessive, recondition the engine for good operation.

Even though the engine develops its rated horsepower using a normal amount of fuel, a tune-up may still improve its efficiency. Consider both hours of operation and the conditions under which the engine has been operated. It is far more economical in the long run to tune the engine *before* a lack of performance makes it mandatory.

Most manufacturers suggest a regular period of operation between tune-ups.

Litho in U.S.A.

ENGINE TUNE-UP

For proper engine tune-up, use the sequence listed in the following chart.

A detailed description of each operation follows the chart.

NOTE: The numbers in the "Item" column refer to numbered groups in the detailed instructions. For best results and time savings, follow the sequence given in the tune-up charts.

TUNE-UP CHART

Item No.	*Operation*	*Gasoline*	*Diesel*	*LP-Gas*
1	**AIR INTAKE AND EXHAUST SYSTEM**			
	Clean out precleaner (if used)	X	X	X
	Remove and clean air cleaner	X	X	X
	Swab out inlet pipe in air cleaner body	X	X	X
	Inspect exhaust system and muffler	X	X	X
	Check crankcase ventilating system for restrictions	X	X	X
2	**BASIC ENGINE**			
	Recheck air intake for restrictions	X	X	X
	Check radiator for air bubbles or oil indicating compression or oil leaks	X	X	X
	Cylinder head gasket leakage	X	X	X
	Retighten cylinder head cap screws	X	X	X
	Adjust valve clearance	X	X	X
	Check compression pressure in each cylinder	X	X	X
3	**IGNITION SYSTEM (Spark-Ignition Engines)**			
	Spark Plugs			
	Clean, test, and adjust spark plugs	X		X

Litho in U.S.A.

TUNE-UP CHART (Continued)

Item No.	*Operation*	*Gasoline*	*Diesel*	*LP-Gas*
	Ignition Coil			
	Test ignition coil	X		X
	Check for proper coil connections	X		X
	Distributor			
	Check the following items:			
	Cap and rotor	X		X
	Condenser	X		X
	Breaker Points	X		X
	Point gap (cam dwell)	X		X
	Breaker point spring tension	X		X
	Lubrication	X		X
	Distributor timing	X		X
4	**FUEL SYSTEMS**			
	Check fuel lines for leaks or restrictions	X	X	X
	Clean fuel pump sediment bowl	X	X	X
	Test fuel pump pressure	X	X	
	Check and clean LP-gas fuel-lock strainer (observe caution)			X
	Drain carburetor sump			X
	Check radiator for LP-gas leaking from converter into cooling system			X
	Drain gasoline carburetor, clean inlet strainer	X		
	Check carburetor choke disk operation	X		X
	Check speed control linkage	X	X	X
	Service diesel fuel filters		X	
	Check diesel injection pump		X	

TUNE-UP CHART (Continued)

Item No.	*Operation*	*Gasoline*	*Diesel*	*LP-Gas*
	Check and clean diesel injection nozzles		X	
	Bleed diesel fuel system		X	
	Check diesel injection pump timing		X	
5	**LUBRICATION SYSTEM**			
	Check operation of pressure gauge or light	X	X	X
	Service oil filter	X	X	X
	Check condition of crankcase oil	X	X	X
	Check engine oil pressure	X	X	X
6	**COOLING SYSTEM**			
	Check water pump for leaks and excessive shaft endplay	X	X	X
	Inspect radiator hoses	X	X	X
	Clean and flush cooling system	X	X	X
	Test thermostat and pressure cap	X	X	X
	Test radiator for leaks	X	X	X
	Check condition of fan belt	X	X	X
7	**ELECTRICAL SYSTEM**			
	Battery			
	Clean battery, cables, terminals, and battery box	X	X	X
	Tighten battery cables and battery hold-down	X	X	X
	Apply petroleum jelly to battery posts and cable clamps	X	X	X
	Check specific gravity of electrolyte and add water to proper level	X	X	X
	Make "light load" test of battery condition	X	X	X

Litho in U.S.A.

TUNE-UP CHART (Continued)

Item No.	*Operation*	*Gasoline*	*Diesel*	*LP-Gas*
	Generator or Alternator			
	Check belt tension	X	X	X
	Test alternator or generator output	X	X	X
	Starting Circuit			
	Check safety starter switch	X	X	X
	Check current draw of starting motor	X	X	X
8	**CLUTCH FREE TRAVEL**			
	Check free travel at clutch pedal or lever	X	X	X
9	**DYNAMOMETER TEST**			
	Use a dynamometer to check engine performance	X	X	X
	Use an exhaust analyzer while engine is on test to check for accurate carburetor adjustment	X		X

DETAILED TUNE-UP PROCEDURES

The following pages have detailed instructions for the checks and adjustments given in the tune-up chart.

The chart is divided into groups and each group is numbered. This number matches the number given in the detailed procedures which follow.

Where specifications are required, always refer to the engine Technical Manual for details.

1. AIR INTAKE AND EXHAUST SYSTEM

Pre-Cleaner

Check the pre-cleaner bowl for collected foreign matter (Fig. 4). Clean the bowl whenever three-fourths inch of dirt collects in the bowl.

Check the pre-cleaner screen (if used) and clean if necessary.

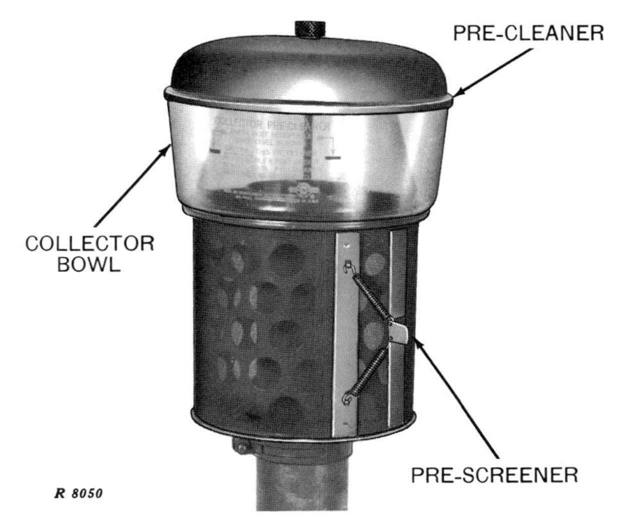

Fig. 4 — Pre-Cleaner and Screen

Air Cleaner (Dry Type)

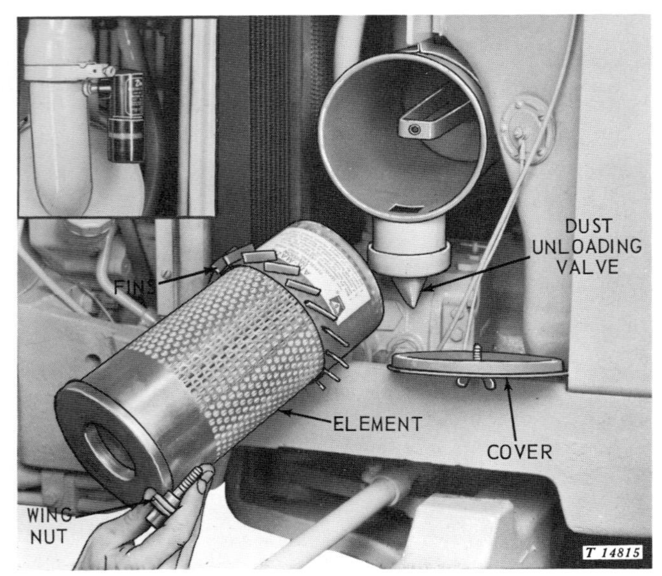

Fig. 5 — Dry-Type Air Cleaner

Remove and check the dry cleaner element (Fig. 5). If the element is ruptured or the seal is damaged, replace it.

Most dry elements can be recleaned a few times before they must be replaced. Two cleaning methods are used: One is for *dusty elements.* The other is for *oily or sooty elements.* See the machine operator's manual for details.

IMPORTANT: Never wash dry elements in gasoline, fuel oil, or solvents. Do not oil the element unless instructed.

Check the operation of the restriction indicator and dust unloading valve (if used). Flex the valve to clean out debris. Also wipe out the inside of the air cleaner housing. If the cleaner has a dust cup, clean it.

Install the cleaner element (usually fins first as shown) and *tighten it securely.* This will prevent air leaks past the end of the element. Then install the cover on the air cleaner housing.

Air Cleaner (Oil Bath Type)

Remove the oil cup from the air cleaner (Fig. 6). Pour the oil from the cup and clean the cup.

Clean the tray or oil trap (if used). Swab out the air intake tube.

If the main cleaner element is extremely dirty, remove it and clean by dashing it up and down in a pail of solvent. *Do not use steam cleaning equipment.* It is impossible to force the steam through the element and so the dirt stays at the center of the element.

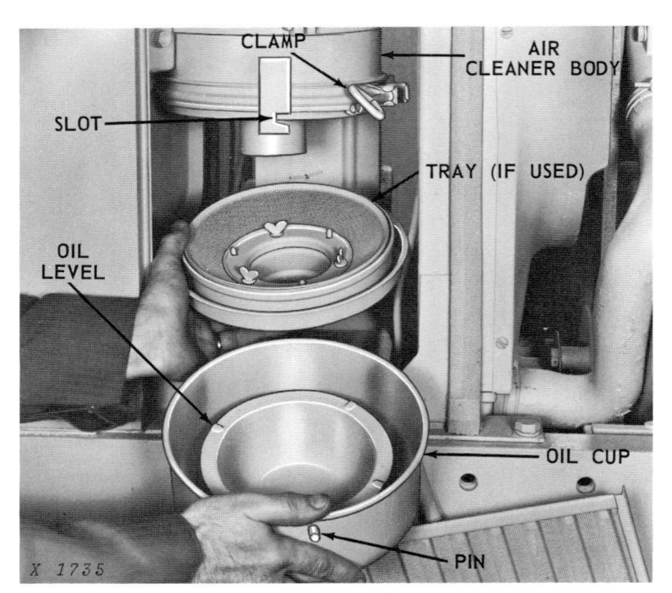

Fig. 6 — Oil Bath Air Cleaner

Install the oil trap or other parts removed. To assure *clean air,* check to see that all connections are tight and leak-proof.

Fill the oil cup to proper level with the correct amount of *new* oil of the same weight as is used in the engine. *Do not overfill the cup.* Install the oil cup and tighten it securely.

IMPORTANT: Never check the air cleaner with the engine running.

Exhaust System

Check the exhaust system for leaks. Leaks at the engine can be dangerous to the operator when the engine is used with an enclosed machine.

Check the exhaust pipe and muffler for restrictions which could prevent the free flow of exhaust gases.

Crankcase Ventilating System

Check for restrictions which could prevent free air flow through the crankcase ventilating system. Lack of ventilation causes sludge to form in the engine crankcase. This can cause clogging of oil passages, filters, and screens, resulting in expensive damage to the engine.

2. BASIC ENGINE

Air Intake System

A further check of the air intake system should now be made. Do this with a water manometer or vacuum gauge calibrated in inches of water.

NOTE: If the diesel engine, on the dynamometer test, produced a reasonable amount of horsepower and the air cleaner appears to have been regularly serviced, this operation can be omitted. However, on diesel engines which have air cleaners equipped with restriction indicators, check the operation of the indicator against the gauge.

INTAKE VACUUM TEST (Diesel Engines)

Test as follows:

1. Warm up the engine.

2. *On engines with restriction indicators,* remove the indicator, install a pipe tee fitting, and reinstall the indicator. Connect the gauge to the tee fitting.

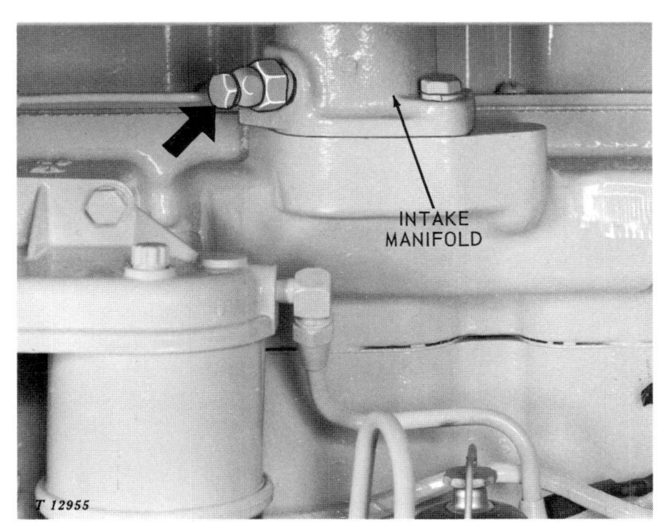

Fig. 7 — Manometer or Gauge Connection Point

3. *On engines without restriction indicators,* connect the gauge to the intake manifold (Fig. 7).

4. Set the diesel engine speed at fast idle and note the reading on the gauge. Too high a reading means that there is a restriction in the air intake system. Check the engine Technical Manual for correct specifications.

5. On engines with restriction indicators, check the operation of the indicator. Use a board or metal plate to very slowly cover the air intake opening. Note the action of the indicator in relation to the reading on the gauge. If the indicator does not operate properly, replace it.

MANIFOLD DEPRESSION TEST (Spark-Ignition Engines)

Much can be learned about the internal condition of a spark-ignition engine by checking the intake manifold depression between the carburetor and the engine. Use a vacuum gauge calibrated in inches of mercury.

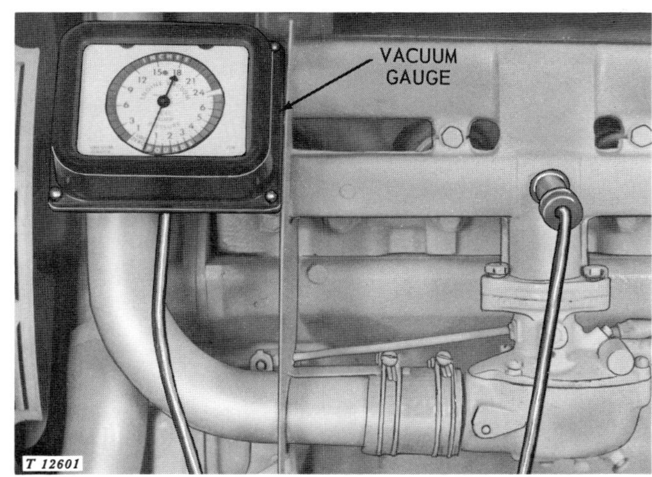

Fig. 8 — Manifold Vacuum Test **(Spark-Ignition Engines)**

1. Connect the vacuum gauge to the intake manifold (Fig. 8).

2. Warm up the engine and operate it at idle speed.

3. Note the reading on the vacuum gauge. Check the engine Technical Manual for exact specifications.

4. Interpret the gauge reading as follows:

- **If the reading is steady and low,** loss of power in all cylinders is indicated. Possible causes are late ignition, bad valve timing, or loss of compression at valves or piston rings. A leaky carburetor gasket will also cause a low reading.

- **If the needle fluctuates steadily,** a partial or complete loss of power in one or more cylinders is indicated. This can be due to an ignition defect, or loss of compression due to stuck piston rings or a leaky cylinder head gasket.

- **Intermittent needle fluctuation** indicates occasional loss of power due to an ignition defect or a sticking valve.

- **Slow needle fluctuation** is usually caused by improper carburetor idle mixture adjustment.

- **A gradual drop in the gauge reading** at idle engine speed indicates back pressure in the exhaust system due to a restriction.

Since the gauge readings can indicate more than one fault, be very careful in analyzing abnormal readings.

Cylinder Head Gasket Leaks

With the engine running at idle speed and at operating temperature, *carefully* remove the radiator cap and check for bubbles or oil at the surface of the coolant.

If bubbles appear, cylinder pressure is leaking past the cylinder head gasket (or LP-gas is leaking from the converter into the engine coolant). Replace the head gasket or repair the converter.

If oil is present, it can be leaking past the cylinder head gasket or leaking from an engine oil cooler. Determine the cause and correct it.

Cylinder Head Cap Screw Tightness

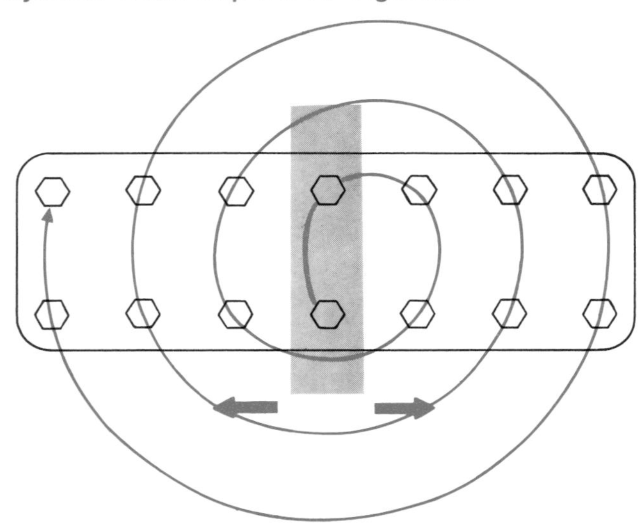

Fig. 9 — Sequence For Tightening Cylinder Head Cap Screws

While the engine is hot, loosen each cylinder head cap screw and retighten to specified torque.

One sequence for tightening cylinder heads is to work from the center outward on alternating sides of the head in a circle as shown in Fig. 9. Figure 10 shows sequence for newer engines.

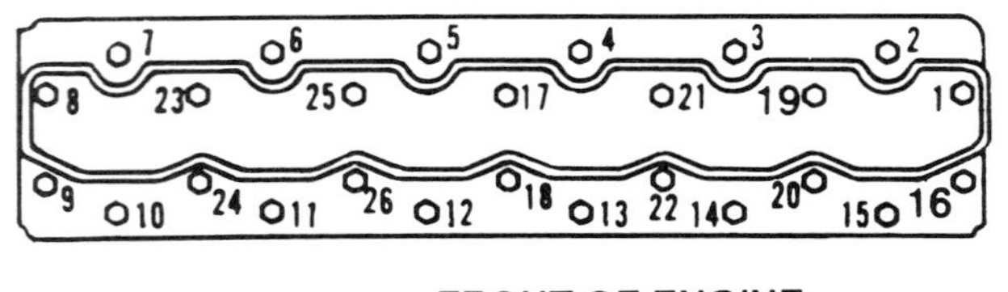

Fig. 10 — Sequence for Tightening Cylinder Heads — Newer Engines

However, use the exact sequence given in the engine specifications.

NOTE: A film of oil under the head of each cap screw will assure better tightening.

IMPORTANCE OF USING A TORQUE WRENCH

Many servicemen do not realize the importance of using a torque wrench to tighten cap screws or nuts. This is particularly true on cylinder head cap screws or nuts.

Unevenly tightened cylinder head cap screws can distort the valve seats (causing leakage) or the cylinder head or block surfaces (which could result in cylinder head gasket failure).

Valve Tappet Clearance

Check the clearance on the engine valves (Fig. 11). This is the distance between the end of the tappet lever and the end of the valve stem when the piston is at "top dead center" of compression stroke.

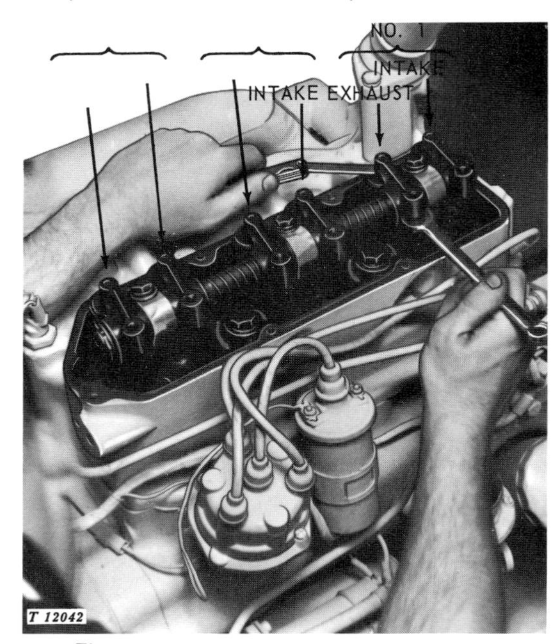

Fig. 11 — Adjusting Valve Tappet Clearance

Turn the engine to "top dead center" on compression stroke of the No. 1 (front) cylinder. Many engines have a timing mark on the front pulley or a timing pin on the flywheel to set the engine at "TDC."

Adjust the clearance on the two valves for the No. 1 cylinder to specifications. Use a feeler gauge as shown in Fig. 10 to check the setting.

Turn the engine until the next piston comes up to "top dead center" and adjust the valve clearances for that cylinder.

Keep on turning the engine until all the valves are adjusted. The sequence will vary for 3-, 4-, and 6-cylinder engines.

THE IMPORTANCE OF ACCURATE VALVE CLEARANCE

The importance of accurate valve adjustment cannot be too highly stressed.

If there is not enough valve clearance, the valve may not entirely close, resulting in loss of compression or burned exhaust valves.

If the valve clearance is too great, the valve opens late and closes early. Thus the valve does not remain open long enough to allow the full charge of air or fuel-air mixture to enter the cylinder. This results in a loss of engine efficiency.

Compression Test

Weak compression results in loss of horsepower and poor engine performance. This is even more true in diesel engines which depend upon the heat of compression to ignite the fuel in the cylinders.

Check the engine compression as follows:

1) Warm up the engine to operating temperature.

2) Remove the spark plugs or injection nozzles.

3) Connect a pressure gauge to the cylinder port (Fig. 12).

4) *On spark-ignition engines,* be sure that both carburetor throttle and choke valves are in the wide open position. Disconnect the ignition coil-to-distributor wire at the distributor. Ground the wire by placing it in contact with some part of the engine.

5) *On diesel engines,* the air intake system has no throttle valve, so merely be sure that the engine speed control is in the "stop" position.

6) Turn the engine with the starter until the pressure gauge registers no further rise in pressure. It is a good practice to count the number of compression strokes (indicated by movement of the gauge needle) and check each cylinder with the same number of strokes. The engine must be at full cranking speed to get a good reading.

7) Check the pressure reading against the engine Technical Manual.

8) If the compression is very low, apply oil to the ring area of the piston through the openings. Do not use too much oil to avoid getting oil on the valves. Then check the compression again.

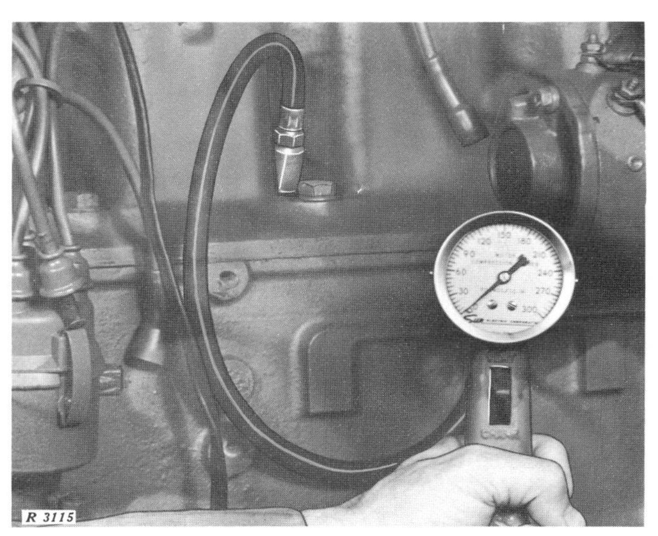

Fig. 12 — Checking Engine Compression

9) Judge the compression readings as follows.

JUDGING THE ENGINE COMPRESSION READINGS

All cylinder pressures should be approximately alike. There should be less than 50 psi (340 k Pa) difference between cylinder pressures.

If compression is too low, it's possible that valves are worn or sticking, the cylinder head gasket is leaking or the piston and its rings are damaged.

If compression is higher, carbon build up could be indicated.

Altitude affects compression pressures. Normally there is about a 4 percent loss for every 1000 feet (300 m) of altitude above sea level. Pressures given are normally for about 1000 feet (300 m) above sea level.

Compression pressures are affected by the **cranking speed** of the engine. Therefore, be sure the batteries are in good condition and fully charged when making a compression test.

HOW COMPRESSION AFFECTS ENGINE PERFORMANCE

Horsepower in an engine is created by the expansion of gases in the cylinder caused by the burning of a mixture of fuel and air.

For example, in a spark-ignition engine, peak firing pressures are roughly four times higher than compression pressures. In an engine with 160 psi (1100 kPa) compression pressure, peak firing pressure will be approximately 640 psi (4400 kPa).

If the pressure is only 150 psi (1000 kPa) in each cylinder when checked, there is approximately a nine percent loss in horsepower. This, of course, is a loss of engine efficiency and results in higher operating costs.

In diesel engines, peak firing pressures do not rise as high, comparatively, as in spark-ignition engines. They are about double that of the compression pressure.

This is because diesel engines only take in air. The fuel is then sprayed into the hot air in the cylinder.

There is less pressure rise but the peak pressure is maintained over a longer period of time. This gives the diesel engine its greater "lugging" ability.

These principles are shown in Fig. 13. Note that while firing pressure rises very rapidly in the spark-ignition engine, it also falls off very rapidly. In diesel engines, the pressure rises, is maintained over a longer period of time, and then falls off.

Causes of Poor Compression

1. *Worn cylinders*
2. *Worn piston rings*
3. *Leaking gaskets*
4. *Worn pistons*
5. *Leaking valves which can be caused by:*
   a. *seats burned*
   b. *valves burned*
   c. *valve springs weak or broken*
   d. *valve stem warped or sticking in valve guide*
6. *Loose spark plug*
7. *Loose cylinder head bolts*
8. *Warped cylinder head*
9. *Broken connecting rod*
10. *Not enough valve tappet clearance*

For remedies, see Chapter 2 of this manual.

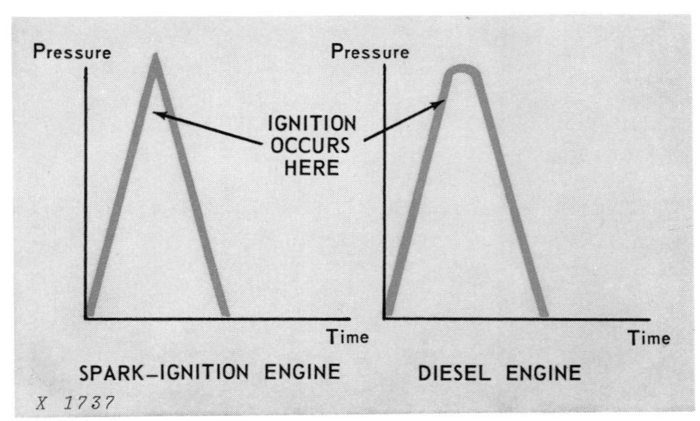

Fig. 13 — Pressure Curves for Engine Compression

3. IGNITION SYSTEM (SPARK-IGNITION ENGINES ONLY)

An oscilloscope is an instrument used to electronically check the performance of the ignition system. It is fast, accurate, and often pinpoints troubles which might be overlooked.

Details on servicing the ignition system are given in the FOS-20 Manual on "Electrical Systems." Here we will only outline the ignition services as they relate to engine tune-up.

Spark Plugs

Close inspection of the spark plugs can reveal much about the condition of the engine.

Black sooty plugs may be caused by:

1) *Engine operated at too low a temperature*

2) *Fuel-air mixture too rich*

3) *Heat range of plug too cold*

4) *If plug is oily—engine consuming oil*

Eroded plugs—insulator chipped or blistered from heat.

May be caused by:

1) *Heat range of plug too hot.*

2) *Carburetor adjusted too lean, resulting in too much heat in the engine cylinders.*

Good plugs *will be relatively free of both deposits and erosion.* This means that the plug is operating properly and is serviceable.

It is a good practice to replace all the spark plugs during tune-up if there is any question of their condition. Engine operation can be seriously affected by dirty or eroded plugs.

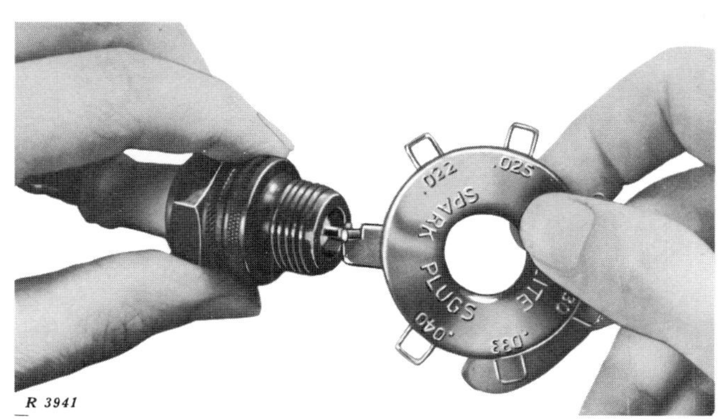

Fig. 14 — Setting The Spark Plug Gap

If the plugs are serviceable, file the electrode surfaces flat and square and adjust the point gap (Fig. 14) as specified in the engine Technical Manual.

After the spark plugs have been cleaned and adjusted, check them with a spark plug tester. Replace any which are doubtful.

When installing plugs, use a new gasket and tighten to exactly the specified torque. *This is important.*

If plugs are not tight enough, the plug can overheat because heat cannot be transferred to the cooling system.

If too tight, the thread area of the plug may stretch. This increases the point gap, resulting in hard starting and poor engine operation under load.

Ignition Coil

Use a reliable meter to check the ignition coil.

If the performance is doubtful, replace the coil. Refer to the Technical Manual for tests and specifications.

Check the coil for proper connections.

Use this rule of thumb:

On negative-ground system, the negative (—) primary terminal is connected to the distributor.

On positive-ground systems, the postitive (+) primary terminal is connected to the distributor.

Distributor

TESTING

If possible, the distributor should be checked out on a distributor tester. Testing the distributor often locates troubles which would otherwise not be found.

CAP AND ROTOR

Remove the distributor cap, rotor, and housing cover. Wipe clean with a rag. Make sure any vent holes in the cap are open. Inspect the posts in the cap for erosion.

Check the rotor for severe erosion at the end of the metal contact. Erosion at this point widens the gap between the end of the rotor and the posts in the distributor cap. This puts an excessive load on the ignition coil and can cause the coil to fail. It also affects engine starting and operation.

Detach the wires from the distributor cap. The ends of the wires may have been burned away, causing a gap which could result in failure of plugs to fire. This makes the engine hard to start.

When installing the wires into the cap, be sure they are pushed all the way in.

Replace any spark plug wires which have cracked, burned, or worn insulation. Here again, leakage of the high-tension current can affect engine operation.

CONDENSER

Use a reliable meter to check the condenser. Refer to the engine Technical Manual for tests and specifications.

BREAKER POINTS

Examine the condition of the breaker points. If they appear burned, replace them. If there is a spur on one contact surface, a corresponding pit will be found on the other.

If the surfaces are clean and are a light gray in color, merely use a point contact file to true up the surfaces. It is not necessary to remove the pit from one surface, but the spur should be removed.

After the breaker points have been filed or when new points are installed, be sure that the contact surfaces meet squarely. This assures a full flow of primary current when the points are in contact.

 Litho in U.S.A.

If the points do not meet squarely, resistance to the flow of primary (low-tension) current results. This can reduce the primary flow until the coil output is affected, resulting in poor engine operation.

NOTE: Before installing the points, check the operation of the automatic spark advance mechanism (if used). Turn the breaker cam in the normal direction of rotation. It should operate freely without evidence of binding. If not, find the cause and correct it. See the engine Technical Manual for information.

Apply a small amount of cam lubricant to the breaker cam. **Avoid using too much lubricant.** If the grease gets on the contacts, they may be burned.

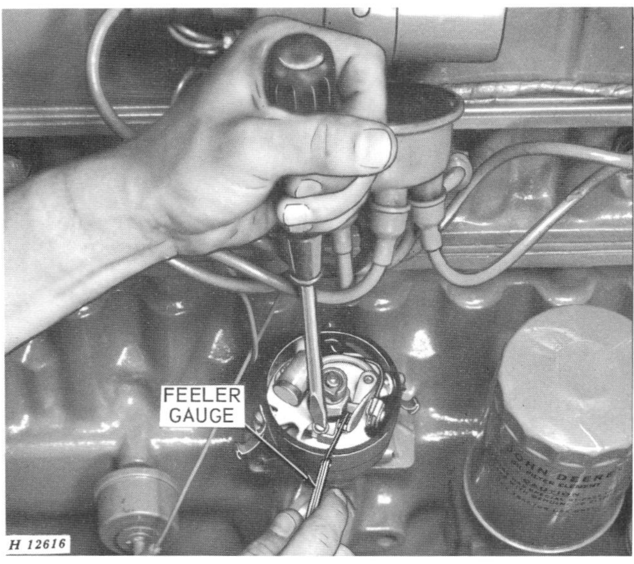

Fig. 15 — Adjusting The Distributor Point Gap

Adjust the distributor breaker point gap (Fig. 15). Follow the instructions in the engine Technical Manual. Also use a reliable meter to check the cam dwell.

Cam dwell is the period of time the breaker points remain closed until they open to create the spark. If the point gap is too wide, the points may not remain closed long enough for complete build-up of the magnetic field in the ignition coil. This can result in a weak spark and poor engine performance. If the point gap is too small, the points will not remain open long enough for complete collapse of the field. This can cause burning of the points and also a weak spark.

Check the tension of the breaker point spring. The tension should be within the limits given in the engine Technical Manual. If necessary, adjust by bending the spring.

Breaker point tension is very important. If the tension is too great, the rubbing block on the movable point will wear too fast. This will change the cam dwell and affect engine operation. If the tension is too weak, the movable point will bounce, causing erratic engine operation.

On distributors which require lubrication, lubricate according to instructions in the engine Technical Manual.

Reassemble the distributor. Be sure all connections are tight and that the spark plug wires are pushed all the way into their sockets.

DISTRIBUTOR TIMING

Check the timing of the distributor to the engine. Use a distributor timing light.

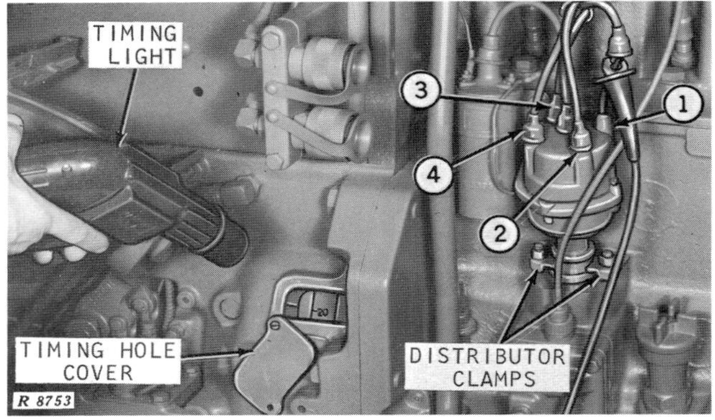

Fig. 16 — Timing The Distributor

Here is a general guide to distributor timing:

1. Start the engine and warm it up.

2. Stop the engine and connect the timing light as instructed in the engine Technical Manual. Also locate the timing marks on the engine.

3. Start the engine and operate it as specified.

4. Observe the flash of the timing light when the spark occurs. It should happen when the timing marks on the engine are lined up.

5. If necessary, loosen the distributor mounting and rotate the distributor until the flash occurs at the right instant.

Rotating the distributor in the direction of cam rotation will delay the spark, while rotating it in reverse will advance the spark.

6. Recheck the timing after the distributor mount is tightened again.

Timing the distributor to the engine should be done with great accuracy. The reason for this can be seen in Fig. 17.

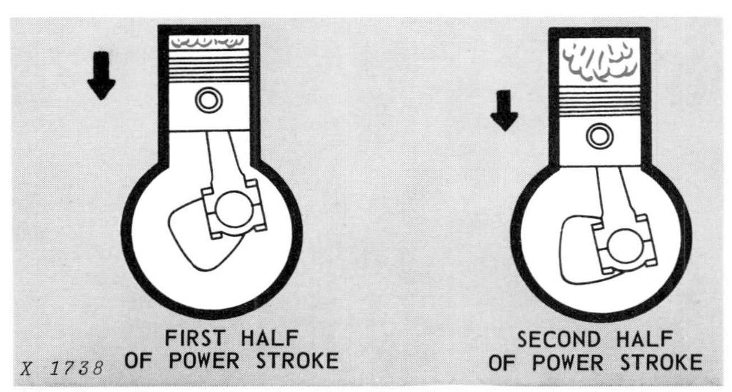

Fig. 17 — Connecting Rod Position In Relation To Crankshaft

In a spark-ignition engine the most power from combustion occurs while the piston is traveling the *first* half if its power stroke. This is because the connecting rod is angled in relation to the crankshaft. The left-hand drawing in Fig. 17 shows why this is true. Notice how the connecting rod drives the crankshaft more directly than during the second half of the power stroke (shown at the right).

If the ignition timing is late, full expansion of the gases will not occur while the piston is within the most effective area of its stroke. This results in a loss of horsepower and greater operating costs.

If the timing is too early, too much expansion takes place before the piston reaches top dead center of its compression stroke and the engine literally works against itself. The result is excessive "pinging", loss of horsepower, and higher operating costs.

4. FUEL SYSTEMS

Fuel Lines

Check fuel lines for leaks or restrictions. Leaking fuel not only is a fire hazard but it gathers dirt and wastes fuel.

CHECKING LP-GAS SYSTEM FOR LEAKS

Use a soap solution or liquid leak detector to coat all parts of the LP-gas system to check for leaks. Pressurize the system and look for bubbles which indicate a gas leak.

CAUTION: Observe all safety rules when working on LP-gas fuel systems (see Chapter 4).

Fuel Transfer Pump

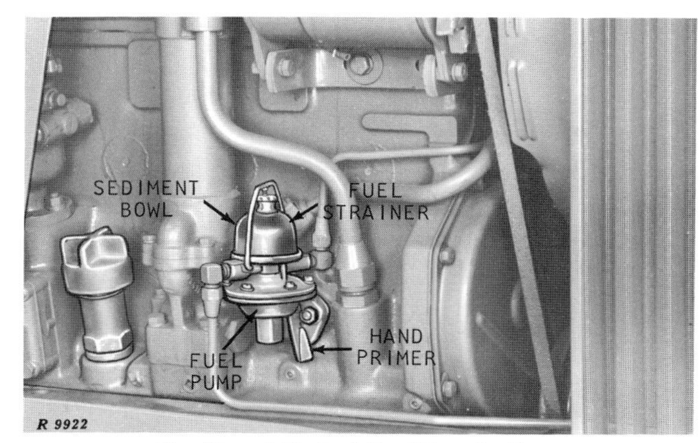

Fig. 18 — A Typical Fuel Transfer Pump

Remove and clean the fuel pump sediment bowl and strainer (Fig. 18).

TESTING FUEL TRANSFER PUMP PRESSURE

Install a tee fitting at the fuel pump outlet and connect a low-pressure test gauge. Then operate the engine and check the fuel pressure.

NOTE: On diesel engines, the hand primer lever can be used to build up the pump pressure and to fill the system after service.

Be careful to avoid too-high pressures. In spark-ignition engines there is a tendency, especially on rough terrain, for the fuel to be forced by the carburetor float valve. This causes the fuel level in the carburetor to be too high, resulting in excess fuel consumption.

If the fuel pump has a line filter, service it as instructed in the engine Technical Manual.

Check and clean the LP-Gas fuel-lock strainer using the precautions in Chapter 4.

Carburetor Sump (LP-Gas)

Drain the LP-gas carburetor sump using precautions (see Chapter 4).

Converter (LP-Gas)

Test the LP-gas converter for leaks as follows:

 Litho in U.S.A.

Turn on the starter switch and open the vapor withdrawal valve. Then remove the radiator cap and check the coolant for bubbles caused by a leaky converter gasket. If bubbles are present, service the converter.

Carburetor (Gasoline)

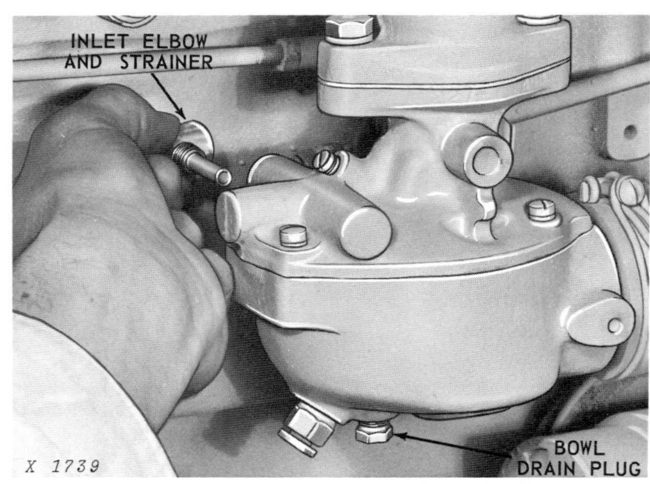

Fig. 19 — Gasoline Carburetor

Remove the drain plug at the bottom of the carburetor bowl (Fig. 19) and drain out any water or sediment in the bowl. After draining, replace the drain plug.

Remove and clean the carburetor fuel inlet strainer. Flush the strainer with gasoline to remove any dirt which may have worked through it.

If the engine has been in long or severe service, remove and clean the carburetor. Disassemble and clean the metal parts by immersing them in a carburetor cleaning solution. When clean, blow out all passages and dry the parts with compressed air. When reassembling, use a new carburetor repair kit.

In operation, the carburetor float needle and seat wear. This raises the fuel level in the carburetor bowl and increases fuel consumption. It is false economy to "make do" when new parts will soon pay for themselves in reduced operating costs.

After the carburetor is installed, check the choke disk for proper operation. When the choke is applied, the disk should be centered in the bore and should entirely close the opening. When the choke is open, the disk should be parallel with the opening.

Carburetor fuel adjustment should also be checked later when the engine is tested on the dynamometer.

Speed Control Linkage

Check the adjustment of the linkage between the speed control lever and the injection pump or carburetor.

Be sure that all linkage operates freely through its entire range. Any binding will cause erratic operation of the engine.

Fuel Filters (Diesel)

Diesel fuel must be clean and free from water. Many cases of injection pump or nozzle failure can be traced to dirt or water in the fuel.

Take special care to avoid contamination of diesel fuel—from the supplier, through storage, to the fuel tank, and the injection system.

Loss of engine horsepower can also be traced to dirty fuel filters.

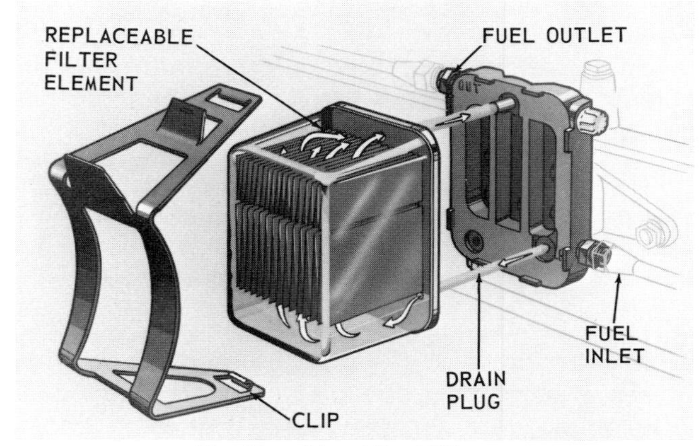

Fig. 20 — Diesel Fuel Filter

Special fuel filters are installed in most diesel fuel systems (Fig. 20) but they cannot perform well if they are not properly serviced.

Replace or clean the fuel filters during tune-up.

If the filter has a drain plug (Fig. 20), loosen it and drain out any water or sediment.

If water is present, be sure to service the filter. Also find out and eliminate the source of water in the fuel.

On dual-stage filters, if the first-stage filter is extremely dirty or water-soaked, also service the second-stage filter.

Injection Pump (Diesel)

During major engine service, the injection pump should be removed, cleaned, inspected, and calibrated on an injection pump test stand.

However, it is a good practice to remove the pump and check it during a complete tune-up, if a test stand is available.

Testing the pump assures that it is operating properly and helps to locate malfunctions which affect engine performance and cause high fuel consumption. It can also reveal whether or not the pump calibration has been tampered with to increase engine horsepower.

Never calibrate the pump to inject more fuel than specified in the engine Technical Manual. The engine is designed to produce its rated horsepower at a certain rate of fuel consumption. Any change in this rate puts an undue strain on the working parts of the engine. This can cause early engine failure and always results in higher operating costs.

Excessive smoke from the exhaust is one sign that the injection pump is calibrated too high.

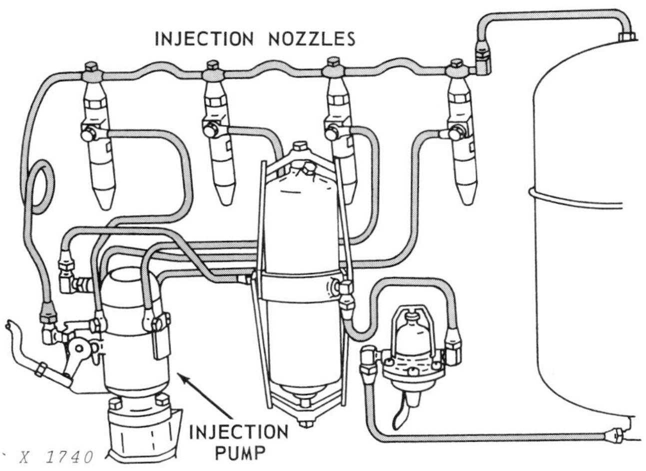

Fig. 21 — Diesel Fuel Injection Pump and Nozzles

Injection Nozzles (Diesel)

During major engine service, the injection nozzles should be removed, cleaned, and tested using a nozzle tester (see Chapter 5).

During a complete tune-up, it is also a good practice to remove and service the nozzles.

Three factors are of prime importance in nozzle operation:

1) **Cracking pressure.** All nozzles should be about equal. Too low—fuel will not be atomized.

2) **Condition of spray tips.** The orifices should be clean and not eroded.

3) **Spray pattern.** The fuel should be finely atomized and spread in an even spray pattern. A bad spray pattern can actually erode metal from the top of the piston.

Engines operated with bad nozzles often have low horsepower and excessive exhaust smoke.

Bleeding Air From The Fuel System (Diesel)

After servicing, always bleed air from the diesel fuel system. This will prevent "lock" in the high-pressure injection system.

Usually the fuel transfer pump has a primer lever which helps force air bubbles out of the system at various bleed plugs and loosened connections (see Chapter 5).

When all air is bled from the system, the engine will start normally and run without missing.

Timing The Injection Pump (Diesel)

In a diesel engine the start of fuel injection occurs *before* the piston reaches "top dead center" of its compression stroke. This is necessary because it takes time for the burning fuel to build up pressure.

Therefore, the pump is timed so the expanding gases reach peak pressure at about the time the piston reaches "top dead center".

If injection is timed too early, expansion occurs before the piston reaches top dead center and the engine is literally working against itself. This results in horsepower loss and decreased engine efficiency.

If injection is timed too late, expansion occurs late, and much power is lost because the connecting rod has a poor angle to the crankshaft. (See the drawing at right in Fig. 22.)

Litho in U.S.A.

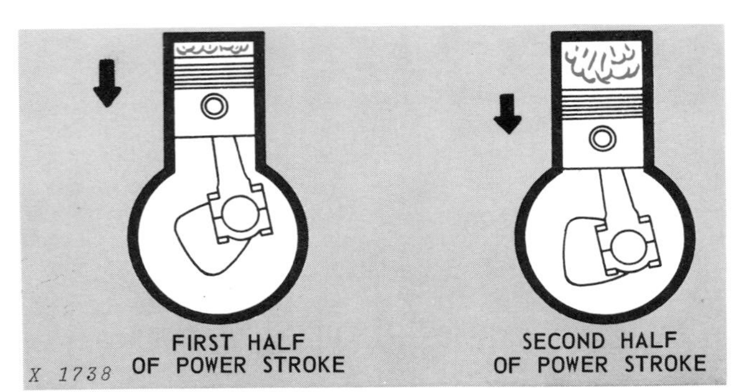

Fig. 22 — Connecting Rod Position In Relation To Crankshaft

When the injection is correctly timed, the connecting rod drives the crankshaft more directly and gets full power as shown at left in Fig. 22.

5. LUBRICATION SYSTEM

Pressure Gauge or Indicator Light

Check the operation of the pressure gauge or light. Turn on the starter switch to check, then start the engine and check during operation. See the engine Technical Manual for service information.

Oil Filter

Replace the engine oil filter during tune-up.

If the filter is extremely dirty, consider these likely problems:

1. The crankcase oil has not been changed often enough.

When oil is used too long in an engine, much of its detergent qualities are neutralized and it can no longer do its job. Either change the oil more often or use a higher quality of oil.

2. The oil is of the wrong quality.

Check the engine Technical Manual for the correct quality and weight of oil to use.

3. Water or antifreeze is in the oil.

Water in the oil is one cause of excessive sludge. Water can leak into the crankcase from around the cylinders through a sand hole in the cylinder block casting, or past the cylinder head gasket. If water is present in the oil, determine the cause and correct it.

Fig. 23 — A Typical Internal Oil Filter Which Is Replaced During Service

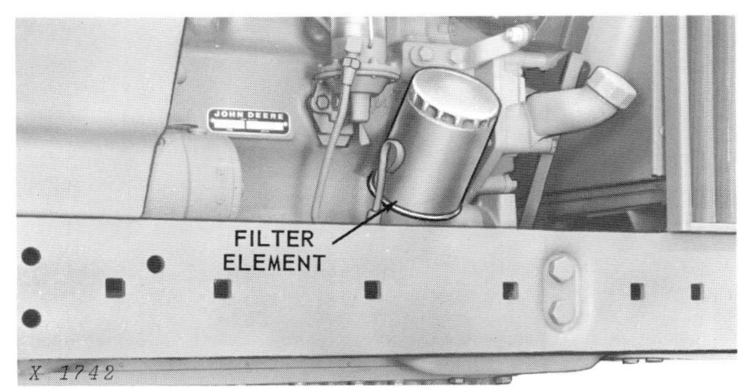

Fig. 24 — A Typical Screw-On Oil Filter Which Is Replaced During Service

Anti-freeze can enter the crankcase in any of the above ways. However, anti-freeze poses much more of a problem than plain water.

- *It forms a deposit on the pistons and rings, causing rings to stick and pistons to score.*
- *It can cause main and connecting rod bearings to fail.*
- *It forms a deposit on all surfaces throughout the engine and can cause severe damage to all moving parts.*

If the engine has been run too long with anti-freeze in the oil, disassemble the engine. Clean all parts thoroughly and replace any that are damaged.

Crankcase Oil

Remove the oil dipstick and examine the condition of the crankcase oil.

Oil in the engine has a variety of functions:

1) Oil's prime purpose is to provide lubrication between the moving parts. This prevents wear and possible damage.

Litho in U.S.A.

2) Oil assists in carrying off heat generated within the engine. The oil is cooled to some extent by the flow of air around the crankcase. Some engines are provided with special coolers to cool the oil. The cooled oil is pumped around the bearing surfaces and thrown around the inside of the engine. This helps to dissipate the engine heat.

3) Oil acts as a seal around the piston rings. Oil not only lubricates the pistons and rings, it also acts as a seal between the piston and rings, preventing loss of compression.

4) Oil neutralizes acids and prevents corrosion. Modern oils contain additives which neutralize acids formed during combustion and so prevent corrosion within the engine.

5) Oil prevents depositing of sludge in the engine. Additives in the oil also enable it to hold in suspension dirt and grit which otherwise would be deposited within the engine as sludge.

From this summary we can see how important it is to use the correct oil.

If the crankcase oil is extremely dirty or contains water or antifreeze, etc., drain and replace it, then correct the other causes.

Refer to the engine Technical Manual for draining procedures and for the correct quality and weight of oil to use.

Oil Pressure

Check the engine oil pressure using a master gauge.

1. Operate the engine until the oil is warmed up to normal. (Also be sure the filter is clean.)

2. Stop the engine and connect a master gauge (Fig. 25) to the engine block (normally where the pressure sending unit is attached.)

3. Run the engine at the specified speed and check the oil pressure against the listing in the Technical Manual.

4. If oil pressure is too high or low, adjust it as instructed in the Technical Manual.

PROBLEMS WITH LOW OIL PRESSURE

Operate the engine at slow idle speed. If the pressure is *very low,* check for one of these three problems:

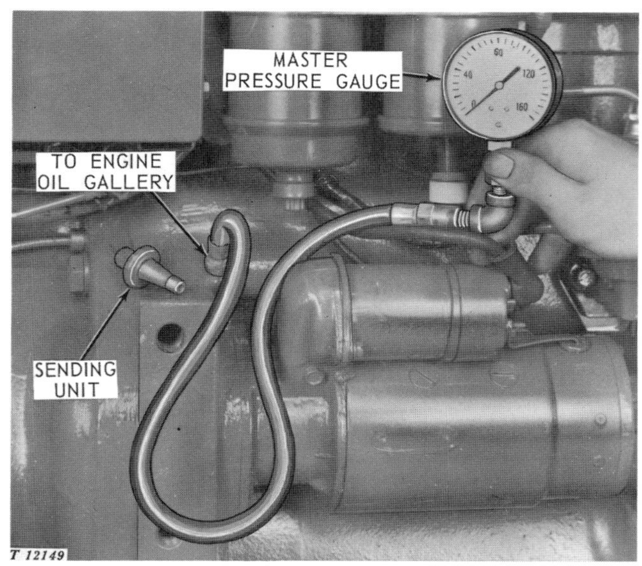

Fig. 25 — Checking Engine Oil Pressure With A Master Gauge

1) Using too-light oil. Check for correct weight.

2) Stuck oil pressure relief valve. Dirt on the valve seat can prevent it from closing. Check and clean the valve and seat.

3) Worn-out bearings on the crankshaft or camshaft. Too much oil escapes past worn parts, lowering oil pressure. In this case, recondition the bearings.

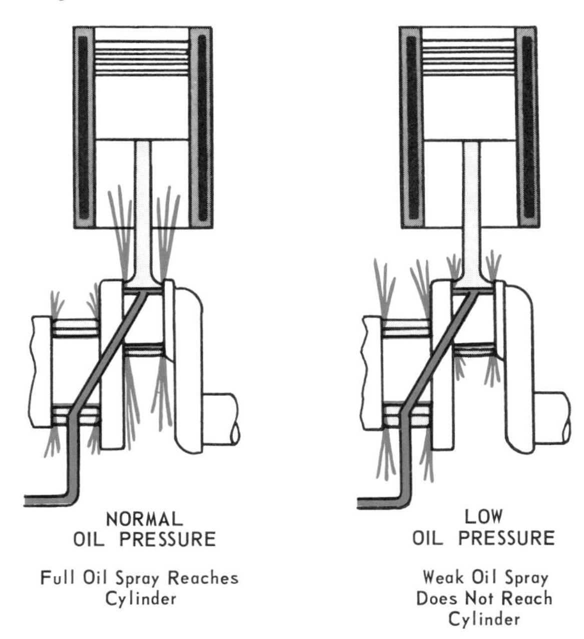

Fig. 26 — Low Oil Pressures Mean Lack of Lubrication

Too-low oil pressures can cause the pistons and bearings to wear faster. This is because not enough oil reaches the cylinders for good lubrication (Fig. 26).

6. COOLING SYSTEM

Water Pump

Fig. 27 — Water Pump and Thermostat Housing

Inspect the water pump (Fig. 27). If the pump is leaking or has too much endplay in its shaft, remove the pump and repair it as instructed in the engine Technical Manual.

Radiator Hoses

Inspect radiator hoses for signs of leakage or rot. Be sure that the inside lining of hoses has not become "mushy". This restricts the flow of water. Replace hoses as necessary.

Cleaning and Flushing the Cooling System

1. Run the engine long enough to stir up any rust or sediment in the cooling system.

2. Drain the cooling system. Detach the thermostat housing and remove the thermostat. Reinstall the housing.

3. Fill the cooling system with water and run the engine long enough to warm the water and stir up any rust or sediment.

4. Stop the engine and drain the system at once before the rust or sediment settles.

5. Fill the system with a solution of water and a good commercial radiator cleaner.

6. Install the radiator cap and run the engine until the solution is thoroughly warmed. Place a cardboard over the front of the radiator core for faster warm-up.

7. After several minutes, drain out the solution, fill with clean soft water, start the engine, and let the water circulate for a few minutes.

8. Condition the system with a recommended antifreeze (winter) or a coolant conditioner which inhibits rust (summer). Also add a water pump lubricant (when recommended).

Thermostat

Fig. 28 — Testing The Thermostat

Test the action of the thermostat (Fig. 28). If it does not open at the temperature given in the engine Technical Manual or is otherwise defective, replace it.

The thermostat is the key factor in cooling the engine.

When the engine is cold, the thermostat is closed. This confines the circulation of the coolant to within the cylinder block, resulting in faster warm-up.

When the engine is warmed up, the thermostat opens to bypass just enough coolant into the radiator core to maintain the proper temperature in the engine cylinder block.

If the thermostat ever sticks shut or fails to open, coolant in the cylinder block will soon overheat which may damage the engine.

If the thermostat opens at too low a temperature or does not close, the engine will not warm up properly and its efficiency will be lost.

Radiator

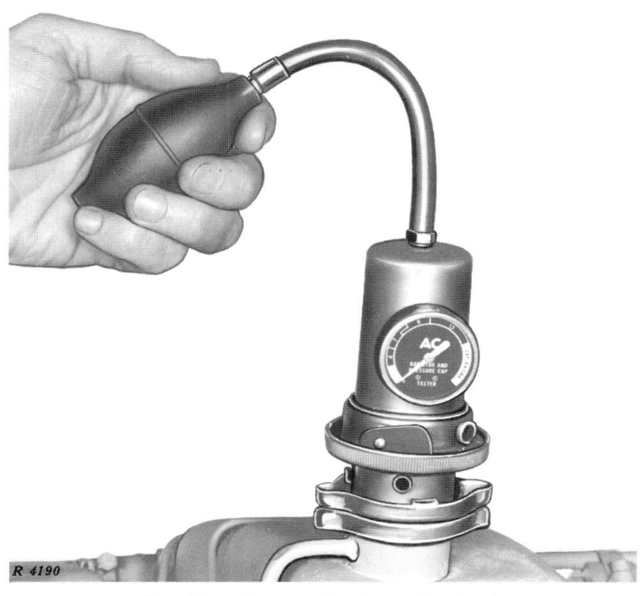

Fig. 29 — Testing Radiator For Leaks

Check the radiator for leaks using a pressure tester as illustrated in Fig. 29. *Avoid applying too much pressure.* A maximum of 15 to 25 psi (100 to 170 kPa) should be sufficient.

If any leaks are observed, repair them. See Chapter 8 for details.

Examine all air passages in the radiator core. Blow out any chaff or dirt and straighten any bent fins.

Clean all chaff or dirt from the radiator screens.

Radiator Cap

Use a reliable tester to test the radiator cap (Fig. 30). The pressure valve should open at a certain pressure (see engine Technical Manual for specifications).

If the pressure valve does not open as specified or leaks before reaching opening pressure, replace the gasket or the whole cap as necessary.

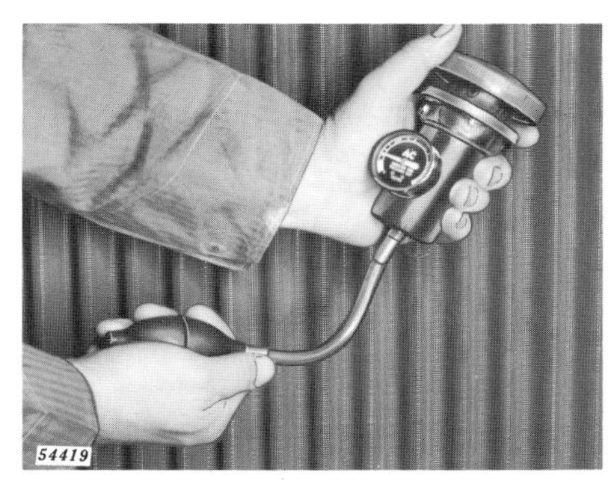

Fig. 30 — Testing Pressure Cap

Be sure that the cap and radiator are free from leaks. The pressure cap permits pressure to build up in the cooling system and so raises the boiling point of the coolant to approximately 230°F (110°C). This prevents boiling away of the coolant which would occur at a lower temperature. The pressure cap also prevents loss of coolant through the overflow when traveling over rough terrain.

Fan Belt

Check the fan belt for excessive wear, cracks, or other signs of damage. Replace the belt if any of these conditions exist.

Fan belt tension is checked later during generator inspection.

7. ELECTRICAL SYSTEM

Servicing of electrical systems is covered in the FOS-20 manual on "Electrical Systems". Refer to that manual for details on the tune-up procedures given here.

Batteries

CLEANING THE BATTERIES

Use a stiff brush and a water and soda solution to thoroughly clean the batteries. Rinse off the batteries with clean water.

If the battery terminals are corroded, disconnect and clean them. Clean the battery posts and the insides of the connectors so they make good electrical contact.

Be sure the batteries are properly installed.

Use petroleum jelly on the battery posts and connectors to prevent corrosion.

TESTING THE SPECIFIC GRAVITY

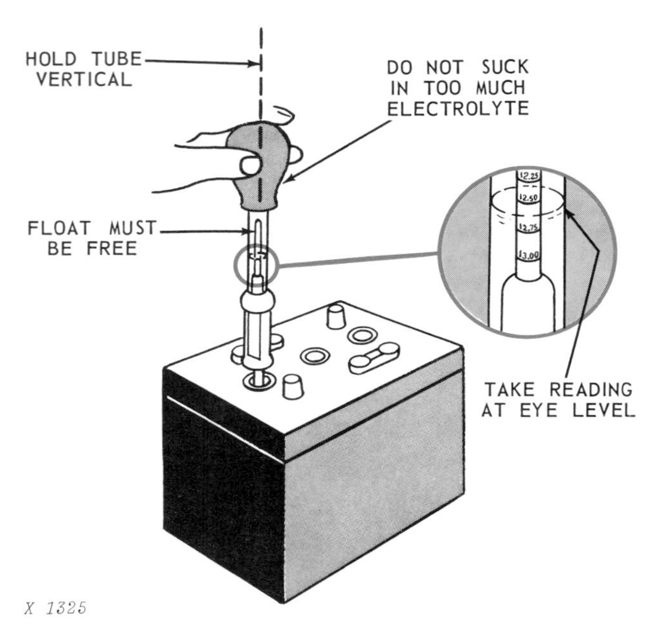

Fig. 31 — Testing The Battery Specific Gravity

Use an accurate battery hydrometer to check the specific gravity of the electrolyte in each battery cell (Fig. 31).

If the specific gravity of the electrolyte is low, it suggests undercharging.

If the specific gravity shows the battery is fully charged but the electrolyte level is low, the battery may have been charged at too high a rate. This will be discussed later.

ADDING WATER TO BATTERIES

Fill each battery cell to the proper level. Battery waters are listed below in the order of their quality.

1. *Distilled water*
2. *Water from a dehumidifier or from defrosting*
3. *Rain water*
4. *Tap water*

The first two waters are best since they contain no minerals.

The third is not so good because rain water can collect minerals from the air.

The fourth is least desirable since it contains chlorine and carbonates which can shorten the life of the battery.

CAUTION: Avoid adding too much water to the batteries during freezing weather unless the engine is run long enough to mix the water with the electrolyte. Unmixed water can freeze and burst the battery.

LIGHT LOAD TEST OF BATTERIES

Make a "light load" test of the batteries to determine their condition.

Recharge or replace the batteries if they fail any part of the "light load" test.

Alternator or Generator

CHECKING BELT TENSION

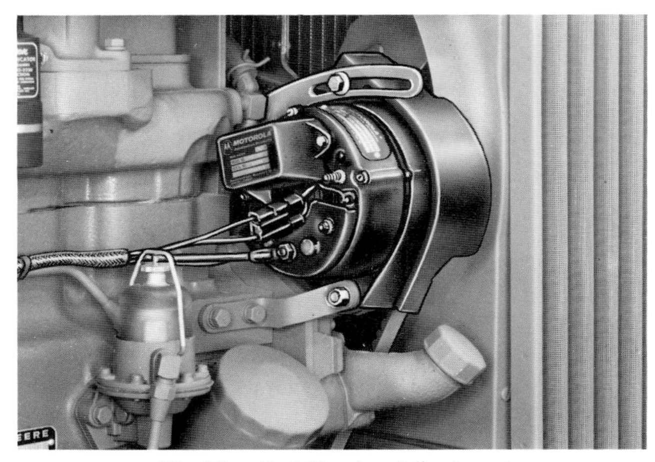

Fig. 32 — Belt Tension Adjustment

Check belt for proper tension. Fig. 32 shows a typical belt installation. Refer to the engine Technical Manual for the correct tension and how to adjust it.

TESTING GENERATING OUTPUT

Check the output of the alternator or generator. Refer to the engine Technical Manual for details.

IMPORTANT: Follow the manual procedure exactly. Making a wrong connection can severely damage an alternator.

Starting Circuit

SAFETY STARTER SWITCH

Some machines use a safety starter switch to prevent starting the engine when the machine is in gear.

Check the operation of the switch by operating the gear shift and trying to start the engine. Refer to the engine Technical Manual and replace or adjust the safety switch as necessary.

PREVENT MACHINE RUNAWAY

Avoid possible injury or death from machinery runaway.

Do not start the engine by shorting across the starter terminals. The machine will start in gear if normal circuitry is bypassed.

NEVER start the engine while standing on the ground. Start the engine only from the operator's seat, with the transmission in neutral or park.

Fig. 33 — Never Bypass The Neutral Start Switch

STARTING MOTOR

With the engine warmed up, check the ampere draw of the starting motor.

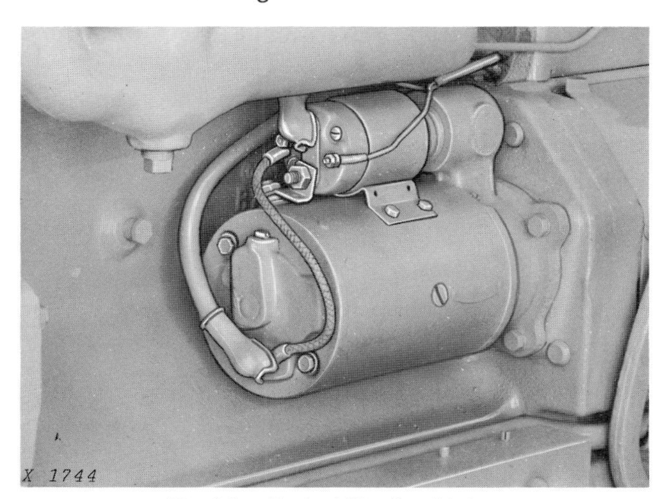

Fig. 34 — Typical Starting Motor

Disconnect the battery cable from the starting motor and use an ammeter connected as instructed in the engine Technical Manual. Record the amperage draw and check it against the specifications.

If the ampere draw is too high, look for worn starter bearings or some drag in the engine.

8. CLUTCH FREE TRAVEL

Clutch free travel is the distance the pedal travels before it starts to disengage the clutch from the engine flywheel.

Too little clutch free travel may cause the clutch release bearing to operate continuously, resulting in early failure. It can also cause slippage.

Too much free travel may prevent the clutch from completely disengaging and so wear it out.

To check free travel, pull the pedal down until you feel a contact at the flywheel. Measure this distance as shown in Fig. 35 and check it against the engine Technical Manual. If necessary, adjust as specified.

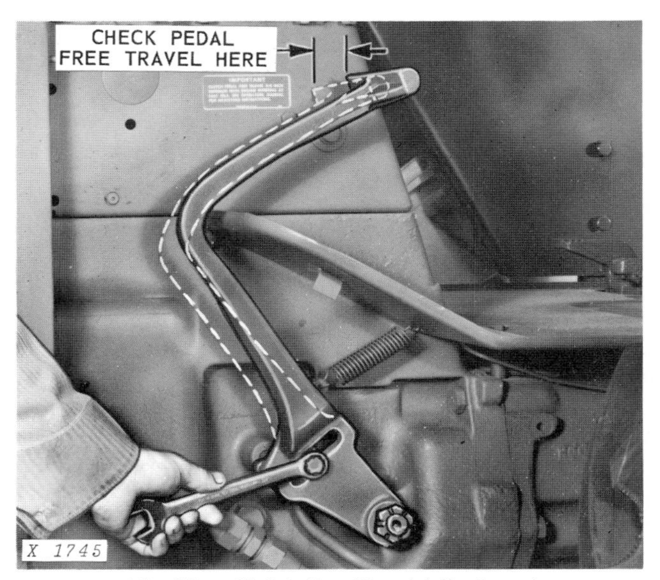

Fig. 35 — Clutch Free Travel Adjustment

9. DYNAMOMETER TEST

The dynamometer test is the final check of overall engine performance. It will tell you whether the tune-up has been adequate. Compare it with the dynamometer test made *before* tune-up.

Test for the following things:

1. **Engine Horsepower**
2. **Exhaust Analysis**

   Smoke analysis (diesel)

   Carburetor adjustment (spark-ignition)
3. **Fuel Consumption**
4. **Crankcase Blow-By**

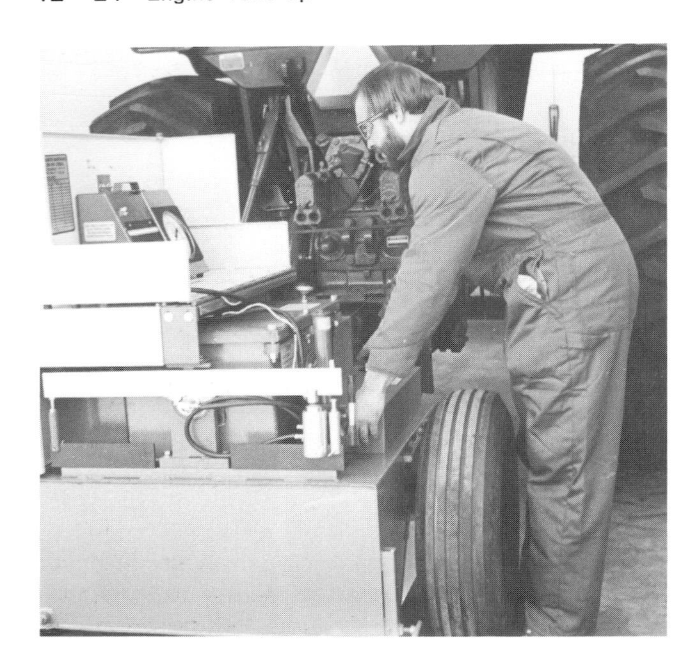

Fig. 36 — A Typical Dynamometer Test

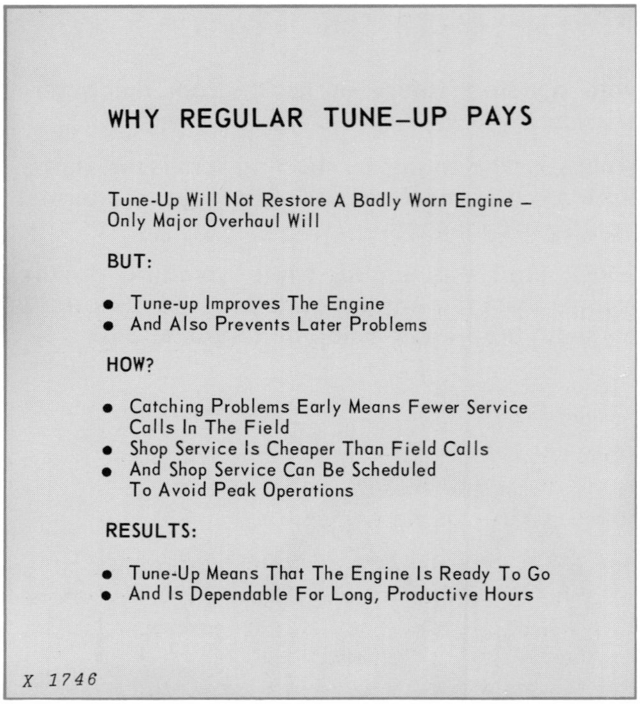

Most dynamometer manufacturers have instruments to be used with the dynamometer for checking the above items.

Use the engine Technical Manual for procedures and specifications.

NOTE: If the engine fails to produce the desired horsepower, and an air cleaner restriction test was not made at the beginning, make one now. It is possible that an air restriction is causing the loss of horsepower.

A completely tuned engine should pass the dynamometer test with no problems.

However, the engine should not put out *more* horsepower than it was designed for. Tampering with the engine to get extra horsepower will shorten engine life and raise operating costs. It may also void the engine warranty.

SUMMARY: ENGINE TUNE-UP

Tune-up of an engine may seem like a long and expensive ordeal.

Actually, most of the items can be checked in a minute or two.

But why check out so many items if the engine has not actually failed?

WHY REGULAR TUNE–UP PAYS

Tune-Up Will Not Restore A Badly Worn Engine – Only Major Overhaul Will

BUT:

- Tune-up Improves The Engine
- And Also Prevents Later Problems

HOW?

- Catching Problems Early Means Fewer Service Calls In The Field
- Shop Service Is Cheaper Than Field Calls
- And Shop Service Can Be Scheduled To Avoid Peak Operations

RESULTS:

- Tune-Up Means That The Engine Is Ready To Go
- And Is Dependable For Long, Productive Hours

X 1746

Fig. 37 — Why Regular Tune-Up Pays

The answer is that **tune-up is preventive maintenance.**

Before the engine fails, we keep it tuned up so that causes are corrected early, and possible causes are prevented.

Tune-up catches the problems early—in the shop, not in the field.

Shop costs are much cheaper than field costs, and by scheduling the tune-up during a lull in operation, costly downtime at peak periods is prevented.

Tune-up means that the engine is ready to go and the operator can depend on it for some long and productive hours on the job.

TEST YOURSELF

QUESTIONS

1. True or false? "A badly worn engine can be restored by a complete tune-up."
2. Before tuning up an engine, what should be done?
3. When should dynamometer tests be made?

(Answers in back of text.)

 Litho in U.S.A.

GLOSSARY

A

ABRASION—Wearing or rubbing away of a part.

ACTUATOR SOLENOID—The solenoid in the actuator housing on the back of the injection pump which moves the control rack as commanded by the engine controller unit.

ADDITIVE—A substance added to oil to give it certain properties. For example, a material added to engine oil to lessen its tendency to congeal or thicken at low temperatures.

AFTERCOOLER—A heat exchanger that cools the intake air, coming from the turbocharger, before it enters the cylinder. This creates more power, better economy, and quieter combustion.

AIR CLEANER—A device for filtering, cleaning, and removing dust from the air admitted to an engine.

ANTIFREEZE—A material such as alcohol, glycerin, etc., added to water to lower its freezing point.

ANTIFRICTION BEARING—A bearing constructed with balls, rollers or the like between the journal and the bearing surface to provide rolling instead of sliding friction.

ARC WELDING—A method of utilizing the heat of an electric current jumping an air gap to provide heat for welding metal.

ASME—American Society of Mechanical Engineers.

ATMOSPHERIC PRESSURE — The weight of the air at sea level; about 14.7 lbs. per square inch (100 kPa); less at higher altitudes.

AUXILIARY SPEED SENSOR—Engine speed sensor located on engine timing gear cover. Serves as back-up to primary engine speed sensor.

B

BACK-FIRE—Ignition of the mixture in the intake manifold by flame from the cylinder such as might occur from a leaking inlet valve.

BACKLASH—The clearance or "play" between two parts, such as meshed gears.

BACK-PRESSURE—A resistance to free flow, such as a restriction in the exhaust line.

BAFFLE OR BAFFLE PLATE—An obstruction for checking or deflecting the flow of gases or sound.

BALL BEARING—An anti-friction bearing consisting of a hardened inner and outer race with hardened steel balls interposed between the two races.

BEARING—A part in which a journal, pivot or the like turns or moves.

B.H.P. (Brake Horsepower)—A measurement of the power developed by an engine in actual operation. It subtracts the F.H.P. (friction losses) from the I.H.P. (pure horsepower).

BLOW-BY—A leakage or loss of compression past the piston ring between the piston and the cylinder.

BOILING POINT—The temperature at which bubbles or vapors rise to the surface of a liquid and escape.

BORE—The diameter of a hole, such as a cylinder; also to enlarge a hole as distinguished from making a hole with a drill.

BRAZE—To join two pieces of metal using a comparatively high-melting-point material. An example is to join two pieces of steel by using brass or bronze as a binder.

BREAK-IN—The process of wearing in to a desirable fit between the surfaces of two new or reconditioned parts.

BURNISH—To smooth or polish using a sliding tool under pressure.

BUSHING—A removable liner for a bearing.

BY-PASS—An alternate path for a flow of air or liquid.

C

CALIBRATE—To determine or adjust the graduation or scale of any measuring instrument.

CALORIFIC VALUE—A measure of the heating value of fuel.

Litho in U.S.A.

CALORIMETER—An instrument to measure the amount of heat given off by a substance when burned.

CALORY—The metric measurement of the amount of heat required to raise 1 gram of water from zero degrees to 1 degree Centigrade.

CAM-GROUND PISTON—A piston ground to a slightly oval shape which under the heat of operation becomes round.

CAMSHAFT—The shaft containing lobes or cams to operate the engine valves.

CARBON MONOXIDE—Gas formed by incomplete combustion. Colorless, odorless, and very poisonous.

CARBONIZE—The process of carbon formation within an engine, such as on the spark plugs and within the combustion chamber.

CARBURETOR—A device for automatically mixing gasoline fuel in the proper proportion with air to produce a combustible vapor.

CARBURETOR "ICING"—A term used to describe the formation of ice on a carburetor throttle plate during certain atmospheric conditions.

CETANE—Measure of ignition quality of diesel fuel—at what pressure and temperature the fuel will ignite and burn.

CHAMFER—A bevel or taper at the edge of a hole.

CHASE—To straighten up or repair damaged threads.

CHOKE—A device such as a valve placed in a carburetor air inlet to restrict the volume of air admitted.

COMBUSTION—The process of burning.

COMBUSTION CHAMBER—The volume of the cylinder above the piston with the piston on top center.

COMPRESSION—The reduction in volume or the "squeezing" of a gas. As applied to metal, such as a coil spring, compression is the opposite of tension.

COMPRESSION RATIO—The volume of the combustion chamber at the end of the compression stroke as compared to the volume of the cylinder and chamber with the piston on bottom center. Example: 8 to 1.

CONDENSATION—The process of a vapor becoming a liquid; the reverse of evaporation.

CONNECTING ROD—Rod that connects the piston to the crankshaft.

CONTRACTION—A reduction in mass or dimension; the opposite of expansion.

CONVECTION—A transfer of heat by circulating heated air.

CONVERTER—As used in connection with LP-Gas, a device which converts or changes LP-Gas from a liquid to a vapor for use by the engine.

CORRODE—To eat away gradually as if by gnawing, especially by chemical action, such as rust.

COUNTERBORE—To enlarge a hole to a given depth.

COUNTERSINK—To cut or form a depression to allow the head of a screw to go below the surface.

CRANKCASE—The lower housing in which the crankshaft and many other parts of the engine operate.

CRANKCASE DILUTION—When unburned fuel finds its way past the piston rings into the crankcase oil, where it dilutes or "thins" the engine lubricating oil.

CRANKSHAFT—The main drive shaft of an engine which takes reciprocating motion and converts it to rotary motion.

CRANKSHAFT COUNTER-BALANCE—A series of weights attached to or forged integrally with the crankshaft to offset the reciprocating weight of each piston and rod.

CROSSHEAD — Part of the valve train in an engine with two intake and two exhaust valves per cylinder. Permits two valves in the same cylinder to be opened at the same time.

CRUDE OIL—Liquid oil as it comes from the ground.

CU. IN.—Cubic Inch.

CYLINDER—A round hole having some depth bored to receive a piston; also sometimes referred to as "bore" or "barrel."

CYLINDER BLOCK—The largest single part of an engine. The basic or main mass of metal in which the cylinders are bored or placed.

CYLINDER HEAD—A detachable portion of an engine fastened securely to the cylinder block which contains all or a portion of the combustion chamber.

 Litho in U.S.A.

CYLINDER LINER—A sleeve or tube interposed between the piston and the cylinder wall or cylinder block to provide a readily renewable wearing surface for the cylinder.

D

DEAD CENTER—The extreme top or bottom position of the crankshaft throw at which the piston is not moving in either direction.

DENSITY—Compactness; relative mass of matter in a given volume.

DETERGENT—A compound of a soap-like nature used in engine oil to remove engine deposits and hold them in suspension in the oil.

DETONATION—A too-rapid burning or explosion of the mixture in the engine cylinders. It becomes audible through a vibration of the combustion chamber walls and is sometimes confused with a "ping" or spark "knock."

DIAGNOSIS—In engine service, the use of instruments to "trouble shoot" the engine parts to locate the cause of a failure.

DIAGNOSTIC CODE—A number which represents a problem detected by the engine control unit. Transmitted for use by on-board displays or a diagnostic reader, so operator or technician is aware there is a problem and where it can be found in the fuel injection system.

DIESEL ENGINE—Named after its developer, Dr. Rudolph Diesel. This engine ignites fuel in the cylinder from the heat generated by compression. The fuel is an "oil" rather than gasoline, and no spark plug or carburetor is required.

DILUTION—See Crankcase Dilution.

DISPLACEMENT—See Piston Displacement.

DISTORTION—A warpage or change in form from the original shape.

DOWEL PIN—A pin inserted in matching holes in two parts to maintain those parts in fixed relation one to the other.

DOWN-DRAFT—A type of carburetor in which the fuel-air mixture flows downward to the engine.

DRAWBAR HORSEPOWER—Measure of the pulling power of a machine at the drawbar hitch point.

DYNAMOMETER—A test unit for measuring the actual power produced by an engine.

E

ECCENTRIC—One circle within another circle but with a different center of rotation. An example of this is a driving cam on a camshaft.

ECONOMIZER—A device installed in a carburetor to control the amount of fuel used under certain conditions.

ECU (Engine Controller Unit)—An electronic module which controls fuel delivery, diagnostic outputs, back-up operation and communications with other electronic modules.

ENERGY—The capacity for doing work.

ENGINE—The prime source of power generation used to propel the machine.

ENGINE DISPLACEMENT—The sum of the displacement of all the engine cylinders. See Piston Displacement.

EVAPORATION—The process of changing from a liquid to a vapor, such as boiling water to produce steam. Evaporation is the opposite of condensation.

EXHAUST GAS ANALYZER—An instrument for determining the efficiency with which an engine is burning fuel.

EXHAUST MANIFOLD—The passages from the engine cylinders to the muffler which conduct the exhaust gases away from the engine.

EXPANSION—An increase in size. For example, when a metal rod is heated it increases in length and perhaps also in diameter. Expansion is the opposite of contraction.

F

F.H.P. (Friction Horsepower)—A measure of the power lost to the engine through friction or rubbing of parts.

FILTER—(Oil, Water, Gasoline, Etc.)—A unit containing an element, such as a screen of varying degrees of fineness. The screen or filtering element is made of various materials depending upon the size of the foreign particles to be eliminated from the fluid being filtered.

FLASH POINT—The temperature at which an oil, when heated, will flash and burn.

FLOATING PISTON PIN—A piston pin which is not locked in the connecting rod or the piston, but is free to turn or oscillate in both the connecting rod and the piston.

FLOAT LEVEL—The height of the fuel in the carburetor bowl, usually regulated by means of a suitable valve and float.

"FLUTTER" OR "BOUNCE"—In engine valves, refers to a condition where the valve is not held tightly on its seat during the time the cam is not lifting it.

FLYWHEEL—A heavy wheel in which energy is absorbed and stored by means of momentum.

FOOT-POUND (Ft.-Lb.)—This is a measure of the amount of energy or work required to lift one pound a distance of one foot.

FOUR-STROKE-CYCLE ENGINE—Also known as Otto cycle, where an explosion occurs every other revolution of the crankshaft. These strokes are (1) intake stroke; (2) compression stroke) (3) power stroke; (4) exhaust stroke.

FUEL KNOCK—Same as Detonation.

G

GAS—A substance which can be changed in volume and shape according to the temperature and pressure applied to it. For example, air is a gas which can be compressed into smaller volume and into any shape desired by pressure. It can also be expanded by the application of heat.

GEAR RATIO—The number of revolutions made by a driving gear as compared to the number of revolutions made by a driven gear of different size. For example, if one gear makes three revolutions while the other gear makes one revolution, the gear ratio would be 3 to 1.

GLAZE—As used to describe the surface of the cylinder, an extremely smooth or glossy surface such as a cylinder wall highly polished over a long period of time by the friction of the piston pin.

GLAZE BREAKER—A tool for removing the glossy surface finish in an engine cylinder.

GOVERNOR—A device to control and regulate speed. May be mechanical, hydraulic, or electrical.

GRIND—To finish or polish a surface by means of an abrasive wheel.

H

HEAT EXCHANGER—Sometimes used to describe a Vaporizer.

HEAT TREATMENT—A combination of heating and cooling operations timed and applied to a metal in a solid state in a way that will produce desired properties.

HONE—An abrasive tool for correcting small irregularities or differences in diameter in a cylinder.

HORSEPOWER (HP)—The energy required to lift 550 lbs. one ft. in one second.

HOT SPOT—Refers to a comparatively thin section or area of the wall between the inlet and exhaust manifold of an engine, the purpose being to allow the hot exhaust gases to heat the comparatively cool incoming mixture. Also used to designate local areas of the cooling system which have attained above average temperatures.

I

I.D.—Inside diameter.

IDLE—Refers to the engine operating at its slowest speed with a machine under no load.

I.H.P. (Indicated Horsepower)—"Pure" horsepower as measured in the combustion chamber before friction and other losses are subtracted.

IN.—Inch.

INERTIA—A physical law which tends to keep a motionless body at rest or also tends to keep a moving body in motion; effort is thus required to start a mass moving or to retard or stop it once it is in motion.

INHIBITOR—A material to restrain some unwanted action, such as a rust inhibitor which is a chemical added to cooling systems to retard the formation of rust.

INJECTION PUMP (Diesel)—A device by means of which the fuel is metered and delivered under pressure to the injector.

INJECTOR (Diesel)—An assembly which receives a metered charge of fuel from another source at relatively low pressure, then is actuated to inject the charge of fuel into a cylinder or chamber at high pressure and at the proper time.

INPUT SHAFT—The shaft carrying the driving gear, such as in a transmission by which the power is applied.

INTAKE MANIFOLD—The passages which conduct the fuel-air mixture from the carburetor to the engine cylinders.

INTAKE VALVE—A valve which permits a fluid or gas to enter a chamber and seals against exit.

INTEGRAL—The whole made up of parts.

INTERNAL COMBUSTION—The burning of a fuel within an enclosed space.

J

JOURNAL—A part or support within which a shaft operates.

K

KEY—A small block inserted between the shaft and hub to prevent circumferential movement.

KEYWAY OR KEYSEAT—A groove or slot cut for inserting a key to hold a part on a shaft, etc.

KNOCK—A general term used to describe various noises occurring in an engine; may be used to describe noises made by loose or worn mechanical parts, preignition, detonation, etc.

KNURLED—Displacing metal on the skirt of a piston to increase the diameter of the piston (on automotive engines).

L

LACQUER—A solution of solids in solvents which evaporate with great rapidity.

LAPPING—The process of fitting one surface to another by rubbing them together with an abrasive material between the two surfaces.

L-HEAD ENGINE—An engine design in which both valves are located on one side of the engine cylinder.

LINER—Usually a thin section placed between two parts, such as a replaceable cylinder liner in an engine.

LP-GAS, LIQUEFIED PETROLEUM GAS—Made usable as a fuel for internal combustion engines by compressing volatile petroleum gases to liquid form. When so used, must be kept under pressure or at low temperature in order to remain in liquid form, until used by the engine.

M

MANIFOLD—A pipe or casting with multiple openings used to connect various cylinders to one inlet or outlet.

MANOMETER—A device for measuring a vacuum. It is a U-shaped tube partially filled with fluid. One end of the tube is open to the air and the other is connected to the chamber in which the vacuum is to be measured. A column of Mercury 30 in. high equals 14.7 lbs. per square in. which is atmospheric pressure at sea level. Readings are given in terms of inches of Mercury.

MECHANICAL EFFICIENCY (Engine)—The ratio between the indicated horsepower and the brake horsepower of an engine.

MICROMETER—A measuring instrument for either external or internal measurement in thousandths and sometimes tenths of thousandths of inches.

MISFIRING—Failure of an explosion to occur in one or more cylinders while the engine is running; may be a continuous or intermittent failure.

MOTOR—This term should be used in connection with an electric motor and should not be used when referring to the *engine* of a machine.

MPH—Miles per hour.

MUFFLER—A chamber attached to the end of the exhaust pipe which allows the exhaust gases to expand and cool. It is usually fitted with baffles or porous plates and serves to subdue much of the noise created by the exhaust.

N

NEEDLE BEARING—An antifriction bearing using a great number of thin rollers.

O

OCTANE—Measurement which indicates the tendency of a fuel to detonate or knock.

O.D.—Outside diameter.

OIL PUMPING—A term used to describe an engine which is using an excessive amount of lubrication oil.

OTTO CYCLE—Also called four-stroke cycle. Named after the man who adopted the principle of four cycles of operation for each explosion in an engine cylinder. They are (1) intake stroke, (2) compression stroke, (3) power stroke, (4) exhaust stroke.

P

PEEN—To stretch or clinch over by pounding with the rounded end of a hammer.

PETROLEUM—A group of liquid and gaseous compounds composed of carbon and hydrogen which are removed from the earth.

PINION—A small gear having the teeth formed on the hub.

PISTON—A cylindrical part closed at one end which is connected to the crankshaft by the connecting rod. The force of the expansion in the cylinder is exerted against the closed end of the piston, causing the connecting rod to move the crankshaft.

Litho in U.S.A.

PISTON COLLAPSE—A condition describing a collapse or a reduction in diameter of the piston skirt due to heat or stress.

PISTON DISPLACEMENT—The volume of air moved or displaced by moving the piston from one end of its stroke to the other.

PISTON HEAD—That part of the piston above the rings.

PISTON LANDS—Those parts of a piston between the piston rings.

PISTON PIN—The journal for the bearing in the small end of an engine connecting rod which also passes through piston walls; also known as a wrist pin.

PISTON RING—An expanding ring placed in the grooves of the piston to seal off the passage of fluid or gas past the piston.

PISTON RING EXPANDER—A spring placed behind the piston ring in the groove to increase the pressure of the ring against the cylinder wall.

PISTON RING GAP—The clearance between the ends of the piston ring.

PISTON RING GROOVE—The channel or slots in the piston in which the piston rings are placed.

PISTON SKIRT—That part of the piston below the rings.

PORT—The openings in the cylinder block for valves, exhaust and inlet pipes, or water connections. In two-cycle engines, the openings for intake and exhaust.

PREIGNITION—Ignition occurring earlier than intended. For example, the explosive mixture being fired in a cylinder as by a flake of incandescent carbon before the electric spark occurs.

PRESS-FIT—Also known as a force-fit or drive-fit. This term is used when the shaft is slightly larger than the hole and must be forced into place.

PRIMARY SPEED SENSOR—A magnetic pickup which generates voltage pulses to the ECU as the teeth on the speed wheel pass by the tip of the sensor. Located within the actuator housing.

PSI—A measurement of pressure in Pounds per Square Inch.

PUSH ROD—A connecting link in an operating mechanism, such as the rod interposed between the valve lifter and rocker arm on an overhead valve engine.

R

RACE—As used with reference to bearings; a finished inner and outer surface in which or on which balls or rollers operate.

RATED HORSEPOWER—Value used by the engine manufacturer to rate the power of his engine, allowing for safe loads, etc.

RATIO—The relation or proportion of one number or quantity to another.

REAM—To finish a hole accurately with a rotating fluted tool.

RECIPROCATING MOTION—A back and forth movement, such as the action of a piston in a cylinder.

ROCKER ARM—In an engine a lever located on a fulcrum or shaft, one end on the valve stem, the other on the push rod.

ROLLER BEARING—An inner and outer race upon which hardened steel rollers operate.

ROTARY MOTION—A circular movement, such as the rotation of a crankshaft.

ROTOR—Rotating valve or conductor for carrying fluid or electrical current from a central source to the individual outlets as required.

RPM—Revolutions per minute.

RUNNING-FIT—Where sufficient clearance has been allowed between the shaft and journal to allow free running without overheating.

S

S.A.E.—Society of Automotive Engineers. This group sets the standards for much engine design.

SAFETY FACTOR—The degree of surplus strength over and above normal requirements which serves as insurance against failure.

SAND BLAST—To clean a surface by means of sand propelled by compressed air.

SCALE—A flaky deposit occurring on steel or iron. Ordinarily used to describe the accumulation of minerals and metals accumulating in an engine cooling system.

SCORE—A scratch, ridge or groove marring a finished surface.

SEAT—A surface, usually machined, upon which another part rests or seats. For example, the surface upon which a valve face rests.

SHIM—Thin sheets used as spacers between two parts, such as the two halves of a journal bearing.

SHRINK-FIT—Where the shaft or parts are slightly larger than the hole in which it is to be inserted. The outer part is heated above its normal operating temperature or the inner part chilled below its normal operating temperature and assembled in this condition; upon cooling an exceptionally tight fit is obtained.

SLIDING-FIT—Where sufficient clearance has been allowed between the shaft and journal to allow free running without overheating.

SLIP-IN BEARING—A liner made to precise measurements which can be used for replacement without additional fitting.

SLUDGE—A composition of oxidized petroleum products along with an emulsion formed by the mixture of oil and water. This forms a pasty substance and clogs oil lines and passages and interferes with engine lubrication.

SOLID INJECTION—The system used in diesel engines where fuel as a fluid is injected into the cylinder rather than a mixture of fuel and air.

SOLVENT—A solution which dissolves some other material. For example, water is a solvent for sugar.

SPLINE—A long keyway.

SPOT WELD—To attach in spots by localized fusion of the metal parts with the aid of an electric current.

SQ. FT.—Square feet.

SQ. IN.—Square inch.

STRESS—The force or strain to which a material is subjected.

STROKE—The distance moved by the piston.

STUDS—A rod with threads cut on both ends, such as a cylinder stud which screws into the cylinder block on one end and has a nut placed on the other end to hold the cylinder head in place.

SUCTION—Suction exists in a vessel when the pressure is lower than the atmospheric pressure; also see Vacuum.

SUPERCHARGER—A blower or pump which forces air into the cylinders at higher-than-atmospheric pressure. The increased pressure forces more air into the cylinder, thus enabling more fuel to be burned and more power produced.

SYNCHRONIZE—To cause two events to occur at the same time. For example, to time a mechanism so that two or more sparks willl occur at the same instant.

T

TACHOMETER—A device for measuring and showing the rotating speed of an engine.

TAP—To cut threads in a hole with a tapered, fluted, threaded tool.

TAPPET—The adjusting device for varying the clearance between the valve stem and the cam. May be built into the valve lifter in an engine or may be installed in the rocker arm on an overhead-valve engine.

T.D.C.—Top dead center (of a piston).

T-HEAD ENGINE—An engine design wherein the inlet valves are placed on one side of the cylinder and the exhaust valves are placed on the other.

THERMAL EFFICIENCY—A gallon of fuel contains a certain amount of potential energy in the form of heat when burned in the combustion chamber. Some of this heat is lost and some is converted into power. The thermal efficiency is the ratio of work accomplished to the total quantity of heat in the fuel.

THERMOSTAT—A heat-controlled valve used in the cooling system of an engine to regulate the flow of water between the cylinder block and the radiator.

THROW—The distance from the center of the crankshaft main bearing to the center of the connecting rod journal.

TIMING GEARS—Any group of gears which are driven from the engine crankshaft to cause the valves, ignition and other engine-driven accessories to operate at the desired time during the engine cycle.

Litho in U.S.A.

TOLERANCE—A permissible variation between the two extremes of a specification of dimensions. Used in the precision fitting of mechanical parts.

TORQUE—The effort of twisting or turning.

TORQUE WRENCH—A special wrench with a built-in indicator to measure the applied turning force.

TROUBLE SHOOTING—A process of diagnosing or locating the source of the trouble or troubles from observation and testing. Also see Diagnosis.

TUNE-UP—A process of accurate and careful adjustments to obtain the best engine performance.

TURBINE—A series of angled blades located on a wheel against which fluids or gases are impelled to rotate a shaft.

TURBULENCE—A disturbed, irregular motion of fluids or gases.

TWO-CYCLE ENGINE—An engine design permitting a power stroke once for each revolution of the crankshaft.

U

UP-DRAFT—A carburetor type in which the mixture flows upward to the engine.

V

VACUUM—A perfect vacuum has not been created as this would involve an absolute lack of pressure. The term is ordinarily used to describe a partial vacuum; that is, a pressure less than atmospheric pressure; in other words a depression.

VACUUM GAUGE—An instrument designed to measure the degree of vacuum existing in a chamber.

VALVE—A device for opening and sealing the cylinder intake and exhaust ports.

VALVE CLEARANCE—The air gap allowed between the end of the valve stem and the valve lifter or rocker arm to compensate for expansion due to heat.

VALVE FACE—That part of a valve which mates with and rests upon a seating surface.

VALVE GRINDING—Also called valve lapping. A process of lapping or mating the valve seat and valve face usually performed with the aid of an abrasive.

VALVE HEAD—The portion of the valve upon which the valve face is machined.

VALVE-IN-HEAD ENGINE—Same as Overhead Valve Engine.

VALVE LIFTER—A push rod or plunger placed between the cam and the valve on an engine; is often adjustable to vary the length of the unit.

VALVE MARGIN—On a poppet valve, the space or rim between the surface of the head and the surface of the valve face.

VALVE SEAT—The matched surface upon which the valve face rests.

VALVE SPRING—A spring attached to a valve to return it to the seat after it has been released from the lifting or opening means.

VALVE STEM—That portion of a valve which rests within a guide.

VALVE STEM GUIDE—A bushing or hole in which the valve stem is placed which allows lateral motion only.

VANES—Any plate, blade or the like attached to an axis and moved by or in air or a liquid.

VAPORIZER—A device for transforming or helping to transform a liquid into a vapor; often includes the application of heat.

VAPOR LOCK—A condition wherein the fuel boils in the fuel system forming bubbles which retard or stop the flow of fuel to the carburetor.

VENTURI—Two tapering streamlined tubes joined at their small ends so as to reduce the internal diameter.

VIBRATION DAMPER—A device to reduce the torsional or twisting vibration which occurs along the length of the crankshaft used in multi-cylinder engines; also known as a harmonic balancer.

VISCOSITY—The resistance to flow of an oil.

VOLATILITY—The tendency for a fluid to evaporate rapidly or pass off in the form of vapor. For example, gasoline is more volatile than kerosene as it evaporates at a lower temperature.

VORTEX—A whirling movement of mass of liquid or air.

W

WRIST PIN—The journal for the bearing in the small end of an engine connecting rod which also passes through piston walls; also known as a Piston Pin.

WEIGHTS AND MEASURES—METRIC TO U.S.

Metric System / **U.S. Equivalent**

LENGTH

Unit	*Abbreviation*	*Number of Meters*	U.S. Equivalent
Kilometer	km	1,000	0.62 mile
Hectometer	hm	100	109.36 yards
Decameter	dkm	10	32.81 feet
Meter	m	1	39.37 inches
Decimeter	dm	0.1	3.94 inches
Centimeter	cm	0.01	0.39 inch
Millimeter	mm	0.001	0.04 inch

AREA

Unit	*Abbreviation*	*Number of Square Meters*	U.S. Equivalent
Square Kilometer	sq km or km^2	1,000,000	0.3861 square mile
Hectare	ha	10,000	2.47 acres
Are	a	100	119.60 square yards
Centare	ca	1	10.76 square feet
Square Centimeter	sq cm or cm^2	0.0001	0.155 square inch

VOLUME

Unit	*Abbreviation*	*Number of Cubic Meters*	U.S. Equivalent
Stere	s	1	1.31 cubic yards
Decistere	ds	0.10	3.53 cubic feet
Cubic Centimeter	cu cm or cm^3 also cc	0.000001	0.061 cubic inch

CAPACITY

Unit	*Abbreviation*	*Number of Liters*	*Cubic*	*Dry*	*Liquid*
Kiloliter	kl	1,000	1.31 cubic yards		
Hectoliter	hl	100	3.53 cubic feet	2.84 bushels	
Decaliter	dkl	10	0.35 cubic foot	1.14 pecks	2.64 gallons
Liter	l	1	61.02 cubic inches	0.908 quart	1.057 quarts
Deciliter	dl	0.10	6.1 cubic inches	0.18 pint	0.21 pint
Centiliter	cl	0.01	0.6 cubic inch		0.338 fluidounce
Milliliter	ml	0.001	0.06 cubic inch		0.27 fluidram

MASS AND WEIGHT

Unit	*Abbreviation*	*Number of Grams*	U.S. Equivalent
Metric Ton	MT or t	1,000,000	1.1 tons
Quintal	q	100,000	220.46 pounds
Kilogram	kg	1,000	2.2046 pounds
Hectogram	hg	100	3.527 ounces
Decagram	dkg	10	0.353 ounce
Gram	g or gm	1	0.035 ounce
Decigram	dg	0.10	1.543 grains
Centigram	cg	0.01	0.154 grain
Milligram	mg	0.001	0.015 grain

POWER

Unit	*Abbreviation*	U.S. Equivalent
Kilowatt	kW	1.34 horsepower

HEAT

Unit	*Abbreviation*	U.S. Equivalent
Watt	W	3.41 Btu

PRESSURE OR VACUUM

Unit	*Abbreviation*	U.S. Equivalent
Kilopascal	kPa	0.145 pounds per sq. inch

TORQUE

Unit	*Abbreviation*	U.S. Equivalent
Newton-meter	N•m	142.86 ounce inch 8.85 pound inch 0.74 pound foot

TEMPERATURE

Unit	*Abbreviation*	U.S. Equivalent
Degrees Celsius	°C	$(°C \times 1.8) + 32 = °F$

 Litho in U.S.A.

WEIGHTS AND MEASURES—U.S. TO METRIC

U.S. System			Metric Equivalent
LENGTH			
Unit	*Abbreviation*	*Equivalents In Other Units*	
Mile	mi	5280 feet, 320 rods, 1760 yards	1.609 kilometers
Rod	rd	5.50 yards, 16.5 feet	5.029 meters
Yard	yd	3 feet, 36 inches	0.914 meters
Foot	ft. or ′	12 inches, 0.333 yards	30,480 centimeters
Inch	in or ″	0.083 feet, 0.027 yards	2.540 centimeters
AREA			
Unit	*Abbreviation*	*Equivalents In Other Units*	
Square Mile	sq mi or m²	640 acres, 102,400 square rods	2.590 square kilometers
Acre	a	4840 square yards, 43,560 square feet	0.405 hectares, 4047 square meters
Square Rod	sq rd or rd²	30.25 square yards, 0.006 acres	25.293 square meters
Square Yard	sq yd or yd²	1296 square inches, 9 square feet	0.836 square meters
Square Foot	sq ft or ft²	144 square inches, 0.111 square yards	0.093 square meters
Square Inch	sq in or in²	0.007 square feet, 0.00077 square yards	6.451 square centimeters
VOLUME			
Unit	*Abbreviation*	*Equivalents In Other Units*	
Cubic Yard	cu yd or yd³	27 cubic feet, 46,656 cubic inches	0.765 cubic meters
Cubic Foot	cu ft or ft³	1728 cubic inches, 0.0370 cubic yards	0.028 cubic meters
Cubic Inch	cu in or in³	0.00058 cubic feet, 0.000021 cubic yards	16.387 cubic centimeters
MASS AND WEIGHT			
Unit	*Abbreviation*	*Equivalents In Other Units*	
Ton	tn (seldom used)		
short ton		20 short hundredweight, 2000 pounds	0.907 metric tons
long ton		20 long hundredweight, 2240 pounds	1.016 metric tons
Hundredweight	cwt		
short hundredweight		100 pounds, 0.05 short tons	45.359 kilograms
long hundredweight		112 pounds, 0.05 long tons	50.802 kilograms
Pound	lb or lb av also #	16 ounces, 7000 grains	0.453 kilograms
Ounce	oz or oz av	16 drams, 437.5 grains	28.349 grams
Dram	dr or dr av	27.343 grains, 0.0625 ounces	1.771 grams
Grain	gr	0.036 drams, 0.002285 ounces	0.0648 grams
CAPACITY			
Unit	*Abbreviation*	*U.S. Liquid Measure*	
Gallon	gal	4 quarts (231 cubic inches)	3.785 liters
Quart	qt	2 pints (57.75 cubic inches)	0.946 liters
Pint	pt	4 gills (28.875 cubic inches)	0.473 liters
Gill	gi	4 fluidounces (7.218 cubic inches)	118.291 milliliters
Fluidounce	fl oz	8 fluidrams (1.804 cubic inches)	29.573 milliliters
Fluidram	fl dr	60 minims (0.225 cubic inches)	3.696 milliliters
Minim	min	1/60 fluidram (0.003759 cubic inches)	0.061610 milliliters
		U.S. Dry Measure	
Bushel	bu	4 pecks (2150.42 cubic inches)	35.238 liters
Peck	pk	8 quarts (537.605 cubic inches)	8.809 liters
Quart	qt	2 pints (67.200 cubic inches)	1.101 liters
Pint	pt	½ quart (33.600 cubic inches)	0.550 liters
POWER			
Unit	*Abbreviation*		
Horsepower	hp		0.746 kilowatt
HEAT			
Unit	*Abbreviation*		
British thermal unit	Btu		0.293 watt
PRESSURE OR VACUUM			
Unit	*Abbreviation*		
Pounds per square inch	psi		6.895 kilopascal
TORQUE			
Unit	*Abbreviation*		
ounce inch	oz/in.		0.007 newton-meters
ounce inch	lb/in.		0.113 newton-meters
pound foot	lb/ft		1.346 newton-meters
TEMPERATURE			
Unit	*Abbreviation*		
Fahrenheit	°F		(°F - 32) 0.556 = °C

 Litho in U.S.A.

UNIFIED INCH BOLT AND CAP SCREW TORQUE VALUES

SAE Grade and Head Markings	NO MARK	5 5.1 5.2	8 8.2
	1 or 2[b]		
SAE Grade and Nut Markings	NO MARK	5	8
	2		

Size	Grade 1				Grade 2[b]				Grade 5, 5.1, or 5.2				Grade 8 or 8.2			
	Lubricated[a]		Dry[a]		Lubricated[a]		Dry[a]		Lubricated[a]		Dry[a]		Lubricated[a]		Dry[a]	
	N·m	lb-ft	N·m	lb-ft	N·m	lb-ft	N·m	lb-ft	N·m	lb-ft	N·m	lb-ft	N·m	lb-ft	N·m	lb-ft
1/4	3.7	2.8	4.7	3.5	6	4.5	7.5	5.5	9.5	7	12	9	13.5	10	17	12.5
5/16	7.7	5.5	10	7	12	9	15	11	20	15	25	18	28	21	35	26
3/8	14	10	17	13	22	16	27	20	35	26	44	33	50	36	63	46
7/16	22	16	28	20	35	26	44	32	55	41	70	52	80	58	100	75
1/2	33	25	42	31	53	39	67	50	85	63	110	80	120	90	150	115
9/16	48	36	60	45	75	56	95	70	125	90	155	115	175	130	225	160
5/8	67	50	85	62	105	78	135	100	170	125	215	160	215	160	300	225
3/4	120	87	150	110	190	140	240	175	300	225	375	280	425	310	550	400
7/8	190	140	240	175	190	140	240	175	490	360	625	450	700	500	875	650
1	290	210	360	270	290	210	360	270	725	540	925	675	1050	750	1300	975
1-1/8	470	300	510	375	470	300	510	375	900	675	1150	850	1450	1075	1850	1350
1-1/4	570	425	725	530	570	425	725	530	1300	950	1650	1200	2050	1500	2600	1950
1-3/8	750	550	950	700	750	550	950	700	1700	1250	2150	1550	2700	2000	3400	2550
1-1/2	1000	725	1250	925	990	725	1250	930	2250	1650	2850	2100	3600	2650	4550	3350

DO NOT use these values if a different torque value or tightening procedure is given for a specific application. Torque values listed are for general use only. Check tightness of fasteners periodically.

Shear bolts are designed to fail under predetermined loads. Always replace shear bolts with identical grade.

Fasteners should be replaced with the same or higher grade. If higher grade fasteners are used, these should only be tightened to the strength of the original.

Make sure fasteners threads are clean and that you properly start thread engagement. This will prevent them from failing when tightening.

Tighten plastic insert or crimped steel-type lock nuts to approximately 50 percent of the dry torque shown in the chart, applied to the nut, not to the bolt head. Tighten toothed or serrated-type lock nuts to the full torque value.

[a] *"Lubricated" means coated with a lubricant such as engine oil, or fasteners with phosphate and oil coatings. "Dry" means plain or zinc plated without any lubrication.*

[b] *Grade 2 applies for hex cap screws (not hex bolts) up to 152 mm (6-in.) long. Grade 1 applies for hex cap screws over 152 mm (6-in.) long, and for all other types of bolts and screws of any length.*

METRIC BOLT AND CAP SCREW TORQUE VALUES

Property Class and Head Markings	4.8	8.8 / 9.8	10.9	12.9
Property Class and Nut Markings	5	10	10	12

Size	Class 4.8				Class 8.8 or 9.8				Class 10.9				Class 12.9			
	Lubricated[a]		Dry[a]		Lubricated[a]		Dry[a]		Lubricated[a]		Dry[a]		Lubricated[a]		Dry[a]	
	N·m	lb-ft	N·m	lb-ft	N·m	lb-ft	N·m	lb-ft	N·m	lb-ft	N·m	lb-ft	N·m	lb-ft	N·m	lb-ft
M6	4.8	3.5	6	4.5	9	6.5	11	8.5	13	9.5	17	12	15	11.5	19	14.5
M8	12	8.5	15	11	22	16	28	20	32	24	40	30	37	28	47	35
M10	23	17	29	21	43	32	55	40	63	47	80	60	75	55	95	70
M12	40	29	50	37	75	55	95	70	110	80	140	105	130	95	165	120
M14	63	47	80	60	120	88	150	110	175	130	225	165	205	150	260	190
M16	100	73	125	92	190	140	240	175	275	200	350	225	320	240	400	300
M18	135	100	175	125	260	195	330	250	375	275	475	350	440	325	560	410
M20	190	140	240	180	375	275	475	350	530	400	675	500	625	460	800	580
M22	260	190	330	250	510	375	650	475	725	540	925	675	850	625	1075	800
M24	330	250	425	310	650	475	825	600	925	675	1150	850	1075	800	1350	1000
M27	490	360	625	450	950	700	1200	875	1350	1000	1700	1250	1600	1150	2000	1500
M30	675	490	850	625	1300	950	1650	1200	1850	1350	2300	1700	2150	1600	2700	2000
M33	900	675	1150	850	1750	1300	220	1650	2500	1850	3150	2350	2900	2150	3700	2750
M36	1150	850	1450	1075	2250	1650	2850	2100	3200	2350	4050	3000	3750	2750	4750	3500

DO NOT use these values if a different torque value or tightening procedure is given for a specific application. Torque values listed are for general use only. Check tightness of fasteners periodically.

Shear bolts are designed to fail under predetermined loads. Always replace shear bolts with identical property class.

Fasteners should be replaced with the same or higher property class. If higher property class fasteners are used, these should only be tightened to the strength of the original.

[a] *"Lubricated means coated with a lubricant such as engine oil, or fasteners with phosphate and oil coatings. "Dry means plain or zinc plated without any lubrication.*

Make sure fasteners threads are clean and that you properly start thread engagement. This will prevent them from failing when tightening.

Tighten plastic insert or crimped steel-type lock nuts to approximately 50 percent of the dry torque shown in the chart, applied to the nut, not to the bolt head. Tighten toothed or serrated-type lock nuts to the full torque value.

Litho in U.S.A.

ENGINE FLYWHEEL HOUSINGS—SAE J617c

SAE Standard

Report of Flywheel Division approved March 1916 and last revised by Construction Machinery Technical Committee June 1976.

Scope—This SAE standard defines flywheel housing and mating flange configuration for industry standardization, interchangeability, and compatibility.

Dimensions and tolerances shown are millimeter (inch).

Table 1 and the figure give dimensions for flywheel housings. The figure also shows spacing for eight, twelve, and sixteen bolt flange mounting patterns.

Mating Housing Flanges—The capscrew holes on the mating housing flanges shall be 1.19 (0.047) larger than the nominal diameter of the capscrews used on the flywheel housing.

The diameter of the pilot on the flange of the mating housing shall be the same as the nominal diameter of the bore in the flywheel housing; the tolerances shall be +0.000 and −0.13 (0.005), and the maximum eccentricity shall be 0.064 (0.0025) [indicated runout 0.13 (0.005)].

The mating housing flange pilot diameter shall be 6.4 (0.25) long, and its lead-in chamfer shall not exceed 2.0 (0.08) in length. The fillet radius between the mounting flange face and the pilot diameter shall not exceed 1.0 (0.04) R.

The maximum variation of the face of the mating housing flange from its true position, when rotated about its axis, shall be 0.064 (0.0025) [indicated runout 0.13 (0.005)].

CONSTRUCTION AND INDUSTRIAL EQUIPMENT

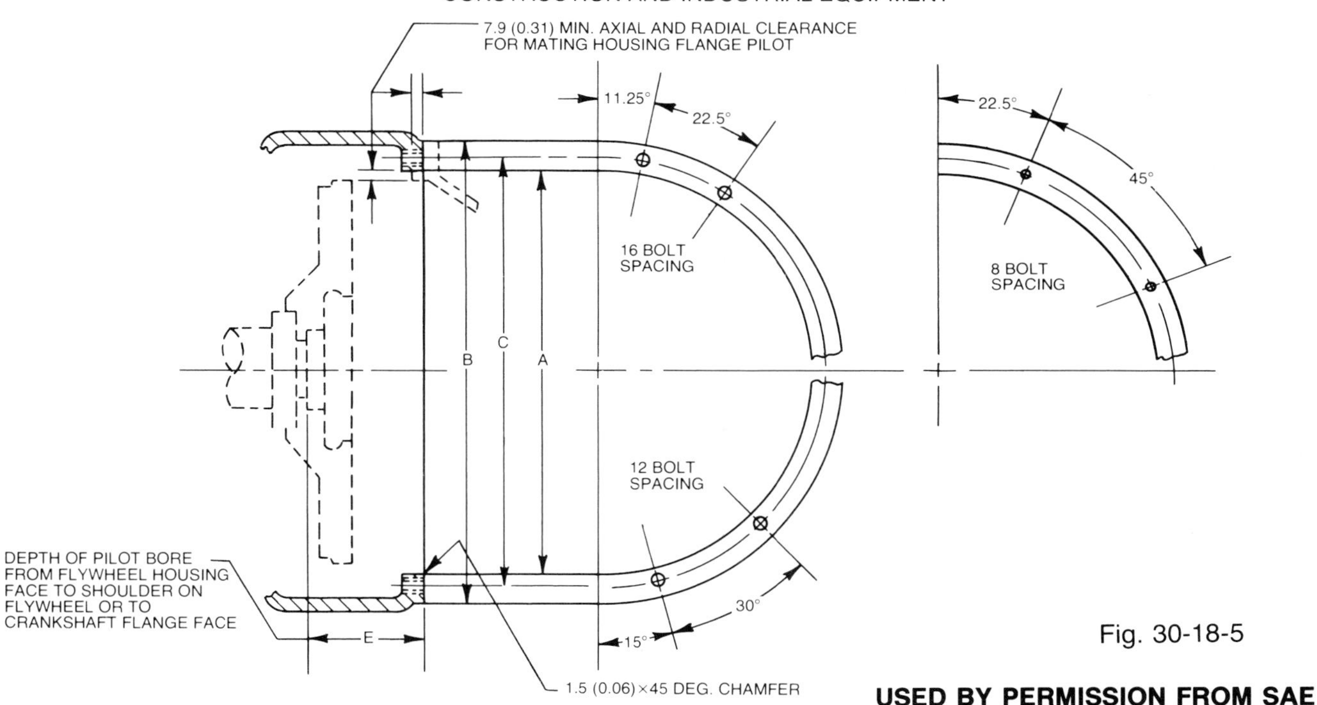

Fig. 30-18-5

USED BY PERMISSION FROM SAE

TABLE 1—FLYWHEEL HOUSING DIMENSIONS, mm (in)

SAE No.	A	Tolerance[a] Bore Diameter −0.000	Tolerance[a] Bore Eccentricity	Tolerance[a] Face Deviation	B	C	E[b]	Tapped Holes No.	Tapped Holes Size[c]
00	787.40 (31.000)	+0.25 (0.010)	0.30 (0.012)		882.6 (34.75)	850.90 (33.500)	100.1 (3.94)	16	1/2-13
0	647.70 (25.500)	+0.25 (0.010)	0.25 (0.010)		711.2 (28.00)	679.45 (26.750)	100.1 (3.94)	16	1/2-13
1/2	584.20 (23.000)	+0.20 (0.008)	0.25 (0.010)		647.7 (25.50)	619.12 (24.375)	100.1 (3.94)	12	1/2-13
1[d]	511.18 (20.125)	+0.13 (0.005)	0.20 (0.008)		552.4 (21.75)	530.22 (20.875)	100.1 (3.94)	12	7/16-14
2	447.68 (17.625)	+0.13 (0.005)	0.20 (0.008)		489.0 (19.25)	466.72 (18.375)	100.1 (3.94)	12	3/8-16
3	409.58 (16.125)	+0.13 (0.005)	0.20 (0.008)		450.8 (17.75)	428.62 (16.875)	100.1 (3.94)	12	3/8-16
4	361.95 (14.250)	+0.13 (0.005)	0.15 (0.006)		403.4 (15.88)	381.00 (15.000)	100.1 (3.94)	12	3/8-16
5	314.32 (12.375)	+0.13 (0.005)	0.15 (0.006)		355.6 (14.00)	333.38 (13.125)	71.4 (2.81)	8	3/8-16
6	266.70 (10.500)	+0.13 (0.005)	0.15 (0.006)		307.8 (12.12)	285.75 (11.250)	71.4 (2.81)	8	3/8-16

[a] Suggested tolerances are to be measured on the assembled engine mounted on its supports. For measuring procedure, see SAE J1033. Figures shown for bore eccentricity and face deviation are total indicator readings.

[b] An "E" dimension of 133.4 (5.25) is optional for multiple-plate clutches in the No. 00, 0, ½, and 1 flywheel housings. An "E" dimension of 71.4 (2.81) is required with No. 6½ or 7½ overcenter clutch when used with the No. 4 housing. An "E" dimension of 100.1 (3.94) is required with No. 8 overcenter clutch when used with the No. 5 housing.

[c] Tapped holes shall be threaded in accordance with UNC Class 2B tolerances of ANSI B1.1 screw threads, and the minimum length of thread engagement shall be 1.5 times the nominal diameter for gray iron housings and 2 times the nominal diameter for aluminum housings. Tapped holes shall be spaced equally on each side of the vertical centerline.

[d] Prior to 1953, the No. 1 SAE housing specified holes for 3/8-16 screws. When service requires differentiation, the new SAE No. 1 as adopted since 1953 may be specified as SAE No. 1 (7/16) and the obsolete housing may be specified as SAE No. 1 (3/8).

Litho in U.S.A.

FLYWHEELS FOR INDUSTRIAL ENGINES USED WITH INDUSTRIAL POWER TAKE-OFFS EQUIPPED WITH DRIVING-RING TYPE OVERCENTER CLUTCHES AND ENGINE-MOUNTED MARINE GEARS AND SINGLE BEARING ENGINE-MOUNTED POWER GENERATORS—SAE J620 OCT88

SAE Standard

Report of the Construction and Industrial Machinery Technical Committee and Engine Committee approved January 1953 and completely revised by the Clutch, Flywheel, and Housing Committee October 1988.

1. Scope—This SAE standard defines flywheel configuration for industry standardization, interchangeability, and compatibility.

Table 1 and Fig. 1 give the dimensions for the flywheels.

For dimensions of industrial power take-offs with driving-ring type over-center clutches, see SAE J621.

For flywheel dimensions for engine mounted torque converters without front disconnect clutch, see SAE J927.

TABLE 1—DIMENSIONS OF FLYWHEELS, mm (in)

Clutch Size	A	B[a,b]	C	D	E[e,k]	F
165 (6-1/2)	184.2 (7.25)	215.90 (8.500)	200.02 (7.875)	127.0 (5.00)	71.4 (2.81)	63.5 (2.50)
190 (7-1/2)	206.2 (8.12)	241.30 (9.500)	222.25 (8.750)	— —	71.4 (2.81)	63.5 (2.50)
200 (8)	225.6 (8.88)	263.52 (10.375)	244.48 (9.625)	— —	100.1 (3.94)	76.2 (3.00)
255 (10)[f]	276.4 (10.88)	314.32 (12.375)	295.28 (11.625)	196.8 (7.75)	100.1 (3.94)	76.2 (3.00)
290 (11-1/2)[g]	314.5 (12.38)	352.42 (13.875)	333.38 (13.125)	203.2 (8.00)	100.1 (3.94)	— —
355 (14)[h]	409.4 (16.12)	466.72 (18.375)	438.15 (17.250)	222.2 (8.75)	100.1 (3.94)	101.6 (4.00)
405 (16)	460.2 (18.12)	517.52 (20.375)	488.95 (19.250)	254.0 (10.00)	100.1 (3.94)	104.6 (4.12)
460 (18)[i]	498.3 (19.62)	571.50 (22.500)	542.92 (21.375)	— —	100.1 (3.94)	104.6 (4.12)
530 (21)[j]	584.2 (23.00)	673.10 (26.500)	641.35 (25.250)	— —	100.1 (3.94)	146.0 (5.75)
610 (24)	644.7 (25.38)	733.42 (28.875)	692.15 (27.250)	— —	100.1 (3.94)	146.0 (5.75)

Clutch Size	G[e]	H	J	K[c,k]	L[b,c,k]	Tapped Holes[d] No.	Tapped Holes[d] Size
165 (6-1/2)	30.2 (1.19)	12.7 (0.50)	9.7 (0.38)	17.5 (0.69)	52.000 (2.0472)	6	(5/16-18)
190 (7-1/2)	30.2 (1.19)	12.7 (0.50)	12.7 (0.50)	17.5 (0.69)	52.000 (2.0472)	8	(5/16-18)
200 (8)	62.0 (2.44)	12.7 (0.50)	12.7 (0.50)	19.0 (0.75)	62.000 (2.4409)	6	(3/8 -16)
255 (10)	53.8 (2.12)	15.7 (0.62)	12.7 (0.50)	28.4 (1.12)	72.000 (2.8346)	8	(3/8 -16)
290 (11-1/2)	39.6 (1.56)	28.4 (1.12)	22.4 (0.88)	31.8 (1.25)	72.000 (2.8346)	8	(3/8 -16)
355 (14)	25.4 (1.00)	28.4 (1.12)	22.4 (0.88)	38.1 (1.50)	80.000 (3.1496)	8	(1/2 -13)
405 (16)	15.7 (0.62)	28.4 (1.12)	22.4 (0.88)	44.4 (1.75)	100.000 (3.9370)	8	(1/2 -13)
460 (18)	15.7 (0.62)	31.8 (1.25)	31.8 (1.25)	44.4 (1.75)	100.000 (3.9370)	6	(5/8 -11)
530 (21)	0.0 (0.00)	31.8 (1.25)	31.8 (1.25)	57.2 (2.25)	130.000 (5.1181)	12	(5/8 -11)
610 (24)	0.0 (0.00)	31.8 (1.25)	31.8 (1.25)	57.2 (2.25)	130.000 (5.1181)	12	(3/4 -10)

NOTE: Suggested tolerances are to be measured on assembled engine; for measuring procedure, see SAE J1033.

[a]Diameter tolerance of driving-ring pilot bore 'B' is + 0.13 (0.005), − 0.000; maximum eccentricity is 0.13 (0.005) total indicator reading (see footnote b); face runout maximum total indicator reading is 0.0005 times the measured diameter. Diameter tolerance for mating driving ring, etc. pilot diameter is + 0.000, −0.13 (0.005).

[b]Eccentricity between driving-ring pilot bore 'B' and pilot bearing bore 'L' is not to exceed 0.20 (0.008) total indicator reading.

[c]'K' is length of bore for pilot bearing; 'L' is nominal diameter of bearing. Diameter and fit are to suit installation. Maximum eccentricity is 0.13 (0.005) total indicator reading. (See footnote b.)

[d]Tapped holes shall be threaded in accordance with UNC Class 2B tolerances of ANSI B1.1 screw threads, and the minimum length of thread engagement shall be 1.5 times the nominal diameter.

[e]Tolerances for dimensions 'G' and 'E' not to exceed the tolerance for 'E' as defined in SAE J617.

[f]Identical to flywheel No. 500 in SAE J162 (cancelled) except for 'G' dimension which was 25.4 (1.00), and 'H' dimension which was 28.4 (1.12). A spacer can be used between flywheel and coupler for equipment designed to the old specification.

[g]Identical to flywheel No. 511 in SAE J162 (cancelled).

[h]Identical to flywheel No. 514 in SAE J162 (cancelled).

[i]Identical to flywheel No. 518 in SAE J162 (cancelled).

[j]Identical to flywheel No. 521 in SAE J162 (cancelled).

[k]Pilot bearing bores, as defined by dimensions 'E', 'K', and 'L' are not typically required on flywheels for single bearing engine mounted power generators, and marine gears.

TAPPED HOLES EQUALLY SPACED

PHANTOM LINES INDICATE CLEARANCE REQUIRED FOR CLUTCH AND MARINE GEARS; CRANKSHAFT BOLTS; ETC. MUST NOT PROJECT BEYOND THIS LINE.

DEPTH OF PILOT BORE FROM FLYWHEEL HOUSING FACE TO SHOULDER ON FLYWHEEL OR TO CRANKSHAFT FLANGE FACE.

0.8 (0.03) R MAX.

0.8 (0.03) × 45 DEG. CHAMFER

6.4 (0.25) MIN.

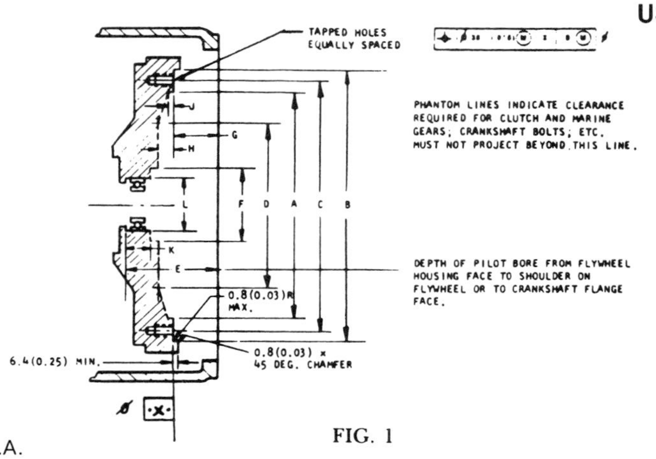

FIG. 1

 Litho in U.S.A.

INDEX

F

G

H

I

L

M

O

P

 Litho in U.S.A.

P (Cont.) **Page**

R

S

T **Page**

V

W

ANSWERS TO CHAPTER 1 QUESTIONS

1. First blank — "heat". Second blank — "Mechanical."

2. Air, fuel, and combustion (or ignition).

3. It heats up. This gives the heat necessary for good ignition of the fuel.

4. Vapor.

5. a — 2, b — 1.

6. First blank — "reciprocating" or "up-and-down". Second black — "rotary."

7. 8 to 1.

8. Intake, compression, power, exhaust.

9. False. The crankshaft turns **two** complete revolutions during a four-stroke cycle.

10. a — 1. b — 3. c — 2.

11. Liquid and air.

12. Diesel.

13. Because more heat is needed for fuel combustion without a spark in diesel engines. (The more air is compressed, the more it heats up.)

ANSWERS TO CHAPTER 2 QUESTIONS

1. Start at the center of the head and work out on both sides.

2. Upsets metal on the inside wall, making the guide bore smaller. This is a way to recondition some valve guides.

3. By turning the valves, the rotators **knock off deposits**. This keeps the valves sealing better, running cooler, and cuts down on corrosion.

4. 1/2

5. First blank — "top dead center." Second blank — "compression".

6. "Wet" liners contact the engine coolant, while "dry" liners do not.

7. a. Top inch of ring travel.

8. False. Measure liners **before** removing them.

9. Because of heat expansion. The thicker top part of the skirt must be slightly narrower since it expands more from the heat of combustion.

10. a — 2, b — 3, c — 1.

CHAPTER 2 ANSWERS CONTINUED

11. False. Always replace every piston ring you remove, regardless of its condition.

12. To be sure they are installed in the same cylinder. This gives a better fit, since the parts are already "wear fitted" to their own cylinders.

13. Because of the heavy throws that must be used as counterweights to balance the engine during rotation.

14. The flywheel.

15. True. The lighter flywheel allows faster acceleration and deceleration between various speeds.

16. 1) Cylinder head bolts; 2) valve tappets; 3) engine timing.

ANSWERS TO CHAPTER 3 QUESTIONS

1. Fuel tank, fuel pump, carburetor

2. To deliver fuel to the carburetor.

3. To mix the proper amounts of fuel and air under all operating conditions.

4. False

5. Natural draft, updraft, downdraft.

6. To provide a richer fuel mixture during starting.

7. To provide extra fuel momentarily during rapid acceleration.

8. The greater the air velocity, the **lower** the air pressure.

ANSWERS TO CHAPTER 4 QUESTIONS

1. False. LP-gas is a vapor when it reaches the carburetor, but gasoline is a liquid.

2. True.

3. True.

4. False! Never fill more than 80% full to allow room for vapor.

5. True.

6. False. Ventilate areas around LP-gas equipment, or use fans and blowers.

7. False. The converter changes liquid to vapor but it **lowers** the pressure.

Litho in U.S.A.

ANSWERS TO CHAPTER 5 QUESTIONS

1. False. No spark is used in a diesel engine. Instead, air is compressed until it heats up and ignites the injected shot of fuel.

2. Because one pump can serve all the cylinders.

3. By the diesel fuel circulating through it during operation.

4. First blank — "spring". Second blank — "fuel".

5. Use a **brass** wire brush. Never use a steel brush as it will scratch the precision tips and erode the spray orifices.

6. In **series** filters, all the fuel goes through one filter, then through the other. In **parallel** filters, part of the fuel goes through each filter.

7. The **series** filters clean fuel the best because the second filter can pick up dirt missed by the first one.

8. Balancing coil and thermostatic.

ANSWERS TO CHAPTER 6 QUESTIONS

1. Cylinder feed and crankcase feed.

2. Roots and centrifugal. The Roots type is a positive displacement pump and provides the same amount of air per revolution, regardless of engine speed. The centrifugal type is most efficient at high speeds.

3. Manifold, exhaust pipe, muffler.

4. Removes heat, muffles sound, carries away exhaust gases.

ANSWERS TO CHAPTER 7 QUESTIONS

1. Any three of these: 1) Reduces friction; 2) absorbs heat; 3) seals the piston rings; 4) cleans and flushes moving parts; 5) helps deaden the noise.

2. Bypasses oil around the filter if the filter clogs. This assures a supply of lubricating oil to the engine at all times.

3. Additives.

4. False. Oil loses many of its good lubricating qualities as it gets dirty and its additives wear out.

ANSWERS TO CHAPTER 8 QUESTIONS

1. Radiator, radiator cap, fan and fan belt, water pump, engine water jacket, thermostat, connecting hoses, liquid or coolant.

2. True.

3. The BELLOWS-type consists of a short length of circular corrugated copper tube closed at both ends and filled with a liquid having a low boiling point.

The BIMETALLIC-type of thermostat consists of a spiral of bimetallic strip. This is a strip of steel welded to a strip of bronze.

4. Water pump.

5. False.

ANSWERS TO CHAPTER 9 QUESTIONS

1. Governors can do any of three jobs: maintain a selected speed, limit the slow and fast speeds, or shut down the engine when it overspeeds.

2. Speed droop is the change in governor speed required to cause the throttle rod to move from full-open to full-closed or full-closed to full-open.

3. Constant speed governors are isochronous governors. An isochronous governor is able to maintain a constant speed without correcting it.

ANSWERS TO CHAPTER 10 QUESTIONS

1. **Testing** tools and **servicing** tools.

2. **Testing** tools, which locate the trouble which the servicing tools are then used to help correct.

3. **Special tools** designed for servicing each model of engine.

ANSWERS TO CHAPTER 11 QUESTIONS

1. Know the system, 2) Ask the operator, 3) Inspect the engine, 4) Operate the engine, 5) List the possible causes, 6) Reach a conclusion, 7) Test your conclusion.

2. During **none** of these steps. Do all these things **before** you begin repairing the engine.

3. The dynamometer test.

4. Fuel-air mixture, compression, and ignition.

Litho in U.S.A.

ANSWERS TO CHAPTER 12 QUESTIONS

1. False. Tune-up can restore an engine which needs minor checks and adjustments, but it cannot make up for overdue repairs. Only a major overhaul can do that.

2. A **visual inspection** should be made to find out the condition of the engine and to see whether a tune-up or an overhaul is necessary.

3. Both **before** and **after** engine tune-up. Testing before tune-up gives an idea of the engine's condition and whether tune-up will do the job. Testing after tune-up gives a final check of engine performance to see if the tune-up has been successful.

Litho in U.S.A.